D1552802

Methods in Enzymology

Volume 195

ADENYLYL CYCLASE, G PROTEINS, AND GUANYLYL CYCLASE

METHODS IN ENZYMOLOGY

EDITORS-IN-CHIEF

John N. Abelson   Melvin I. Simon

DIVISION OF BIOLOGY
CALIFORNIA INSTITUTE OF TECHNOLOGY
PASADENA, CALIFORNIA

FOUNDING EDITORS

Sidney P. Colowick and Nathan O. Kaplan

Methods in Enzymology

Volume 195

Adenylyl Cyclase, G Proteins, and Guanylyl Cyclase

EDITED BY

Roger A. Johnson

DEPARTMENT OF PHYSIOLOGY AND BIOPHYSICS
SCHOOL OF MEDICINE, HEALTH SCIENCES CENTER
STATE UNIVERSITY OF NEW YORK AT STONY BROOK
STONY BROOK, NEW YORK

Jackie D. Corbin

DEPARTMENT OF MOLECULAR PHYSIOLOGY AND BIOPHYSICS
HOWARD HUGHES MEDICAL INSTITUTE
VANDERBILT UNIVERSITY
NASHVILLE, TENNESSEE

ACADEMIC PRESS, INC.
Harcourt Brace Jovanovich, Publishers
San Diego New York Boston
London Sydney Tokyo Toronto

This book is printed on acid-free paper. ∞

Academic Press, Inc.
San Diego, California 92101

United Kingdom Edition published by
Academic Press Limited
24-28 Oval Road, London NW1 7DX

Library of Congress Catalog Card Number: 54-9110

ISBN 0-12-182096-3 (alk. paper)

Printed in the United States of America
91 92 93 94 9 8 7 6 5 4 3 2 1

Table of Contents

Section I. Adenylyl Cyclase

A. Determination of Adenylyl Cyclase Activity

B. Preparation of Materials Useful in Purification of Components of Hormonally Responsive Adenylyl Cyclase Systems

C. Purification of Adenylyl Cyclases

Section II. Guanine Nucleotide-Dependent Regulatory Proteins

A. Purification and Characterization of G Proteins

B. Labeling and Quantitating of G Proteins

C. Reconstitution

Section III. Guanylyl Cyclase

A. Assay of Guanylyl Cyclase

B. Purification and Characterization of Guanylyl Cyclase Isozymes

C. Regulation of Guanylyl Cyclases

Contributors to Volume 195

Article numbers are in parentheses following the names of contributors.
Affiliations listed are current.

J. Kelley Bentley (43), *Pharmacology Department, University of Washington Health Sciences Center, Seattle, Washington 98195*

Lutz Birnbaumer (15), *Department of Cell Biology, Baylor College of Medicine, Houston, Texas 77030*

David A. Bobak (21), *Departments of Medicine and Microbiology, University of Virginia School of Medicine, Charlottesville, Virginia 22908*

Eycke Böhme (35), *Institut für Pharmakologie, Freie Universität Berlin, Universitätsklinikum Charlottenburg, D-1000 Berlin 33, Federal Republic of Germany*

Theodor Braun (11), *Section of Cell Growth, Regulation, and Oncogenesis, Duke University Medical Center, Chicago, Illinois 60611*

Donna J. Carty (15, 26), *Department of Pharmacology, Mount Sinai School of Medicine, City University of New York, New York, New York 10029*

Patrick J. Casey (27), *Department of Pharmacology, University of Texas Southwestern Medical Center, Dallas, Texas 75235*

Richard A. Cerione (29), *Department of Pharmacology, Cornell University, New York State College of Veterinary Medicine, Ithaca, New York 14853*

Juan Codina (15), *Department of Cell Biology, Baylor College of Medicine, Houston, Texas 77030*

Steven E. Domino (30), *Department of Obstetrics & Gynecology, University of Michigan, Ann Arbor, Michigan 41809*

Maura G. Donovan (12), *Department of Pharmacology, University of Washington School of Medicine, Seattle, Washington 98195*

Kenneth M. Ferguson (16, 28), *ICOS Corporation, Bothell, Washington 98021*

Gabriela Fischer (35), *Max-Planck-Institut für Biochemie, D-8033 Martinsried bei München, Federal Republic of Germany*

Michael Freissmuth (17), *Pharmacology Institute, University of Vienna, Vienna, Austria*

David L. Garbers (30, 33, 39), *Howard Hughes Medical Institute, Department of Pharmacology, University of Texas Southwestern Medical Center, Dallas, Texas 75235*

Rupert Gerzer (31, 34), *Medizinische Klinik Innenstadt, Universität München, D-8000 München 2, Federal Republic of Germany*

D. Michael Gill (23), *Department of Molecular Biology and Microbiology, Tufts University School of Medicine, Boston, Massachusetts 02111*

Alfred G. Gilman (17, 18, 19, 27), *Department of Pharmacology, University of Texas Southwestern Medical Center, Dallas, Texas 75235*

Michael P. Graziano (17), *Department of Molecular Pharmacology and Biochemistry, Merck and Company, Rahway, New Jersey 07065*

Mary K. Gross (8), *Department of Pharmacology, University of Washington School of Medicine, Seattle, Washington 98195*

Pavel Hamet (40, 42), *Laboratory of Molecular Pathophysiology, Clinical Research Institute of Montreal, Montreal, Quebec H2W 1R7, Canada*

Tsutomu Higashijima (16, 28), *Department of Pharmacology, University of*

Texas Southwestern Medical Center, Dallas, Texas 75235

KLAUS-DIETER HINSCH (35), *Zentrum für Dermatologie und Andrologie, Universität Giessen, D-6300 Giessen, Federal Republic of Germany*

PETER HUMBERT (35), *Institut für Pharmakologie, Freie Universität Berlin, Universitätsklinikum Charlottenburg, D-1000 Berlin 33, Federal Republic of Germany*

CÉLINE HUOT (40), *Laboratory of Molecular Pathophysiology, Clinical Research Institute of Montreal, Montreal, Quebec H2W 1R7, Canada*

TADASHI INAGAMI (38), *Department of Biochemistry, Vanderbilt University, School of Medicine, Nashville, Tennessee 37232*

RAVI IYENGAR (15, 26), *Department of Pharmacology, Mount Sinai School of Medicine, City University of New York, New York, New York 10029*

ROGER A. JOHNSON (1), *Department of Physiology and Biophysics, School of Medicine, Health Sciences Center, State University of New York at Stony Brook, Stony Brook, New York 11794*

RICHARD A. KAHN (20), *Laboratory of Biological Chemistry, DCT, National Cancer Institute, National Institutes of Health, Bethesda, Maryland 20892*

SUSANNE KLUMPP (44), *Pharmazeutisches Institut, Universität Tübingen, D-7400 Tübingen, Federal Republic of Germany*

CAROLINE KOCH (40), *Laboratory of Molecular Pathophysiology, Clinical Research Institute of Montreal, Montreal, Quebec H2W 1R7, Canada*

DORIS KOESLING (35), *Institut für Pharmakologie, Freie Universität Berlin, Universitätsklinikum Charlottenburg, D-1000 Berlin 33, Federal Republic of Germany*

GREGORY S. KOPF (22), *Division of Reproductive Biology, Department of Obstetrics and Gynecology, University of Pennsylvania School of Medicine, Philadelphia, Pennsylvania 19104*

PUSHKARAJ J. LAD (32), *Genencor Inc., South San Francisco, California 94080*

ANTONIO LAURENZA (5), *Division of Chemotherapy, GLAXO Inc., Research Triangle Park, North Carolina 27709*

DALE C. LEITMAN (36, 37, 41), *Metabolic Research Unit, University of California, School of Medicine, San Francisco, California 94143*

STEPHEN H. LEPPLA (13), *Laboratory of Microbial Ecology, National Institute of Dental Research, National Institutes of Health, Bethesda, Maryland 20892*

MAURINE E. LINDER (18), *Department of Pharmacology, University of Texas Southwestern Medical Center, Dallas, Texas 75235*

H. ROBERT MASURE (12), *Laboratory of Microbiology, Rockfeller University, New York, New York 10021*

BERND MAYER (35), *Institut für Pharmakologie, Freie Universität Berlin, Universitätsklinikum Charlottenburg, D-1000 Berlin 33, Federal Republic of Germany*

STEFAN MOLLNER (7, 10), *Physiologisch Chemisches Institut der Univeristät Würzberg, D-8700 Würzburg, Federal Republic of Germany*

MALCOLM C. MOOS, JR. (3), *Biochemical Pharmacology, Center for Biologics Evaluation and Research, National Institutes of Health, Bethesda, Maryland 20892*

JOEL MOSS (21), *Laboratory of Cellular Metabolism, National Heart, Lung, and Blood Institute, National Institutes of Health, Bethesda, Maryland 20892*

ALEXANDER MÜLSCH (31,34), *Department of Applied Physiology, University of Freiburg, D-7800 Freiburg im Breisgau, Federal Republic of Germany*

SUSANNE M. MUMBY (19), *Department of Pharmacology, University of Texas Southwestern Medical Center, Dallas, Texas 75235*

FERID MURAD (36, 37, 41), *Pharmaceutical Products Research and Development, Abbott Laboratories, Abbott Park, Illinois 60064*

FERAYDOON NIROOMAND (35), *Institut für Pharmakologie, Freie Universität Berlin,*

Universitätsklinikum Charlottenburg, D-1000 Berlin 33, Federal Republic of Germany

STEFAN OFFERMANNS (25), *Institut für Pharmakologie, Freie Universität Berlin, D-1000 Berlin 33, Federal Republic of Germany*

IOK-HOU PANG (27), *Research and Development, Alcon Laboratories, Inc., Fort Worth, Texas 76134*

ELKE PFEUFFER (7, 14), *Physiologisch Chemisches Institut der Universität Würzburg, D-8700 Würzburg, Federal Republic of Germany*

THOMAS PFEUFFER (4, 7, 10, 14, 24), *Physiologisch Chemisches Institut der Universität Würzburg, D-8700 Würzburg, Federal Republic of Germany*

MICHEL POTIER (40), *Department of Medical Genetics, St. Justine's Hospital, Montreal, Quebec H3T 1C5, Canada*

RICHARD T. PREMONT (26), *Department of Pharmacology, Mount Sinai School of Medicine, City University of New York, New York, New York 10029*

S. RUSS PRICE (21), *Laboratory of Cellular Metabolism, National Heart, Lung, and Blood Institute, National Institutes of Health, Bethesda, Maryland 20892*

CHODAVARAPU S. RAMARAO (33), *Department of Pharmacology, Vanderbilt University Medical Center, Nashville, Tennessee 37232*

GARY B. ROSENBERG (9), *Department of Pharmacology, University of Washington School of Medicine, Seattle, Washington 98195*

WALTER ROSENTHAL (25), *Institut für Pharmakologie, Freie Universität Berlin, D-1000 Berlin 33, Federal Republic of Germany*

ELLIOTT M. ROSS (29), *Department of Pharmacology, University of Texas Southwestern Medical Center, Dallas, Texas 75235*

YORAM SALOMON (1, 2), *Department of Hormone Research, Weizmann Institute of Science, Rehovot 76100, Israel*

GÜNTER SCHULTZ (25, 35), *Institut für Pharmakologie, Freie Universität Berlin, Universitätsklinikum Charlottenburg, D-1000 Berlin 33, Federal Republic of Germany*

JOACHIM E. SCHULTZ (44), *Pharmazeutisches Institut, Universität Tübingen, 7400 Tübingen, Federal Republic of Germany*

KENNETH B. SEAMON (5), *Laboratory of Molecular Pharmacology, Division of Biochemistry and Biophysics Center for Biologics Evaluation and Research, Food and Drug Administration, Bethesda, Maryland 20892*

SUJAY SINGH (39), *Department of Pharmacology, Vanderbilt University Medical Center, Nashville, Tennessee 37232*

RUDOLF M. SNAJDAR (38), *Cleveland Clinic Research Institute Cleveland, Ohio 44195*

DANIEL R. STORM (6, 9, 12), *Department of Pharmacology, University of Washington School of Medicine, Seattle, Washington 98195*

RYOICHI TAKAYANAGI (38), *The Third Department of Internal Medicine, Kyushu University School of Medicine, Fukuoka, Japan*

ROLF THOMAS (24), *Physiologisch Chemisches Institut der Universität, Würzburg, D-8700 Würzburg, Federal Republic of Germany*

WILLIAM A. TOSCANO, JR. (8), *Division of Environmental and Occupational Health (Toxicology), University of Minnesota, Minneapolis, Minnesota 55455*

JOHANNE TREMBLAY (40, 42), *Laboratory of Molecular Pathophysiology, Clinical Research Institute of Montreal, Montreal, Quebec H2W 1R7, Canada*

SU-CHEN TSAI (21), *Laboratory of Cellular Metabolism, National Heart, Lung, and Blood Institute, National Institutes of Health, Bethesda, Maryland 20892*

D. JANETTE TUBB (30), *Howard Hughes Medical Institute, University of Texas Southwestern Medical Center, Dallas, Texas 75235*

MARTHA VAUGHAN (21), *Laboratory of Cellular Metabolism, National Heart, Lung, and Blood Institute, National Institutes of Health, Bethesda, Maryland 20892*

SCOTT A. WALDMAN (36, 37, 41), *Division of Clinical Pharmacology, Department of Medicine, Thomas Jefferson University, Philadelphia, Pennsylvania 19107*

TIMOTHY F. WALSETH (3), *Department of Pharmacology, University of Minnesota, Minneapolis, Minnesota 55455*

PING WANG (6, 9), *Department of Pharmacology, University of Washington School of Medicine, Seattle, Washington 98195*

ARNOLD A. WHITE (32), *Department of Biochemistry and the John M. Dalton Research Center, University of Missouri-Columbia, Columbia, Missouri 65211*

MARILYN J. WOOLKALIS (22, 23), *Department of Pharmacology, University of Pennsylvania School of Medicine, Philadelphia, Pennsylvania 19104*

PETER S. T. YUEN (3), *Howard Hughes Medical Institute, University of Texas Southwestern Medical Center, Dallas, Texas 75235*

Preface

Since the first volume of *Methods in Enzymology* on cyclic nucleotides (Vol. XXXVIII) was published in 1974, substantial progress has been made in cyclic nucleotide research, particularly on the enzymes involved in their synthesis, degradation, and mode of action. Cyclic nucleotide-dependent protein kinases, cyclic nucleotide phosphodiesterases, and quantitative assays of cAMP and cGMP levels were updated in Volumes 99 and 159. This volume emphasizes methods for the assay, purification, and characterization of adenylyl cyclases, guanine nucleotide-dependent regulatory proteins (G proteins), and guanylyl cyclases. Research in each of these areas has grown rapidly in the past sixteen years, especially recently with the application of molecular biological approaches that augment biochemical techniques.

One consequence of the rapid growth is that it has become impossible to have an absolutely current book describing these advances. Although adenylyl and guanylyl cyclases have been purified and characterized from numerous sources, it is becoming clear that each is, in fact, a family of enzymes. For adenylyl cyclase this is most easily recognized by differences in function and distribution of mammalian forms that are either sensitive or insensitive to calmodulin. These are dealt with in depth in this volume. Additional members of the adenylyl cyclase family are currently being purified and/or cloned from numerous prokaryotic and eukaryotic sources. These are only briefly discussed. The soluble and particulate forms of guanylyl cyclase, while appearing diverse due to differences in their distribution and in their sensitivities to specific peptide hormones and to nitrous oxide, also form a growing family of enzymes. The "snapshot" of the field presented suggests substantial future developments. For both adenylyl and guanylyl cyclases and for the G proteins it is becoming clear that in addition to the established modes of regulation, e.g., of the cyclases through G proteins or hormones, there are likely other mechanisms through which cells may regulate the activities of these important enzymes. Prominent among these are covalent modifications, e.g., phosphorylation–dephosphorylation, as well as allosteric regulation, e.g., inhibition of adenylyl cyclases by specific cell-derived adenine nucleotides. Thus, future direction of research will certainly include additional details of the number and structure of the various members of the families of adenylyl and guanylyl cyclases and of the modes of their regulation.

The impact of rapid growth in a research area is most obvious with the G proteins. Given that important aspects of our current interest and understanding of G proteins derive from investigations on their role in the regulation of adenylyl cyclases, it is obviously imperative that a section on G proteins be included in any volume dealing with adenylyl cyclases. In part, G proteins were discovered due to the effects of GTP to mediate hormonal activation of adenylyl cyclases initially described by Rodbell and co-workers. G proteins were later found to be involved also in mediating hormonal inhibition of this enzyme and to be targets for ADP-ribosylation by cholera and pertussis toxins. However, the explosion in G-protein research, in the number and variety of G proteins, and in the myriad of actions they mediate force a limitation on coverage. The emphasis in this volume is limited to the purification and quantification of those G proteins mediating stimulatory (G_S) and inhibitory (G_i) effects on adenylyl cyclases and to the low molecular weight proteins that enhance the actions of cholera toxin on G_S. While G proteins mediate the effects of stimulatory and inhibitory hormones on the activity of adenylyl cyclases, we thought it beyond the scope of this volume to deal with their interactions with each of the numerous hormone receptors with which they are known to interact. We have limited the treatment of these interactions simply to general aspects of hormone receptor–G protein–adenylyl cyclase reconstitution. Similarly, the actual mechanisms by which G proteins mediate activation or inhibition of adenylyl cyclases are skirted since they are presently very poorly understood. Another volume on this topic will become necessary as our understanding of these enzymes and their regulation develops.

We are grateful to the authors for their excellent contributions, and we apologize to those who have made many contributions to these fields but whose work may not be adequately recognized here. The inevitable omissions have been due to editorial oversight, to potential authors being already overcommitted, and to the rapid rate at which research in these areas has occurred and, hence, to timing.

This volume is dedicated to Dr. Martin Rodbell for his very many contributions to our understanding of the hormonal regulation of adenylyl cyclases.

Roger A. Johnson
Jackie D. Corbin

METHODS IN ENZYMOLOGY

VOLUME I. Preparation and Assay of Enzymes
Edited by SIDNEY P. COLOWICK AND NATHAN O. KAPLAN

VOLUME II. Preparation and Assay of Enzymes
Edited by SIDNEY P. COLOWICK AND NATHAN O. KAPLAN

VOLUME III. Preparation and Assay of Substrates
Edited by SIDNEY P. COLOWICK AND NATHAN O. KAPLAN

VOLUME IV. Special Techniques for the Enzymologist
Edited by SIDNEY P. COLOWICK AND NATHAN O. KAPLAN

VOLUME V. Preparation and Assay of Enzymes
Edited by SIDNEY P. COLOWICK AND NATHAN O. KAPLAN

VOLUME VI. Preparation and Assay of Enzymes (*Continued*)
Preparation and Assay of Substrates
Special Techniques
Edited by SIDNEY P. COLOWICK AND NATHAN O. KAPLAN

VOLUME VII. Cumulative Subject Index
Edited by SIDNEY P. COLOWICK AND NATHAN O. KAPLAN

VOLUME VIII. Complex Carbohydrates
Edited by ELIZABETH F. NEUFELD AND VICTOR GINSBURG

VOLUME IX. Carbohydrate Metabolism
Edited by WILLIS A. WOOD

VOLUME X. Oxidation and Phosphorylation
Edited by RONALD W. ESTABROOK AND MAYNARD E. PULLMAN

VOLUME XI. Enzyme Structure
Edited by C. H. W. HIRS

VOLUME XII. Nucleic Acids (Parts A and B)
Edited by LAWRENCE GROSSMAN AND KIVIE MOLDAVE

VOLUME 81. Biomembranes (Part H: Visual Pigments and Purple Membranes, I)
Edited by LESTER PACKER

VOLUME 82. Structural and Contractile Proteins (Part A: Extracellular Matrix)
Edited by LEON W. CUNNINGHAM AND DIXIE W. FREDERIKSEN

VOLUME 83. Complex Carbohydrates (Part D)
Edited by VICTOR GINSBURG

VOLUME 84. Immunochemical Techniques (Part D: Selected Immunoassays)
Edited by JOHN J. LANGONE AND HELEN VAN VUNAKIS

VOLUME 85. Structural and Contractile Proteins (Part B: The Contractile Apparatus and the Cytoskeleton)
Edited by DIXIE W. FREDERIKSEN AND LEON W. CUNNINGHAM

VOLUME 86. Prostaglandins and Arachidonate Metabolites
Edited by WILLIAM E. M. LANDS AND WILLIAM L. SMITH

VOLUME 87. Enzyme Kinetics and Mechanism (Part C: Intermediates, Stereochemistry, and Rate Studies)
Edited by DANIEL L. PURICH

VOLUME 88. Biomembranes (Part I: Visual Pigments and Purple Membranes, II)
Edited by LESTER PACKER

VOLUME 89. Carbohydrate Metabolism (Part D)
Edited by WILLIS A. WOOD

VOLUME 90. Carbohydrate Metabolism (Part E)
Edited by WILLIS A. WOOD

VOLUME 91. Enzyme Structure (Part I)
Edited by C. H. W. HIRS AND SERGE N. TIMASHEFF

VOLUME 92. Immunochemical Techniques (Part E: Monoclonal Antibodies and General Immunoassay Methods)
Edited by JOHN J. LANGONE AND HELEN VAN VUNAKIS

Volume 106. Posttranslational Modifications (Part A)
Edited by Finn Wold and Kivie Moldave

Volume 107. Posttranslational Modifications (Part B)
Edited by Finn Wold and Kivie Moldave

Volume 108. Immunochemical Techniques (Part G: Separation and Characterization of Lymphoid Cells)
Edited by Giovanni Di Sabato, John J. Langone, and Helen Van Vunakis

Volume 109. Hormone Action (Part I: Peptide Hormones)
Edited by Lutz Birnbaumer and Bert W. O'Malley

Volume 110. Steroids and Isoprenoids (Part A)
Edited by John H. Law and Hans C. Rilling

Volume 111. Steroids and Isoprenoids (Part B)
Edited by John H. Law and Hans C. Rilling

Volume 112. Drug and Enzyme Targeting (Part A)
Edited by Kenneth J. Widder and Ralph Green

Volume 113. Glutamate, Glutamine, Glutathione, and Related Compounds
Edited by Alton Meister

Volume 114. Diffraction Methods for Biological Macromolecules (Part A)
Edited by Harold W. Wyckoff, C. H. W. Hirs, and Serge N. Timasheff

Volume 115. Diffraction Methods for Biological Macromolecules (Part B)
Edited by Harold W. Wyckoff, C. H. W. Hirs, and Serge N. Timasheff

Volume 116. Immunochemical Techniques (Part H: Effectors and Mediators of Lymphoid Cell Functions)
Edited by Giovanni Di Sabato, John J. Langone, and Helen Van Vunakis

VOLUME 117. Enzyme Structure (Part J)
Edited by C. H. W. HIRS AND SERGE N. TIMASHEFF

VOLUME 118. Plant Molecular Biology
Edited by ARTHUR WEISSBACH AND HERBERT WEISSBACH

VOLUME 119. Interferons (Part C)
Edited by SIDNEY PESTKA

VOLUME 120. Cumulative Subject Index Volumes 81–94, 96–101

VOLUME 121. Immunochemical Techniques (Part I: Hybridoma Technology and Monoclonal Antibodies)
Edited by JOHN J. LANGONE AND HELEN VAN VUNAKIS

VOLUME 122. Vitamins and Coenzymes (Part G)
Edited by FRANK CHYTIL AND DONALD B. MCCORMICK

VOLUME 123. Vitamins and Coenzymes (Part H)
Edited by FRANK CHYTIL AND DONALD B. MCCORMICK

VOLUME 124. Hormone Action (Part J: Neuroendocrine Peptides)
Edited by P. MICHAEL CONN

VOLUME 125. Biomembranes (Part M: Transport in Bacteria, Mitochondria, and Chloroplasts: General Approaches and Transport Systems)
Edited by SIDNEY FLEISCHER AND BECCA FLEISCHER

VOLUME 126. Biomembranes (Part N: Transport in Bacteria, Mitochondria, and Chloroplasts: Protonmotive Force)
Edited by SIDNEY FLEISCHER AND BECCA FLEISCHER

VOLUME 127. Biomembranes (Part O: Protons and Water: Structure and Translocation)
Edited by LESTER PACKER

VOLUME 128. Plasma Lipoproteins (Part A: Preparation, Structure, and Molecular Biology)
Edited by JERE P. SEGREST AND JOHN J. ALBERS

VOLUME 129. Plasma Lipoproteins (Part B: Characterization, Cell Biology, and Metabolism)
Edited by JOHN J. ALBERS AND JERE P. SEGREST

VOLUME 130. Enzyme Structure (Part K)
Edited by C. H. W. HIRS AND SERGE N. TIMASHEFF

VOLUME 131. Enzyme Structure (Part L)
Edited by C. H. W. HIRS AND SERGE N. TIMASHEFF

VOLUME 132. Immunochemical Techniques (Part J: Phagocytosis and Cell-Mediated Cytotoxicity)
Edited by GIOVANNI DI SABATO AND JOHANNES EVERSE

VOLUME 133. Bioluminescence and Chemiluminescence (Part B)
Edited by MARLENE DELUCA AND WILLIAM D. MCELROY

VOLUME 134. Structural and Contractile Proteins (Part C: The Contractile Apparatus and the Cytoskeleton)
Edited by RICHARD B. VALLEE

VOLUME 135. Immobilized Enzymes and Cells (Part B)
Edited by KLAUS MOSBACH

VOLUME 136. Immobilized Enzymes and Cells (Part C)
Edited by KLAUS MOSBACH

VOLUME 137. Immobilized Enzymes and Cells (Part D)
Edited by KLAUS MOSBACH

VOLUME 138. Complex Carbohydrates (Part E)
Edited by VICTOR GINSBURG

VOLUME 139. Cellular Regulators (Part A: Calcium- and Calmodulin-Binding Proteins)
Edited by ANTHONY R. MEANS AND P. MICHAEL CONN

VOLUME 140. Cumulative Subject Index Volumes 102–119, 121–134

VOLUME 141. Cellular Regulators (Part B: Calcium and Lipids)
Edited by P. MICHAEL CONN AND ANTHONY R. MEANS

VOLUME 142. Metabolism of Aromatic Amino Acids and Amines
Edited by SEYMOUR KAUFMAN

VOLUME 143. Sulfur and Sulfur Amino Acids
Edited by WILLIAM B. JAKOBY AND OWEN GRIFFITH

VOLUME 144. Structural and Contractile Proteins (Part D: Extracellular Matrix)
Edited by LEON W. CUNNINGHAM

VOLUME 145. Structural and Contractile Proteins (Part E: Extracellular Matrix)
Edited by LEON W. CUNNINGHAM

VOLUME 146. Peptide Growth Factors (Part A)
Edited by DAVID BARNES AND DAVID A. SIRBASKU

VOLUME 147. Peptide Growth Factors (Part B)
Edited by DAVID BARNES AND DAVID A. SIRBASKU

VOLUME 148. Plant Cell Membranes
Edited by LESTER PACKER AND ROLAND DOUCE

VOLUME 149. Drug and Enzyme Targeting (Part B)
Edited by RALPH GREEN AND KENNETH J. WIDDER

VOLUME 150. Immunochemical Techniques (Part K: *In Vitro* Models of B and T Cell Functions and Lymphoid Cell Receptors)
Edited by GIOVANNI DI SABATO

VOLUME 151. Molecular Genetics of Mammalian Cells
Edited by MICHAEL M. GOTTESMAN

VOLUME 152. Guide to Molecular Cloning Techniques
Edited by SHELBY L. BERGER AND ALAN R. KIMMEL

VOLUME 153. Recombinant DNA (Part D)
Edited by RAY WU AND LAWRENCE GROSSMAN

VOLUME 154. Recombinant DNA (Part E)
Edited by RAY WU AND LAWRENCE GROSSMAN

VOLUME 155. Recombinant DNA (Part F)
Edited by RAY WU

VOLUME 156. Biomembranes (Part P: ATP-Driven Pumps and Related Transport: The Na,K-Pump)
Edited by SIDNEY FLEISCHER AND BECCA FLEISCHER

VOLUME 183. Molecular Evolution: Computer Analysis of Protein and Nucleic Acid Sequences
Edited by RUSSELL F. DOOLITTLE

VOLUME 184. Avidin–Biotin Technology
Edited by MEIR WILCHEK AND EDWARD A. BAYER

VOLUME 185. Gene Expression Technology
Edited by DAVID V. GOEDDEL

VOLUME 186. Oxygen Radicals in Biological Systems (Part B: Oxygen Radicals and Antioxidants)
Edited by LESTER PACKER AND ALEXANDER N. GLAZER

VOLUME 187. Arachidonate Related Lipid Mediators
Edited by ROBERT C. MURPHY AND FRANK A. FITZPATRICK

VOLUME 188. Hydrocarbons and Methylotrophy
Edited by MARY E. LIDSTROM

VOLUME 189. Retinoids (Part A: Molecular and Metabolic Aspects)
Edited by LESTER PACKER

VOLUME 190. Retinoids (Part B: Cell Differentiation and Clinical Applications)
Edited by LESTER PACKER

VOLUME 191. Biomembranes (Part V: Cellular and Subcellular Transport: Epithelial Cells)
Edited by SIDNEY FLEISCHER AND BECCA FLEISCHER

VOLUME 192. Biomembranes (Part W: Cellular and Subcellular Transport: Epithelial Cells)
Edited by SIDNEY FLEISCHER AND BECCA FLEISCHER

VOLUME 193. Mass Spectrometry
Edited by JAMES A. MCCLOSKEY

VOLUME 194. Guide to Yeast Genetics and Molecular Biology
Edited by CHRISTINE GUTHRIE AND GERALD R. FINK

VOLUME 195. Adenylyl Cyclase, G Proteins, and Guanylyl Cyclase
Edited by ROGER A. JOHNSON AND JACKIE D. CORBIN

Section I

Adenylyl Cyclase

A. Determination of Adenylyl Cyclase Activity
Articles 1 and 2

B. Preparation of Materials Useful in Purification of Components of Hormonally Responsive Adenylyl Cyclase Systems
Articles 3 through 5

C. Purification of Adenylyl Cyclases
Articles 6 through 13

[1] Assay of Adenylyl Cyclase Catalytic Activity

By ROGER A. JOHNSON and YORAM SALOMON

Adenylyl cyclase (ATP pyrophosphate-lyase, cyclizing, EC 4.6.1.1, adenylate cyclase) is a family of membrane-bound enzymes that exhibit inactive and active configurations resulting from the actions of a variety of agents, acting indirectly and directly on the enzyme. Enzyme activity may be increased or decreased by stimulatory or inhibitory hormones acting via specific hormone receptors coupled to the catalytic unit of the enzyme by the respective guanine nucleotide-dependent regulatory proteins (G_s and G_i, respectively). The G proteins are also activated by aluminum fluoride and are targets for ADP-ribosylation by specific bacterial toxins. The catalytic moiety of adenylyl cyclase from most tissues is stimulated by the diterpene forskolin and is inhibited by analogs of adenosine, the most potent of which is 2′-deoxyadenosine 3′-monophosphate, and the enzyme from some tissues is also stimulated directly by Ca^{2+}/calmodulin. The nature of some of these agents and the range of resulting activities can influence the assay conditions used for determining the catalytic activity of the enzyme.

The catalytic activity of adenylyl cyclase is determined by methods that rely on the measurement on cAMP formed from unlabeled substrate, with cAMP-binding proteins or radioimmunoassay procedures, or by methods that rely on the use of radioactively labeled substrate followed by isolation and determination of the radioactively labeled product. The two different approaches have different purposes, different sensitivities, and different ease of use. The method of choice will depend in part on the facilities and orientation of a given laboratory. The procedures described here focus on the use of radioactively labeled substrate and isolation of the labeled product. Additional detailed considerations for the assay of adenylyl cyclase by these procedures can be found in the review by Salomon.[1]

Considerations for Establishing Reaction Conditions

Requirements for Metal-ATP and Divalent Cations

Both ATP and divalent cation (Mg^{2+} or Mn^{2+}) are required for adenylyl cyclase-catalyzed formation of cAMP.[2,3] The enzyme conforms to a bireac-

[1] Y. Salomon, *Adv. Cyclic Nucleotide Res.* **10,** 35 (1979).
[2] T. W. Rall and E. W. Sutherland, *J. Biol. Chem.* **232,** 1065 (1958).
[3] E. W. Sutherland, T. W. Rall, and T. Menon, *J. Biol. Chem.* **237,** 1220 (1962).

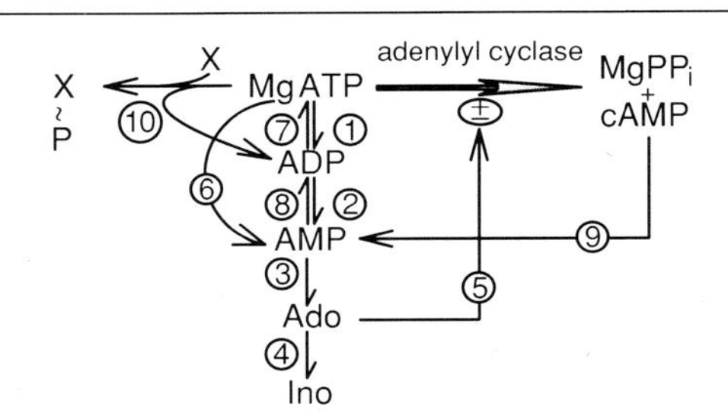

FIG. 1. Reactions involving adenine nucleotides in membrane preparations.

tant sequential mechanism in which metal-ATP^{2-} is substrate and free divalent cation is a requisite cofactor.[4] Thus, adenylyl cyclase requires divalent cation in excess of the ATP concentration, and for determining kinetic constants the concentrations of free Mg^{2+} or Mn^{2+} should be kept essentially constant by maintaining the total metal concentration at a predetermined fixed concentration above the total ATP concentration.[4,5] Kinetic constants can then be calculated by linear regression analysis of the slopes and intercepts of secondary plots as suggested by Cleland.[5] Buffers that can significantly affect concentrations of free divalent cation (especially Mn^{2+}) in the reaction mixture should be avoided. Triethanolamine hydrochloride does not have this problem, whereas Tris-Cl is particularly poor.[4] Examples of approximate K_m values for adenylyl cyclases have been reported as follows: detergent-dispersed enzyme from rat brain, K_m(MnATP) 7–9 μM, $K_m(Mn^{2+})$ 2–3 μM, K_m(MgATP) 30–60 μM, $K_m(Mg^{2+})$ 800–900 μM[4]; human platelets, $K_m(Mg^{2+})$ 1100 μM, K_m(MgATP) 50 μM[6]; liver, K_m(MnATP) and K_m(MgATP) were similar (~50 μM).[7]

Contaminating Enzyme Activities

Adenylyl cyclase is an enzyme that constitutes a very small percentage of membrane-bound protein and as such exists in an environment rich in contaminating enzyme activities, including a number of nucleotide phosphohydrolases, cyclic nucleotide phosphodiesterases, and ATP-utilizing kinases. Thus, adenylyl cyclase in membrane preparations competes with other enzymes for ATP (Fig. 1) and the cAMP formed is readily hydrolyzed to 5′-AMP. Analogously, GTP, which is required for hormone-induced

[4] D. L. Garbers and R. A. Johnson, *J. Biol. Chem.* **250,** 8449 (1975).

[5] W. W. Cleland, *in* "The Enzymes" P. E. Boyer, ed.), 3rd Ed., Vol. 2, p. 1. Academic Press, New York, 1970.

[6] R. A. Johnson, W. Saur, and K. H. Jakobs, *J. Biol. Chem.* **254,** 1094 (1979).

[7] C. Londos and M. S. Preston, *J. Biol. Chem.* **252,** 5957 (1977).

activation or inhibition of adenylyl cyclases, is also metabolized. Consequently, the use of regenerating systems to counteract this alternative metabolism of ATP and/or GTP and the use of inhibitors of cAMP phosphodiesterases are nearly unavoidable.

Cyclic Nucleotide Phosphodiesterases. cAMP is effectively inactivated through the hydrolysis of its 3′-phosphate bond, yielding 5′-AMP (Step 9 in Fig. 1 above). Since cyclic nucleotide phosphodiesterase activity is substantial in most membrane preparations, these enzymes must be inhibited to measure accurately the rate of formation of cAMP by adenylyl cyclase. This is usually accomplished by the use of unlabeled cAMP in the reaction mixture or by the use of inhibitors of the enzyme, such as papaverine or alkylxanthines (e.g., 3-isobutyl-1-methylxanthine; IBMX). IBMX is also useful through its additional action of effectively blocking adenosine receptors. One must be cautious in the selection of a phosphodiesterase inhibitor, though, because some agents do not block all cAMP phosphodiesterases. For example, the sole use of the imidazolidinone derivative Ro 20-1724 [4-(3-butoxy-4-methoxybenzyl)-2-imidazolidinone] is not recommended. This compound has been useful in the study of adenosine receptor-mediated effects on adenylyl cyclases because it is not an adenosine receptor antagonist as are the alkylxanthines. Although it may substantially suppress the hydrolysis of cAMP in preparations from some tissues (pig coronary arteries), it does so incompletely in others (e.g., platelets). In preparations of human platelet membranes, for example, the addition of a second phosphodiesterase inhibitor is required. In our hands the most effective combinations of agents for inhibition of hydrolysis of labeled cAMP produced by adenylyl cyclase have been the following: 1 m*M* IBMX; cAMP (100 μM) with or without papaverine (100 μM); 100 μM anagrelide [6,7-dichloro-1,5-dihydroimidazol[2,1-*b*]quinazolinone monohydrochloride (BL-4162A); or anagrelide (100 μM) plus Ro 20-1724 (500 μM).[8]

ATP-Regenerating Systems. The accurate determination of adenylyl cyclase activities is adversely affected by the hydrolysis of ATP between the β- and γ-phosphates (Step 1 in Fig. 1), due to various membrane-bound ATPases, nonspecific phosphohydrolases, and flux through membrane-bound kinases (Step 10, Fig. 1) and phosphatases. Cleavage of the bond between the α- and β-phosphates occurs for ADP (Step 2, Fig. 1) by membrane phosphohydrolases and for ATP by nucleotide pyrophosphatase (Step 6, Fig. 1). Whether by (Step 2 or by Step 6) (Fig. 1), the result is 5′-AMP, which is rapidly hydrolyzed by 5′-nucleotidase (Step 3, Fig. 1) to adenosine (Ado). Adenosine can stimulate or inhibit (Step 5, Fig. 1) adenylyl cyclase, either indirectly via inhibitory (A_1) or stimulatory (A_2)

[8] E. A. Martinson and R. A. Johnson, unpublished observations (1986).

receptors, or inhibit the enzyme directly via the P-site. These reactions can be counteracted and their influence minimized by an ATP-regenerating system. The most commonly used systems have utilized creatine kinase and creatine phosphate or pyruvate kinase and phosphoenolpyruvate to counteract hydrolysis of ATP between the β- and γ-phosphates by catalyzing the conversion of ADP to ATP (Step 7, Fig. 1), but to counteract hydrolysis between the α- and β-phosphates myokinase (adenylate kinase) is used to convert 5′-AMP to ADP (Step 8, Fig. 1). Since the action of 5-nucleotidase cannot be reversed, the influence of the formed adenosine can be minimized by the use of adenosine deaminase (Step 4, Fig. 1), as the product inosine is without effect on adenylyl cyclase. In adenylyl cyclase reaction mixtures effective concentrations of these enzymes are as follows: myokinase, 100 μg/ml (Boehringer Mannheim, Indianapolis, IN, ammonium sulfate suspension from rabbit muscle); adenosine deaminase, 5 U/ml (Sigma, St. Louis, MO, ammonium sulfate suspension, Type VIII from calf intestinal mucosa).

The influence of nucleotide pyrophosphatase and to a lesser extent 5′-nucleotidase can be further minimized by pretreatment of membranes with 5 m*M* EDTA and 3 m*M* dithiothreitol.[9,10] The necessity for the additions to the assay or the effectiveness of the membrane-pretreatment with chelator and/or dithiothreitol depends on the source and purity of the adenylyl cyclase being studied. An example of the effectiveness of adenosine deaminase and myokinase to enhance hepatic adenylyl cyclase is shown in Table I for an assay conducted at a low (10 μM) concentration of ATP. Myokinase enhanced adenylyl cyclase activity and helped maintain ATP levels during the reaction, whereas adenosine deaminase did not affect the stability of ATP but enhanced enzyme activity, presumably by the removal of inhibitory levels of adenosine (Table I).

Although both creatine kinase and creatine phosphate or pyruvate kinase and phosphoenolpyruvate have been utilized as the basis of ATP-regenerating systems, neither is without its pitfalls. Both of the enzymes bind and utilize adenosine phosphates, and both substrates form weak complexes with divalent cations. Moreover, phosphoenolpyruvate has been shown to cause both stimulatory and inhibitory effects on the enzyme from liver and to inhibit the enzyme from heart[11] and contaminants in creatine phosphate have been found to cause both stimulatory and inhibitory effects on adenylyl cyclases.[9] For these reasons the preferable ATP-

[9] R. A. Johnson, *J. Biol. Chem.* **255,** 8252 (1980).

[10] R. A. Johnson and J. Welden, *Arch. Biochem. Biophys.* **183,** 2176 (1977).

[11] R. A. Johnson and E. L. Garbers, *in* "Receptors and Hormone Action" B. W. O'Malley and L. Birnbaumer, eds.), Vol. 1, p. 549. Academic Press, New York, 1977.

TABLE I
EFFECTS OF MYOKINASE AND ADENOSINE DEAMINASE ON ADENYLYL CYCLASE ACTIVITY AND ON ATP REGENERATION IN LIVER PLASMA MEMBRANES[a]

	Basal		NaF (10 mM)		Glucagon (1 μM) + GTP (10 μM)	
Additions	AC	ATP	AC	ATP	AC	ATP
None	8.5	0.57	22	0.50	35.5	0.60
Adenosine deaminase	11.5	0.52	31	0.52	52	0.62
Myokinase	10.5	0.88	33.5	0.81	57	0.86
Adenosine deaminase + myokinase	11	0.74	40.5	0.79	78.5	0.80

[a] Adenylyl cyclase activity (AC) is expressed as pmol cAMP (min·mg protein)$^{-1}$. Activity was determined with 10 μM MnATP, 400 μM excess $MnCl_2$, 10 mM creatine phosphate, 100 μg/ml creatine kinase, 1 mM IBMX, and 50 mM glycylglycine, pH 7.5, without additions (basal) or with 10 mM NaF or 1 μM glucagon and 10 μM GTP. Residual ATP (ATP) is the quantity (ATP with membranes)/(ATP without membranes), determined at the end of the 2-min reaction. ATP concentrations were determined by luciferase luminescence. Adenosine deaminase was 5 U/ml myokinase was 100 μg/ml. Adapted from Ref. 9.

regenerating system is creatine kinase (100 μg/ml; Boehringer Mannheim, from rabbit muscle) and creatine phosphate. Creatine phosphate should be used at concentrations low enough (e.g., 2 mM) to minimize the influence of the contaminants, or, preferably, be purified before use, for example, by anion-exchange chromatography.[9]

Enzyme Concentrations, Reaction Times, and Temperatures

Three additional factors that obviously influence adenylyl cyclase-catalyzed formation of cAMP interdependently are the concentration of enzyme, the time, and the incubation temperature.

Enzyme Concentration. Adenylyl cyclase is more active in crude membrane preparations from some tissues than from others, and the levels of the various contaminating enzymes that utilize adenine nucleotides (Fig. 1) also vary. Consequently, under any given set of reaction conditions it is imperative to establish (1) that sufficient enzyme is used to catalyze the formation of measurable amounts of [^{32}P]cAMP; (2) that concentrations of both ATP and creatine phosphate are adequate for sustaining stable ATP concentrations; and (3) that formation of cAMP is linear with respect to enzyme concentration. Formation of measurable amounts of [^{32}P]cAMP is improved with increasing specific radioactivity of [α-^{32}P]ATP and increasing enzyme concentration. However, increasing concentrations of crude membrane preparations of adenylyl cyclase typically also require increased amounts of creatine phosphate or proportionally decreased incubation times. Hence, enzyme concentration, ATP specific radioactivity, and incubation times must be adjusted to allow linear formation of cAMP with respect to both protein concentration and time. Useful ranges of specific radioactivity of [α-^{32}P]ATP are 10 to 200 cpm/pmol, depending principally on enzyme source. For studies of enzyme kinetics with respect to metal-ATP, the range of [α-^{32}P]ATP specific activity will be greater than this. Examples of, but not necessarily upper limits for, protein concentrations yielding linear product formation of 30° with 0.1 m*M* ATP and 5 m*M* creatine phosphate, 100 μg/ml creatine kinase, 100 μg/ml myokinase, 5 U/ml adenosine deaminase, and 10 m*M* $MnCl_2$ or $MgCl_2$ would be as follows (in mg/ml): heart, 0.6; liver, 1.2; kidney, 0.7; skeletal muscle, 0.3; adipocytes, 0.2; spleen, 0.5; human platelets, 0.2; bovine sperm particles, 1.0; washed particles from brain, 0.2; detergent-solubilized brain, 0.2.

Time. The incubation time for adenylyl cyclase reactions is dictated by a balance between rates of formation of cAMP from ATP, hydrolysis of cAMP to 5′-AMP by contaminating cyclic nucleotide phosphodiesterases, hydrolysis of ATP by a number of membrane-bound phosphohydrolytic enzymes (Fig. 1), inactivation of adenylyl cyclase by regulatory components, and denaturation of the enzyme. It is essential that linearity of product formation with respect to time be established. The reaction is typically linear with respect to time for crude membrane preparations, with the conditions given above, for 2 to 15 min and for purified enzyme for 60 min.

Incubation Temperature. Temperature can be used to advantage to exhibit certain characteristics of behavior of adenylyl cyclases as well as to modify rates of alternative substrate utilization and enzyme denaturation. Formation of cAMP is linear with respect to time for a longer period at 30° than at a more physiological 37°. However, the catalytic moiety is readily inactivated in the absence of protective agents by exposure to heat for short periods of time. For example, exposure of adenylyl cyclases from platelets and from S49 lymphoma wild-type and *cyc*$^-$ cells for 8 min at 35° causes 70 to 75% inactivation.[12,13] Comparable inactivation of adenylyl cyclases from bovine sperm and detergent-dispersed porcine brain occurred by exposure at 45° for 8 and 4 min, respectively.[12] Partial protection against thermal inactivation is afforded by forskolin (200 μ*M*), metal-ATP (submillimolar), the P-site agonist 2′,5′-dideoxyadenosine (250 μ*M*), guanine nucleotides (micromolar), and, for the Ca^{2+}/calmodulin-sensitive

[12] J. A. Awad, R. A. Johnson, K. H. Jakobs, and G. Schultz, *J. Biol. Chem.* **258,** 2960 (1983).
[13] V. A. Florio and E. M. Ross, *Mol. Pharmacol.* **24,** 195 (1983).

form of adenylyl cyclase, by Ca^{2+}/calmodulin (50 μM/5 μM).[12-16] In addition, adenylyl cyclase reactions conducted at different temperatures can be used to enhance selective regulatory properties of the enzyme. For example, inhibition of adenylyl cyclase mediated by guanine nucleotide-dependent regulatory protein (G_i), whether by hormone or stable guanine nucleotide, is more readily shown experimentally at lower temperatures (e.g., 24°), whereas activation, mediated by the stimulatory G protein (G_s), is evident at higher temperatures (e.g., 30°).[17]

Guanine Nucleotides

GTP is required for G_s and G_i mediation of hormone-induced activation and inhibition of adenylyl cyclases.[18-22] The more stable GTP analogs guanosine 5′-(β,γ-imino)triphosphate [GPP(NH)P] and guanosine 5′-*O*-(3-thiotriphosphate) (GTPγS) can substitute for GTP. Effects of these analogs are evident after a distinct lag phase, and preincubation of the enzyme with them will result in persistently activated or inhibited enzyme, depending on incubation conditions. In addition, the effectiveness of G_s and G_i in regulating adenylyl cyclase is further influenced by divalent cation, type, and concentration and by membrane perturbants (e.g., Mn^{2+} and detergents obliterate G_i-mediated inhibition). Half-maximal stimulation of adenylyl cyclase by hormones is usually observed with 10 to 50 nM GTP, or with 50 to 100 nM GPP(NH)P or GTPγS, with optimal stimulation between 1 and 10 μM GTP, GPP(NH)P, or GTPγS. Half-maximal inhibition occurs with 100 to 500 nM GTP, 10 to 100 nM GPP(NH)P, or 1 to 10 nM GTPγS, with optimal inhibition occurring with GTPγS over 100 nM, GPP(NH)P over 10 nM, and GTP over 1 μM. Consequently, even in relatively pure membrane preparations, enzyme activity may be increased somewhat by stimulatory hormones due to endogenously present GTP (e.g., as a contaminant of ATP; GTP-free ATP can be purchased from Sigma). The addition of GTP enhances stimulation further. By comparison, GTP must

[14] M. A. Brostrom, C. O. Brostrom, and D. J. Wolff, *Arch. Biochem. Biophys.* **191,** 341 (1978).
[15] R. S. Salter, M. H. Krinks, C. B. Klee, and E. J. Neer, *J. Biol. Chem.* **256,** 9830 (1981).
[16] J. P. Harwood, H. Löw, and M. Rodbell, *J. Biol. Chem.* **248,** 6239 (1973).
[17] D. M. F. Cooper and C. Londos, *J. Cyclic Nucleotide Res.* **5,** 289 (1979).
[18] M. Rodbell, L. Birnbaumer, S. L. Pohl, and H. M. J. Krans, *J. Biol. Chem.* **246,** 1877 (1971).
[19] K. H. Jakobs, W. Saur, and G. Schultz, *FEBS Lett.* **85,** 167 (1978).
[20] C. Londos, D. M. F. Cooper, W. Schlegel, and M. Rodbell, *Proc. Natl. Acad. Sci. U.S.A.* **75,** 5362 (1978).
[21] D. M. F. Cooper, W. Schlegel, M. C. Lin, and M. Rodbell, *J. Biol. Chem.* **254,** 8927 (1979).
[22] E. Perez-Reyes and D. M. F. Cooper, *J. Neurochem.* **46,** 1508 (1986).

be added for hormonal inhibition of adenylyl cyclase, and GTP-dependent inhibition is often best elicited with enzyme that has been stimulated by forskolin or a stimulatory hormone. In addition, the concentrations of guanine nucleotides necessary for regulation of adenylyl cyclase activity are dependent on enzyme source and incubation temperature and are influenced by the relative activities and abundances of G_s and G_i.

Radioactive Substrates: [^{3}H]ATP versus [α-^{32}P]ATP

[^{3}H]ATP and [α-^{32}P]ATP are commonly used as labeled substrate for measuring adenylyl cyclase catalytic activity. The use of each has both advantages and disadvantages, some of which are described below.

[^{3}H]ATP Advantages

The one main advantage to the use of [^{3}H]ATP as labeled substrate is its long half-life (~12.3 years). This allows the nearly complete usage of purchased isotope without regard to loss through decay. Low usage rates may adequately compensate for its being initially substantially more expensive than [α-^{32}P]ATP.

[^{3}H]ATP Disadvantages

There are several significant disadvantages to the use of [^{3}H]ATP as substrate in adenylyl cyclase reactions. First, tritium-labeled adenine nucleosides and nucleotides are chemically unstable in that the C-8 tritium exchanges with water, especially under alkaline conditions. This results in a continuous loss of tritium to ^3H_2O that can occur at the rate of several percent per month. Consequently, for accurate estimations of substrate specific activity the ^3H_2O must be removed periodically, either chromatographically or by lyophilization. Both procedures necessitate undue handling of and exposure to moderate quantities of isotope and have the potential of major isotope spills in a laboratory environment.

The low energy of the beta decay of tritium necessitates the use of scintillation cocktails to detect [^{3}H]cAMP, and the long half-life of tritium means that large volumes of liquid radioactive waste, which necessarily also contains large quantities of organic solvents, must be disposed rather than be allowed to dissipate through radioactive decay as would be the case with ^{32}P. Disposal of radioactive waste, especially mixed with scintillation cocktail, is an expensive and undesirable consequence of the use of tritium-labeled substrate. The low energy of tritium decay also makes it more difficult to detect if there are inadvertant spills or contamination in a laboratory and could thereby lead to undue exposure of laboratory personnel to low-energy radiation.

Breakdown products of [^{3}H]ATP or [^{3}H]cAMP include various nucleotides and nucleosides, as well as xanthine, hypoxanthine, and others, that are also labeled. Chromatographic techniques must take this into consideration. These various breakdown products and the continuous formation of ^3H_2O from tritium-labeled adenine nucleotides contribute to blank values [counts per minute (cpm) for samples in the absence of enzyme], with the Dowex 50/Al_2O_3 column system described below being substantially higher than those obtained with [α-^{32}P]ATP as substrate. With adenylyl cyclases of low specific activity in crude membrane preparations, such high blank values may constitute a substantial percentage of the [^{3}H]cAMP formed enzymatically.

[α-^{32}P]ATP Advantages

There are several important advantages to the use of [α-^{32}P]ATP as substrate for adenylyl cyclase reactions. The specificity of labeling of [α-^{32}P]ATP, which is dictated by the enzymatic means typically used for its synthesis (as per Ref. 23), means that only α-phosphates are labeled and, since the α-phosphate of ATP is not readily transferred to other compounds, that only nucleotides derivable from ATP will be labeled. Additional products that could result from contaminating activities in crude membrane preparations, for example, [α-^{32}P]ADP, [α-^{32}P]AMP, [α-^{32}P]IMP, and [^{32}P]P_i, due to differences in their ionic properties are all readily separated from [^{32}P]cAMP.

^{32}P is a high-energy beta emitter that allows detection by Cerenkov radiation and obviates the use of scintallation cocktails, that is, can be detected in aqueous solutions with efficiencies approaching that of tritium in scintillation cocktails but with little influence of agents that typically quench detection of tritium. The high energy of the beta emission also allows easy detection of inadvertant spills with a Geiger/Müller detector and thereby actually enhances laboratory safety. The short half-life of ^{32}P (~14.3 days) allows waste to be decayed off before disposal, effectively eliminating expensive or awkward disposal of radioactive waste, whether solid or liquid.

[α-^{32}P]ATP Disadvantages

The short half-life of ^{32}P also implies that the usefulness of the isotope is often lost to decay before the [α-^{32}P]ATP is fully utilized. Consequently, if usage rates are low the decay of the isotope may result in [α-^{32}P]ATP actually being more expensive than [^{3}H]ATP. Blank values can depend on the quality of substrate, even with the double column procedures described below. The quality can vary substantially among different suppliers and in different batches from a given supplier. Blank values for the adenylyl cyclase assay should be supplied on the product data sheet. The quality of [α-^{32}P]ATP can be assured by its purification before use or through its laboratory synthesis from carrier-free $^{32}P_i$ by published procedures,[23] a process which also includes its purification.

Stopping the Reaction

There are several good methods for stopping adenylyl cyclase reactions, the choice depending in part on whether [^{3}H]ATP or [α-^{32}P]ATP is used as substrate and on the method used for estimating loss of labeled cAMP during its purification. Some of these considerations are dealt with below.

Stopping with Zinc Acetate/Na_2CO_3/cAMP or /[^{3}H]cAMP

The use of coprecipitation or adsorption of nucleotides with inorganic salts dates from an early assay for adenylyl cyclase developed by Krishna *et al.*,[24] who used a combination of column chromatography on Dowex 50 and precipitation with $ZnSO_4$ and $Ba(OH)_2$, yielding the insoluble salts of $BaSO_4$ and $Zn(OH)_2$, which adsorb phosphomonoesters and polyphosphates but not cyclic nucleotides. A disadvantage in the use of $ZnSO_4$/$Ba(OH)_2$ is that cAMP may be formed nonenzymatically from ATP at alkaline pH, especially at elevated temperatures, leading to variable and high blank values. This problem is circumvented by the use of other salt combinations or by the use of acidic inactivation of adenylyl cyclase. The effectiveness of a variety of combinations of inorganic salts, for example, $ZnSO_4/Na_2CO_3$, $CdCl_2/Na_2CO_3$, $ZnSO_4/BaCl_2$, $BaCl_2/Na_2CO_3$, to bind labeled ATP, ADP, AMP, cAMP, and adenosine has been cataloged previously in this series.[25] Since comparable separation of ATP and cAMP can be achieved with columns packed with $ZnCO_3$[25] or Al_2O_3,[26] it is likely that adsorption to the insoluble inorganic salts, rather than coprecipitation with them, is the basis of the separation of cAMP from the multivalent nucleotides and hence the basis of their usefulness in assays of adenylyl or guanylyl cyclases. It is important to emphasize that since none of these salt combinations will separate cAMP from adenosine, [α-^{32}P]ATP is

[23] R. A. Johnson and T. F. Walseth, *Adv. Cyclic Nucleotide Res.* **10,** 135 (1979).
[24] G. Krishna, B. Weiss, and B. B. Brodie, *J. Pharm. Exp. Ther.* **163,** 379 (1968).
[25] P. S. Chan and M. C. Lin, this series, Vol. 38, p. 38.
[26] A. A. White and T. V. Zenser, *Anal. Biochem.* **41,** 372 (1971).

preferred to [^{3}H]ATP as substrate, to avoid the possibility that the [^{3}H]cAMP formed through adenylyl cyclase may be contaminated with [^{3}H]adenosine, [^{3}H]inosine, or other labeled compounds deriving from adenine nucleotides.

The characteristics of the insoluble inorganic salts are taken advantage of in the following procedure adapted from Jakobs *et al.*[27]

Reagents

Zinc acetate/cAMP: 120 m*M* $Zn(C_2H_3O_2)_2 \cdot 2H_2O$ (FW 219.49) is prepared in deionized, double-distilled, or Millipore grade water that has been boiled and then cooled to remove dissolved carbon dioxide, and cAMP is then added (165 mg/liter, to 0.5 m*M*); this is kept refrigerated and tightly capped between uses

Zinc acetate/[^{3}H]cAMP: prepare as above except add tritiated cAMP, from which ^3H_2O has been removed, to an amount of zinc acetate needed for a given assay to yield approximately 10,000 to 20,000 cpm/ml, when counted in the same volume of eluate (see below) used for samples

Sodium carbonate: 144 m*M* Na_2CO_3, anhydrous (FW 106.0)

Both zinc acetate and Na_2CO_3 solutions are stored in and dispensed from glass repipetting dispensers.

Procedure

1. Adenylyl cyclase reactions, typically 50 to 200 μl in 1.5-ml plastic Eppendorf tubes, are terminated by the addition of 0.6 ml of 120 m*M* zinc acetate/cAMP or zinc acetate/[^{3}H]cAMP. Aliquots of these stopping solutions are taken for determining absorbance ($A_{259\ nm}$) or radioactivity, as appropriate; the values are to be used for quantitating sample recovery (see below).
2. One-half milliliter of 144 m*M* Na_2CO_3 is added to precipitate $ZnCO_3$ and multivalent adenine nucleotides.
3. Samples are placed on ice, or they can be kept refrigerated or frozen overnight. The $ZnCO_3$ precipitate is sedimented by centrifugation in a benchtop centrifuge. Pellets of frozen samples are smaller and heavier than those of unfrozen samples.
4. The supernatant fractions are decanted onto columns for purification of sample cAMP.
5. Assay blanks are prepared by substituting enzyme buffer for enzyme.
6. A potential disadvantage of this method is that if [^{3}H]cAMP is

[27] K. H. Jakobs, W. Saur, and G. Schultz, *J. Cyclic Nucleotide Res.* **2,** 381 (1976).

used, it becomes necessary to use, and hence eventually dispose of, scintillation cocktails for quantitating [^{32}P]cAMP and its recovery.

7. An advantage of this procedure that has lead to its use in many laboratories is that over 98% of all multivalent nucleotides, namely, substrate [α-^{32}P]ATP, [α-^{32}P]ADP, [^{32}P]AMP, as well as any [^{32}P]P_i, are retained in the capped Eppendorf assay tubes in the $ZnCO_3$ precipitate. The waste radioactivity is thus highly confined, occupies little volume, and can be allowed to decay off and then be dealt with as normal solid waste.

Stopping with ATP/SDS/cAMP with or without [^{3}H]cAMP

The following method, adapted from Salomon *et al.*,[1,28] relies on sodium dodecyl sulfate to inactivate adenylyl cyclase and on unlabeled ATP and unlabeled cAMP to overwhelm adenylyl cyclase and cAMP phosphodiesterases with substrate and thereby effectively prevent the further formation or degradation of [^{32}P]cAMP.

Reagents. "Stopping solution" contains 2% (w/v) sodium dodecyl sulfate (SDS), 40 m*M* ATP, 1.4 m*M* cAMP, pH 7.5, and approximately 100,000 cpm/ml [^{3}H]cAMP, to monitor recovery of [^{32}P]cAMP. Alternatively, [^{3}H]cAMP could be omitted from the stopping solution and added separately.

Procedure

1. Adenylyl cyclase reactions, typically 50 to 200 μl in 13 $\times$ 65 mm glass or plastic tubes or in 1.5-ml plastic Eppendorf tubes, are terminated by the addition of 100 μl of the stopping solution.
2. To achieve full membrane solubilization in cases of high membrane content, it is advisable to boil the test tubes for 3 min at this stage. This also facilitates the rate of chromatography. Hence, use heat-stable tubes.
3. The mixtures in the reaction tubes are then diluted and decanted onto chromatography columns for purification of sample cAMP.
4. Assay blanks are prepared by omitting enzyme or by adding enzyme after the stopping solution.
5. A disadvantage of this procedure is that all radioactive compounds, including unused substrate [α-^{32}P]ATP, [α-^{32}P]ADP, [^{32}P]AMP, [^{32}P]P_i, as well as degradation products of [^{3}H]cAMP, are passed with the labeled cAMP onto the chromatography column and are typically eluted in a fall through fraction that must be collected and

[28] Y. Salomon, C. Londos, and M. Rodbell, *Anal. Biochem.* **58,** 541 (1974).

then dealt with as liquid radioactive waste. To minimize this waste, see below.

6. A disadvantage of either stopping procedure when [^{3}H]cAMP is used to monitor recovery of sample [^{32}P]cAMP is that scintillation cocktails must be used and consequently disposed.

Chromatographic Alternatives

The characteristic property of alumina and other insoluble inorganic salts to bind multivalent nucleotides but not cAMP is the central feature of a number of variations of assays for adenylyl and guanylyl cyclases. White and Zenser[26] passed reaction mixtures over columns of neutral alumina that were equilibrated and then developed with neutral buffer. However, assay blanks with this procedure were variable and depended highly on the radiochemical purity of the α-^{32}P-labeled substrate. Salomon *et al.*[28] and later Wincek and Sweat[29] showed that sequential chromatography on Dowex 50 and alumina produced an assay for adenylyl cyclase that was more consistent than alumina columns alone, a combination that has also been utilized for the assay of guanylyl cyclase.[30] Additional variations on this procedure have been reported by a number of investigators. For example, nearly quantitative separation of cAMP from ATP was achieved by a combination of precipitation with inorganic salts (zinc acetate/Na_2CO_3) followed by chromatography on alumina.[27] Another approach to lower blank values has been to inactivate adenylyl cyclase by stopping the reaction with the addition of acid (e.g., 1 ml of 1 *M* HCl) followed by heating (90° for 8–10 min) and then precipitation with $ZnSO_4/Ba(OH)_2$[31] or, subsequent to acidic heat inactivation (4 min at 95° in 0.165 *N* HCl), chromatography on a selected alumina column.[32] Thus, to minimize the influence of variations in the quality of [α-^{32}P]ATP, the method of choice has become sequential chromatography with Dowex 50 and then alumina columns.[1,28] The procedure is as follows.

Reagents

Dowex 50, H^+ form [e.g., Bio-Rad (Richmond, CA) AG50-X8, 100–200 mesh]. Before use the Dowex 50 is washed sequentially with approximately 6 volumes each of 0.1 *N* NaOH, water, 1 *N* HCl, and water. The Dowex 50, in an approximately 2 : 1 slurry, is

[29] T. J. Wincek and F. W. Sweat, *Anal. Biochem.* **64,** 631 (1975).

[30] J. A. Nesbitt III, W. B. Anderson, Z. Miller, I. Pastan, T. R. Russell, and D. Gospodarowicz, *J. Biol. Chem.* **251,** 2344 (1976).

[31] C. Nakai and G. Brooker, *Biochim. Biophys. Acta* **391,** 222 (1975).

[32] R. Counis and S. Mongongu, *Anal. Biochem.* **84,** 179 (1978).

then poured into columns (~0.6 × 4 cm). After each use, Dowex 50 columns are regenerated by washing with 5 ml of 1 *N* HCl, then stored until reused. Before use the columns are washed 3 times with 10 ml of water. Columns can be reused dozens of times. If flow rates decrease, columns should be regenerated with NaOH, water, and HCl as above.

Neutral alumina [e.g., Bio-Rad AG7, 100–200 mesh; Sigma WN-3; ICN (Costa Mesa, CA) Alumina N, Super I]: The source of Al_2O_3 is less critical with the two-column procedure than if it is used alone. The alumina (0.6 g) may be poured dry into columns (e.g., with a plastic scoop or a large disposable plastic syringe from which the alumina is allowed to drain).

Elution buffers: The original method of Salomon *et al.*[28] calls for 100 m*M* imidazole, pH 7.5. An equally effective and less expensive alternative is 100 m*M* Tris-Cl, pH 7.5. The purpose of the buffer is to elute cyclic nucleotides. Since eluate from the Dowex 50 columns is acidic, which enhances adsorption of cyclic nucleotides to alumina, elution of cyclic nucleotides is achieved principally through an increase in the pH of the buffer rather than through increased ionic strength. Consequently, there is probably wide latitude in buffer choice here.

Before initial use alumina columns *must* be washed once with elution buffer, either 10 ml of 100 m*M* Tris-Cl or 10 ml of 1 *M* imidazole, pH 7.5, otherwise the procedure does not work right away. After each use the columns are washed with 10 ml of 100 m*M* Tris-Cl or 10 ml of 100 m*M* imidazole, pH 7.5. Once poured both alumina and Dowex 50 columns may be reused virtually indefinitely, though they may need to be topped off occasionally.

Apparatus. Rapid flow rates for the alumina columns, and consequently short chromatography times, are achieved with glass columns with a large cross-sectional area and a coarse sintered glass plug to retain the alumina (Fig. 2). Satisfactory dimensions are alumina to approximately 1 cm in a column 11 mm i.d. by 9 cm attached to a 4-cm glass funnel (24 mm i.d.). [Smaller columns (e.g., ~0.6 × 2 cm, alumina) while allowing satisfactory chromatographic performance are slow.)

Both Dowex 50 and alumina columns are most conviently used if they are mounted in racks (e.g., plastic) with spacing identical to that of the racks of scintillation vials to be used. The design of the racks supporting the Dowex 50 columns should be such that the columns can be conveniently mounted aboved the alumina columns so that the eluate of the Dowex 50 columns can drip directly onto the alumina columns. Similarly, the design

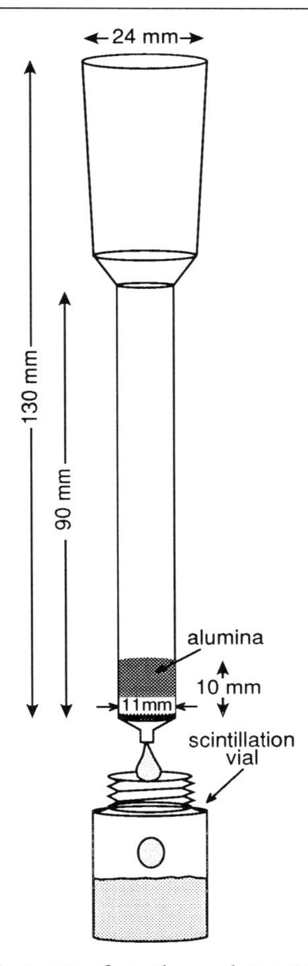

FIG. 2. Apparatus for column chromatography.

of the racks supporting the alumina columns should allow the eluate from them to drip directly into scintillation vials.

Carbon filters for the removal of adsorbable radioactive materials from column eluates are Gelman Ann Arbor, MI (#12011) Carbon Capsules, each containing 100 g activated charcoal.

Procedure. Two alternative procedures are described, the choice of which depends on the quality of ^{32}P-labeled substrate. The first procedure should be adequate with all but the poorest quality substrate and the second procedure should lower blank values further if necessary.

Water elution of Dowex 50

1. Whether reactions are stopped by the zinc acetate/Na_2CO_3 or the ATP/SDS/cAMP method, the resulting solutions are decanted directly onto the Dowex 50 columns.
2. The Dowex 50 columns are then washed with about 3 ml water. (The actual volume necessary for this step may vary slightly from batch to batch or with the age of the Dowex 50 and should be determined.[1] The eluate from this wash contains [^{32}P]P_i, [α-^{32}P]ATP, and [α-^{32}P]ADP and should be disposed of by the procedure described below.
3. The Dowex 50 columns are then mounted above a comparable number of alumina columns so that the eluate drips directly onto the alumina. Dowex 50 columns are washed with 8 ml water. The eluate from the Dowex 50 columns is slightly acidic and causes cAMP to be retarded on the alumina column.
4. After the eluate from the Dowex 50 columns had dripped onto and through the alumina columns, the alumina columns are placed over scintillation vials.
5. cAMP is eluted from the alumina columns directly into scintillation vials. The volume of elution buffer used depends on whether unlabeled cAMP or [^{3}H]cAMP is used to quantitate sample recovery and on the types of vials used in the scintillation counter. It is important to use sufficient buffer to elute all the cAMP as well as to optimize counter efficiency, which is dictated by the geometry of the counter phototubes. For example, if recovery is monitored with unlabeled cAMP and [^{32}P]cAMP is determined by Cerenkov radiation in 20 ml counting vials, [^{32}P]cAMP is eluted from alumina columns with 8 ml of 100 m*M* Tris-Cl. Smaller volumes do not give optimal counting efficiency. Following counting, the absorbance at 259 nm is determined on an aliquot of the sample to quantitate recovery of unlabeled cAMP. If recovery is monitored with [^{3}H]cAMP, both [^{3}H]cAMP and [^{32}P]cAMP are eluted with 4 ml of 100 m*M* imidazole into 10-ml vials to which 5 ml scintillation cocktail is then added. The smaller vials spare expensive cocktail. Sample recovery is determined from dual channel counting.
6. Scintillation counting of ^{32}P by Cerenkov radiation is achieved with a single channel with wide open windows.
7. Counting of samples containing both ^{3}H and ^{32}P can be achieved in a two-channel scintillation counter with windows adjusted such that there is zero crossover of ^{3}H into the ^{32}P window and small but measureable crossover of ^{32}P into the ^{3}H window.

Acid elution of Dowex 50. The following procedure is adapted from White and Karr.[33]

1. Before use the Dowex 50 columns are washed with 10 ml of 10 m*M* $HClO_4$.
2. Whether reactions are stopped by the zinc acetate/Na_2CO_3 or the ATP/SDS/cAMP method, the resulting solutions are decanted directly onto the Dowex 50 columns.
3. The Dowex 50 columns are then washed with 6 ml of 10 mM $HClO_4$. (The actual volume necessary for this step may vary from batch to batch or with the age of the resin and should be determined.) The eluate from this wash contains $[^{32}P]P_i$, $[\alpha\text{-}^{32}P]ATP$, and $[\alpha\text{-}^{32}P]ADP$ and should be disposed of by the procedure described below.
4. The Dowex 50 columns are mounted directly above a comparable number of alumina columns and then washed with 8 ml of 10 m*M* $HClO_4$ that is allowed to drain through both columns.
5. The alumina columns are then washed with 10 ml of water. This eluate is discarded.
6. The alumina columns are mounted above a rack of scintillation vials and the cAMP is eluted as above.

Disposal of waste isotope. To avoid contamination of wastewater, all radioactive waste from the above procedures is collected and pooled. It is poured onto a large Büchner funnel, containing perhaps 250 g alumina (e.g., Fisher, Al_2O_3, anhydrous), attached in series to two parallel 100-g carbon filters and a flask attached to a water aspirator. (The purpose of the flask is to allow an aliquot of the filtered waste to be monitored for radioactivity before the waste is discarded down the drain.) By use of alumina and carbon filters, virtually no ^{32}P radioactivity is discarded in wastewater, though some $^{3}H_2O$ will be lost if tritiated nucleotides are used ($[^{3}H]ATP$ or $[^{3}H]cAMP$), and any radioactive waste can be treated as compact solid waste. This is especially useful for tritiated nucleosides and nucleotides. The ^{32}P-labeled solid waste can be allowed to decay off. The alumina can be used almost indefinitely, whereas the carbon filters tend to clog with prolonged use and need to be replaced periodically (e.g., annually).

Data Analysis

Calculation of adenylyl cyclase activities determined with radioactive substrates is straightforward but must take into consideration the inevitable loss of sample cAMP that occurs during chromatographic purification.

[33] A. A. White and D. B. Karr, *Anal. Biochem.* **85,** 451 (1978).

For this reason sample recovery is determined either with unlabeled cAMP or with cAMP that is labeled with a second isotope. In addition, since many adenylyl cyclases exhibit low activities, especially under basal assay conditions, the radioactivity measured in the sample in the absence of enzyme (no enzyme blank) can represent sizable percentage of that measured with enzyme. Consequently, it becomes important to consider how this value is to be treated in the calculation of activity. If there is measurable nonenzymatic formation of cAMP from ATP, as may be the case under alkaline assay conditions, especially in the presence of manganese, the labeled cAMP must be corrected for sample loss during purification. However, if it can be established that the sample radioactivity in the absence of enzyme is due to ^{32}P-labeled contaminants in the sample, that is, ^{32}P-labeled compounds not adsorbed by alumina[33] or as determined through alternative chromatographic techniques, the blank value should not be corrected for sample recovery. Such a correction would give rise to an erroneously high blank value, and the apparent enzyme activity would be lower than it should be. Both sample recovery and assay blank adjustments to the determination are readily made with programmable calculators, though they are more conveniently done with a computer since programs can be written to accommodate variable amounts of protein, substrate concentrations, assay times, and volumes, can be extended to the computation of enzyme kinetic constants, and can be interfaced with graphic plotters. In addition, scintillation counters may be attached directly to such computers to enhance data acquisition and processing. Examples of these calculations are given below.

Example 1. For the calculation of adenylyl cyclase activities with [^{3}H]cAMP used for sample recovery, the assumption is made that the windows for the ^{32}P and ^{3}H channels of the scintillation spectrometer have been set so there is zero crossover of ^{3}H cpm into the ^{32}P channel. The calculation compensates for crossover of ^{32}P cpm into the ^{3}H channel.

$$\begin{aligned}\text{Velocity} = {} & (\text{sample } ^{32}\text{P cpm} - \text{no enzyme } ^{32}\text{P cpm}) \\ & \times \text{ATP concentration} \times \text{reaction volume/fraction of} \\ & \text{sample counted/}([\alpha\text{-}^{32}\text{P}]\text{ATP cpm} - \text{no enzyme } ^{32}\text{P cpm}) \\ & \times [^{3}\text{H}]\text{cAMP std cpm/(sample } ^{3}\text{H cpm} - \{(\text{sample } ^{32}\text{P cpm} \\ & - \text{no enzyme } ^{32}\text{P cpm}) \times {}^{32}\text{P cpm in } ^{3}\text{H channel/} \\ & [\alpha\text{-}^{32}\text{P}]\text{ATP-cpm})\}\text{/time/protein}\end{aligned}$$

[^{3}H]cAMP std cpm is the value that would represent 100% recovery of the added [^{3}H]cAMP, for example, the total ^{3}H cpm in the 0.6 ml of zinc acetate containing [^{3}H]cAMP used to stop the reaction, counted under comparable quench conditions used to count the samples.

Example 2. An analogous though simpler calculation is used for activities when unlabeled cAMP is used for sample recovery and is the same whether [α-^{32}P]ATP or [^{3}H]ATP is used as substrate tracer.

Velocity = (sample cpm − no enzyme cpm)
× ATP concentration × reaction volume/fraction of sample counted/(substrate cpm − no enzyme cpm) × cAMP standard A_{259}/sample A_{259}/time/protein

cAMP standard A_{259} is the optical density at 259 nm that would represent 100% recovery of the added unlabeled cAMP. This value usually also includes a factor to compensate for the volume of the final sample. In the example given here for samples chromatographed first on Dowex 50 then on Al_2O_3 columns, samples are 8 ml. For example, the optical density (A_{259}) of the 0.6 ml of zinc acetate containing unlabeled cAMP is typically determined on an aliquot diluted 40-fold in 100 m*M* Tris-C pH 7.5, and gives a value of approximately 0.2. Hence, in this example, $0.2 \times 40 \times 0.6$ ml/8 ml → 0.6 for cAMP standard A_{259}.

For either procedure the determinations of velocity assume that ^{32}P cpm observed in the absence of enzyme is *not* cAMP, and no correction is made for loss during purification of those samples. This is an important assumption only in instances when enzyme activity is low and the cpm observed in the absence of enzyme represents a sizable percentage of sample cpm. For both calculations the value for "fraction of sample counted" is usually 1 and would be less than 1 only if an aliquot of the sample were used for some other purpose. The velocities obtained are in nanomoles cAMP formed per minute per milligram protein when substrate concentration is entered as micromolar, time as minutes, protein as micrograms per tube, and reaction volume as microliters. If protein is not known or if it is not desirable to normalize to protein, a value of 1 is used and velocities are picomoles cAMP formed per minute per tube.

Acknowledgments

Yoram Salomon is the Charles and Tillie Lubin Professor of Hormone Research. Research in the laboratory of R.A.J. was supported by Grant DK38828 from the National Institutes of Health.

[2] Cellular Responsiveness to Hormones and Neurotransmitters: Conversion of [^{3}H]Adenine to [^{3}H]cAMP in Cell Monolayers, Cell Suspensions, and Tissue Slices

By YORAM SALOMON

Changes in intracellular cyclic adenosine 3′,5′-monophosphate (cAMP) levels reflect the state of responsiveness of cells to respective hormones/neurotransmitters. Thus, by comparing the intracellular levels of cAMP in stimulated vis-à-vis unstimulated cells, the response of the subject cells to a given substance can be accurately assessed, and the presence of putative receptors can be deduced. The method described here provides a simple and rapid means for assessing responsiveness and sensitivity of cells to different stimulants, as well as for screening the receptor repertoire of the cell.

Assay Method

The following procedure enables the determination of cAMP levels in intact stimulated cells (monolayers/suspensions) or tissue slices relative to unstimulated controls. The cells are initially preincubated with [^{3}H]adenine, in order to label the adenine nucleotide pool of the cell.[1,2] Subsequently, following a wash to remove extracellular [^{3}H]adenine, the cells are further incubated in fresh medium containing the test substance(s) and appropriate controls. During the incubation step, the cells are permitted to respond to the extracellular challenge. The incubation is terminated by the addition of perchloric acid (PCA), which inactivates the metabolism of the cell, denatures the cellular proteins, and permits extraction of all cellular soluble nucleotides, including the [^{3}H]adenine nucleotides. The clear, protein-free PCA extract is then neutralized, and [^{3}H]cAMP is isolated by the Dowex/alumina double-column technique,[2,3] as described in [1], this volume.

As desired, the procedure enables the determination of relative intracellular cAMP levels in treated and untreated cells or tissues, in cells plus the conditioned medium, or in the conditioned medium only.

Reagents. For experiments with cultured cells, use appropriate culture

[1] J. L. Humes, M. Rounbehler, and F. L.Kuel, Jr., *Anal. Biochem.* **32,** 210 (1969).
[2] Y. Salomon, *Adv. Cyclic Nucleotide Res.* **10,** 35 (1979).
[3] Y. Salomon, C. Londos, and M. Rodbell, *Anal. Biochem.* **58,** 541 (1974).

medium. For experiments with dissociated cells or tissue slices, use appropriate physiological medium. *Note:* Media should not contain unlabeled adenine.

1-Methyl-3-isobutylxanthine (IBMX), 10 m*M*: dissolve in water, in a boiling water bath (keep frozen and thaw for use by heating; stable for several months)

Perchloric acid (PCA), 25% (keep in the dark; stable)

KOH, approximately 4.2 *N* standardized (v/v) with 25% PCA reagent (stable)

Cyclic AMP, 0.1 *M*, neutralized with Tris to pH 7.5 (stable frozen)

[^{3}H]Adenine, 1 mCi/ml

Hormones, stimulants, and test substances:

PCA 2.5%, cAMP 0.1 m*M* (to be prepared freshly from stock solutions and to be used ice cold)

PCA 5%, cAMP 0.2 m*M* (to be prepared freshly from stock solutions and to be used ice cold)

PCA 25%, cAMP 1 m*M* (to be prepared freshly from stock solutions and to be used ice cold)

For the use of cell monolayers, cell suspensions, or tissue slices, continue to the appropriate section below.

Procedure for Monolayers

The procedure described below is for cell monolayers cultured in 24-well plates. It is recommended that the experiment be performed in triplicate wells for each test condition.

Solutions

Preincubation medium: [^{3}H]Adenine, 5–10 μCi/ml in appropriate culture medium (0.5 ml/well); keep a 0.1-ml sample for counting, to calculate total radioactivity added (TA)

Incubation medium: Appropriate culture medium without serum but containing 0.1 mg/ml bovine serum albumin (BSA) and 0.1 m*M* IBMX or other inhibitor of cyclic nucleotide phosphodiesterases

Hormones and stimulants, dissolved freshly in incubation medium; calculate 0.5 ml/well

Preincubation. At this stage, cells are preloaded with [^{3}H]adenine. Aspirate the medium from the wells and replace with warm (37°) preincubation medium, 0.5 ml/well. Incubate (37°) for 90–120 min* in a CO_2 incubator.

* The conditions cited are typical for M2R mouse melanoma cells, grown in Dulbecco's modified Eagle's medium: Ham's F12 medium (DMEM : F12) containing 10% horse serum,

Incubation. At this stage, prelabeled cells are washed and stimulated to form [^{3}H]cAMP. Place the dish on the table on insulating paper and aspirate the medium. (*Caution:* Medium is radioactive! Dispose of carefully.) Add 0.5 ml of test substance (37°), dissolved in incubation medium. Prepare zero time controls at this step (see next paragraph). Incubate at 37°* in a CO_2 incubator, for a time appropriate for the test substance.

Stop Reaction. For determination of intracellular [^{3}H]cAMP (cells only), aspirate the incubation medium (be careful not to lose cells) and add 1 ml of 2.5% PCA, 0.1 m*M* cAMP. Continue to next paragraph. For the zero time control, add 1 ml of 2.5% PCA, 0.1 m*M* cAMP to *one* set of triplicate wells. Continue to next paragraph. For determination of [^{3}H]cAMP in cells plus medium, add 0.5 ml of 5% PCA, 0.2 m*M* cAMP. Cells usually do not come off, but if the BSA/serum of the medium shows up as turbidity, leave for 30 min at 4° for nucleotide extraction. Then transfer the medium to test tubes and centrifuge to obtain a clear, protein-free supernatant. Continue to next paragraph. For a zero time control, add 0.5 ml of 5% PCA, 0.2 m*M* cAMP, followed by 0.5 ml incubation medium.

Keep culture dishes cold for 30 min, with occasional agitation for efficient [^{3}H]cAMP extraction. Then transfer 0.7–0.9 ml (MX) of the cell-free, clear PCA extract to a set of marked test tubes and prepare the extracts for chromatography (see below).

Procedure for Cell Suspensions

The following procedure is appropriate for cells grown in suspension or cells grown as monolayers, loaded with [^{3}H]adenine, and suspended for the experiment. It is recommended that the experiment be performed in triplicate.

Solutions

Preincubation medium: As described above, 1 ml/0.5–1 × 10^6 cells
Incubation medium: As described above, but containing 0.2 m*M* IBMX (0.25 ml/point) or other inhibitor of cyclic nucleotide phosphodiesterases
Hormones, etc., dissolved in incubation medium

Preincubation. At this stage, cell suspensions are preloaded with [^{3}H]-adenine. For cells grown in suspension, wash the cells free of culture

as used in the author's laboratory.[4–6] These times are reasonable as starting conditions for various other cell types, but conditions must be verified experimentally for each system.

[4] J. E. Gerst, J. Sole, J. P. Mather, and Y. Salomon, *Mol. Cell. Endocrinol.* **46,** 137 (1986).

[5] J. E. Gerst, J. Sole, and Y. Salomon, *Mol. Pharmacol.* **31,** 81 (1987).

[6] J. E. Gerst and Y. Salomon, *J. Biol. Chem.* **263,** 7073 (1988).

medium by centrifugation. Resuspended cell pellet to a concentration of 0.5–1 × 10^6 cells/ml in preincubation medium and incubate (37°) for 90–120 min* in a CO_2 incubator/spinner. Take a sample of 0.1 ml of this medium for counting, to calculate total radioactivity of [^{3}H]adenine added (TA). Centrifuge cells and aspirate medium. (*Caution:* Medium is radioactive! Dispose of carefully.) Resuspend cells in warm (37°) culture medium (2–4 × 10^5 cells/250 μl/point). Continue to the incubation step.

For cells grown as monolayers, use, in principle, the same conditions. At this stage, cells are preloaded with [^{3}H]adenine. Aspirate the culture medium and add preincubation medium (1 ml/0.5–1 × 10^6 cells) as described above and incubate for 90–120 min at 37°.* Aspirate the medium. (*Caution:* Radioactive! Dispose of carefully.) Then scrape or shake cells off† and resuspend them in warm (37°) culture medium (2–4 × 10^5 cells/250 μl/point).

Incubation. At this stage, prelabeled cells are stimulated to form [^{3}H]cAMP. Add 0.25 ml of the cell suspension (2–4 × 10^5 cells) to test tubes containing the test substance in 0.25 ml. Prepare zero time controls at this step (see next paragraph). Incubate at 37°* for a time appropriate for the test substance.

Stop Reaction. For determination of intracellular [^{3}H]cAMP (cells only), centrifuge to sediment cells, then decant supernatant; add 1 ml of 2.5% PCA, 0.1 m*M* cAMP to extract labeled nucleotides from the cell pellet. Resuspend the cell pellets for efficient extraction of nucleotides, and continue to next paragraph. For a zero time control, add 0.25 ml cell suspension to 0.25 ml incubation medium and centrifuge immediately to sediment cells. Decant the supernatant and add 1 ml 2.5% PCA, 0.1 m*M* cAMP. Resuspend for efficient extraction of nucleotides, then continue to next paragraph. For determination of [^{3}H]cAMP in cells plus medium, add to each test-tube 0.5 ml of 5% PCA, 0.2 m*M* cAMP. Continue to next paragraph. For a zero time control, add 0.25 ml cell suspension, followed by 0.5 ml of 5% PCA, 0.2 m*M* cAMP and 0.25 ml incubation medium.

Keep test tubes cold for 30 min, with occasional shaking, and centrifuge to obtain a clear, protein-free PCA supernatant containing the extracted ^{3}H-labeled nucleotides. Transfer 0.7–0.9 ml (MX) of the clear PCA supernatant to a set of marked test tubes and prepare extracts for chromatography (see below).

Procedure for Tissue Slices

It is recommended that the experiment be performed in triplicate.

† The use of trypsin, at this stage, should be avoided due to potential damage to receptors and peptide hormones.

Solutions

Preincubation medium: [^{3}H]adenine, 50–200 μCi/ml, in Krebs–Ringer bicarbonate (KRB)‡

KRB medium containing 0.1 mM IBMX, or other inhibitor of cyclic nucleotide phosphodiesterases, and 1 mg/ml BSA (prepare fresh)

Hormones and stimulants freshly dissolved in KRB.

Preincubation. At this stage, the entire pool of tissue slices prepared for the experiment is preloaded with [^{3}H]adenine. Prepare tissue slices 0.1–0.5 mm thick (25–100 mg wet weight/point) under tissue-specific conditions in KRB. Keep good contact with air or bubble with 95% O_2/5% CO_2 (v/v) at 37°. Wash the slices gently 2 or 3 times with warm KRB, decanting blood and cell debris. Replace with preincubation medium (see above), 250–1000 mg wet weight/ml medium. Incubate for 90–120 min,§ keeping the tissue under the best metabolic conditions at 37°. Keep a 0.1-ml sample of the medium for counting, to calculate total [^{3}H]adenine added (TA).

Decant the radioactive medium. (*Caution:* Medium radioactive! Dispose of carefully.) Wash once or twice with fresh medium at 37° to remove the excess [^{3}H]adenine. Put all the labeled slices on a sheet of Parafilm placed on an insulating layer of tissue paper and rapidly divide the slices into equal portions (25–100 mg/experimental point) with a spatula (make sure that the tissue does not dry by keeping sufficient medium for wetting on the Parafilm sheet). Each portion (0.5–1 rat lacrimal gland, as used in the author's laboratory) represents one experimental point. To quickly achieve easy division, divide the total material into two equal portions and each half in two equal portions, etc. (*Caution:* Tissue is radioactive!)

Incubation. Transfer the tissue slice portions, one each, into glass scintillation vials containing the test substance in 1 ml. Flush with a 95% O_2/5% CO_2 (v/v) mixture and close the lid. Prepare zero time controls at this step (see next paragraph). Incubate for 20 min with gentle agitation at 37° and oxygenate if needed.

Stop Reaction. Add 0.11 ml of 25% PCA, 1 mM cAMP to each vial and place vials on ice. Homogenize each system (tissue plus medium), using a Polytron homogenizer equipped with a microprobe; transfer homogenate into a set of marked test tubes and spin to separate protein from extracted radioactive nucleotides. Keep protein pellets for protein determination, which may be required for standardization of the results. Continue to next paragraph. For a zero time point, add to one triplicate

‡ KRB is suitable for most mammalian tissues but may be substituted by any appropriate physiological solution.

§ Conditions used for rat lacrimal tissue in the author's laboratory.

of test tubes 1 ml of KRB medium containing 0.1 m*M* IBMX and 0.11 ml of 25% PCA, 1 m*M* cAMP.

Transfer 0.7–0.8 ml (MX) of the clear, protein-free PCA supernatant to a set of marked test tubes.

Preparation of PCA Extracts for Chromatography

Extracts, as obtained above, are 2.5% PCA final and contain 0.1 m*M* cAMP as carrier. To each PCA supernatant (optimally 0.7–0.9 ml), add one-tenth volume of 4.2 N KOH (e.g., to 0.85 ml PCA extract, add 1 equivalent of KOH in 85 μl) and mix. A sediment of potassium perchlorate appears shortly. The procedure may be discontinued at this stage. Store test tubes at 4° overnight or at $-20°$ for longer periods. To proceed, continue with chromatographic separation (below).

Chromatographic Separation of $[^3H]$cAMP

Add water to each test tube to a final volume of 1.3 ml. Mix and let crystals settle. Sample 0.1 ml from each extract to a set of marked scintillation vials (for counting of total $[^3H]$adenine uptake, TU). Subject the rest, 1.2 ml, to double-column chromatography, as outlined in Chapter [1]. Decant test tubes slowly (the crystals will mostly stay in the test tubes).

Calculations

Definitions. PE is the volume of PCA extract. MX is the volume of PCA extract transferred for neutralization with KOH. TA (total $[^3H]$adenine added) is calculated by counting replicas (0.1 ml each) of the original $[^3H]$adenine-containing preincubation medium.

$$\text{TA} = (\text{cpm sample})/0.1 \times [\text{preincubation volume taken per experimental point (ml)}]$$

TU (total $[^3H]$adenine uptake/experimental point) is calculated from a single 0.1-ml sample taken from each test tube prior to chromatography.

$$\text{TU} = (\text{cpm sample}) \times 13 \times \text{PE/MX}$$

CA ($[^3H]$cAMP produced) is calculated from the radioactivity eluted as cAMP from the alumina columns:

$$\text{CA} = (\text{cpm sample}) \times 13/12 \times \text{PE/MX}$$

PC is the $[^3H]$cAMP percent conversion.

$$\text{PC} = \text{CA/TU} \times 100$$

AU is the percent uptake of $[^3H]$adenine.

$$AU = TU/TA \times 100$$

Results are conveniently expressed as percent conversion (PC).

Remarks

The time required for maximal [^{3}H]adenine uptake differs from system to system and must be determined empirically under the experimental conditions selected. This can be achieved by counting PCA cell/tissue extracts at the end of the preincubation step. The amount of cells/tissue to be used per experimental point and the incubation time required for [^{3}H]cAMP accumulation may vary, depending on the test system and must, therefore, be adapted in early experiments. The amount of [^{3}H]adenine used can be adapted to each system and experimental need by varying the concentration of [^{3}H]adenine in the preincubation medium. Percent conversion may be lower than 1%. Thus, appropriate uptake levels of [^{3}H]adenine per experimental point must be obtained.

Experiments are best run in triplicate and should yield good repetition; moreover, the use of [^{32}P]cAMP in a recovery mixture can be omitted. If desired, [^{32}P]cAMP can be added as internal standard when the incubation is terminated, together with the PCA and before the extraction step. By double-isotope counting, the recovery of the individual test systems may be calculated in a manner similar to that described in Chapter [1].[2,3]

This method provides a good, rapid (several hours), and sensitive measure for tissue responsiveness to external stimuli and is especially convenient for establishing the pharmacological profile or receptor repertoire of given cells or tissues. The method, however, provides only relative levels of cAMP and is not suitable for determination of absolute tissue cAMP levels.

Note: If cell culture media contain adenine, this method cannot be applied. Care must be taken as per international and local regulations concerning handling and waste disposal of radioactive materials.

Acknowledgments

I wish to thank Rachel Benjamin for devoted secretarial assistance, Josepha Schmidt-Sole for technical help, and Nira Garty for enthusiasm and help. The author is the Charles W. and Tillie K. Lubin Professor for Hormone Research at The Weizmann Institute of Science.

[3] Preparation of α-^{32}P-Labeled Nucleoside Triphosphates, Nicotinamide Adenine Dinucleotide, and Cyclic Nucleotides for Use in Determining Adenylyl and Guanylyl Cyclases and Cyclic Nucleotide Phosphodiesterase

By TIMOTHY F. WALSETH, PETER S. T. YUEN, and MALCOLM C. MOOS, JR.

Introduction

Nucleotides radiolabeled in specific positions have become indispensable for the study of signal transduction pathways, enzyme kinetics, and gene structure. We present here a simple, economical, and efficient procedure for the enzymatic synthesis of α-^{32}P-labeled nucleoside triphosphates and cyclic nucleotides from ^{32}P-labeled inorganic phosphate. Although this procedure can be used to prepare a wide variety of γ- and α-^{32}P-labeled ribo- and deoxyribonucleotides,[1,2] this chapter focuses on the preparation of labeled substrates for use in the assay of adenylyl and guanylyl cyclases (EC 4.6.1.1, adenylate cyclase; EC 4.6.1.2, guanylate cyclase) and cyclic nucleotide phosphodiesterases and points out improvements and modifications made to the original method.[1,2]

The conversion of [^{32}P]P$_i$ to α-labeled nucleoside triphosphates or cyclic nucleotides proceeds through the series of enzymatic reactions indicated schematically below.

Stage 1: [^{32}P]P$_i$ $\rightarrow$ [γ-^{32}P]ATP
Stage 2: [γ-^{32}P]ATP + 3′-AMP $\rightarrow$ 3′-[5′-^{32}P]ADP
Stage 3: 3′-[5′-^{32}P]ADP $\rightarrow$ [5′-^{32}P]AMP
Stage 4: [5′-^{32}P]AMP $\rightarrow$ [α-^{32}P]ATP
Stage 5: [α-^{32}P]ATP $\rightarrow$ [^{32}P]cAMP

Modifications necessary to prepare [α-^{32}P]GTP and [^{32}P]cGMP are noted below. Stage 1 produces [γ-^{32}P]ATP through substrate level phosphorylation of ADP during the enzymatic conversion of L-α-glycerophosphate to 3-phosphoglycerate. The two dehydrogenase steps in this conversion are rendered irreversible by the inclusion of lactate dehydrogenase and pyruvate to maintain NAD$^+$ levels. The synthesis of [γ-^{32}P]ATP is very efficient; conversion yields of over 99% are obtained routinely. Polynucleotide kinase is utilized to transfer the γ-phosphoryl of [γ-^{32}P]ATP to

[1] T. F. Walseth and R. A. Johnson, *Biochim. Biophys. Acta* **562,** 11 (1979).
[2] R. A. Johnson and T. F. Walseth, *Adv. Cyclic Nucleotide Res.* **10,** 135 (1979).

the 5′-hydroxyl of 3′-AMP in Stage 2. The resulting 3′-[5′-^{32}P]ADP is converted to [5′-^{32}P]AMP by nuclease P1 in Stage 3. Adenylate kinase and pyruvate kinase are used to generate [α-^{32}P]ATP from [5′-^{32}P]AMP in Stage 4. The [α-^{32}P]ATP is converted to [^{32}P]cAMP in Stage 5. Substitutions necessary to synthesize [α-^{32}P]GTP or [^{32}P]cGMP are 3′-GMP for 3′-AMP in Stage 2, GMP kinase for adenylate kinase in Stage 4, and guanylyl cyclase for adenylyl cyclase in Stage 5. The synthesis of [^{32}P]NAD from [α-^{32}P]ATP is also possible after Stage 4. All the enzymes necessary are commercially available except the adenylyl or guanylyl cyclase preparations employed for generating [^{32}P]cAMP or cGMP in Stage 5. Yields of [α-^{32}P]ATP or [α-^{32}P]GTP are generally 90% or better with respect to [^{32}P]P_i. Yields of ^{32}P-labeled cyclic nucleotides vary depending on the preparation of the cyclase used.

Major advantages of this procedure are that no purification of intermediate products is necessary and all stages are conducted in the same vial in which the [^{32}P]P_i is shipped, eliminating any transfer of radioactivity until the final purification of the [α-^{32}P]nucleoside triphosphate or ^{32}P-labeled cyclic nucleotide. This procedure can also generate products of very high specific activity if the reagents used are free of contaminating inorganic phosphate and nucleotides.[1,2] Since high specific activity nucleotides are not required for adenylyl and guanylyl cyclase or cyclic nucleotide phosphodiesterase assays, requirements to produce very high specific activity products will not be given in detail.

General Information

Caution should be exercised whenever radioisotopes are concerned. Shielding and eye protection should be used at all times when handling ^{32}P. Proper techniques for safe handling of ^{32}P have recently been discussed by Zoon[3] in this series. Presented below are several specific recommendations for anyone setting up this procedure.

1. Identify an isolated place, not subject to routine laboratory traffic, for the preparation.
2. Dedicate equipment such as a microfuge and automatic pipettes (Pipetman) for use with isotopes as these items are sure to become contaminated. It is also wise to have a Lucite disk (6 inches in diameter, 0.5 inch thick) fitted onto the barrel of a Pipetman to provide shielding to the hand while pipeting into the vial containing ^{32}P.
3. Wear disposable surgical gowns and two pairs of gloves. The inner pair of gloves should be surgical gloves that extend up the arm to overlap

[3] R. A. Zoon, this series, Vol. 152, p. 25.

the arms of the surgical gown and provide protection to the wrist. The outer layer of gloves should be of the common latex or vinyl laboratory type and changed often to minimize contamination.

4. Absorbent pads should be taped to the floor immediately in front of the working area. Disposable surgical booties should be worn over shoes when working on the absorbent pads and removed before leaving the working area. This precaution will prevent isotope from being tracked away from the preparation area on shoes.

5. All liquid scintillation counting can be done by Cerenkov radiation, eliminating the need for scintillation cocktail. We find that counting samples in 10 ml of water (when using 20-ml counting vials) works well.

Analytical Procedure

Stage 1: Synthesis of [γ-^{32}P]ATP

Materials. Carrier-free [^{32}P]P_i should be ordered with instructions to the supplier that the material be shipped in the smallest possible volume (without dilution). If the [^{32}P]P_i arrives in a greater volume (1 ml is standard) the procedure can still be carried out by scaling up the volumes of reagents and enzymes listed below. We use [^{32}P]P_i in 20 m*M* HCl from New England Nuclear (Boston, MA) routinely. The reagents necessary for preparing a 2.5× Stage 1 reaction mixture are listed in Table I. The enzymes necessary are listed in Table II. The reagents should be prepared just prior to initiating Stage 1. An option for combining Stages 1 and 2 is also listed in this section. This option is particularly useful when only α-^{32}P-labeled nucleotides or cyclic nucleotides are to be synthesized.

Stage 1 Protocol. The volume of the [^{32}P]P_i (the volume of 10 mCi of [^{32}P]P_i is usually about 50 μl) dictates the volume in which Stage 1 and other stages are carried out. The 20 m*M* HCl in [^{32}P]P_i must be neutralized with Tris base before proceeding. This is accomplished by adjusting the volume of [^{32}P]P_i to 100 μl with a concentration of Tris base resulting in a final Tris concentration of 25–30 m*M*. For example, if the volume of [^{32}P]P_i is 55 μl, 45 μl of 60 m*M* Tris base would be added, resulting in a volume of 100 μl and a Tris concentration of 27 m*M*. The ratio of volumes for components necessary for Stage 1 are listed below. The volumes in parentheses indicate the actual volume if the Tris-neutralized [^{32}P]P_i is 100 μl.

Tris base-neutralized [^{32}P]P_i	1 volume (100 μl)
2.5× Stage 1 reaction mixture (Table I)	0.8 volume (80 μl)
Enzyme–DTT mixture (Table II)	0.2 volume (20 μl)
Total	2 volumes (200 μl)

TABLE I
2.5× STAGE 1 REAGENT MIXTURE

Component[a]	Stock concentration (mM)	Stock volume used (μl)	Final concentration in Stage 1 (mM)
Tris-HCl (pH 8.0)	1000	25	25
$MgCl_2$	480	25	12
Dithiothreitol (DTT)	120	50	6
Spermine	36	50	1.8
EGTA	4	25	0.1
L-α-Glycerophosphate	10	25	0.25
β-NAD^+	10	25	0.5
ADP[b]	2	25	0.05
Sodium pyruvate[c]	80	25	2
Water	—	125	—
		400	

[a] All reagent stock solutions are stored at −20° unless otherwise noted.
[b] ADP is purified by anion-exchange high-performance liquid chromatography before use by the method described by J. T. Axelson, J. W. Bodley, and T. F. Walseth, *Anal. Biochem.* **116,** 357 (1981). The trifluoroacetic acid in the fractions containing the purified ADP is neutralized with 2 M Tris base, and ADP is quantitated by absorbance (molar extinction coefficient 15,400 at 259 nm) and stored at −20° in 0.25-ml aliquots.
[c] Sodium pyruvate (8.8 mg/ml) is prepared immediately before use.

For addition of reagents and enzymes in Stages 1 in terms of a "volume," it refers to the volume of the Tris-neutralized $[^{32}P]P_i$ used in the Stage 1 reaction.

Stage 1 is carried out at room temperature in the vessel in which the $[^{32}P]P_i$ arrives from the manufacturer. Stage 1 is initiated by neutralizing with Tris base, then adding the 2.5× Stage 1 reaction mixture and finally the enzyme–DTT mixture. Before and immediately after the addition of the enzyme–DTT mixture, the contents of the reaction vessels are thoroughly mixed by carefully drawing the contents up and down several times in a pipette tip with the aid of a Pipetman. The 2.5× Stage 1 reaction mixture (see Table I) and enzyme–DTT mixture (see Table II) are prepared from the reagent stocks or enzyme mix just prior to use. The final concentrations of reagents and enzymes in Stage 1 have been chosen such that complete conversion of $[^{32}P]P_i$ to $[\gamma\text{-}^{32}P]ATP$ occurs with minimal contamination of $[^{32}P]P_i$ with unlabeled P_i. If high specific activity products are not required, it is suggested that the enzyme–DTT mixture (see Table II) be prepared with 3 to 4 times more enzyme mixture (e.g., 15 μl of ammonium

TABLE II
STAGE 1 ENZYME MIXTURE AND ENZYME–DTT MIXTURE

Enzyme[a]	Source and concentration	Stock enzyme mixture[b]		Enzyme DTT mixture[c]	Final concentration in Stage 1[d]	
		μl	mg/ml	(μg/ml)	μg/ml	U/ml
Glycerophosphate dehydrogenase	Rabbit muscle, 60 U/mg, 2 mg/ml	100	1.4	100	10	0.6
Triose-phosphate isomerase	Rabbit muscle, 5000 U/mg, 2 mg/ml	1	0.014	1	0.1	0.5
Glyceraldehyde-3-phosphate dehydrogenase	Rabbit muscle, 80 U/mg, 10 mg/ml	20	1.4	100	10	0.8
3-Phosphoglycerate kinase	Yeast, 450 U/mg, 10 mg/ml	2	0.14	10	1	0.45
Lactate dehydrogenase	Rabbit muscle, 550 U/mg, 5 mg/ml	20	0.7	50	5	2.75

[a] All enzymes are purchased from Boehringer Mannheim (Indianapolis, IN) and are in 3.2 *M* ammonium sulfate.

[b] The enzymes are mixed in the proportions shown and stored at 4°. This stock enzyme mixture is stable for up to 1 year.

[c] The enzyme–DTT mixture is prepared just before use by adding 5 μl of ammonium sulfate-depleted stock enzyme mix to 65 μl of a buffer containing 50 m*M* Tris-HCl, pH 8, and 5 m*M* DTT. Ammonium sulfate is removed from the stock enzyme mix by centrifugation for 2 min in a microfuge and aspiration of the supernatant. The protein pellet is resuspended to the original volume (usually 20 μl) in 50 m*M* Tris-HCl, pH 8, 5 m*M* DTT.

[d] The enzyme–DTT mixture constitutes 10% of the Stage 1 volume.

sulfate-depleted enzyme mixture added to 55 μl of 50 m*M* Tris-HCl, pH 8, 5 m*M* DTT). This modification will quicken the rate of conversion to [γ-^{32}P]ATP. When conversion is complete (see monitoring procedure below), the Stage 1 enzymes are inactivated by placing the reaction vessel into 70° water for 10 min. Stage 2 can be initiated after cooling to room temperature. Stage 1 is usually complete within 30 to 60 min. On occasion it has been noted that the [^{32}P]P$_i$ contains an inhibitor(s) of Stage 1. This inhibition of Stage 1 can be overcome by dilution if the same proportions of Stage 1 enzymes and reagents are maintained. For example, if inhibition occurs in the example above, the reaction can be diluted 2-fold by the addition of 100 μl water, 80 μl of 2.5× Stage 1 reaction mixture, and 20 μl of enzyme–DTT mixture.

Option: Running Stage 1 and 2 Concurrently. For applications in which only α-^{32}P-labeled nucleotides are required, such as those featured in this chapter, it is convenient to combine Stages 1 and 2. The only modifications necessary in order to do this are to include 2.5 m*M* 3′-AMP or 3′-GMP in the 2.5× Stage 1 reaction mixture for [α-^{32}P]ATP or [α-^{32}P]GTP, respectively, and to add 30 to 60 units of polynucleotide kinase (United States Biochemical Corporation, Cleveland, OH) to the enzyme–DTT mixture. This reaction is allowed to run for at least 4 hr and usually overnight. Monitoring is accomplished as described for Stage 2 (see below).

Stage 2: Polynucleotide Kinase Reaction

Materials

Cloned polynucleotide kinase, 30 units/μl (United States Biochemical Corporation), and 10 m*M* 3′-AMP or 10 m*M* 3′-GMP

Stage 2 Protocol. After the Stage 1 enzymes have been inactivated and the reaction contents cooled to room temperature, Stage 2 is initiated by addition of 0.2 volume of 10 m*M* 3′-AMP for [α-^{32}P]ATP or 3′-GMP for [α-^{32}P]GTP and 30 to 60 units of polynucleotide kinase (the volume of polynucleotide kinase is negligible and is usually added to the volume of 3′-nucleotide so only one addition needs to be made). If the volume of Tris-neutralized [^{32}P]P_i in Stage 1 volume is 100 μl (e.g., 1 volume), then Stage 2 would be initiated by adding 20 μl of 10 m*M* 3′-AMP containing 30 to 60 units of polynucleotide kinase. Stage 2 is usually allowed to proceed overnight at room temperature (16 hr) for convenience, although it will usually be complete within 4 hr.

Stage 3: Nuclease P1 Reaction

Materials

Nuclease P1 (Boehringer Mannheim) and 30 m*M* sodium acetate, pH 5.3: 1 mg of nuclease P1 is dissolved in 1 ml of 30 m*M* sodium acetate, pH 5.3, and stored at −20° in 50-μl aliquots.

Stage 3 Protocol. Stage 3 is initiated by adding 0.25 volume of 1 mg/ml nuclease P1 and incubating at room temperature. If the volume of Stage 1 is 200 μl, then 25 μl of nuclease P1 would be added. Complete conversion of the 3′-[5′-^{32}P]nucleoside diphosphates to [5′-^{32}P]-labeled nucleoside monophosphates is usually complete within 15 to 30 min.

Before proceeding to Stage 4 any remaining [γ-^{32}P]ATP can be degraded by adding 0.05 volume each of 20 m*M* glucose and 1 mg/ml hexokinase. This reaction should be done in about 15 min. The hexokinase does

TABLE III
4× REAGENT MIXTURES FOR STAGE 4

Component	Stock concentration	Final concentration in 4× reagent	Stock volumes (μl)		
			[α-^{32}P]ATP preparation	[α-^{32}P]GTP preparation	[^{32}P]cGMP preparation
KCl	1000 m*M*	200 m*M*	80	80	80
Phosphoenolpyruvate	100 m*M*	4 m*M*	16	16	16
Dithiothreitol	120 m*M*	7.5 m*M*	25	25	25
Pyruvate kinase[a]	10 mg/ml	200 μg/ml	8	8	8
Adenylate kinase[a]	5 mg/ml	200 μg/ml	16	—	—
Triethanolamine, pH 7.5	500 m*M*	100 m*M*	—	—	80
$MnCl_2$	500 m*M*	20 m*M*	—	—	16
GMP kinase[a]	2 mg/ml	200 μg/ml	—	40	40
Water	—	—	255	231	265
			400	400	400

[a] The enzymes are all from Boehringer Mannheim. The pyruvate kinase and GMP kinase are in 50% glycerol. The myokinase is in 3.2 *M* ammonium sulfate. The ammonium sulfate is removed by centrifugation before use as described for the Stage 1 enzymes.

not need to be inactivated before proceeding to Stage 4 since it eventually becomes ineffective through the depletion of glucose during Stage 4. If Stage 2 has been allowed to incubate overnight and less than 2% of the [γ-^{32}P]ATP remains (which is usually the case), the hexokinase step can be eliminated.

Stage 4: Conversion of [5′-^{32}P]AMP to [α-^{32}P]ATP

Materials. The reagents and enzymes required for Stage 4 are listed in Table III. Separate 4× reagents are used depending on whether [α-^{32}P]ATP, [α-^{32}P]GTP, or [^{32}P]cGMP is to be prepared.

Stage 4 Protocol. Stage 4 is initiated by adding 0.8 volume of the appropriate 4× Stage 4 reagent to the reaction vessel (80 μl if the volume of Tris-neutralized [^{32}P]P_i in Stage 1 volume was 100 μl). The reaction should be complete within 30 min at room temperature.

Stage 5: Synthesis of [^{32}P]cAMP, [^{32}P]cGMP, or [^{32}P]NAD

[^{32}P]cAMP or [^{32}P]cGMP for cyclic nucleotide phosphodiesterase assays, or [^{32}P]NAD for ADP-ribosylation studies, can easily be prepared after completion of Stage 4 without prior purification of [α-^{32}P]ATP or

[α-^{32}P]GTP. Since some laboratories may want to convert a portion of the Stage 4 products to cyclic nucleotides or NAD, and also purify a portion of the [α-^{32}P]nucleotides from Stage 4 for cyclase assays, the descriptions of additions in terms of a "volume" below no longer refer to the volume of Tris-neutralized [^{32}P]P_i used in Stage 1. They instead refer to the volume of Stage 4 used in each conversion.

Materials. The calmodulin-sensitive adenylyl cyclase from *Bordetella pertussis*[4,5] in the form of a dialyzed urea extract is used to prepare [^{32}P]cAMP and was a generous gift from Dr. Dennis Confer. A partially purified sea urchin sperm guanylyl cyclase[6] is used to prepare [^{32}P]cGMP. NAD pyrophosphorylase for the synthesis of [α-^{32}P]NAD[7] is purchased from Boehringer Mannheim. Bovine brain calmodulin is obtained from Calbiochem (San Diego, CA).

Synthesis of [^{32}P]cAMP. Dialyzed urea extracts of *B. pertussis*[4,5] have been found to be very efficient in converting [α-^{32}P]ATP to [^{32}P]cAMP. These extracts typically contain an adenylyl cyclase activity of over 1 μmol cAMP formed/(min · mg protein). [^{32}P]cAMP synthesis is initiated by adding 2 volumes of dialyzed urea extract that has been adjusted to 3 μM calmodulin to 1 volume of [α-^{32}P]ATP from Stage 4. There is no need to purify the [α-^{32}P]ATP from Stage 4 before initiating this step. We typically obtain over 90% conversion to [^{32}P]cAMP in 2 to 3 hr at room temperature. The reaction is terminated by the addition of an equal volume of 1 M perchloric acid and the [^{32}P]cAMP purified as described below.

Preparation of [^{32}P]cAMP by use of a Lubrol PX extract of rat brain membranes has been previously described.[1,2] The yield and specific activity of the resulting [^{32}P]cAMP are not as high as with the *B. pertussis* extract. An adenylyl cyclase expressed by *Bacillus anthracis*[8,9] should also prove useful in the preparation of [^{32}P]cAMP.

Synthesis of [^{32}P]cGMP. Unpurified [α-^{32}P]GTP from Stage 4 is converted to [^{32}P]cGMP by a partially purified preparation of sea urchin sperm guanylyl cyclase.[6] [^{32}P]cGMP synthesis is started by adding 0.5 volume of guanylyl cyclase (about 20 to 50 nmol cGMP formed/min) to 1 volume of [α-^{32}P]GTP from Stage 4. Depending on the preparation of guanylyl cyclase, 50 to 90% conversion to [^{32}P]cGMP should occur within 3 to 4 hr at room temperature. The reaction is terminated by the addition of an equal

[4] D. L. Confer and J. W. Eaton, *Science* **217,** 948 (1982).

[5] D. L. Confer, A. S. Slungaard, E. Graf, S. S. Panter, and J. W. Eaton, *Adv. Cyclic Nucleotide Res.* **17,** 183 (1984).

[6] G. E. Ward, D. L. Garbers, and V. D. Vacquier, *Science* **227,** 768 (1985).

[7] D. Cassel and T. Pfeuffer, *Proc. Natl. Acad. Sci. U.S.A.* **75,** 2669 (1978).

[8] S. H. Leppla, *Proc. Natl. Acad. Sci. U.S.A.* **79,** 3162 (1982).

[9] S. H. Leppla, *Adv. Cyclic Nucleotide Res.* **17,** 189 (1984).

volume of 1 *M* perchloric acid and the [^{32}P]cGMP purified as described below.

We have used sperm guanylyl cyclase purified on concanavalin A-agarose[6] as well as Lubrol PX-solubilized sperm membranes to prepare [^{32}P]cGMP. If the latter is used, it is recommended that a phosphodiesterase inhibitor (2 m*M* 3-isobutyl-1-methylxanthine) be included in the 4$\times$ Stage 4 reagent (Table III).

Synthesis of [α-^{32}P]NAD. [α-^{32}P]NAD is readily prepared from the unpurified [α-^{32}P]ATP after the completion of Stage 4 by adding an equal volume of a reagent containing 50 m*M* Tris-HCl, 10 m*M* $MgCl_2$, 50 m*M* KCl, 4 m*M* phosphoenolpyruvate, 100 m*M* nicotinamide, 25 m*M* nicotinamide mononucleotide, 100 μg/ml pyruvate kinase, 100 μg/ml myokinase (adenylate kinase), and 1 mg/ml NAD^+ pyrophosphorylase (Boehringer Mannheim). Yields of [α-^{32}P]NAD are routinely above 85% after 1 to 2 hr when incubated at room temperature. Upon completion, the resulting [α-^{32}P]NAD is diluted with water and purified as described below.

Monitoring Procedures

It is strongly recommended that all stages be monitored to ensure that they are working properly and, if not, that the appropriate action can be taken. Listed below are procedures to monitor the progress of each stage. These procedures are also useful in determining when each stage is complete and the next one can begin.

Stage 1. Stage 1 is conveniently monitored by determining the amount of charcoal-adsorbable radioactivity ([γ-^{32}P]ATP) generated with time. A very small aliquot of the Stage 1 reaction is taken by quickly dipping the end of a small piece of surgical silk (size 00) into it. The suture is then dipped into 500 μl of 0.1 *M* H_3PO_4 in a 1.5-ml microfuge tube. After mixing, a 10-μl aliquot is removed for determination of total radioactivity ([^{32}P]P_i and [γ-^{32}P]ATP) by liquid scintillation counting (Cerenkov radiation in 10 ml of water is adequate). Charcoal-adsorbable radioactivity is then removed by the addition of 500 μl of 25 mg/ml Norit A in 0.1 *M* H_3PO_4 and centrifuging in a microfuge for 30 sec. The amount of radioactivity not adsorbed to charcoal ([^{32}P]P_i) is also determined by liquid scintillation counting. As Stage 1 progresses to completion the amount of [^{32}P]P_i diminishes. The percent conversion can be calculated as follows: %[γ-^{32}P]ATP = (total cpm before Norit A $-$ cpm after Norit A $\times$ 2)/total cpm before Norit A. Conversion yields to [γ-^{32}P]ATP are usually over 98%.

Stage 1 can also be monitored by thin-layer chromatography (TLC) on polyethyleneimine (PEI)-cellulose (Brinkmann). Plastic-backed 20 $\times$ 20 cm plates of PEI-cellulose are washed with water and cut into 1 × 5 cm pieces. At the appropriate times string aliquots are removed from the Stage 1 reaction vessel (as described above) and dipped into 100 μl of water contained in 1.5-ml microfuge tubes. One microliter of this is then spotted onto the origin of a 1 × 5 cm PEI-cellulose strip that has been prespotted with 1 μl of 10 m*M* ATP. The strip is then developed in a small beaker containing 0.4 *M* LiCl. After development the ATP is located under UV light and marked. P_i migrates above ATP. The strip is cut just above the ATP region, and ^{32}P in both pieces is determined by liquid scintillation counting in vials containing 10 ml of water. This method is generally applicable to all stages and is referred to as quick TLC (development takes 10 min or less).

Stage 2. Stage 2 is monitored by determining the amount of charcoal-adsorbable radioactivity present before and after treatment with potato apyrase (Sigma). A string aliquot is taken and dipped into a 1.5-ml microfuge tube containing 200 μl of 25 m*M* Tris-HCl, pH 7, and 2.5 m*M* $CaCl_2$. After mixing, a 100-μl aliquot of this sample is transferred to a microfuge tube containing 10 μl of 1 mg/ml potato apyrase. Both aliquots are incubated for 10 min at room temperature and terminated by the addition of 400 μl of 0.1 *M* H_3PO_4. At this point the amount of charcoal-adsorbable radioactivity in each aliquot is determined as described above in the procedure for monitoring Stage 1. Apyrase hydrolyzes [γ-^{32}P]ATP but not 3′-[5′-^{32}P]ADP to [^{32}P]P_i. At the beginning of Stage 2, all the radioactivity should be charcoal adsorbable ([γ-^{32}P]ATP) but converted to [^{32}P]P_i by apyrase. As Stage 2 progresses, the amount of charcoal-adsorbable radioactivity in the apyrase-treated sample should approach that in the control sample. When complete, virtually all the radioactivity should be charcoal adsorbable, both before and after apyrase treatment.

Stage 2 can also be monitored by quick TLC as described above for Stage 1 with either 0.4 *M* LiCl or 2 *M* sodium formate, pH 3.4, to develop the strips. The strips should be prespotted with ATP and 5′,3′-ADP or GDP standards. ATP is marginally resolved from the 5′,3′-ADP or GDP on both systems, so it is recommended that Stage 2 be monitored by the apyrase method.

Stages 3, 4, and 5. Stages 3, 4, and 5 are monitored by quick TLC on PEI-cellulose. In all cases, an aliquot is taken from the reaction vessel with the string technique and dipped into a microfuge tube containing 100 μl of water. One microliter is then spotted onto a PEI-cellulose strip prespotted with the appropriate standard. After development, the areas of the strip corresponding to each nucleotide are cut out and the radioactivity determined by liquid scintillation counting.

The PEI-cellulose strips for stages 3 and 4 are prespotted with 1 μl of either a mixture of 10 mM ATP, ADP, and AMP or 10 mM GTP, GDP, and GMP. The strips are developed with 2 M sodium formate, pH 3.4. The PEI-cellulose strips for the monitoring of [^{32}P]cAMP or [^{32}P]cGMP synthesis are prespotted with 1 μl of a mixture of 10 mM ATP and cAMP or 10 mM GTP and cGMP, respectively, and developed in 0.4 M LiCl. [α-^{32}P]NAD synthesis is monitored on strips prespotted with 1 μl of a mixture containing 10 mM NAD and ATP and developed with 0.1 M formic acid.

Purification Procedures

[α-^{32}P]ATP and [α-^{32}P]GTP

The α-^{32}P-labeled nucleoside triphosphates are purified by anion-exchange chromatography on AG MP-1 (Bio-Rad, Richmond, CA). The trifluoroacetate form of AG MP-1 is generated by batch washing the resin sequentially with 1 M NaOH, water, and 1 M trifluoroacetic acid (TFA). On the day of use the AG MP-1 is defined by washing with water several times for 10 min in a 50-ml centrifuge tube. The sample is applied to a 0.7 $\times$ 3 cm column of AG MP-1 that has been prewashed with 10 ml of water. If the [α-^{32}P]nucleotide being purified is going to be used only in cyclase assays, dilution of the specific activity by adding 5 nmol of unlabeled nucleotide per millicurie will enhance the stability of the purified product. After allowing the sample to drain into the column, the column is washed twice with 1 ml of water. Inorganic phosphate, ADP, or GDP is removed from the column by washing with 10 ml of either 30 mM TFA or 75 mM TFA for [α-^{32}P]ATP or [α-^{32}P]GTP, respectively. The eluate up to this point is discarded. Elution of [α-^{32}P]ATP or [α-^{32}P]GTP is accomplished by washing the column with 8 ml of 150 mM TFA. The 150 mM TFA eluate is collected in 1-ml fractions in plastic test tubes, each containing 10 μl of 100 mM DTT. The [α-^{32}P]ATP or GTP generally elutes in the first six fractions. The peak fractions are pooled and neutralized with 2 M Tris base (about 50 μl/fraction) and stored in 0.5-ml aliquots at $-80°$. The pooled peak is diluted such that it does not exceed 3 μCi/μl. We have found that ^{32}P-labeled nucleotides stored at this concentration at $-80°$ in the presence of 1 mM DTT remain stable for at least 2 months.

High-performance liquid chromatography (HPLC) should also be helpful in the purification of ^{32}P-labeled nucleotides (see also chapter [25] by Offermanns *et al.*, this volume). Many HPLC systems for the purification of nucleotides have been developed and could presumably be applicable to the purification of labeled nucleotides. The major drawback of HPLC purification of ^{32}P-labeled nucleotides is the contamination of the HPLC equipment. To minimize the contamination we have developed a system in which a suitably equilibrated HPLC column is connected to a peristaltic pump with Peek tubing and plastic fittings (Upchurch, Oak Harbor, WA). The ^{32}P-labeled sample is pumped onto the column at a flow rate of about 0.1 ml/min. After sample application has been completed, the column is connected to an HPLC system and elution is initiated with the appropriate gradient. The outlet of the column is usually connected directly to a fraction collector, bypassing the detector in order to avoid contamination. This sample loading system minimizes contamination of HPLC components except for the column. In some cases the specific activity of the nucleotide can be determined by allowing the effluent to pass through the UV detector.

[α-^{32}P]ATP and [α-^{32}P]GTP from Stage 4 are successfully purified by this sample loading system using a 30 × 4.6 mm MF-PLUS column (Alltech, Deerfield, IL) packed with AG MP-1 and eluted with a TFA gradient (0 to 150 m*M* over 40 min at 1 ml/min) as described by Axelson *et al.*[10] Fractions containing the purified [α-^{32}P]ATP or [α-^{32}P]GTP are neutralized and stored in DTT at −80° as described above. The column will be contaminated by this procedure and should be stored behind suitable screening when not in use.

[^{32}P]cAMP and [^{32}P]cGMP

The [^{32}P]cAMP and [^{32}P]cGMP are purified by sequential chromatography on acidic alumina and Dowex 50 columns as described by Jakobs *et al.*,[11] with the modification that the Dowex 50 columns are 10 and 15 cm instead of 4 cm for cAMP and cGMP, respectively. The acidified sample is applied to a 0.7 × 4 cm column of alumina (Bio-Rad, Neutral Alumnia AG7) that has been prewashed with 15 ml of 0.5 *M* perchloric acid. After the sample has drained into the column, the sample vessel is rinsed with 1 ml of 0.5 *M* perchloric acid (PCA). The rinse is applied to the column and allowed to drain in. The column is then washed with 5 ml of 0.5 *M* perchloric acid, followed by 10 ml of water. All eluate from the alumina column up to this point is discarded. At this point the alumina column is placed over 0.7 × 10 ([^{32}P]cAMP) or 15 cm ([^{32}P]cGMP) column of Dowex 50 (Bio-Rad, AG50-X8, 100–200 mesh, H^+ form) that had been prewashed with 20 ml of water. The cyclic nucleotides are eluted onto the Dowex 50 columns by washing the alumina columns with 4 ml of 0.2 *M* ammonium formate, pH 7.5. The ammonium formate is allowed to drain into the Dowex 50 column, and then the cyclic nucleotides are eluted by washing

[10] J. T. Axelson, J. W. Bodley, and T. F. Walseth, *Anal. Biochem.* **116,** 357 (1981).

[11] K. H. Jakobs, E. Böhme, and G. Schultz, *in* "Eukaryotic Cell Function and Growth" (J. E. Dumont, B. L. Brown, and N. J. Marshall, eds.), p. 295. Plenum, New York, 1976.

these columns with water. One-milliliter fractions are collected in plastic tubes containing 10 μl of 100 m*M* DTT starting after the elution of the first 4 ml from the Dowex 50 column (i.e., at the start of the water wash). [^{32}P]cAMP elutes in the third to sixth water fraction, and [^{32}P]cGMP elutes in the fifth to eighth water fraction. The peak fractions are pooled, neutralized with 2 *M* Tris base (about 75 μl/fraction), diluted to 3 μCi/μl or less, and stored at $-80°$ in 0.5-ml aliquots.

Further purification of the purified [^{32}P]cGMP may be necessary if the product is to be used in the determination of high affinity cGMP binding sites by conventional binding assays. To accomplish this the neutralized pooled eluate from the Dowex 50 column is applied to a 0.7 $\times$ 2 cm column of AG MP-1 (trifluoroacetate form). This column is developed as described for the purification of [α-^{32}P]ATP above, except that the cGMP is eluted from the column with 8 ml of 75 m*M* TFA. The cGMP elutes in the first 5 ml of the 75 m*M* eluate and is pooled, neutralized, and stored as described above. The HPLC adaptation of this purification on AG MP-1 which is described above can also be employed at this step.

The specific activity of the purified [^{32}P]cAMP or [^{32}P]cGMP can be determined by analyzing an aliquot of the purified product by reversed-phase HPLC on a 5-μm Ultrasphere column (0.46 $\times$ 15 cm, Altex, Berkeley, CA). Solvent A consists of 10 m*M* potassium phosphate, pH 7.1, and solvent B is 20% (v/v) methanol. Cyclic nucleotides are eluted by gradient elution at a flow rate of 1 ml/min. The solvent composition is held at 100% A for the first 6 min. The composition of B is increased linearly to 100% from 6 to 16 min. Fractions of 0.5 ml are collected, and ^{32}P is determined by liquid scintillation counting. The amount of cyclic nucleotide is determined by comparing the peak area (from UV detection) to a standard curve generated by injecting known amounts of cyclic nucleotide standards. Specific activities have been determined with this technique by injection of 100 to 200 μCi of purified cyclic nucleotide.

[α-^{32}P]NAD

The sample is applied to a 0.7 $\times$ 4 cm column of AG1-X8 (Bio-Rad, 100–200 mesh, formate form) that has been prewashed with 15 ml of water. The sample is allowed to drain into the column, and the sample vessel is rinsed with 1 ml of water. The rinse is applied to column, and, after it has drained in, the column is washed with 15 ml of 0.1 formic acid. One-milliliter fractions are collected in plastic test tubes containing 10 μl of 100 m*M* DTT beginning with the sample application. The [α-^{32}P]NAD elutes in a broad peak from about the third to the tenth fraction. The peak is pooled, aliquoted into 1.5-ml Eppendorf tubes as 0.5-ml aliquots and concentrated to dryness on a Savant Speed-Vac concentrator. The dried aliquots are stored at $-80°$ and resuspended in water just before use.

Adenylyl and Guanylyl Cyclase Assays

There are many satisfactory methods for the radiochemical assay of adenylyl and guanylyl cyclases. For detailed descriptions of assays for adenylyl and guanylyl cyclases, see Johnson and Salomon [1] and Domino *et al.* [30] in this volume. α-^{32}P-Labeled nucleoside triphosphates are the substrates of choice for these assays. Here we describe a simple procedure to assay both adenylyl and guanylyl cyclase based on the method described by Garbers and Murad.[12] The use of [α-^{32}P]ATP or [α-^{32}P]GTP generated by the enzymatic procedure detailed above in this assay results in negligible blank values. The main advantages of this assay are as follows: (1) the alumina columns employed for the separation of ^{32}P-labeled cyclic nucleotide from labeled substrate are reusable, (2) the majority of labeled substrate is precipitated by $ZnCO_3$, thus eliminating the need to chromatograph large amounts of radioactivity, and (3) the radioactivity in the purified cyclic nucleotide fraction can be determined by Cerenkov radiation, eliminating the need for scintillation cocktail.

The setup of adenylyl and guanylyl cyclase assays has been adequately dealt with elsewhere.[12,13] Our basic assay is done in a 100-μl volume containing the appropriate concentrations of buffer, metal ion, nucleoside triphosphate, phosphodiesterase inhibitor, nucleoside triphosphate-regenerating system, etc., and 0.5 to 2 μCi of [α-^{32}P]ATP or GTP. The reactions are initiated by adding enzyme and incubated for the desired time at 37°. The reaction is terminated by addition of 0.5 ml of 36 m*M* Na_2CO_3 and 0.6 ml of 30 m*M* zinc acetate containing 2 m*M* cyclic AMP or cyclic GMP in rapid succession, vortexed, and kept on ice. The samples can be stored frozen at this point and the purification steps carried out later with a modest increase in the blank. The zinc carbonate precipitate, which contains over 98% of the [α-^{32}P]ATP or GTP and less than 5% of the cyclic nucleotide product, is removed by centrifugation at 2000 rpm for 10 min in a tabletop centrifuge. The ^{32}P-labeled cyclic nucleotide contained in the supernatant is purified by chromatographing the supernatant on acidic alumina columns. The columns (0.6 × 4 cm) are prepared from AG7 alumina (Bio-Rad) and washed with 10 ml of water and 3 ml of 1 *M* PCA before use. The column procedure is as follows:

1. Acidify the sample by adding 3 ml of 1 *M* PCA to the column and applying the zinc carbonate supernatant directly to the acid (the zinc

[12] D. L. Garbers and F. Murad, *Adv. Cyclic Nucleotide Res.* **10,** 57 (1979).
[13] Y. Salomon, *Adv. Cyclic Nucleotide Res.* **10,** 35 (1979).

carbonate pellet is sufficiently solid such that the supernatant can be poured directly to the column).

2. Allow the acidified samples to drain into the columns.
3. Wash the columns with 3 ml of 1 *M* PCA.
4. Wash the columns with 10 ml of water. All effluent up to this point should be discarded appropriately.
5. Place the columns over scintillation vials and elute the ^{32}P-labeled cyclic nucleotides with 3 ml of 0.2 *M* ammonium formate, pH 7.5.
6. Determine the absorbance at 259 nm of a 0.5-ml aliquot of the 0.2 *M* ammonium formate fractions for recovery.
7. Add 10 ml of water to the scintillation vials and determine the amount of ^{32}P by Cerenkov radiation.

The alumina columns are reusable and can be recycled after use by washing first with 10 ml of water and then with 3 ml of 1 *M* PCA. The recovery of cyclic AMP and cyclic GMP is 70–80 and 50–60%, respectively. The need to read the absorbance of an aliquot of the 0.2 *M* ammonium formate fraction for the determination of the recovery of cyclic nucleotide during purification can be circumvented by adding ^{3}H-labeled cyclic AMP or cyclic GMP in the 30 m*M* zinc acetate solution (about 5000 cpm/ml). However, if this method is followed, the 0.2 *M* ammonium formate fraction must be counted in an appropriate scintillation cocktail to allow for the determination of both ^{32}P and ^{3}H. Care should be taken to determine total counts of ^{32}P and ^{3}H and spillover of each of these isotopes into the counting channel of the other under conditions identical to those used for the determination of radioactivity in samples eluted from the alumina columns.

Cyclic Nucleotide Phosphodiesterase Assay

The use of ^{32}P-labeled cyclic nucleotides simplifies greatly the determination of cyclic nucleotide phosphodiesterase activity. The assay presented here is a modified version of the cyclic AMP phosphodiesterase assay described by Schönhöfer *et al.*[14] The major modifications are that the [5′-^{32}P]AMP product is converted to [^{32}P]P_i by 5′-nucleotidase instead of a snake venom preparation, and separation of [^{32}P]P_i from labeled substrate is accomplished by charcoal adsorption of the substrate instead of ammonium molybdate precipitation of the [^{32}P]P_i.

The basic assay is done in a 100-μl volume containing the appropriate concentrations of buffer, metal ion, cyclic nucleotide, etc., and 0.01 to 0.1 μCi of [^{32}P]cAMP or [^{32}P]cGMP. The assays are initiated by the addition

[14] P. S. Schönhöfer, L. F. Skidmore, H. R. Bourne, and G. Krishna, *Pharmacology* **7**, 65 (1972).

of enzyme and incubated at the desired time and temperature. The reactions are terminated by the addition of 200 μl of ice-cold 35 m*M* HCl containing 1 m*M* unlabeled cyclic AMP or cyclic GMP, depending on the assay. The samples are then neutralized by adding 30 μl of 0.6 *M* sodium HEPES supplemented with 10 m*M* $MgCl_2$. Conversion of the labeled 5′-AMP or 5′-GMP product to $[^{32}P]P_i$ and adenosine or guanosine is accomplished by adding 5 μl of 0.2 unit/ml 5′-nucleotidase (Sigma, St. Louis, MO, Cat. No. N-5880) and incubating for 30 min at 37°. The 5′-nucleotidase does not hydrolyze cyclic nucleotides. Separation of $[^{32}P]P_i$ from labeled substrate is achieved by adding 1 ml of a slurry of 25 mg/ml Norit A (Sigma) in 0.1 *M* phosphoric acid. The remaining labeled cyclic nucleotide substrate will be adsorbed by Norit A, while the $[^{32}P]P_i$ will not be adsorbed. Norit A is sedimented by centrifugation for 10 min at 2000 *g*, and an aliquot (0.8 ml) of the supernatant (containing the $[^{32}P]P_i$) is quantitated by Cerenkov radiation in a scintillation vial containing 10 ml of water.

This assay is applicable to the determination of both cyclic AMP and cyclic GMP phosphodiesterase activity. ^{32}P-Labeled cyclic nucleotides prepared by the procedure described above usually have blanks of about 0.1 to 0.3% of the total counts used in the assay. The phosphodiesterase assay described here is convenient because there is no column step and no loss of product needs to be taken into account. The assay is also fast, and no scintillation cocktail is necessary.

[4] Synthesis of Forskolin-Agarose Affinity Matrices

By THOMAS PFEUFFER

Introduction

Understanding signal transfer via hormonally regulated adenylyl cyclase at the molecular level requires characterization, isolation, and purification of the individual components. Considerable progress has been made with respect to stimulatory and inhibitory receptors and GTP-binding proteins (G_s and G_i)[1] for which cDNAs have been cloned so that

[1] MOPS, 3-(*N*-Morpholino)propane sulfonate; GPP(NH)P, guanosine 5′-(β,γ-imido)triphosphate; GTPγS, guanosine 5′-(γ-thio)triphosphate; TLC, thin-layer chromatography; G

their primary structures are now available.[2–5] Isolation of the catalytic component has lagged behind, most likely because of the absence of suitable ligands for affinity techniques which seem to be indispensable for the isolation of low-abundance proteins. A hypotensive plant constituent, forskolin, has been shown to stimulate the enzyme from a broad variety of tissues and species.[6,7] Unlike other effectors of adenylyl cyclase tested so far, forskolin seems to act through the catalytic portion itself. Forskolin stimulates the $G_{s\alpha}$-deficient S49 mutant *cyc*$^-$, although to a lesser degree than the wild-type S49 cell.[8]

We have begun to synthesize bifunctional derivatives of forskolin suitable for affinity chromatography of detergent-solubilized adenylyl cyclase. Since hydroxyl groups of the forskolin molecule seem to be most easily modified, a variety of hemisuccinyl derivatives have been prepared and tested. Direct succinylation of forskolin afforded a single product in good yield (most likely 1-succinylforskolin), which barely activated myocardial (and other) adenylyl cyclase (EC 4.6.1.1), and exhibited only marginal affinity ($K_a > 2 \times 10^{-4}$ *M*) (Table I). The most promising route to obtain a potent analog of forskolin, however, was suggested through a replacement of the acetyl group via deacetylforskolin. Although at least three hydroxyl groups could potentially be acylated we aimed at the preferential reactivity of the 7-hydroxyl group toward acylating agents.[9] Although this proved to be correct, considerable amounts of a second product (most likely 1,7-bissuccinyldeacetylforskolin) were formed. However, the convenient one-step method justified the direct (i.e., without protecting

proteins, GTP-binding proteins; C, catalytic component of adenylyl cyclase; G_s, guanine nucleotide-binding regulatory component responsible for stimulation of adenylyl cyclase; G_i, guanine nucleotide-binding regulatory component responsible for inhibition of adenylyl cyclase.

[2] R. A. F. Dixon, B. K. Kobilka, D. J. Strader, J. L. Benovic, H. G. Dohlmann, T. Frielle, M. A. Bolanowski, C. D. Bennett, D. Rands, R. E. Diehl, R. A. Mumford, E. E. Slater, I. S. Sigal, M. G. Caron, R. J. Lefkowitz, and C. D. Strader, *Nature (London)* **321,** 75 (1986).

[3] J. D. Robishaw, D. W. Russel, B. A. Harris, M. D. Smigel, and A. G. Gilman, *Proc. Natl. Acad. Sci. U.S.A.* **83,** 1251 (1986).

[4] T. Nukada, T. Tanabe, H. Takahashi, M. Noda, K. Haga, T. Haga, A. Ichiyama, M. Hiranaga, H. Matsuo, and S. Numa, *FEBS Lett.* **197,** 305 (1986).

[5] K. A. Sullivan, Y.-C. Liao, A. Alborzi, B. Beidermann, F.-H. Chang, S. B. Masters, A. D. Levinson, and H. R. Bourne, *Proc. Natl. Acad. Sci. U.S.A.* **83,** 6687 (1986).

[6] H. Metzger and E. Lindner, *IRCS Med. Sci.* **9,** 99 (1981).

[7] K. B. Seamon and J. W. Daly, *J. Biol. Chem.* **256,** 9799 (1981).

[8] K. B. Seamon, W. Padgett, and J. W. Daly, *Proc. Natl. Acad. Sci U.S.A.* **78,** 3363 (1981).

[9] S. V. Bhat, B. S. Bajwa, H. Dornauer, and N. J. De Souza, *Tetrahedron Lett.* **19,** 1669 (1977).

TABLE I
BIOCHEMICAL PROPERTIES OF FORSKOLIN DERIVATIVES

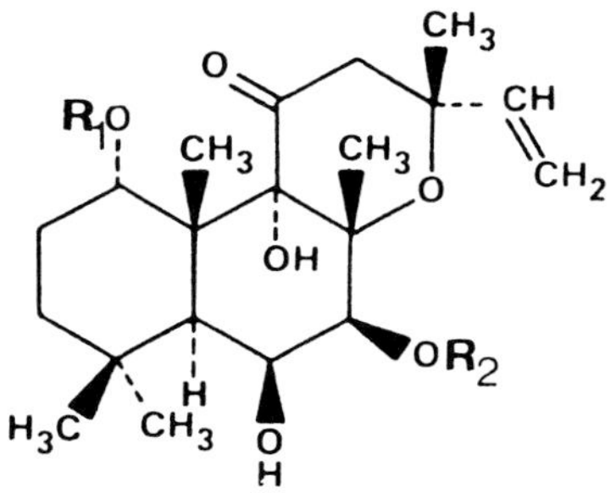

Derivative	R_1	R_2	K_a[a] (μM)	V_{max}[a] (pmol mg^{-1} min^{-1}
Deacetylforskolin	H	H	18	295
Forskolin	H	$\mathrm{-C(=O)-CH_3}$	3.2	610
1-Succinylforskolin	$\mathrm{-C(=O)-CH_2CH_2COOH}$	$\mathrm{-C(=O)-CH_3}$	>200	85
7-Succinyl-7-deacetylforskolin	H	$\mathrm{-C(=O)-CH_2CH_2COOH}$	10.4	362
1,7-Bissuccinyl-7-deacetylforskolin	$\mathrm{-C(=O)-CH_2CH_2COOH}$	$\mathrm{-C(=O)-CH_2CH_2COOH}$	>500	97

[a] K_a and V_{max} are determined by measuring adenylyl cyclase activity in rabbit myocardial membrane (100 μg per assay) in the presence of at least six different concentrations of these derivatives which are dissolved in ethanol. The final concentration of the vehicle in the assay is 1%. Vehicl alone causes an activity of 30 pmol mg^{-1} min^{-1}. The adenylyl cyclase assay is performed in 2 mM MOPS buffer, pH 7.4, containing 5 mM $MgCl_2$, 3 mM theophylline, 10 mM creatine phosphate 20 μg/ml creatine kinase, and 0.5 mM [α-^{32}P]ATP (0.5 Bq/pmol).

agents) approach, since both products could be easily separated on silica gel.

Synthesis of 1-Succinylforskolin

One-half gram succinic anhydride (Serva, Heidelberg), finely divided in a mortar, is dissolved in 2 ml of pyridine (dried over KOH) by heating to 90°. The solution is cooled to 60°, 0.2 g forskolin is added, and the mixture is maintained for 20 hr at 60° under exclusion of moisture. Inspection of the reaction by thin-layer chromatography (see below) reveals almost quantitative formation of 1-succinylforskolin. The work up of the reaction mixture and the isolation of the product by chromatography on silica gel are performed analogously with the procedure described below for the preparation of 7-succinyl-7-deacetylforskolin.

Synthesis of 7-Succinyl-7-deacetylforskolin and 1,7-Bissuccinyl-7-deacetylforskolin

Preparation of Deacetylforskolin

Forskolin (0.25 g) is dissolved in 50 ml of a methanolic Na_2CO_3 solution (saturated solution of Na_2CO_3 in 90% methanol v/v) and refluxed for 16 hr until the reaction is complete (check by TLC). Methanol is largely removed by reduced pressure during which the crystalline deacetyl derivative precipitates. The mixture is left for 6 hr in a refrigerator, and the crystals formed are collected by filtration and washed with ice-cold water until the filtrate becomes neutral. The product is dried *in vacuo* over phosphorus pentoxide. Yield, 80%.

Succinylation of Deacetylforskolin[10,11]

Deacetylforskolin (0.2 g) is dissolved in a solution of 0.7 g of succinic anhydride in 3 ml of anhydrous pyridine (dried over KOH). The mixture is heated at 40° for 80 hr, after which the reaction is controlled by TLC (silica sheets, Merck, Darmstadt, developed in toluene/ethyl acetate, 1 : 1, containing 1% formic acid; detection is by charring with concentrated H_2SO_4). The first product detected has an R_f value of 0.4 [7-succinyl-7-deacetylforskolin (**I**)], followed by a derivative with an R_f of 0.22, most likely 1,7-bissuccinyl-7-deacetylforskolin (**II**). After this period 80–85% of the deacetylforskolin disappears, while the formation of **I** reaches a maximum (35–40%). Prolongation of the reaction time causes the conversion of **I** to **II**.

The reaction is stopped following addition of 1 ml distilled water and left for 2 hr at 0°. Following evaporation of solvents at 40° under reduced pressure, the resulting solid residue is extracted 4 times with 10 ml each of ethyl acetate under vigorous stirring. The combined extracts are dried with anhydrous Na_2SO_4, evaporated, and chromotagraphed on a 50-ml

[10] T. Pfeuffer and H. Metzger, *FEBS Lett.* **146,** 369 (1982).

[11] E. Pfeuffer, R.-M. Dreher, H. Metzger, and T. Pfeuffer, *Proc. Natl. Acad. Sci. U.S.A.* **82,** 3086 (1985).

silica gel column with chloroform/ethyl acetate/formic acid (at a ratio of 68 : 32 : 0.5, 250 ml) as eluant. The inclusion of formic acid prevents smearing of the succinates, most likely by suppressing their dissociation. The 7-succinyl-7-deacetyl derivative elutes after residual (if any) deacetylforskolin and appears as a colorless solid following evaporation of solvents (yield 35%, based on deacetylforskolin). When the chromatography is continued, compound **II** (1,7-bissuccinyl-7-deacetylforskolin) is eluted. Its bissuccinate nature is also suggested by the fact that it can be converted to compound **I** (and further to deacetylforskolin) by treatment with methanolic KOH.

The identity of compound **I** as 7-succinyl-7-deacetylforskolin is obvious from the 90-MHz ^1H NMR spectrum (Bruker, EM 390). When compared with that of forskolin, the spectrum of **I** lacks the ^3H singlet at 2.18 ppm (acetyl group) but otherwise reveals almost identical δ and J values for carbinol protons and those adjacent to the acyl groups. ^1H NMR spectra (in $CDCl_3$, tetramethylsilane as standard): 7-Succinyl-7-deacetylforskolin: $2H$, m at 4.55 ppm (1β-H, 6α-H); $1H$, d at 5.4 ppm, J = 4 Hz (7β-H). Forskolin: $2H$, d of d at 4.4 and 4.55 ppm, $W_{1/2}$ = 10 Hz (1β-H, 6α-H); $1H$ at 5.48 ppm, J = 4 Hz (7β-H).

Coupling of Forskolin Derivatives to Insoluble Supports[10,11]

Forskolin-Sepharose

Sepharose CL-4B (Pharmacia, Uppsala, Sweden), 18 g, is activated with 0.5 g carbonyldiimidazole (Merck) and coupled to ethylenediamine (3.4 g, pH 10.0) as described by Bethell *et al.*[12] Activation of 7-succinyl-7-deacetylforskolin (7.52 mg, 16 μmol) is achieved by treatment with 2.75 mg (24 μmol) *N*-hydroxysuccinimide and 4.95 mg dicyclohexylcarbodiimide (24 μmol, Merck) in 200 μl of anhydrous acetonitrile. After standing for 4 hr at 22°, the dicyclohexylurea formed is removed by centrifugation and extracted 3 times each with 200 μl of acetonitrile. The combined supernatants are added to 10 g (wet) aminoethyl-Sepharose CL-4B in 15 ml of dry dimethylformamide. The mixture is gently agitated for 16 hr at 22°, following which the resin is collected by filtration and washed successively with 30 ml dimethylformamide, 20 ml ethanol, 30 ml of 50% ethanol, and 50 ml of distilled water.

The affinity resin prepared this way contains positively charged groups due to residual NH_3^+ functions, which makes the use of higher ionic strength buffers (>400 m*M* NaCl) necessary. In order to use the resin at

[12] G. S. Bethell, J. S. Ayers, and W. S. Hancock, *J. Biol. Chem.* **254,** 2572 (1979).

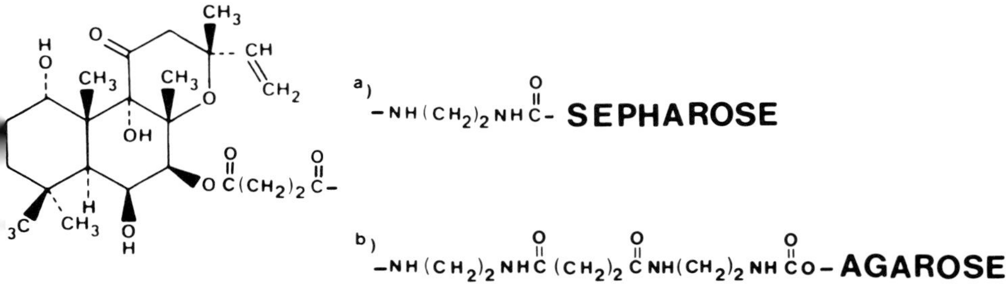

FIG. 1. Structural formulas of immobilized forskolin derivatives. (a) Forskolin-Sepharose; (b) forskolin-Affi-Gel 10.

lower NaCl concentrations (e.g., 100 m*M*) free NH_3^+ groups are blocked by acetylation as follows. Wet forskolin-Sepharose (10 g) is resuspended in 30 ml distilled water, and acetic anhydride is added in three portions (50 μl each) under rigorous stirring at 22°, while the pH is maintained at 6.0 by addition of 1 *M* NaOH. When the pH remains constant the resin is washed with 50 ml water and 50 ml of 70% ethanol, which also serve to store the affinity support at −20°. The disappearance of NH_3^+ groups is indicated by the negative test with trinitrobenzenesulfonic acid.

Forskolin-Affi-Gel 10

For the isolation of brain adenylyl cyclase a longer spacer proves to be more useful.[13] The activated agarose derivative Affi-Gel 10 (Bio-Rad, Richmond, CA) is reacted with 0.8 ml ethylenediamine (Merck), applied to the original 25-ml container by means of a syringe. The mixture is shaken for 12 hr at 22° and washed with 200 ml each of 0.1 *M* HCl and distilled water. The Affi-Gel 10–ethylenediamine derivative is reacted with the *N*-hydroxysuccinimide ester of 7-succinyl-7-deacetylforskolin as described above for the derivatization of aminoethyl-Sepharose CL-4B. The affinity resin is stored in 70% ethanol v/v at −20° (Fig. 1).

Discussion

The potent bifunctional forskolin derivative 7-succinyl-7-deacetylforskolin can be readily synthesized in one step from deacetylforskolin in reasonable yields. It is conveniently coupled to aminoalkyl agaroses via

[13] E. Pfeuffer, S. Mollner, and T. Pfeuffer, *EMBO J.* **4,** 3675 (1985).

N-hydroxysuccinimide esters (Fig. 1). The 7-succinyl-7-deacetylforskolin is now commercially available from Calbiochem (San Diego, CA).

Immobilized forskolin can be regenerated by treatment with urea or sodium dodecyl sulfate (SDS), and it may be used many times with negligible loss of active ligand. Binding of adenylyl cyclase to forskolin-agarose is highly specific; release of bound cyclase from the affinity support is strictly dependent on the presence of a forskolin derivative, following the same order known for the stimulation of adenylyl cyclase: forskolin > 7-succinyl-7-deacetylforskolin = deacetylforskolin ≫ 1-succinylforskolin. 1,9-Dideoxyforskolin, completely devoid of adenylyl cyclase-activating activity, likewise fails to desorb the enzyme from a forskolin-agarose matrix. Despite the specificity of the diterpene, all chromatographic steps require the continuous presence of detergent, that is, no activity could be released by forskolin alone. Nonionic detergents (Triton, Lubrol, Tween) as low as 1 m*M* are usually sufficient to achieve maximal release in the presence of 100 μM forskolin.

The fact that forskolin also activates adenylyl cyclase from S49 cyc^- membranes argues for the binding site of the diterpene being the catalytic moiety.[8] This is also corroborated by results obtained from affinity chromatography of detergent-solubilized adenylyl cyclase. When the soluble enzyme from rabbit myocardial membranes is treated with the immobilized forskolin derivative, 70% of the activity applied remains matrix bound following removal of nonrelevant proteins by washing. After desorption by excessive forskolin and removal of the latter by gel filtration, a preparation is obtained which lacks stimulation by agents known to act through G_s, such as $[AlF_4]^-$ or guanine nucleotides. However, the purified enzyme is activated by forskolin or by activated G_s (Table II).[14] These data are best interpreted in terms of removal of G_s (or part of it) following affinity chromatography. However the α component of G_s can be coisolated with the catalytic subunit as a $C\alpha_s$ complex when the enzyme is preactivated by $[AlF_4]^-$ or the nonhydrolyzable GTP analogs GPP(NH)P or GTPγS.[11,14] This $C\alpha_s$ complex has considerably higher affinity for forskolin and congeners. In combination with its superior stability, this species allows for higher yields on purification by affinity chromatography. This is exploited for the isolation of the GPP(NH)P-stabilized adenylyl cyclase complex from rabbit myocardium.[11]

A 10,000- to 40,000-fold purification by forskolin-agarose is necessary to reach homogeneity of the catalytic subunit. This further underscores the extraordinary selectivity of the hypotensive drug forskolin. However, forskolin binding sites other than adenylyl cyclase have also been demon-

[14] T. Pfeuffer, B. Gaugler, and H. Metzger, *FEBS Lett.* **164,** 154 (1983).

TABLE II
REFRACTORINESS OF FORSKOLIN-SEPHAROSE-PURIFIED MYOCARDIAL ADENYLYL CYCLASE TOWARD GUANINE NUCLEOTIDES AND NaF[a]

Preparation	Adenylyl cyclase activity (nmol mg^{-1} min^{-1}) in presence of					
	None	Mn^{2+}	GPP(NH)P	NaF	Forskolin	G_s
Crude, solubilized	0.13	0.53	0.57	0.65	1.56	1.71
Forskolin-Sepharose eluate	59.2	239.1	58.8	53.7	536	497

[a] Lubrol PX-solubilized adenylyl cyclase from rabbit hearts is chromatographed on forskolin-Sepharose as described in Ref. 10. Bound cyclase is released with 100 μM forskolin, which is removed for adenylyl cyclase assay on a Sephadex G-25 fine column.[10] Twenty-nine micrograms of crude solubilized enzyme or 0.5 μg of purified enzyme is tested for cyclase activity. The adenylyl cyclase assay is performed as described in the footnote to Table I. Activators are added at the following final concentrations: $MnCl_2$, 2 mM; GPP(NH)P, 100 μM; NaF, 10 mM; forskolin, 100 μM; G_s protein [prepared from duck erythrocytes,[14] GPP(NH)P form], 30 μg. Reprinted in part from Pfeuffer and Metzger[10] with the permission of Elsevier Biomedical Press.

strated, namely, the insulin-dependent and -independent glucose carriers.[15] Since their affinity for the diterpene seems to be in the same range as that of adenylyl cyclase, the immobilized forskolin derivatives presented here should also be of value for the isolation of these carriers.

Acknowledgments

We thank Mrs. E. Pfeuffer and Mr. M. Brenner for expert technical assistance. The generous supply of forskolin from Drs. Metzger, Schöne, and Schorr (Hoechst Company, Frankfurt) is gratefully acknowledged. This work was supported by grants from the Deutsche Forschungsgemeinschaft (Sonderforschungsbereich 176) and by the Fonds der Chemischen Industrie.

[15] S. Sergeant and H. D. Kim, *J. Biol. Chem.* **260,** 14677 (1985).

[5] High-Affinity Binding Sites for [^{3}H]Forskolin

By ANTONIO LAURENZA and KENNETH B. SEAMON

Introduction

Forskolin was originally discovered as a stimulator of adenylyl cyclase (EC 4.6.1.1, adenylate cyclase) and has been shown to interact directly with the catalytic subunit of adenylyl cyclase.[1] Forskolin activates adenylyl cyclase in intact cells and tissues, membranes, and detergent-solubilized and purified preparations of the enzyme.[2] The EC_{50} for forskolin activation of adenylyl cyclase in most preparations is between 1 and 20 μM. Forskolin synergistically stimulates adenylyl cyclase in combination with hormones that activate adenylyl cyclase via the G_s protein. The synergistic interactions of forskolin occur at concentrations (10 nM to 0.1 μM) that are lower than those required for the direct activation of adenylyl cyclase. The ability of forskolin to interact with adenylyl cyclase with high affinity suggested that there might be high-affinity binding sites for forskolin that could be detected with a suitable binding assay. Tritiated forskolin was synthesized and used to detect high-affinity binding sites for forskolin in membranes from a number of tissues. The detection of forskolin binding sites can be affected by agents that activate adenyl cyclase the G_s protein. It was therefore suggested that the high-affinity forskolin-binding sites are associated with complexes of the catalytic subunit and the G_s protein.[2]

Forskolin affects the function of a number of membrane transport proteins other than adenylyl cyclase,[3] including the glucose transporter,[4] the nicotinic acetylcholine receptor,[5] the γ-aminobutyric acid (GABAa) receptor,[6] voltage-dependent K^+ channels,[7] and possibly the P-glycoprotein multidrug transporter.[8] The inhibition of these transport proteins by forskolin occurs at concentrations of forskolin that are similar to those

[1] K. B. Seamon, W. Padgett, and J. W. Daly, *Proc. Natl. Acad. Sci. U.S.A.* **78,** 3363 (1981).

[2] K. B. Seamon and J. W. Daly, *Adv. Cyclic Nucleotide Protein Phosphorylation Res.* **20,** 1 (1986).

[3] A. Laurenza, E. McHugh-Sutkowski, and K. B. Seamon, *Trends Pharmacol. Sci.* **10,** 442 (1989).

[4] A. Kashiwagi, T. P. Hueckstadt, and J. E. Foley, *J. Biol. Chem.* **258,** 13685 (1983).

[5] E. McHugh and R. M. McGee, Jr., *J. Biol. Chem.* **261,** 3103 (1986).

[6] G. Heuschneider and R. D. Schwartz, *Proc. Natl. Acad. Sci. U.S.A.* **86,** 2938 (1989).

[7] T. Hoshi, S. S. Garber, and R. W. Aldrich, *Science* **240,** 1652 (1988).

[8] S. Wadler and P. H. Wiemick, *Cancer Res.* **48,** 539 (1988).

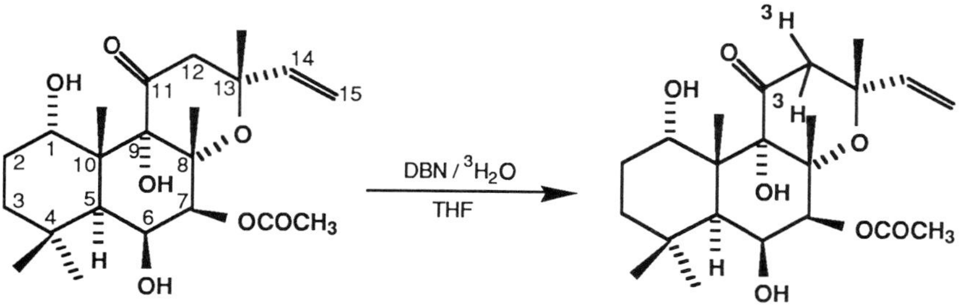

FIG. 1. Synthesis of [12-³H]forskolin.

required for the activation of adenylyl cyclase. Although the interaction of forskolin with these membrane transport proteins is not so well characterized as its interaction with adenylyl cyclase, forskolin has been demonstrated to bind directly to the glucose transporter.[9,10] The ability of forskolin to interact with a number of different membrane-associated proteins makes it difficult to assume that all effects of forskolin are due to activation of adenylyl cyclase and similarly that all forskolin binding sites are associated with the interaction of forskolin with adenylyl cyclase. It is therefore crucial to define characteristics of the binding of forskolin to adenylyl cyclase that can distinguish this from the interaction of forskolin with other proteins. Thus, the characteristics of the high-affinity forskolin binding sites will be described and the ability to uniquely associate this binding with adenylyl cyclase discussed.

Methods

[12-³H]Forskolin. [12-³H]Forskolin ([³H]forskolin) is synthesized by the base-catalyzed exchange of protons on the C-12 methylene group, which is adjacent to the carbonyl group[11] (Fig. 1). Forskolin is incubated with the base 1,5-diazobicyclo[4.3.0]non-5-ene (DBN) in tetrahydrofuran and a stoichiometric amount of ^3H_2O at 45° for 3 hr. The solution is then passed through a silica column equilibrated with ethyl acetate to remove the DBN. The solvent is removed under N_2, and the residue is taken up in ethanol in order to exchange the tritium at the hydroxyl groups. The ethanol is evaporated, and the residue is dissolved in ethanol and stored at −80°. [³H]Forskolin is purified by flash chromatography on a silica gel

[9] S. Sergeant and H. D. Kim, *J. Biol. Chem.* **260,** 14677 (1985).
[10] V. R. Lavis, D. P. Lee, and S. Shenolikar, *J. Biol. Chem.* **262,** 14571 (1987).
[11] K. B. Seamon, R. Vaillancourt, M. W. Edwards, and J. W. Daly, *Proc. Natl. Acad. Sci. U.S.A.* **81,** 5081 (1984).

column with dichloromethane/ethyl acetate (9/1) as solvent. Fractions are collected from the column, and radioactivity is counted. The forskolin is identified by chromatography on silica gel plates with dichloromethane/ethyl acetate (9/1) as solvent and detected by use of I_2 vapor. Forskolin has an R_f of 0.4 with this chromatography system. Forskolin-containing fractions are pooled, diluted with ethanol to a concentration of 15 mCi/ml, and stored at $-80°$. The specific activity of the tritiated forskolin is 32 Ci/mmol.

The site of the exchange is determined by carrying out the reaction with ^2H_2O in place of ^3H_2O and monitoring the reaction by NMR spectroscopy. Quantitative exchange of the protons at the 12-position of forskolin is verified by loss of the proton intensities in the NMR spectrum. There is no hydrolysis of the 7-acetyl group of forskolin under the exchange conditions. The radiochemical purity of [^{3}H]forskolin is determined by using reverse phase high-performance liquid chromatography [C18 column, 4.6 mm × 25 cm; solvent CH_3CN/H_2O (7/3)] and monitoring the radioactivity with a Radiomatic HS flow detector. Purified [^{3}H]forskolin is radiochemically pure (>99%) by this procedure.

The tritium on forskolin does not exchange with solvent under basic conditions with moderate heat (pH 8.5, 60°, 3 hr). [^{3}H]Forskolin is equipotent with forskolin in activating rat brain adenylyl cyclase. [^{3}H]Forskolin is kept as a 500 μM stock solution in ethyl acetate. A slight decomposition of forskolin is observed after 4–5 months when it is kept in ethanol at room temperature. [^{3}H]Forskolin prepared by the above procedure is available from New England Nuclear (Boston, MA).

Nonspecific binding of [^{3}H]forskolin to tubes is determined as the percentage of total [^{3}H]forskolin that can be extracted from tubes by 0.2 *M* NaOH. Tubes are exposed to [^{3}H]forskolin and rinsed 3 times with cold 50 m*M* Tris-HCl, pH 7.5, before extraction with 1 ml of 0.2 *M* NaOH. Polycarbonate and polysulfone tubes have high nonspecific binding, about 1%, while glass tubes and polypropylene tubes have the lowest nonspecific binding, <0.05%. Glass tubes are therefore used in all experiments. For binding studies a solution of [^{3}H]forskolin is prepared by placing an aliquot of the stock [^{3}H]forskolin into a glass tube, removing the organic solvent under N_2, and adding 50 m*M* Tris-HCl, pH 7.5, to the desired volume. It is crucial to allow at least 20 min for the [^{3}H]forskolin to go completely into solution. It is important to note that a small percentage of [^{3}H]forskolin will precipitate on filters in the presence of high concentrations of unlabeled forskolin (>30 μM). This is due to the limited solubility of forskolin (<100 μM) in aqueous solutions and can lead to anomalous results when displacement experiments are carried out with unlabeled forskolin or with analogs of forskolin having similar solubility characteristics.

Rat Brain Membranes. Forebrains from two rats sacrificed by decapitation are removed and homogenized in a motor-driven homogenizer in 20 ml of ice cold 0.32 *M* sucrose. The homogenate is centrifuged at 1000 *g* for 10 min and the pellet discarded. The supernatant is centrifuged at 10,000 *g* for 10 min. The pellet is resuspended in 10 ml of 50 m*M* Tris-HCl, pH 7.5, at a concentration of 2.5 mg of protein/ml, and is homogenized in a Dounce homogenizer.

Human Platelet Membranes. A 1/50 volume of 0.1 *M* EDTA is added to 5- to 10-day-old platelet concentrates obtained from local blood banks and left on ice for 10 min. The suspension is then centrifuged at 3500 *g* for 15 min and the supernatant discarded. The pellet is resuspended with a plastic pipette in 10 m*M* Tris-HCl, pH 7.4, containing 1.1 m*M* EDTA, 90 m*M* glucose, and 100 m*M* NaCl. The suspension is centrifuged at 1000 *g* for 5 min to remove erythrocytes. The supernatant is then centrifuged at 3500 *g* for 10 min. The pellet is resuspended in 12 ml of the same buffer and incubated for 10 min at 20° with 3.8 ml of a 60% glycerol solution. The cell suspension is then centrifuged at 5000 *g* for 10 min and the pellet resuspended in 20 ml of ice-cold 10 m*M* Tris-HCl pH 7.4, 25 m*M* sucrose, 2 m*M* EDTA. This suspension is vortexed vigorously for 2 min, homogenized with 5 strokes in a Dounce glass homogenizer, and centrifuged at 3500 *g* for 10 min. The pellet is centrifuged twice at 25,000 *g* for 10 min in 10 m*M* Tris-HCl, pH 7.4, 1 m*M* EDTA and resuspended in the same buffer at a protein concentration of 2.5 mg/ml.

Preparation of Solubilized Proteins. Frozen (−70°) bovine brain (300 g) is thawed in 10 m*M* Tris-HCl, pH 7.5, containing 0.9% NaCl and minced. The buffered saline is replaced with 800 ml of 10 m*M* Tris-HCl, pH 7.5, containing 0.32 *M* sucrose. Minced tissue is homogenized with a Brinkman Polytron at setting 5 for 60 sec and centrifuged at 3000 *g* for 15 min at 4°. The resulting supernatant is centrifuged at 25,000 *g* for 15 min at 4°, and the pellet is resuspended in 50 m*M* Tris-HCl at pH 7.5 and centrifuged as above. The last washing step is repeated twice, and the membranes are suspended in 50 m*M* Tris-HCl buffer at pH 7.5. After the final centrifugation the membranes are suspended in 100 ml of 10 m*M* Tris-HCl, pH 7.5, containing 5 m*M* $MgCl_2$, 0.3% Lubrol PX, 1 m*M* dithiothreitol (DTT), and 1 m*M* EDTA and stirred at 4° for 60 min. The suspension is then centrifuged at 100,000 *g* for 90 min; the supernatant is removed and stored in liquid nitrogen.

Membrane Binding Assay. Incubations are carried out in 13 × 100 mm glass test tubes in a total volume of 0.4 ml with 50 m*M* Tris-HCl buffer, pH 7.4, at 20° for 60 min. The membranes (0.5 mg protein/tube) are incubated with 10 n*M* [^{3}H]forskolin in the presence of 5 m*M* $MgCl_2$ and 100 μM GPP(NH)P or 10 m*M* NaF where indicated. The assay is terminated by rapid filtration over Whatman GF/C filters with a Brandel cell harvester (Gaithersburg, MD). The filters are quickly washed 3 times with 4 ml of ice-cold buffer and placed in scintillation vials for the determination of radioactivity. Specific [^{3}H]forskolin binding is calculated as the difference between total binding in the absence of unlabeled forskolin and nonspecific binding in the presence of 20 μM forskolin. The nonspecific binding is about 10% of the total binding and is slightly higher in membranes from platelets that are older than 10 days and from membranes that had been frozen.

Centrifugation Assay. After incubation, assay tubes are centrifuged at 28,000 *g* for 15 min at 4° to pellet the membranes. The supernatant is carefully aspirated so as not to disturb the membrane pellet, and the tubes are washed 3 times with 3 ml of ice-cold 50 m*M* Tris-HCl, pH 7.5. Washing of the tubes is carried out carefully with a Brandel cell harvester so that the pellet is not disturbed. The pellet is then dissolved in 1 ml of 0.2 *M* NaOH, and the radioactivity is determined by liquid scintillation counting.

Binding to Solubilized Proteins. Solubilized proteins (600 μg/tube) are incubated in 25 m*M* Tris-HCl buffer at pH 7.5 containing 20 m*M* $MgCl_2$, 0.25 m*M* DTT, 0.15% Lubrol PX, and [^{3}H]forskolin at 20° for 45 min in a total volume of 0.4 ml. After incubation the tubes are placed in an ice bath, and 2 mg of bovine γ-globulin (Cohn fraction II) in 100 μl is added to each tube and vortexed. Five hundred microliters of a 32% solution of polyethylene glycol (PEG, MW 8000) is added to each tube, vortexed, and incubated on ice for 10 min. Four milliliters of PEG (5%) is then added to each tube, and the solutions are filtered over Whatman GF/B filter paper which is washed with 8 ml of PEG (5%). There is no specific binding of [^{3}H]forskolin in the absence of added γ-globulin. The number of specific binding sites is maximal in the presence of 2 mg of γ-globulin and accounts for 50% of the total number of binding sites.

Comments

High-Affinity Binding Sites. The use of a filtration assay to separate bound from free forskolin was very effective in developing a binding assay for detecting the high-affinity binding of forskolin with K_d values that are less than 100 n*M*.[11] Low-affinity binding of [^{3}H]forskolin is more difficult to determine using a filtration assay to separate bound from free forskolin; however, low-affinity binding sites can be detected using a centrifugation assay.[11] High-affinity binding will be considered as that binding with a K_d less than 50 n*M* and where a filtration assay is appropriate for separating the bound from free forskolin. Using a filtration assay, a complete binding isotherm can be produced by incubating membranes with increasing amounts of [^{3}H]forskolin, with nonspecific binding being determined in the presence of 20 μM unlabeled forskolin. The nonspecific binding is generally less than 10% of the total binding. The assay is very reproducible and the standard deviation (S.D.) for triplicate determinations is generally less than 10% of the mean. The K_d and B_{max} for forskolin binding can be determined using different protocols. Binding can be determined in the presence of increasing concentrations of [^{3}H]forskolin; alternatively, the amount of [^{3}H]forskolin can be kept constant and the binding determined in the presence of increasing amounts of unlabeled forskolin. The number of binding sites and the affinity of forskolin for these sites are the same regardless of which protocol is used. The binding parameters can be analyzed with the Ligand program of Munson and Rodbard.[12] The affinity for forskolin binding to membranes which have a high specific activity of adenylyl cyclase, such as those from brain and platelets, is about 20 nM. The IC_{50} for forskolin to inhibit binding is about 32 nM in the presence of 10 nM [^{3}H]forskolin.

The B_{max} for forskolin binding depends on the tissue being studied and the conditions of the assay (Table I). For example, in the presence of 5 mM $MgCl_2$, rat brain and human platelet membranes have a B_{max} of 270 and 125 fmol/mg, respectively, while rat striatum has a B_{max} of 630 fmol/mg. It is pertinent that forskolin is more efficacious in stimulating striatal adenylyl cyclase than either human platelet or rat brain adenylyl cyclase.[2] The majority of studies that have detected high-affinity binding sites for forskolin have been carried out with membrane preparations from brain or platelets. Fewer data are available for the binding of forskolin to peripheral tissues. Binding sites for forskolin have been detected in liver, lung, and heart membranes; however, these sites have much lower affinity for forskolin than the binding sites in brain (see Table I).

High-Affinity Binding to Solubilized Proteins. [^{3}H]Forskolin binds to solubilized proteins from bovine brain with a K_d of about 10 nM and a B_{max} in the presence of $MgCl_2$ of 30 fmol/mg, which is increased about 3-fold in the presence of GPP(NH)P. Preactivation of membranes with GPP(NH)P prior to solubilization produces a solubilized preparation which binds [^{3}H]forskolin with a B_{max} of about 100 fmol/mg, and this value is not increased in the presence of GPP(NH)P. This is consistent with the requirement for an activated G_s protein for maximal binding of [^{3}H]forskolin. The number of binding sites per milligram of protein in solubilized membranes from brain is less than that for brain membranes. The decreased amount of binding observed with the solubilized membranes in comparison to intact membranes could be due to the presence of the detergent; high

[12] P. J. Munson and D. Rodbard, *Anal. Biochem.* **107,** 220 (1980).

TABLE I

BINDING PARAMETERS FOR [^{3}H]FORSKOLIN BINDING SITES[a]

Source	K_d	B_{max}	Conditions	Ref.[d]
Membranes				
Rat brain	15 nM	270 fmol/mg	$MgCl_2$	1,3
	1.1 μM	4.2 pmol/mg	$MgCl_2$[b]	1,3
	26 nM	400 fmol/mg	$MgCl_2$[b]	1,3
	15 nM	220 fmol/mg	$MgCl_2$	2
	15 nM	380 fmol/mg	NaF	2
Human platelets	20 nM	125 fmol/mg	$MgCl_2$	4
	20 nM	455 fmol/mg	NaF	4
Rat myocardium	250 nM	5 pmol/mg	$MgCl_2$	5
	650 nM	5 pmol/mg	$MgCl_2$[c]	5
Rat lung	$>100\ \mu M$	ND	$MgCl_2$	6
Rat heart	740 nM	ND	$MgCl_2$	6
Rat liver	1.43 μM	5.2 pmol/mg	$MgCl_2$	6
Rat liver	600 nM	114 pmol/mg	$MgCl_2$	7
Rat striatum	27 nM	95 fmol/mg	No additions	8
	28 nM	120 fmol/mg	$MgCl_2$	8
	15 nM	190 fmol/mg	NaF	8
Rabbit ciliary body	17 nM	184 fmol/mg	NaF	9
Rat brain	150 nM	3.2 pmol/mg	$MgCl_2$[c]	10
Rat liver	1.6 μM	179 pmol/mg	$MgCl_2$[c]	11
Rat brain	0.8 μM	3.2 pmol/mg	$MgCl_2$[c]	11
Rat striatum	9 nM	888 fmol/mg	$MgCl_2$	12
	1.8 μM	280 pmol/mg	$MgCl_2$	12
Slices				
Gerbil brain	8.2 nM	354 fmol/mg	NaF	13
Rat forebrain	309 nM	12 pmol/mg	GPP(NH)P	14
Rat brain	42 nM	102 fmol/mg	$MgCl_2$	15
Rat striatum	41 nM	630 fmol/mg	$MgCl_2$	16
Rat hippocampus	44 nM	335 fmol/mg	$MgCl_2$	16
Rat cerebellum	48 nM	285 fmol/mg	$MgCl_2$	16
Intact cells				
S49 wild-type lymphoma	ND	1800 sites/cell		17
Solubilized membranes				
Bovine brain	10 nM	38 fmol/mg	$MgCl_2$	18
	12 nM	94 fmol/mg	GPP(NH)P	18

[a] Binding was determined using a filtration assay unless otherwise indicated. ND, Not determined.

[b] Binding was determined using a centrifugation assay.

[c] Binding was determined using 14,15-ditritioforskolin.

[d] *Key to References:* (1) K. B. Seamon, R. Vaillancourt, M. Edwards, and J. W. Daly, *Proc. Natl. Acad. Sci. U.S.A.* **81,** 5081 (1984): (2) K. B. Seamon, R. Vaillancourt, and J. W. Daly, *J. Cyclic Nucleotide Protein Phosphorylation Res.* **10,** 536 (1985); (3) K. B. Seamon and J. W. Daly, *Adv. Cyclic Nucleotide Protein Phosphorylation Res.* **19,** 125 (1985); (4) C. A. Nelson and K. B. Seamon, *J. Biol. Chem.* **261,** 13469 (1986); (5) K. Schmidt and W. R. Kukovetz, *J. Cyclic Nucleotide Protein Phosphorylation Res.* **10,**

concentrations of detergent can inhibit the binding of [^{3}H]forskolin to bovine brain membranes in an apparently noncompetitive manner.[13]

Inhibition of Forskolin Binding. Forskolin analogs have been tested for their ability to active adenylyl cyclase and their ability to inhibit [^{3}H]forskolin binding in membranes and solubilized proteins.[14,15] Analogs of forskolin that do not activate adenylyl cyclase do not inhibit the high-affinity binding of [^{3}H]forskolin. These include 1,9-dideoxyforskolin and other analogs of forskolin that contain large lipophilic groups esterified at the 7-position. There is a good correlation between the potency of forskolin analogs to inhibit high-affinity binding and their potency to stimulate adenylyl cyclase. This evidence suggests that the high-affinity forskolin binding sites are associated with the sites on adenylyl cyclase that are relevant for stimulation by forskolin.

Many other agents have been tested for their ability to inhibit forskolin binding to high-affinity sites in bovine brain membranes (Table II). These include compounds that bind to the erythrocyte glucose transporter such as D-glucose and cytochalasin B. These compounds have no effect on the high-affinity binding of forskolin. Some steroids, lipids, and crude lipid fractions from bovine brain do not have any effect on [^{3}H]forskolin binding (Table II). Drugs that interact with cytoskeletal components, such as cytochalasin B, colchicine, and vinblastine, do not inhibit [^{3}H]forskolin binding (Table II). Calcium channel antagonists, such as verapamil and diltiazem, as well as other neuroactive drugs also do not inhibit [^{3}H]forskolin binding (Table II).

[13] A. Laurenza and K. B. Seamon, unpublished data (1989).

[14] K. B. Seamon, J. W. Daly, H. Metzger, N. J. De Souza, and J. Reden, *J. Med. Chem.* **26,** 436 (1983).

[15] A. Laurenza, Y. Khandelwal, N. J. De Souza, H. R. Rupp, H. Metzger, and K. B. Seamon, *Mol. Pharmacol.* **32,** 133 (1987).

425 (1985); (6) G. P. Jackman and A. Bobik, *Biochem. Pharmacol.* **35,** 2247 (1986); (7) K. Schmidt, H. Baer, A. Shariff, W. A. Ayer, and L. Browne, *Can. J. Physiol. Pharmacol.* **65,** 803 (1987); (8) A. B. Norman, T. J. Wachendorf, and P. R. Sanberg, *Life Sci.* **44,** 831 (1989); (9) M. E. Goldman, P. Mallorga, D. J. Pettibone, and M. F. Sugrue, *Life Sci.* **42,** 1307 (1988); (10) K. Schmidt, R. Munshi, and H. P. Baer, *Naunyn-Schmiedebergs Arch. Pharmacol.* **325,** 153 (1984); (11) K. Schmidt and H. P. Baer, *Eur. J. Pharmacol.* **94,** 337 (1983); (12) J. A. Poat, H. E. Cripps, and L. L. Iversen, *Proc. Natl. Acad. Sci. U.S.A.* **85,** 3216 (1988); (13) K. Tanaka, F. Gotoh, N. Ishihara, S. Gomi, S. Takashima, and B. Mihara, *Brain Res. Bull.* **21,** 693 (1988); (14) D. R. Gehlert, *J. Pharmacol. Exp. Ther.* **239,** 952 (1986); (15) D. R. Gehlert, T. M. Dawson, H. I. Yamamura, and J. K. Wamsley, *Brain Res.* **361,** 351 (1985); (16) D. R. Gehlert, T. M. Dawson, F. M. Filloux, E. Sanna, I. Hanbauer, and J. K. Wamsley, *Neurosci. Lett.* **73,** 1146 (1987); (17) R. Barber, *Second Messengers Phosphoproteins* **12,** 59 (1988); (18) C. A. Nelson and K. B. Seamon, *Life Sci.* **42,** 1375 (1988).

TABLE II
AGENTS THAT DO NOT INHIBIT HIGH-AFFINITY [^{3}H]FORSKOLIN BINDING TO BOVINE BRAIN MEMBRANES

Agent	Concentration
Sugars	
D-Glucose	0.5 *M*
L-Glucose	0.5 *M*
2-Deoxy-D-glucose	0.5 *M*
6-Deoxy-D-glucose	0.5 *M*
6-Fluoro-6-deoxy-D-glucose	0.125 *M*
3-Fluoro-3-deoxy-D-glucose	0.125 *M*
D-Glucosamine	0.5 *M*
N-Acetyl-D-glucosamine	0.5 *M*
3-*O*-Methyl-D-glucose	0.5 *M*
D-Galactose	0.5 *M*
2-Deoxy-D-galactose	0.5 *M*
N-Acetylgalactose	0.5 *M*
N-Acetylgalactosamine	0.5 *M*
Fructose	0.5 *M*
Sucrose	0.5 *M*
Altrose	0.5 *M*
Idose	0.5 *M*
D-Xylose	0.5 *M*
D-Fucose	0.5 *M*
D-Gulose	0.5 *M*
D-Allose	0.5 *M*
D-Mannose	0.5 *M*
D-Sorbitol	0.5 *M*
Saccharose	0.5 *M*
Lipids and lipid fractions	
Folch fraction III from bovine brain	50 μg/ml
Folch fraction I from bovine brain	50 μg/ml
Fraction VIII from bovine brain	50 μg/ml
Fraction X from bovine brain	50 μg/ml
Folch fraction V from bovine brain	50 μg/ml
Fraction VI from bovine brain	50 μg/ml
Fraction VII from bovine brain	50 μg/ml
Monosialoganglioside G_{M1}	50 μg/ml
Ganglioside G_{T1b}	50 μg/ml
Psychosine	1 mg/ml
Sulfatide	1 mg/ml
Galactocerebroside	1 mg/ml
Sphingosine	1 mg/ml
Ceramide III	1 mg/ml
Sialic acid	20 μ*M*
Cardiolipin	200 μ*M*
Phosphatidylinositol	40 μ*M*
Phosphatidylinositol phosphate	40 μ*M*

TABLE II (*continued*)

Agent	Concentration
Steroids	
Diethylstilbestrol	20 μM
Androstane-3β,7β,11α-triol-17-one	20 μM
Androstane-3β,7β,11α-triol-11-one	20 μM
Androstane-7α,9α,11α-triol-3-one	20 μM
Androstane-1β,3α,6α-triol-17-one	20 μM
Androstane-6β,11α,16β-triol-3-one	20 μM
5-Pregnen-3β-ol-20-one sulfate	20 μM
Aldosterone	20 μM
Testosterone	20 μM
17β-Estradiol	20 μM
Dexamethasone	20 μM
Cortisone	20 μM
Androsterone	20 μM
Hydrocortisone	20 μM
Estriol	20 μM
4-Androstene-3,17-dione	20 μM
Estrone	20 μM
Other agents	
Cytochalasin B	100 μM
Cytochalasin E	100 μM
Carbachol	20 μM
Muscimol	20 μM
D-Tubocurare	20 μM
Hexamethonium	20 μM
Chlorpromazine	20 μM
Phencyclidine	20 μM
Histrionicotoxin	20 μM
Pentobarbitol	20 μM
Tolbutamide	20 μM
Furosemide	20 μM
Chlorothiazide	20 μM
Benzthiazide	20 μM
Ouabain	20 μM
Verapamil	20 μM
Diltiazem	20 μM
Clonidine	20 μM
Chloroquine	20 μM
Vinblastine	20 μM
Vincristine	20 μM
Daunomycin	20 μM
Doxorubicin	20 μM
Colchicine	20 μM
Thioridazine	20 μM
Prochlorpromazine	20 μM
Actinomycin D	20 μM
Trifluoperazine	20 μM
Reserpine	20 μM

Detergents and organic solvents inhibit [^{3}H]forskolin binding to bovine brain membranes. Ethanol at a concentration of 0.5% has a minimal effect on binding; however, concentrations of ethanol greater than 1% are associated with a decrease in B_{max} and an increase in the IC_{50} for forskolin. Similar results have been observed with detergents. Octylglucoside at concentrations up to 0.2% has no effect on the amount of [^{3}H]forskolin bound to bovine brain membranes or on the IC_{50} for forskolin; however, concentrations of octylglucoside greater than 0.4% are associated with a dramatic inhibition of binding.[13] This inhibition may be due to detergent solubilization of the binding sites.

Modulation of Forskolin Binding Sites. The binding of [^{3}H]forskolin to membranes is decreased after treating the membranes with proteolytic enzymes, *N*-ethylmaleimide, heat and alkylating derivatives of forskolin.[16,17] The binding is therefore associated with a protein and is not due to a nonspecific association with the membrane.

Forskolin binding is increased in the presence of ligands, such as GPP(NH)P or NaF, that activate the G_s protein. The number of forskolin binding sites in membranes from bovine brain and human platelet membranes is increased 2- and 4-fold, respectively, in the presence of GPP(NH)P or NaF.[11,18] The increase in the binding sites in the presence of GPP(NH)P and NaF requires $MgCl_2$ and is associated with an increase in B_{max} with no change in K_d. Increases in binding sites due to GPP(NH)P and NaF are not additive, indicating a common site of action. These results suggest that the high-affinity sites for forskolin are increased when the G_s protein is activated.

High-affinity binding of [^{3}H]forskolin to membranes from S49 lymphoma cells requires the presence of the G_s protein. [^{3}H]Forskolin binding to membranes from wild-type S49 lymphoma cells is observed only in the presence of 10 m*M* NaF (Fig. 2). In contrast, there is no detectable binding of forskolin to membranes from the *cyc*$^-$ variant of S49 lymphoma cells even in the presence of NaF (Fig. 2). [^{3}H]Forskolin binding in the presence of NaF could be observed in *cyc*$^-$ membranes when such membranes were reconstituted with G_s protein.[19]

Low-Affinity Binding Sites. Low-affinity binding ($K_d > 50$ n*M*) of forskolin has been observed in a number of tissues including brain and platelets and other peripheral tissues (Table I). The low-affinity binding has been observed using centrifugation and filtration assays to separate bound

[16] K. B. Seamon, R. Vaillancourt, and J. W. Daly, *J. Cyclic Nucleotide Protein Phosphorylation Res.* **10,** 535 (1985).

[17] A. Laurenza, D. Morris, and K. B. Seamon, *Mol. Pharmacol.* **37,** 69 (1990).

[18] C. A. Nelson and K. B. Seamon, *FEBS Lett.* **183,** 349 (1985).

[19] K. B. Seamon and R. B. Clark, unpublished data (1985).

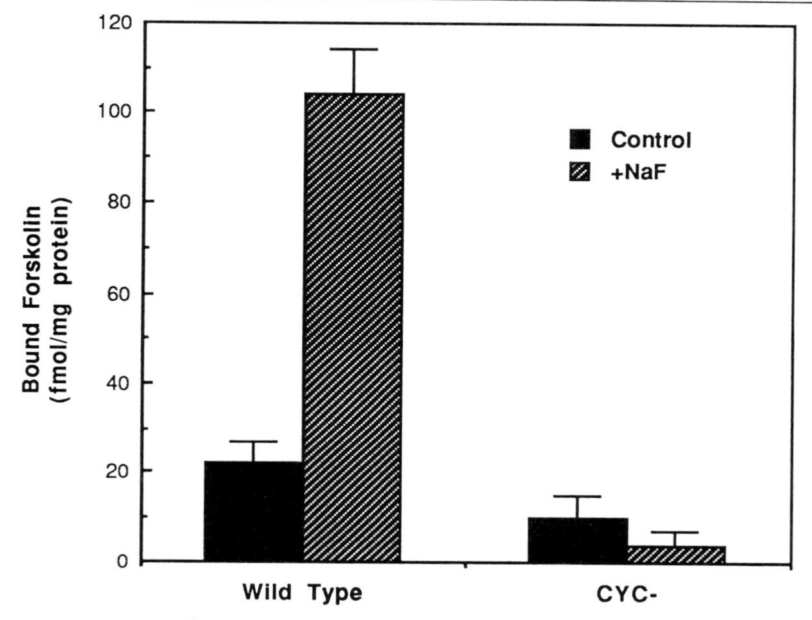

FIG. 2. Binding of [^{3}H]forskolin to wild-type and *cyc*$^-$ S49 lymphoma membranes in the presence and absence of NaF. The binding of [^{3}H]forskolin (10 n*M*) was determined in the presence of 5 m*M* $MgCl_2$ alone or with 10 m*M* NaF as described in the text for high-affinity binding.

from free ligand. The water solubility of forskolin is limited to about 100 μM, and therefore one must be cautious in interpreting binding data with forskolin at concentrations higher than about 40 μM. We have observed that high concentrations of forskolin (>40 μM) can cause an increase in the amount of [^{3}H]forskolin on the filter even in the absence of membranes. This may be due to coprecipitation of [^{3}H]forskolin with forskolin due to the limited solubility of forskolin.

The association of the low-affinity binding sites with adenylyl cyclase is not so well established as that for the high-affinity binding sites. The B_{max} for the low-affinity binding ranges from 5 pmol/mg in rat myocardium to 280 pmol/mg in rat striatum. Peripheral tissues such as heart and liver bound [^{3}H]forskolin only with low affinity (K_d > 100 n*M*) and had B_{max} values of about 5 pmol/mg.

Discussion

The high-affinity binding sites for forskolin detected with a filtration assay have characteristics that associate them with adenylyl cyclase. The ability of forskolin analogs to compete for the high-affinity binding sites is consistent with their ability to stimulate adenylyl cyclase. Forskolin has been proposed to interact directly with the catalytic subunit of adenylyl cyclase, and there are experimental data to support this proposal. Forskolin activates adenylyl cyclase in the absence of the G_s protein, and forskolin affinity columns can be used to purify the catalytic subunit in the absence of the G_s protein. However, high-affinity binding of [^{3}H]forskolin requires the presence of the G_s protein, and the number of [^{3}H]forskolin binding sites is increased when the G_s protein is activated.[18] This suggests that forskolin binds very tightly to a complex of the adenylyl cyclase catalytic subunit and the activated G_s protein. High-affinity binding of forskolin will therefore depend on the presence of the catalytic subunit and the G_s protein as well as the state of activation of the G_s protein.

Forskolin interacts with other membrane-associated proteins with an EC_{50} that is similar to that for forskolin activation of adenylyl cyclase. It is therefore difficult to distinguish the actions of forskolin based solely on the concentration of forskolin required to produce the effect. However, there are a number of criteria that can be used which demonstrate that the high-affinity binding of [^{3}H]forskolin is associated specifically with adenylyl cyclase.[3] For example, the high-affinity binding of forskolin is not inhibited by D-glucose or cytochalsin B, both of which inhibit the interaction of forskolin with the glucose transporter.[10] Analogs of forskolin can be used to associate the high-affinity binding sites of forskolin with adenylyl cyclase. 1,9-Dideoxyforskolin inhibits the glucose transporter,[20] voltage-dependent potassium channels,[7] and ligand-gated ion channels.[5] However, 1,9-dideoxyforskolin does not inhibit the high-affinity binding of forskolin and does not activate adenylyl cyclase.

Forskolin appears to bind only with low affinity to peripheral tissues, such as liver, heart, and lung (see Table I and references therein). It is more difficult to characterize the low-affinity binding of forskolin, and it is possible that low-affinity sites for forskolin binding may be associated with proteins other than adenylyl cyclase, such as the glucose transporter. Binding of [^{3}H]forskolin to the glucose transporter could be verified by the ability of D-glucose and cytochalasin B to inhibit such binding. However, low-affinity binding sites for forskolin may be associated with other forskolin-binding proteins that are not yet characterized. It is therefore essential that the effects of forskolin and forskolin analogs be determined on the other membrane transport proteins that can be affected by forskolin. This will allow a determination of criteria for distinguishing these different sites of action.

[20] H. G. Joost, A. D. Habberfield, I. A. Simpson, A. Laurenza, and K. B. Seamon, *Mol. Pharmacol.* **33,** 449 (1988).

The high-affinity binding sites for forskolin are easily identified in membranes from brain and human platelets. The absence of high-affinity binding of forskolin to peripheral tissues may also suggest that there might be heterogeneity in adenylyl cyclase. Although forskolin can activate adenylyl cyclase in many peripheral tissues, these tissues do not have easily detectable high-affinity binding sites for [^{3}H]forskolin. It is not clear why these tissues apparently lack high-affinity [^{3}H]forskolin binding sites; however, this may be related to differences in the adenylyl cyclase catalytic subunit or the amounts of G_s protein. Alternatively, there might be a requirement for other proteins to interact with adenylyl cyclase for the formation of the high-affinity bindings sites for forskolin.

[6] Purification and Characterization of Calmodulin-Sensitive Adenylyl Cyclase from Bovine Brain

By PING WANG and DANIEL R. STORM

Introduction

Adenylyl cyclase (EC 4.6.1.1, adenylate cyclase) sensitive to stimulation by Ca^{2+} and calmodulin (CaM)[1] was first reported by Brostrom *et al.*[2] in 1975. Although hormone-stimulated adenylyl cyclase is present in almost every mammalian cell type,[3] CaM-sensitive adenylyl cyclase has been demonstrated only in a limited number of tissues, including brain,[2] pancreatic islet cells,[4] and adrenal medulla.[5] Bovine cerebral cortex contains

[1] CaM, Calmodulin; WGA, wheat germ agglutinin; GPP(NH)P, guanylyl imidodiphosphate; EDTA, ethylenediaminetetraacetic acid; EGTA, ethylene glycol bis(β-aminoethyl ether)-*N,N,N′,N′*-tetraacetic acid; G_s, stimulatory guanyl nucleotide regulatory complex of adenylyl cyclase; SDS, sodium dodecyl sulfate; MOPS, 3-(*N*-morpholino)propanesulfonic acid; DTT, dithiothreitol; Tris-HCl, tris(hydroxymethyl)aminomethane hydrochloride; PMSF, phenylmethylsulfonyl fluoride; CHAPS, 3-[(cholamidopropyl)dimethylammonio]-1-propane sulfonate; PAGE, polyacrylamide gel electrophoresis; BSA, bovine serum albumin.

[2] C. O. Brostrom, Y. C. Huang, B. M. Breckenridge, and D. J. Wolff, *Proc. Natl. Acad. Sci. U.S.A.* **72,** 64 (1975).

[3] E. M. Ross and A. G. Gilman, *Annu. Rev. Biochem.* **49,** 533 (1980).

[4] N. C. LeDonne and C. J. Coffee, *Fed. Proc., Fed. Am. Soc. Exp. Biol.,* Abstr. No. 469 (1979).

[5] I. Valverde, A. Vandermeers, R. Anjaneyula, and W. J. Malaisse, *Science* **206,** 225 (1979).

METHODS IN ENZYMOLOGY, VOL. 195

both CaM-sensitive and CaM-insensitive forms of adenylyl cyclase.[6,7] Characterization of the CaM-sensitive adenylyl cyclases with membranes or unfractionated detergent-solubilized extracts has been difficult because of the presence of CaM-insensitive forms of the enzyme and other CaM-binding proteins. In this chapter, we describe three procedures for the purification of the CaM-sensitive adenylyl cyclase and summarize some of the properties of the enzyme. The purification protocols all employ CaM-Sepharose affinity chromatography as a tool for purification of the enzyme.

Purification of Calmodulin-Sensitive Adenylyl Cyclase by Calmodulin-Sepharose and Heptanediamine-Sepharose Chromatography

The CaM-sensitive adenylyl cyclase was originally purified from bovine brain as a complex between the catalytic subunit and G_s-α[8] using the following procedure.

Adenylyl Cyclase Assay. Adenylyl cyclase is assayed by the general method of Salomon *et al.*[9] using [α-^{32}P]ATP as a substrate and [^{3}H]cAMP to monitor product recovery (see [1], this volume). Assays contain, in a final volume of 250 μl, 20 m*M* Tris-HCl, pH 7.5, 1 m*M* [α-^{32}P]ATP (20 cpm/pmol), 5 m*M* theophylline, and 0.1% bovine serum albumin (BSA). All results are presented as the mean of triplicate assays with standard errors of less than 5%. Sensitivity to GPP(NH)P is determined by preincubation of the enzyme with 0.10 m*M* GPP(NH)P for 30 min at 30° in the presence of 5 m*M* $MgCl_2$ prior to assay. Protein concentrations are determined by the method of Peterson.[10]

Preparation of Calmodulin. Calmodulin is prepared from bovine brain by a modification of the procedure of Dedman *et al.*,[11] as modified by LaPorte *et al.*[12] CaM-Sepharose is prepared from purified CaM and cyanogen bromide-activated Sepharose 4B according to the procedure of Westcott *et al.*[7]

Enzyme Purification. Frozen bovine cerebral cortex (500 g), obtained

[6] C. O. Brostrom, M. A. Brostrom, and D. J. Wolff, *J. Biol. Chem.* **252,** 5677 (1977).

[7] K. R. Westcott, D. C. LaPorte, and D. R. Storm, *Proc. Natl. Acad. Sci. U.S.A.* **76,** 204 (1979).

[8] R. E. Yeager, W. Heideman, G. B. Rosenberg, and D. R. Storm, *Biochemistry* **24,** 3766 (1985).

[9] Y. Salomon, C. Londos, and M. Rodbell, *Anal. Biochem.* **58,** 541 (1974).

[10] G. L. Peterson, *Anal. Biochem.* **83,** 346 (1977).

[11] J. R. Dedman, J. D. Potter, R. L. Jackson, J. D. Johnson, and A. R. Means, *J. Biol. Chem.* **252,** 8415 (1977).

[12] D. C. LaPorte, W. A. Toscano, and D. R. Storm, *Biochemistry* **18,** 2820 (1979).

from a local slaughterhouse, is fractured with a hammer and thawed in phosphate-buffered saline (22.5 mM KH_2PO_4, 75 mM NaCl, and 12.8 mM NaOH, pH 7.2). Thawed cortex is drained and homogenized in a Waring blendor (30 sec) in an equal volume of homogenization buffer [20 mM glycylglycine, pH 7.2, 250 mM sucrose, 5 mM $MgCl_2$, 1 mM EDTA, 3 mM dithiothreitol (DTT), and 1 mM phenylmethylsulfonyl fluoride (PMSF)]. The homogenate is further disrupted with a Polytron homogenizer (30 sec at the maximum setting). The resulting homogenate is centrifuged in a Sorvall RC-3B centrifuge for 30 min at 4500 rpm. The pelleted membranes are resuspended in an equal volume of homogenization buffer, and the Polytron homogenization, centrifugation, and resuspension steps are repeated 3 times.

The washed membrane pellet is detergent extracted by the addition of solubilization buffer [20 mM Tris, pH 7.4, 250 mM sucrose, 5 mM $MgCl_2$, 1 mM EDTA, 1 mM DTT, and 0.5% (w/v) Lubrol PX at a detergent to protein ratio of 2.5 : 1 (w/w)]. In some cases, membranes are pretreated with 0.10 mM GPP(NH)P for 30 min at 30° in the presence of 5 mM $MgCl_2$ prior to solubilization. The mixture is stirred overnight and centrifuged 2 hr at 4500 rpm in a Sorvall RC-3B centrifuge, and the supernatant fluid is decanted.

Two liters of DEAE-Sephacel equilibrated in 50 mM Tris, pH 7.4, 250 mM sucrose, 5 mM $MgCl_2$, 1 mM EDTA, 1 mM DTT, and 0.1% (w/v) Lubrol PX (buffer A) is added to 5 liters of detergent extract and stirred for 45 min. The resin is washed with 4 liters of buffer A containing 50 mM KCl, poured into a 9 × 30 cm column, and eluted with buffer A containing 150 mM KCl. Adenylyl cyclase activity is eluted in a single protein peak. CaM is not eluted from the column under these conditions. The enzyme activity is pooled, diluted with an equal volume of buffer A, and brought to 1.1 mM $CaCl_2$. The KCl concentration in the pooled enzyme solution is diluted from approximately 110 mM to about 55 mM. Addition of Ca^{2+} and the dilution of the KCl in the pool are required for successful CaM-Sepharose chromatography.

The diluted enzyme solution is loaded onto a 2.5 × 26 cm CaM-Sepharose column equilibrated in buffer A containing 1.1 mM $CaCl_2$. The column is washed with this same buffer until the effluent absorbance at 280 nm reaches a steady value, at which time the column is eluted with buffer A.

Adenylyl cyclase eluted from CaM-Sepharose is pooled on the basis of activity, loaded onto a heptanediamine-Sepharose column equilibrated in buffer A, and eluted with a concave 0–500 mM NaCl gradient in buffer A. Fractions are pooled on the basis of adenylate cyclase activity and diluted with 1.5 volumes of buffer A. CaM and $CaCl_2$ are added to the

TABLE I
PURIFICATION OF CALMODULIN-SENSITIVE ADENYLYL CYCLASE BY CALMODULIN-SEPHAROSE AND HEPTANEDIAMINE-SEPHAROSE CHROMATOGRAPHY

Purification step	Total activity[a]	% Yield activity	Total protein (mg)	Specific activity CaM-sensitive adenylyl cyclase[b]	Purification (-fold)
Membranes[c]	3.62	100	45,000	0.00008	1.0
Detergent extract	2.88	80	16,000	0.00018	2.3
DEAE-Sephacel	1.94	54	3800	0.00051	6.4
Calmodulin-Sepharose	1.80	50	34	0.0532	665
Heptanediamine-Sepharose I	1.08	30	9	0.120	1500
Heptanediamine-Sepharose II	0.50	14	2	0.240	3000
Heptanediamine-Sepharose II + Tween 20 + Forskolin	—	—	2	2.70	—

[a] Adenylyl cyclase was assayed in the presence of 10 mM $MnCl_2$ and 5 μM calmodulin. Total activity is expressed as μmol cAMP/min; specific activity is μmol cAMP/min/mg.
[b] The specific activity of calmodulin-insensitive adenylyl cyclase plus calmodulin-sensitive adenylyl cyclase in membranes was 0.00044.
[c] Membranes were pretreated with 0.1 mM GPP(NH)P for 30 min at 30°.

diluted pool to final concentrations of 5 μM and 1.1 mM, respectively. This pool is loaded onto the same heptanediamine-Sepharose column equilibrated in buffer A containing 1.1 mM $CaCl_2$ (buffer B). The column is eluted with a linear 0–500 mM NaCl gradient in loading buffer. Fractions are assayed for adenylyl cyclase activity, frozen in liquid nitrogen, and stored at −70° for future use. All operations described in the purification procedure are at 4°.

It was discovered that the yield of adenylyl cyclase activity and the levels of purification are greatly improved by pretreatment of the membranes with GPP(NH)P. However, this treatment renders the final preparation insensitive to GPP(NH)P. The purification reported in Table I is for GPP(NH)P-treated membranes. DEAE-Sephacel affords some purification; however, the major function of this column is to remove endogenous CaM. As illustrated in Fig. 1, CaM is well separated from adenylyl cyclase activity on DEAE-Sephacel when Ca^{2+} is chelated with EDTA. The

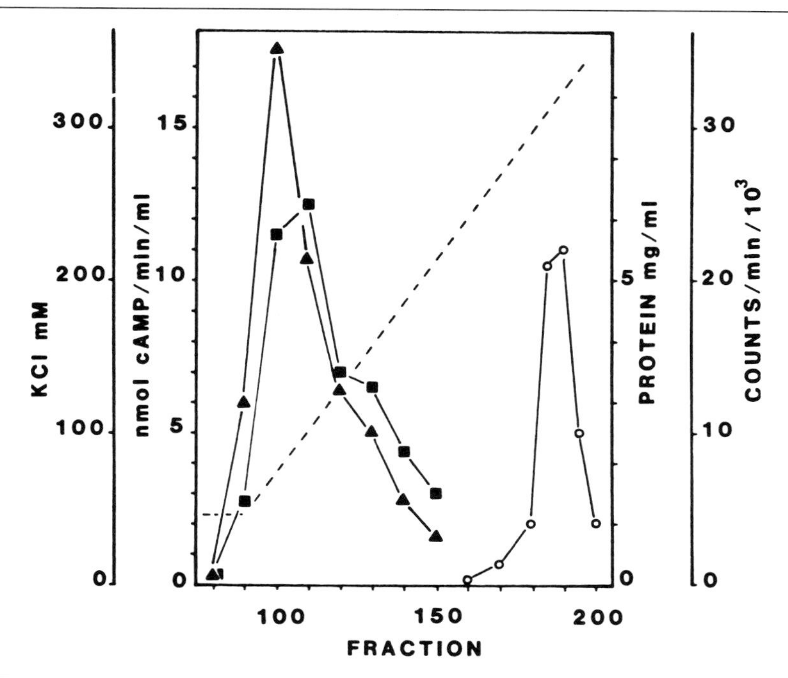

FIG. 1. DEAE-Sephacel chromatography of Lubrol PX detergent extract. Detergent-solubilized adenylyl cyclase was applied to DEAE-Sephacel as described under Enzyme Purification. The column (9 × 30 cm) was washed with buffer A containing 50 m*M* KCl (4 liters) and eluted with a 50–600 m*M* KCl gradient (---) in buffer A. Fractions (15 ml) were collected and assayed for adenylyl cyclase activity (▲) and protein (■). Radioiodinated calmodulin (○) was added to monitor the elution of calmodulin.

DEAE-Sephacel elution profile is highly reproducible; therefore, the 50–600 m*M* KCl gradient shown in Fig. 1 is routinely replaced with a 150 m*M* KCl step gradient. Attempts to remove endogenous CaM from the starting membranes with repeated washes of chelator-containing buffers were unsuccessful. This is apparently due to the presence of neuromodulin in the membranes, a protein which has significant affinity for CaM even in the presence of excess EDTA or EGTA.[13]

When the CaM-depleted preparation is applied to CaM-Sepharose in Ca^{2+}-containing buffer, approximately 50% of the total adenylyl cyclase activity absorbs to the column and is eluted with EDTA-containing buffer

[13] T. J. Andreasen, C. W. Luetje, W. Heideman, and D. R. Storm, *Biochemistry* **22,** 4615 (1983).

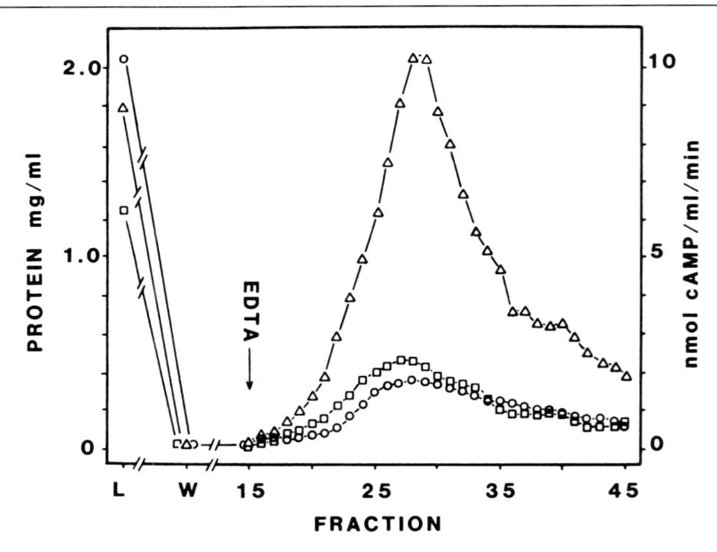

FIG. 2. CaM-Sepharose chromatography. Adenylyl cyclase pooled from DEAE-Sephacel was applied to a 2.5 × 26 cm calmodulin-Sepharose column as described under Enzyme Purification. The column was washed with calcium-containing buffer (buffer B) and eluted with buffer containing EDTA and no added calcium (buffer A). Loaded material flowing through the column and the material flowing through during the calcium wash were collected as single bulk fractions. Fractions (7 ml) were collected during the EDTA elution and assayed for adenylyl cyclase activity in the presence (△) and absence (□) of calmodulin and for protein (○).

(Fig. 2). This column provides significant purifications of the CaM-stimulated adenylyl cyclase with excellent yields (Table I). When the adenylyl cyclase pooled from CaM-Sepharose is reapplied to CaM-Sepharose, 95% of the applied activity is recovered by EDTA elution. Purification of the enzyme by CaM-Sepharose is not improved by the use of chelator gradients or attempts to remove nonspecifically absorbed proteins by washing the column with high salt or increased Lubrol PX concentrations.

The major contamination of the adenylyl cyclase purified through CaM-Sepharose is a CaM-sensitive phosphodiesterase which is removed by heptanediamine-Sepharose chromatography (Fig. 3). The elution profile from this column consists of two major protein peaks. The CaM-sensitive phosphodiesterase is present in the first protein peak, and the adenylyl cyclase is associated with the second. Further purification is accomplished by addition of CaM and Ca^{2+} to the pooled enzyme and reapplication to heptanediamine-Sepharose. Binding of CaM to the adenylyl cyclase causes a shift in the elution position of the adenylyl cyclase to higher salt concen-

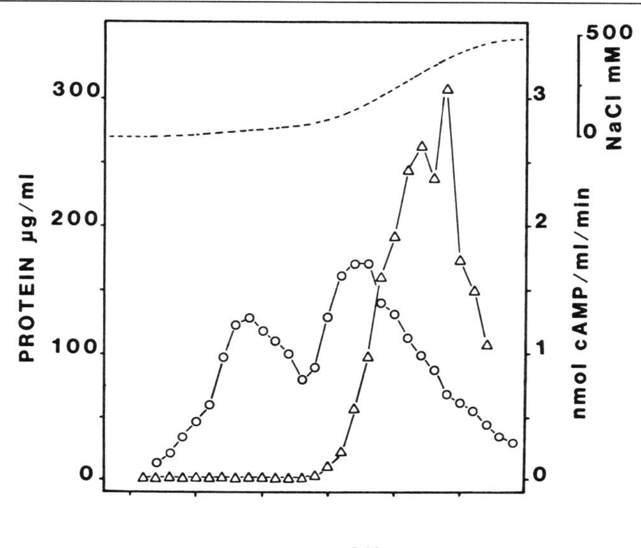

FIG. 3. Heptanediamine-Sepharose chromatography I. CaM-sensitive adenylyl cyclase pooled from the CaM-Sepharose step (80 ml) was applied to a heptanediamine-Sepharose column (10 ml) equilibrated in buffer A as described in Enzyme Purification. The column was eluted with a concave 0–500 m*M* NaCl gradient (---) in buffer A, and fractions were assayed for adenylyl cyclase in the presence of CaM and Mn^{2+} (△) and for protein (○).

trations and results in further purification of the enzyme (Fig. 4). The final specific activity of the pooled preparation is 0.24 μmol/min/mg with a 14% yield. The specific activity in the presence of forskolin and Tween 20 is 2.7 μmol/min/mg. The stability of adenylyl cyclase activity through CaM-Sepharose is excellent, whereas the final preparation is considerably less stable at 4°. The most highly purified preparations are frozen at −80 without loss of activity for 12 months.

Pooled samples from each stage of the purification are submitted to SDS–polyacrylamide gel electrophoresis and stained with Coomassie blue (Fig. 5). Although some protein fractionation is achieved by detergent extraction and DEAE-Sephacel chromatography, the most striking change in polypeptide composition follows CaM-Sepharose chromatography. The major CaM-binding protein eluting from CaM-Sepharose has an apparent molecular weight of 64,000, and it comprises more than 90% of the total protein present in the CaM-Sepharose pool. This protein was subsequently identified as a CaM-stimulated phosphodiesterase. The most highly purified preparation contains a major polypeptide of molecular weight 150,000,

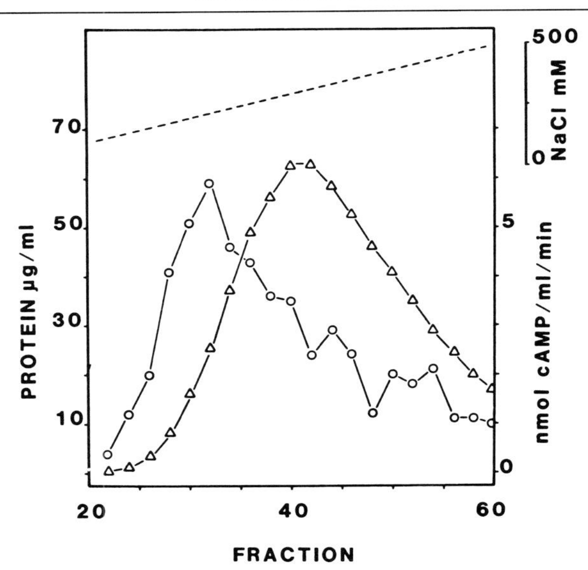

FIG. 4. Heptanediamine-Sepharose chromatography II. Adenylyl cyclase pooled from heptanediamine-Sepharose I was diluted with buffer A, adjusted to 1.1 m*M* calcium and 5 μM CaM, and reloaded onto a heptanediamine-Sepharose column equilibrated in buffer B as described under Enzyme Purification. The column was eluted with a linear 0–500 m*M* NaCl gradient in buffer B (---). Fractions (3.5 ml) were assayed for protein (○) and for adenylyl cyclase activity in the presence of Mn^{2+} and CaM (△).

which is the catalytic subunit of the enzyme, and G_s-α with a molecular weight of 47,000.

Sensitivity to Mn^{2+}, NaF, GPP(NH)P, and Forskolin. The purified enzyme is stimulated 8 to 10-fold by 10 μM CaM and 7-fold by forskolin when Mg^{2+} is the supporting divalent cation. The enzyme prepared without pretreatment of membranes with GPP(NH)P is also stimulated by NaF and GPP(NH)P. The basal activity is approximately 5 times higher when Mn^{2+} is present as the divalent cation compared to Mg^{2+}. Mn^{2+} is also able to replace Ca^{2+} in supporting CaM stimulation of the enzyme; however, in contrast to Ca^{2+}, Mn^{2+} has no inhibitory effect at concentrations up to 20 m*M*. The sensitivity of the partially purified enzyme to NaF and GPP (NH)P indicates functional coupling of the stimulatory guanine nucleotide regulatory complex with the CaM-sensitive adenylyl cyclase.

Molecular Weight of Enzyme Complex. The purified adenylyl cyclase is obtained in 0.1% Lubrol PX, and the complex would be expected to contain a mixture of detergent and protein. Therefore, the general method

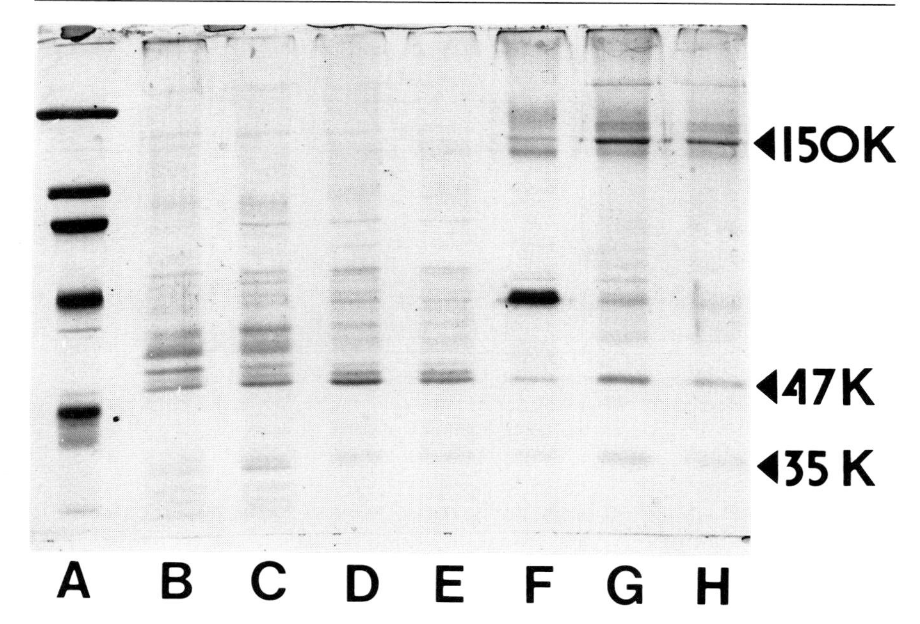

FIG. 5. SDS–polyacrylamide gel electrophoresis of CaM-sensitive adenylyl cyclase after each step in the purification scheme. Samples of adenylyl cyclase (11 μg) from each stage of the purification scheme summarized in Table I were diluted to 0.4 ml in buffer A, brought to 1.0 ml with distilled water, and precipitated with 0.1 ml of 0.15% sodium deoxycholate and 72% trichloroacetic acid. The precipitates were pelleted by centrifugation for 30 min at 3500 rpm in a Sorvall RC-3B centrifuge. The pellets were then washed with 1 ml of cold acetone and recentrifuged for 30 min. The acetone was removed by aspiration, and the residues were dissolved in 150 μl of Laemmli sample buffer. The samples were subjected to SDS–polyacrylamide gel electrophoresis on a 7.5%, 1.5-mm-thick slab gel. The gel was stained with Coomassie Brilliant Blue and photographed wet. (A) Bio-Rad (Richmond, CA) high molecular weight standards (myosin, 200,000; β-galactosidase, 116,250; phosphorylase *b*, 92,500; BSA, 66,200; ovalbumin, 45,000); (B) bovine cerebral membranes; (C) Lubrol PX-solubilized membranes; (D) DEAE-Sephacel pool; (E) flow-through from CaM-Sepharose Ca^{2+} wash; (F) CaM-Sepharose EDTA elution pool; (G) heptanediamine-Sepharose I pool; (H) heptanediamine-Sepharose II pool.

of Clark[14] for determination of the molecular weights of detergent-solubilized membrane proteins is used. It was discovered that the inclusion of Mn^{2+} in buffers used either for gel chromatography or for sucrose gradients converts the enzyme to a much smaller form than that observed without Mn^{2+}. The size of both forms of the enzyme is estimated (Table II). The molecular weights of both forms of the enzyme, with and without

[14] S. J. Clark, *J. Biol. Chem.* **250,** 5459 (1975).

TABLE II
HYDRODYNAMIC PROPERTIES OF CALMODULIN-SENSITIVE ADENYLYL CYCLASE

Property	Large form	Small form[a]
$s_{20,w}$	10.3	7.4
Partial specific volume	0.75	0.80
Lubrol PX binding (%)	7.0	31.0
Stokes radius	75.4	52.5
M_r[b]		
Complex with detergent	353,000	220,000
Adenylyl cyclase	328,000	153,000

[a] The small form was seen in the presence of 10 m*M* Mn^{2+}.
[b] Molecular weight of the protein component was determined assuming a partial specific volume of 0.73.

detergent, are calculated from the data reported in Table II. The apparent molecular weight of the large form is 328,000 and the small form, 153,000.

Purification of Calmodulin-Sensitive Adenylyl Cyclase by Calmodulin-Sepharose, Forskolin-Sepharose, and Wheat Germ Agglutinin-Sepharose

The catalytic subunit of the CaM-sensitive adenylyl cyclase has been purified without associated regulatory subunits with CaM-Sepharose, forskolin-Sepharose, and WGA-Sepharose by the method of Minocherhomjee *et al.*[15] The purification protocol through CaM-Sepharose is identical to that described above except that 0.1% Tween 20 or 6 m*M* CHAPS is substituted for Lubrol in all column steps after the initial solubilization.

Enzyme Purification. The CaM-Sepharose pool is concentrated to approximately 30 ml, and NaCl and CHAPS are added to final concentrations of 0.5 *M* and 6 m*M*, respectively. This preparation is loaded at 30 ml/hr onto a forskolin-Sepharose column (15 ml), equilibrated in 50 m*M* Tris-HCl, pH 7.4, 250 m*M* sucrose, 5 m*M* $MgCl_2$, 1 m*M* EDTA, 1 m*M* DTT, 1 m*M* PMSF, containing 0.5 *M* NaCl, and 6 m*M* CHAPS (buffer C). Forskolin-Sepharose is synthesized by the method of Pfeuffer *et al.*[16] The column is washed with 250 ml of buffer A at 60 ml/hr (forskolin flow-through). Adenylyl cyclase activity is eluted with 100 ml of buffer A containing 75 μM forskolin. The enzyme is pooled on the basis of adenylyl

[15] A. M. Minocherhomjee, S. Selfe, N. J. Flowers, and D. R. Storm, *Biochemistry* **26,** 4444 (1987).
[16] E. Pfeuffer, S. Moller, and T. Pfeuffer, *EMBO J.* **4,** 3675 (1985).

cyclase activity (forskolin pool). The forskolin pool is concentrated to 5–7 ml on a PM10 Amicon ultrafiltration membrane (Danvers, MA), diluted with 3 volumes of buffer A without PMSF and sucrose, and loaded at 30 ml/hr onto a wheat germ agglutinin-agarose column (10 ml) equilibrated in 50 m*M* Tris-HCl, pH 7.4, 250 m*M* sucrose, 5 m*M* $MgCl_2$, 1 m*M* EDTA, 1 m*M* DTT, and 0.1% Tween 20. The column is then washed with 40 ml of equilibrium buffer (WGA flow-through). The enzyme is eluted with 50 ml of the equilibration buffer containing 0.1 M *N*-acetyl-D-glucosamine (WGA pool). Both the WGA flow-through and pool are concentrated on an Amicon PM10 ultrafiltration membrane to approximately 5 ml.

Iodocalmodulin Gel Overlay. Adenylyl cyclase (WGA pool) is subjected to SDS–PAGE (7.5% acrylamide, 0.17% bisacrylamide), fixed with 40% methanol/10% acetic acid, and washed with 10% ethanol overnight. The gel is washed with water and equilibrated with buffer D (50 m*M* imidazole-HCl, pH 7.5, 0.15 *M* NaCl, 1 mg/ml of defatted bovine serum albumin, 1 m*M* $CaCl_2$) for 30 min. The gel is incubated in buffer D containing ^{125}I-labeled CaM with or without WGA (0.5 mg/ml) for 16 hr. The gel is then washed at least 6 times with buffer D to remove nonspecifically bound ^{125}I-labeled CaM. For control, 1 m*M* $CaCl_2$ is replaced by 1 m*M* EGTA. The gel is finally stained with Coomassie blue, dried, and autoradiographed.

Iodo-WGA Gel Overlay. The gel is treated as described above for Iodocalmodulin Gel Overlay; however, buffer D is replaced by buffer E (50 m*M* Tris-HCl, pH 7.5, 0.15 *M* NaCl, 1 mg/ml defatted BSA, and 1 m*M* $CaCl_2$). The gel is incubated with buffer E containing ^{125}I-labeled WGA with or without CaM (10 μM) for 16 hr and washed, stained, dried, and autoradiographed as described above. For a control, 0.2 *M* *N*-acetyl-D-glucosamine is added to buffer E.

This purification protocol yields catalytic subunit unassociated with guanine nucleotide regulatory proteins. This purification protocol is similar to that published by Yeager *et al.*,[8] through CaM-Sepharose, except that Lubrol PX used in chromatography buffers is replaced by Tween 20, and CaM-Sepharose is followed by forskolin-Sepharose and WGA-Sepharose. Tween 20 or CHAPS is used in place of Lubrol PX in order to optimize chromatography on forskolin-Sepharose and WGA-Sepharose. The adenylyl cyclase preparation obtained after CaM-Sepharose chromatography provides significant purification of the enzyme with good yields (>10 mg) (Table III). The catalytic subunit of adenylyl cyclase in the CaM-Sepharose pool is detectable as a major band of M_r 135,000 on SDS gels (Fig. 6). The enzyme is further purified using forskolin-Sepharose and WGA-Sepharose chromatography. The final enzyme preparation contains one major band of approximately M_r 135,000 on SDS–PAGE (Fig. 6),

TABLE III
PURIFICATION OF CALMODULIN-SENSITIVE ADENYLYL CYCLASE FROM BOVINE BRAIN MEMBRANES[a] BY CALMODULIN-SEPHAROSE, FORSKOLIN-SEPHAROSE, AND WHEAT GERM AGGLUTININ-SEPHAROSE CHROMATOGRAPHY

Purification step	Total activity (μmol/min)	Total activity (%)		Total protein (mg)	Specific activity (nmol/min/mg)	Purificatio (-fold)
		Membrane	CaM pool			
Membrane	14.14	100		20,350.0	0.0007	1.0
Lubrol solubilized	13.90	99		6858.0	0.002	2.9
DEAE-Sephadex	9.18	65		1170.0	0.008	11.3
CaM-Sepharose	1.78	13	100.0	11.3	0.158	227.0
Forskolin						
Flow-through	0.33		19.0	10.0	0.033	47.6
Pool	0.25		14.0	0.08	3.125	4503.0
WGA-Sepharose						
Flow-through	0.13		7.0	0.046	2.826	4072.0
Pool	0.06		3.4	0.032	1.875	2702.0

[a] Membranes were treated with GPP(NH)P (100 μM) for 30 min at 30°. Adenylyl cyclase wa assayed in the presence of $MnCl_2$ and 5 μM CaM.

although the molecular weight of this polypeptide varies considerably with the gel composition (135,000 ± 10,000). There is no detectable G_s-α subunit present in the preparation, and the only contaminating protein is the CaM-sensitive phosphodiesterase of M_r 57,000. The turnover number of the enzyme is greater than 1000/min. The final yield is only a few micrograms of protein compared to milligram quantities obtained after CaM-Sepharose chromatography. The 19-fold increase in specific activity obtained after forskolin-Sepharose chromatography is due in part to stimulation of the enzyme by forskolin, which is used to elute the enzyme from the affinity column. We estimate that the forskolin-Sepharose affinity column gives approximately 3-fold purification, with the addition of forskolin stimulating the enzyme about 6-fold. The advantage of this preparation over that reported by Yeager *et al.*[8] is that the catalytic subunit is obtained unassociated with G_s.

Interaction of Catalytic Subunit with ^{125}*I-Labeled Calmodulin.* The enzyme isolated by the procedure described above is stimulated 3- to 4-fold by CaM. Half-maximal stimulation occurs at approximately 10^{-7} *M* CaM; however, the log concentration–response curve for CaM stimulation of the catalytic subunit extends over more than 3 log units, suggesting heterogeneity of CaM binding sites or negative cooperativity between multiple binding sites.

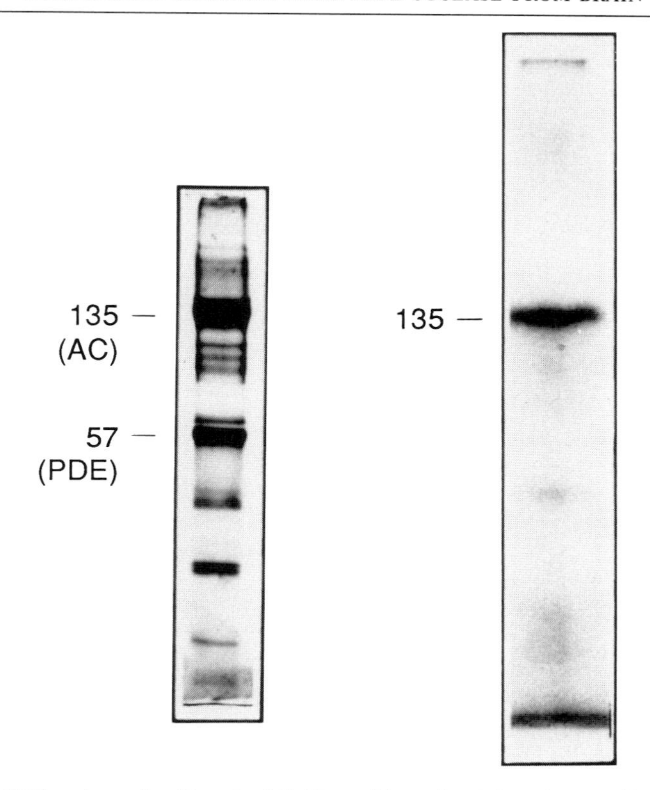

FIG. 6. SDS–polyacrylamide gel of CaM-sensitive adenylyl cyclase purified by CaM-Sepharose, forskolin-Sepharose, and WGA-Sepharose. (Left) Adenylyl cyclase purified through CaM-Sepharose (10 μg) was electrophoresed on 7.5% SDS–polyacrylamide gel and stained by Coomassie blue. (Right) SDS–polyacrylamide gel of adenylyl cyclase purified through WGA-Sepharose (Table III). The WGA pool (200 ng) was iodinated with chloramine-T and electrophoresed on a 7.5% gel (7.5% acrylamide, 0.17% bisacrylamide). The dried gel was exposed to film for 3 hr.

Direct interaction of the catalytic subunit with CaM is shown by the ^{125}I-labeled CaM gel overlay technique described above. The catalytic subunit of CaM-sensitive adenylyl cyclase interacts directly and specifically with CaM in the presence of Ca^{2+} but not in its absence (Fig. 7). The contaminating CaM-binding protein at M_r 57,000 which interacts with ^{125}I-labeled CaM is a CaM-sensitive cAMP phosphodiesterase. The addition of unlabeled WGA (0.5 mg/ml) during the ^{125}I-labeled CaM gel overlay has no effect on the ability of the catalytic subunit to bind ^{125}I-labeled CaM, and WGA has no effect on CaM stimulation of the enzyme. These results

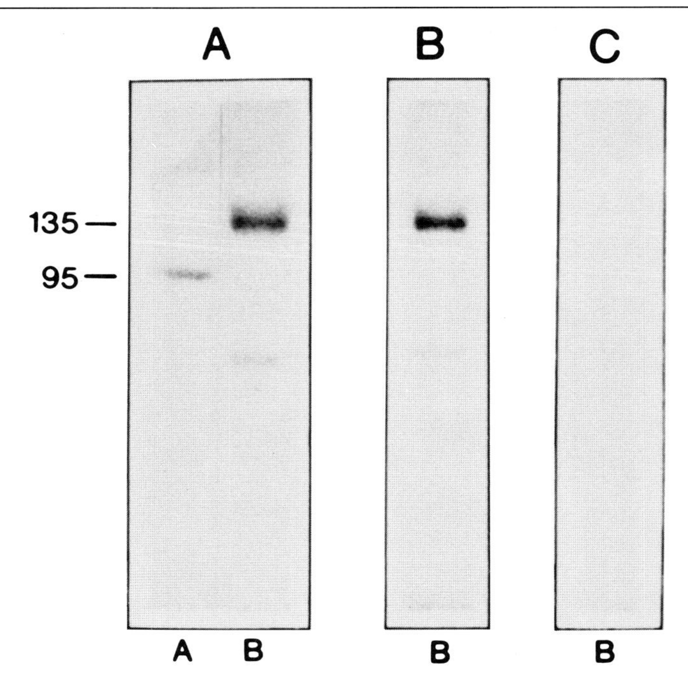

FIG. 7. ^{125}I-Labeled CaM binding to the catalytic subunit of adenylyl cyclase demonstrated by ^{125}I-labeled CaM gel overlay. Adenylyl cyclase (WGA pool) was subjected to SDS–PAGE (7.5% acylamide, 0.17% bisacrylamide), fixed with 40% methanol/10% acetic acid, and washed with 10% ethanol overnight. The gel was washed with water and equilibrated with buffer D (50 m*M* imidazole-HCl, pH 7.5, 0.15 *M* NaCl, 1 mg/ml of defatted BSA, 1 m*M* $CaCl_2$) for 30 min. The gel was incubated in buffer C containing ^{125}I-labeled CaM for 16 hr. The gel was then washed 6 times with buffer C to remove nonspecifically bound ^{125}I-labeled CaM. For control, 1 m*M* $CaCl_2$ was replaced by 1 m*M* EGTA. The gel was finally stained with Coomassie blue, dried, and autoradiographed. (A) Lane A, M_r markers (myosin, 200,000; β-galactosidase, 116,250; phosphorylase *b*, 95,000; bovine serum albumin, 66,000; ovalbumin, 45,000); lane B, purified catalytic subunit with Ca^{2+} present. (B) Purified catalytic subunit with Ca^{2+} and WGA (0.5 mg/ml) present. (C) Purified catalytic subunit with EGTA substituted for Ca^{2+}.

suggest that the CaM-binding domain is not coincident with or overlapping with the lectin-binding domain of the catalytic subunit.

Interaction of the Catalytic Subunit with ^{125}I-Labeled Wheat Germ Agglutinin. Absorption of the enzyme to wheat germ agglutinin-Sepharose suggests that it may be a glycolipid. However, the catalytic subunit may be associated with a glycolipid or a highly glycosylated protein contamination. Furthermore, wheat germ agglutinin has no effect on the activity of the enzyme or its sensitivity to CaM. Therefore, direct interaction between

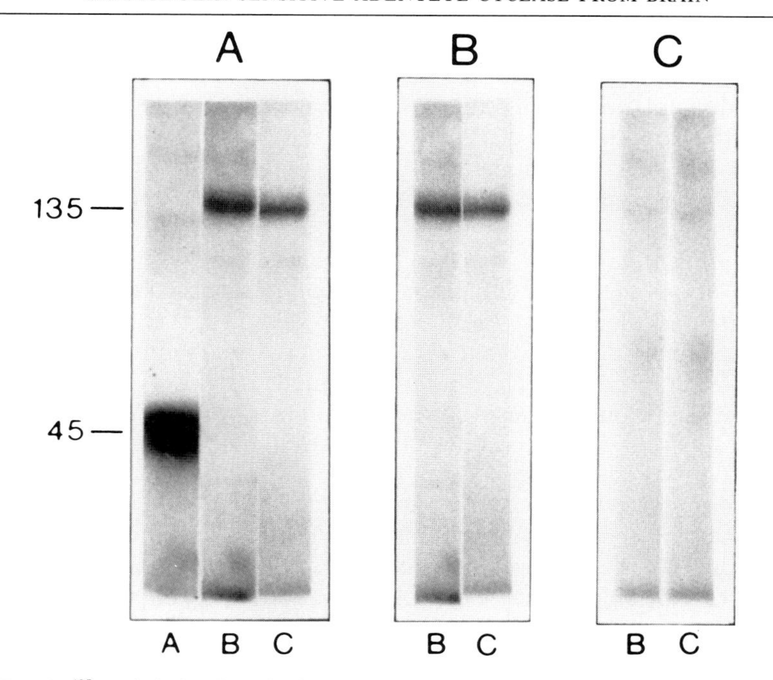

FIG. 8. ^{125}I-Labeled WGA binding to the catalytic subunit of adenylyl cyclase demonstrated by ^{125}I-labeled WGA overlay. The gel was treated as described in Fig. 7; however, buffer B was replaced by buffer C (50 m*M* Tris-HCl, pH 7.5, 0.15 *M* NaCl, 1 mg/ml defatted BSA, and 1 m*M* $CaCl_2$). The gel was incubated with buffer C containing ^{125}I-labeled WGA for 16 hr and washed, stained, dried, and autoradiographed as described previously. For control, 0.2 *M N*-acetyl-D-glucosamine was added to buffer C. (A) Lane A, M_r markers (same as in Fig. 7); lane B, adenylyl cyclase purified through forskolin-Sepharose; lane C, adenylyl cyclase purified through WGA-Sepharose. (B) Lane B, adenylyl cyclase purified through forskolin-Sepharose with 10 μM CaM present; lane C, adenylyl cyclase purified through WGA-Sepharose with 10 μM CaM present. (C) Lane B, adenylyl cyclase purified through forskolin-Sepharose with 0.2 *M N*-acetyl-D-glucosamine present; lane C, adenylyl cyclase purified through WGA-Sepharose with 0.2 *M N*-acetyl-D-glucosamine present.

wheat germ agglutinin and the catalytic subunit of the CaM-sensitive adenylyl cyclase is examined using ^{125}I-labeled WGA and the SDS gel overlay technique described above (Fig. 8). The pure catalytic subunit of adenylyl cyclase activity purified through CaM-Sepharose, forskolin-Sepharose, and WGA-Sepharose columns did interact with ^{125}I-labeled WGA on an SDS gel overlay system. This interaction is blocked by *N*-acetyl-D-glucosamine, indicating that the interaction of the lectin with the catalytic subunit is specific. In addition, the presence of unlabeled CaM during the ^{125}I-labeled WGA gel overlay has no effect on the ability of the catalytic subunit to bind ^{125}I-labeled WGA (Fig. 8B). These data are consistent with the results discussed above which suggested that the CaM-binding domain (presumed to be on the inner membrane surface) does not overlap the lectin-binding domain (presumed to be on the outer membrane surface). It appears that the catalytic subunit of the CaM-sensitive adenylyl cyclase is a glycoprotein and that it therefore may be a transmembrane polypeptide.

Inhibition of Adenylyl Cyclase by Adenosine. There is indirect evidence that "P" type adenosine inhibition of adenylyl cyclase may be due to interactions of adenosine with the catalytic subunit,[17,18] although this has not been directly demonstrated with pure catalytic subunit. The availability of pure catalytic subunit, without associated G_s, allows direct examination of this question. Adenosine directly inhibits the activity of the enzyme, with an apparent K_d of about 1 m*M*. This K_d value is somewhat higher than that observed for adenosine inhibition of the enzyme associated with G_s,[18] and it suggests that G_s may have indirect affects on interactions of adenosine with the catalytic subunit. Wheat germ agglutinin (0.5 mg/ml) did not prevent adenosine inhibition of the purified enzyme. There are at least three molecules which can modify the activity of the CaM-sensitive adenylyl cyclase by interacting directly with the catalytic subunit: CaM, forskolin, and adenosine.

Purification of Calmodulin-Sensitive Adenylyl Cyclase by Calmodulin-Sepharose and Wheat Germ Agglutinin-Sepharose Chromatography

Another variation of the purification schemes described above has been to utilize CaM-Sepharose followed by WGA-Sepharose and then CaM-Sepharose chromatography again. Although this method provides a partially purified enzyme, these two columns are the most reproducible methods available for purification of the enzyme. The total amount of protein and total enzyme activity obtained are superior to the methods described above.

Enzyme Purification. The CaM-sensitive adenylyl cyclase is purified through CaM-Sepharose chromatography by the method of Minocherhomjee *et al.*[15] described above. Adenylyl cyclase eluted from the CaM-Sepharose column is loaded onto a 2.5 × 22 cm WGA-Sepharose column

[17] C. Londos, J. Wolff, and D. M. F. Cooper, *in* "Physiological and Regulatory Functions of Adenosine and Adenine Nucleotides" (H. P. Baer and G. I. Drummond, eds.), p. 271. Raven, New York, 1979.

[18] R. E. Yeager, R. Nelson, and D. R. Storm, *J. Neurochem.* **47**, 139 (1986).

TABLE IV
PURIFICATION OF CALMODULIN-SENSITIVE ADENYLYL CYCLASE FROM BOVINE BRAIN MEMBRANES[a] BY CALMODULIN-SEPHAROSE AND WHEAT GERM AGGLUTININ-SEPHAROSE CHROMATOGRAPHY

Purification step	Total activity[b]	% Total activity	Total protein (mg)	Specific activity[b]	Purification (-fold)
Membrane	8.95	100.0	15,475	0.00058	1.0
Lubrol extract	11.43	127.7	8573.4	0.00133	2.3
DEAE pool	5.68	63.5	1633.2	0.00348	6.0
CaM I pool	2.50	27.9	24.03	0.10400	179.7
WGA pool	1.72	19.3	8.28	0.20800	359.4
CaM II pool	1.00	11.2	3.06	0.32700	565.8

[a] Membranes were pretreated with 0.1 m*M* GPP(NH)P for 30 min at 30° in the presence of 5 m*M* Mg^{2+}.

[b] Adenylyl cyclase was assayed as described under Enzyme Purification in the presence of 10 m*M* Mn^{2+} and 2.4 μM CaM. Total activity is expressed as μmol cAMP/min and specific activity is μmol cAMP/min/mg protein.

washed with 1 *M* NaCl and equilibrated in 50 m*M* Tris-HCl, pH 7.4, 5 m*M* $MgCl_2$, 1 m*M* EDTA, 1 m*M* DTT, 1 m*M* PMSF, and 0.1% Tween 20 containing 0.5 μg/ml leupeptin, 0.7 μg/ml pepstatin, and 0.5 μg/ml aprotinin (buffer A without sucrose plus inhibitors). The column is washed with the same buffer until the effluent absorbance at 280 nm reaches a steady value. Then the column is eluted with 100 ml of buffer A without sucrose plus inhibitors containing 100 m*M* *N*-acetyl-D-glucosamine. The enzyme is pooled on the basis of adenylyl cyclase activity (WGA pool).

$CaCl_2$ is added to the WGA pool to a final concentration of 1.1 m*M*. The second CaM-Sepharose chromatography is performed for the first one except that buffer A without sucrose plus inhibitors containing 1.1 m*M* Ca^{2+} is used to equilibrate the column and wash it after loading the WGA pool. The enzyme is eluted with 300 ml of buffer A without sucrose plus inhibitors containing 2 m*M* EGTA (CaM II pool). All operations described in the purification procedure are at 4°.

A summary of this purification protocol is reported in Table IV. The final specific activity of the unstimulated enzyme is 0.327 μmol cAMP/min/mg, with an overall yield of 11% of total adenylyl cyclase activity or 40% of CaM-stimulated adenylyl cyclase activity. The final enzyme preparation contains several polypeptides, including a major band with an M_r of approximately 165,000 on SDS–PAGE (Fig. 9). This enzyme is stimulated 4-fold by CaM, with half-maximal stimulation seen at 0.1 μM CaM. The enzyme is stimulated 4-fold by 100 μM forskolin.

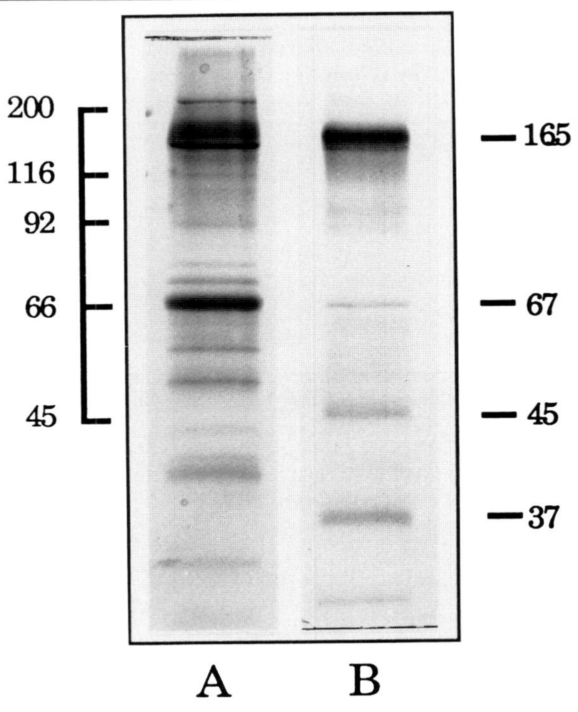

FIG. 9. SDS–polyacrylamide gel of CaM-sensitive adenylyl cyclase purified by CaM-Sepharose and WGA-Sepharose. The samples were subjected to SDS–polyacrylamide gel electrophoresis on a 7.5%, 1.5-mm-thick slab gel, and the gel was stained with Coomassie Brilliant Blue. (A) Enzyme purified through CaM-Sepharose (Table IV). (B) Enzyme purified through a second CaM-Sepharose column.

Summary

The catalytic subunit of the CaM-sensitive adenylyl cyclase can be purified to near homogeneity by several different purification protocols, although the yields of homogeneous catalytic subunit are still very slow. The most reliable purification method for this enzyme is CaM-Sepharose affinity chromatography. WGA-Sepharose and forskolin-Sepharose affinity columns also afford some purification of the enzyme, with WGA-Sepharose columns being more reproducible and reliable than forskolin-Sepharose. The catalytic subunit purified by these methods has an M_r of 150,000 ± 15,000 and is apparently a glycopeptide which interacts directly with CaM and adenosine. This catalytic subunit–G_s complex has been reconstituted *in vitro* with β-adrenergic receptors and muscarinic receptors and G_i.

[7] Purification of Adenylyl Cyclase from Heart and Brain

By ELKE PFEUFFER, STEFAN MOLLNER, and THOMAS PFEUFFER

Introduction

Adenylyl cyclases (EC 4.6.1.1, adenylyl cyclase) are usually membrane-bound enzymes and of low abundance as expected from their function as part of a hormonal signal-amplifying machinery. Compared with certain other membrane-bound enzymes, for example, ATPases, their specific activity in membranes is 100- to 1000-fold lower. The recent isolation of a $[^{35}S]GTP\gamma S \cdot \alpha_s \cdot C$ complex[1] by Pfeuffer and colleagues has verified that the modest specific activity is mostly due to the low molar concentration of the catalytic subunit of the adenylyl cyclase complex in the membrane.[2] Its concentration therefore approaches that of hormone receptors. Proteins of that scarcity usually imply the necessity for affinity techniques if isolation to homogeneity is desired. So far, calmodulin seems to be the only ligand suitable for construction of an affinity support, although the stimulation by Ca^{2+}/calmodulin is restricted to a subspecies of adenylyl cyclases occurring predominantly in neuronal tissues.[3] The second drawback of calmodulin is that it interacts with many proteins. Therefore, attempts to isolate the enzyme through the use of immobilized calmodulin were disappointing in that inhomogeneous preparations with low specific activity were obtained.[4]

Chapter [4] describes the synthesis of novel affinity matrices, based on forskolin, a newly discovered plant constituent endowed with the unique property of binding to the catalytic subunit of adenylyl cyclases.[5] Recent work from our laboratory has established that both isoenzymes of adenylyl cyclase, namely, the Ca^{2+}-sensitive and Ca^{2+}-insensitive forms, can be

[1] MOPS, 3-(*N*-Morpholino)propane sulfonate; CHAPS, 3-[(3-cholamidopropyl)dimethylammonio]propane 1-sulfonate; 4-azidoanilido-GTP, P^3-(4-azidoanilido)-P^1-guanosine triphosphate; GPP(NH)P, guanosine 5′-(β,γ-imido)triphosphate; GTPγS, guanosine 5′-(γ-thio)triphosphate; DEAE, diethylaminoethyl; G proteins, GTP-binding proteins; C, catalytic component of adenylyl cyclase; G_s, guanine nucleotide-binding regulatory component responsible for stimulation of adenylyl cyclase; G_i, guanine nucleotide-binding regulatory component responsible for inhibition of adenylyl cyclase; SDS, sodium dodecyl sulfate.

[2] T. Pfeuffer, B. Gaugler, and H. Metzger, *FEBS Lett.* **164,** 154 (1983).

[3] C. O. Brostrom, Y. C. Huang, B. M. Breckenridge, and D. J. Wolff, *Proc. Natl. Acad. Sci. U.S.A.* **72,** 64 (1977).

[4] R. E. Yeager, W. Heideman, G. B. Rosenberg, and D. R. Storm, *Biochemistry* **24,** 3776 (1985).

[5] T. Pfeuffer and H. Metzger, *FEBS Lett.* **146,** 369 (1982).

isolated by immobilized forskolin.[6,7] In this chapter we describe the isolation of the Ca^{2+}-insensitive enzyme from rabbit myocardium and that of the Ca^{2+}-activatable enzyme from bovine brain, two tissues available in reasonable quantities either commercially or from slaughterhouses.

Isolation of Activated Adenylyl Cyclase from Rabbit Myocardium

Reagents

Buffer A: 10 m*M* MOPS, pH 7.4, 1 m*M* $MgCl_2$, 1 m*M* EDTA, 1 m*M* benzamidine hydrochloride, 10 μM benzethonium chloride, 0.3 mg/liter Trasylol (Bayer, Leverkusen), 8% (w/v) sucrose, 0.02% (w/v) NaN_3

Buffer B: Buffer A containing 1% (w/v) instead of 8% sucrose

Buffer C: 10 m*M* MOPS, pH 7.4, 1 m*M* $MgCl_2$, 1 m*M* EDTA, 1 m*M* benzamidine hydrochloride, 10 μM benzethonium chloride, 0.02% (w/v) NaN_3, 100 m*M* NaCl, 1 m*M* Tween 60 (Serva, Heidelberg)

Buffer D: 10 m*M* MOPS, pH 7.4, 1 m*M* $MgCl_2$, 1 m*M* EDTA, 1 m*M* Tween 60, 500 m*M* NaCl

Preparation of Rabbit Myocardial Membranes. Fifty frozen rabbit hearts (mature, 5 g each, obtained from Pel-Freez Biologicals, Rogers, AR) are thawed, liberated from adherent fat tissue, and chopped with scissors. Following addition of 750 ml of buffer A, the tissue is twice homogenized in a Braun Mixer (setting 3) for 1 min each. The slurry is passed through a double layer of gauze and centrifuged in a Sorvall GSA rotor at 12,000 rpm for 30 min. The pellet is suspended in 300 ml of buffer A and homogenized (3 strokes) with a motor-driven Potter homogenizer (Braun, Melsungen) at setting 1200. Following centrifugation for 30 min at 12,000 rpm in a GSA rotor the supernatant is discarded while the pellet is treated twice more with the homogenizer prior to centrifugation. The crude membranes are adjusted to a protein concentration of 20 mg/ml and either kept frozen at $-80°$ or immediately processed for solubilization. The yield varies between 25 and 30 g.

Activation of Myocardial Adenylyl Cyclase with Guanine Nucleotides or $[AlF_4]^-$ and Solubilization. For nucleotide activation membranes are incubated at 20 mg/ml in buffer A with 30 μM GPP(NH)P (or 10 μM GTPγS) in the presence of 10 μM DL-isoproterenol and 3 m*M* additional $MgCl_2$ for 20 min at 30°. For preparation of larger quantities of enzyme we have activated membranes with $[AlF_4]^-$ because this version is less expensive. For this, membranes (20 mg/ml) in buffer A are treated with 10 m*M* NaF, 6 m*M* $MgCl_2$, and 20 μM $AlCl_3$ for 1 hr at 0°.

[6] E. Pfeuffer, R.-M. Dreher, H. Metzger, and T. Pfeuffer, *Proc. Natl. Acad. Sci. U.S.A.* **82,** 3086 (1985).

[7] E. Pfeuffer, S. Mollner, and T. Pfeuffer, *EMBO J.* **4,** 3675 (1985).

Following activation membranes are centrifuged for 20 min at 12,000 rpm (GSA rotor), the supernatant is removed, and the pellet is suspended in the original volume with buffer B. This procedure is repeated once. The pellet is suspended in one-half the original volume of buffer B containing 20 m*M* Lubrol PX, homogenized in a Potter homogenizer, and adjusted to the original volume with buffer B. The mixture is allowed to stand for 30 min at 0° (with occasional shaking), and solubilized proteins are separated by centrifugation in a Beckman B 19 rotor (19,000 rpm for 60 min). The supernatant, containing 50 to 70% of the membrane-bound adenylyl cyclase activity, is removed, frozen in 200-ml portions in liquid nitrogen, and freeze-dried overnight. The dry residue is taken up in one-tenth of the original volume of membranes by adding distilled water (concentrated crude adenylyl cyclase).

Chromatography on Forskolin-Sepharose. Twenty-five milliliters of forskolin-Sepharose is washed with 100 ml each of 70% (v/v) ethanol, 30% v/v ethanol, water, and finally with buffer C on a Büchner funnel and transferred to a chromatographic column (1.5 cm inner diameter). Eighty milliliters of concentrated crude cyclase (3500 mg) is applied to the affinity column during 3.5 to 4.5 hr. The loaded column is sequentially washed overnight (16 hr) with the following buffers by the use of an automatic multichannel delivery system set to 1.35 ml/min: 600 ml of buffer C, 270 ml of buffer C plus 900 m*M* NaCl, 100 ml of buffer C, 100 ml of buffer C plus 0.6 *M* $MgCl_2$, and finally 300 ml of buffer C. Adenylyl cyclase is eluted with 150 ml of buffer C containing 0.1 m*M* forskolin at a rate of 2 ml/min.

The following DEAE-Sepharose step does not contribute to further purification but rather is designed to obtain the purified enzyme in a reduced volume. For this purpose the forskolin-Sepharose eluate is diluted 3-fold with 10 m*M* MOPS, pH 7.4, 1 m*M* Tween 60 and applied to a 0.8 × 2.5 cm column of DEAE-Sepharose CL-6B equilibrated with 10 m*M* MOPS, pH 7.4, 1 m*M* benzamidine, 10 μ*M* benzethonium chloride, 1 m*M* Tween 60, at a rate of 150 ml/hr. The adenylyl cyclase is eluted in 5 ml of the same buffer containing 500 m*M* NaCl (buffer D) during 1 hr. The enzyme is supplied with 20% w/v glycerol, shock-frozen in liquid nitrogen, and stored at −80°. No loss of activity is observed during several months. The purification of $[AlF_4]^-$-activated myocardial adenylyl cyclase is summarized in Table I. All operations are at 4°.

Isolation of "Basal" Form of Adenylyl Cyclase from Bovine Brain

Reagents

Buffer E: 10 m*M* MOPS, pH 7.4, 1 m*M* $MgCl_2$, 1 m*M* EDTA, 1 m*M* benzamidine hydrochloride, 10 μ*M* benzethonium chloride,

TABLE I
PURIFICATION OF RABBIT MYOCARDIAL ADENYLYL CYCLASE BY FORSKOLIN-SEPHAROSE AFFINITY CHROMATOGRAPHY

Step	Protein (mg)	Total activity[a] (nmol/min)	Specific activity (nmol/mg/min)	Purification (-fold)	
				Per step	Overall
Crude $[AlF_4]^-$-treated membranes	24,878	10,200 (100)	0.41	1	1
Lubrol PX extract	3420	7148 (70)	2.09	5.1	5.1
Concentrated Lubrol extract	3244	4932 (69)	1.52	0.73	3.7
Forskolin-Sepharose	0.108	1617 (31)	14,997	9866	36,580
DEAE-Sepharose	0.098	1440 (89)	14,693	0.98	35,836

[a] Adenylyl cyclase activity in the presence of 100 μ*M* forskolin was estimated as described in the text. Percent recovery of each step is given in parentheses.

1 μg/ml Trasylol (Bayer), 0.2 m*M* phenylmethylsulfonyl fluoride, 2% sucrose

Buffer F: Buffer E without sucrose but with 100 m*M* NaCl and 1 m*M* Tween 60

Preparation of Crude Bovine Cerebral Membranes and Solubilization of Adenylyl Cyclase. Fresh bovine brain cortex (500 g) is transferred to 1.2 liters of 10 m*M* MOPS, pH 7.4, 2 m*M* EDTA, 8% sucrose, 3 m*M* benzamidine hydrochloride, 10 μ*M* benzethonium chloride, 1 m*M* dithioerythritol, 0.2 m*M* phenylmethylsulfonyl fluoride, 5 μg/ml pepstatin, 10 μg/ml Trasylol, 10 μg/ml soybean typsin inhibitor. The mixture is homogenized in portions by a motor-driven Teflon pestle. The homogenate is centrifuged in a Sorvall GSA rotor for 20 min at 12,000 rpm. The supernatant is removed, and the pellet is washed 3 times by repeated homogenization in 10 volumes of buffer E, then centrifuged as described above. The crude membranes (30 g) are adjusted to a protein concentration of 20 mg/ml with buffer E containing 10% glycerol and are immediately frozen in liquid nitrogen and stored at −70°.

For solubilization crude membranes are thawed and washed twice with 2 volumes of buffer E. The membrane pellet is adjusted to a final protein concentration of 20 mg/ml with 20 m*M* Lubrol PX in buffer E and left for 30 min on ice. The mixture is then centrifuged for 20 min at 12,000 rpm in a Sorvall GSA rotor, and the supernatant is saved (Lubrol PX extract). The Lubrol PX extract is immediately lyophilized and the residue resuspended in distilled water at 1/10 of the original volume. The turbid suspension is centrifuged in a Beckman 60 Ti rotor for 30 min at 40,000 rpm. A clear yellow solution, containing essentially all of the adenylyl cyclase activity, is removed by aspiration of the fraction between a thick floating layer of lipid material and a small, lucid pellet. The concentrated Lubrol PX extract is frozen in liquid nitrogen and stored at $-70°$. Recovery of adenylyl cyclase activity following lyophilization is usually between 80 and 90%. All steps are at 4°.

Chromatography of Bovine Brain Adenylyl Cyclase on Forskolin-Affi-Gel 10. For purification of the bovine brain enzyme we use the bifunctional forskolin derivative 7-succinyl-7-deacetylforskolin coupled to Affi-Gel 10 via an ethylenediamine bridge (see Chapter [4]), forming a 19-atom spacer between the matrix and the ligand. The longer spacer proves to be advantageous for the brain enzyme since yields are 30% better than with the shorter spacer. Nonspecific binding is a greater problem than with the myocardial enzyme, and major changes are necessary: (1) the stronger dispersing detergent CHAPS is introduced to remove some tenaciously adhering proteins, and (2) a second cycle of affinity chromatography is used.

Concentrated Lubrol PX extract (45 ml) is treated with 9 ml of forskolin-Affi-Gel 10 in buffer F for 90 min under gentle agitation. The slurry is then poured onto a Büchner funnel, and the fluid phase is removed by suction. The resin is rinsed with 150 ml of buffer F at 5 ml/min, is transferred to a chromatography column (1.6 cm diameter) equipped with a flow adapter, and is successively washed (3 ml/min) with 180 ml of buffer F plus 0.6 *M* $MgCl_2$, 30 ml of buffer F, 200 ml of 6 m*M* CHAPS and 1 *M* NaCl in buffer F, and 200 ml of buffer F. Adenylyl cyclase is released with a gradient (30 ml each) of 0–100 μM forskolin in buffer F during 60 min. Adenylyl cyclase-containing fractions (1 ml) are concentrated, and forskolin is removed by passage over 0.6 ml hydroxyapatite (HTP, Bio-Rad, Richmond, CA), equilibrated with buffer F during 60 min. The column is washed with 30 ml of buffer F at 1 ml/min. Adenylyl cyclase is eluted with 3 ml of 0.5 *M* potassium phosphate, 1 m*M* Lubrol PX, pH 7.4. The eluate is diluted with the same volume of buffer F and is then applied within 60 min to a second forskolin-agarose column (1×2 cm) in buffer F. Following washing of the column with 6 ml of buffer F, 15 ml of 6 m*M* CHAPS, 1 *M* NaCl in buffer F, and 6 ml of buffer F, adenylyl cyclase is desorbed with 15 ml of buffer F plus 100 μM forskolin by reversed flow through the column (60 min), and 1-ml fractions are collected. Active fractions are pooled, adjusted to 20% glycerol (v/v), and stored in liquid nitrogen. All chromatography steps are at 4°. The purification of bovine cerebral adenylyl cyclase is summarized in Table II.

Analytical Procedures

Adenylyl cyclase activity is measured in a medium consisting of 20 m*M* MOPS, pH 7.4, 10 m*M* creatine phosphate, 20 μg/ml creatine kinase,

TABLE II
PURIFICATION OF ADENYLYL CYCLASE FROM BOVINE BRAIN CORTEX[a]

Step	Protein (mg)	Total activity[b] (nmol/min)	Specific activity (nmol/mg/min)	Purification (-fold)	
				Per step	Overall
Crude membranes	8000	4970	0.621	—	—
Lubrol PX extract	1420	4987 (100)	3.512	5.6	5.6
First Forskolin-Affi-Gel 10	0.143	648 (13)	4520	1287.0	7278
Hydroxyapatite	0.087	408 (67)	4710	1.04	7585
Second Forskolin-Affi-Gel 10	0.052	330 (81)	6350	1.35	10,255

[a] Data are representative of 10 preparations starting from 2–8 g crude membranes. Adenylyl cyclase was measured as given in the text, with 100 μM forskolin as activator. Reprinted from Pfeuffer *et al.*[7] with permission of IRL Press Limited.

[b] Percent recovery of each step is given in parentheses.

5 mM $MgCl_2$, 3 mM theophylline, 0.5 mM [α-^{32}P]ATP (0.3–1.5 Bq/pmol). Protein is estimated according to Ref. 8. Purified cyclase and standard proteins for electrophoresis are iodinated according to Greenwood *et al.*[9] SDS–polyacrylamide gel electrophoresis is performed with 5–15% gradient gels according to Laemmli.[10]

Discussion

Two different adenylyl cyclases have been isolated in a homogeneous form by means of an immobilized forskolin derivative and purified between 10,000- to 40,000-fold: the Ca^{2+}-sensitive enzyme from bovine brain and the Ca^{2+}-insensitive enzyme from myocardial tissue. The M_r values of the catalytic units are 115,000 and 150,000, further corroborated by their cross-linking products with [^{32}P]ADP-ribosylated $G_{s\alpha}$.[6,7] Both enzymes seem to be glycoproteins because of their binding to wheat germ agglutinin-agarose.

The myocardial enzyme was isolated in a one-step procedure from solubilized membranes preactivated with $[AlF_4]^-$. The final product appeared on an SDS–polyacrylamide gel as three peptides with M_r values of 150,000, 42,000, and 46,000 (minor band) (Fig. 1A). The latter two bands

[8] R. M. Schultz, J. D. Bleil, and P. M. Wassarman, *Anal. Biochem.* **91,** 354 (1978).

[9] F. C. Greenwood, W. M. Hunter, and J. S. Glover, *Biochem. J.* **89,** 114 (1963).

[10] U. K. Laemmli, *Nature (London)* **227,** 680 (1970).

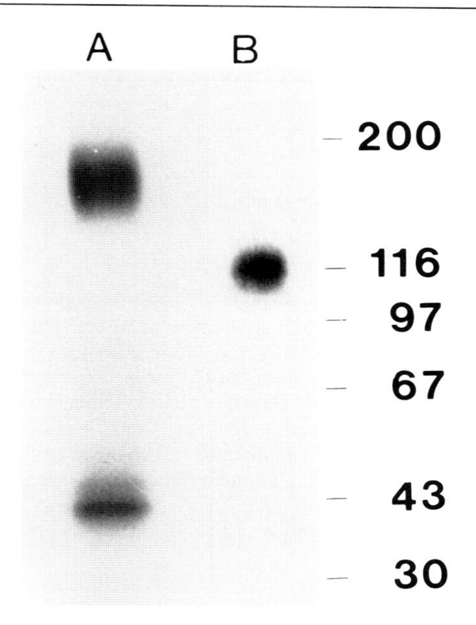

FIG. 1. SDS–polyacrylamide gels of purified ^{125}I-labeled adenylyl cyclases. (A) $C\alpha_s$ complex (21 ng) from rabbit myocardial membranes activated with $[AlF_4]^-$. (B) Catalytic protein (37 ng) from bovine brain cortex.

could be identified by photoaffinity labeling[11] with ^{32}P-labeled azidoanilido-GTP as α subunits of G_s (which occur in two forms in most tissues). The M_r 150,000 band could be shown to bind to forskolin by the use of an azido derivative of the diterpene.[11a] Most notably, the β,γ-portion of the G_s heterotrimer is missing in the purified activated enzyme, suggesting that a $C\alpha_s$ complex is the active species of adenylyl cyclase. The stability of the pure complex is remarkable. As expected, this species is not further activated by stimulators (forskolin, nucleotides, G_s),[6] with the exception of Mn^{2+} (2- to 3-fold stimulation). The disadvantage of this enzyme form is that it is not readily deactivated. Therefore, it is of use where "large" quantities of protein are needed, for instance, for microsequencing or immunization protocols.

Isolation of the brain enzyme was of interest because it was known that this form is stimulated by Ca^{2+}/calmodulin. Indeed, it has been suggested that the brain enzyme is a mixture of Ca^{2+}-sensitive and Ca^{2+}-insensitive forms. This could be verified by establishing monoclonal anti-

[11] T. Pfeuffer, unpublished results (1988).

[11a] E. Pfeuffer and T. Pfeuffer, *FEBS Lett.* **248,** 13 (1989).

TABLE III

RESPONSE TO VARIOUS EFFECTORS OF CRUDE AND PURIFIED BOVINE BRAIN ADENYLYL CYCLASE

Step	Adenylyl cyclase activity (nmol/mg/min)				
	Basal	GPP(NH)P	Forskolin	G_s[GPP(NH)P]	Ca^{2+}/CaM[b]
Lubrol PX extract	0.234	0.819 (3.5)	2.76 (11.8)	2.32 (9.9)	1.83 (7.8
Purified enzyme (second Forskolin-Sepharose)[c]	1726.6	1720.0 (1.0)	6820.0 (3.95)	5350.6 (3.1)	2416.4 (1.4

[a] Data are representative of at least eight individual preparations. Numbers in parentheses indicat (-fold) stimulation. Effectors of adenylyl cyclase in the assay (see text) were 100 μM GPP(NH)P 10 mM NaF, 100 μM forskolin, and 30 ng/ml G_s [purified from duck erythrocyte membranes GPP(NH)P activated] [T. Pfeuffer, B. Gaugler, and H. Metzger, *FEBS Lett.* **164,** 154 (1983) Reprinted from Pfeuffer *et al.*[7] with the permission of IRL Press Limited, Oxford, England.

[b] When assayed for stimulation by Ca^{2+}/calmodulin (CaM) the adenylyl cyclase assay contained 5 μM $CaCl_2$ and 50 μg/ml bovine brain calmodulin (Pharmacia, Uppsala, Sweden). Stimulation b Ca^{2+} alone was 4-fold with the Lubrol PX extract and 1.0-fold with the purified enzyme.

[c] Purified adenylyl cyclase was depleted of forskolin by chromatography on hydroxyapatite a described in the text.

bodies discriminating between these two forms.[12] Isolation of the brain enzyme in its basal form occurred at lower yields (about 5%), which is, however, compensated by the higher concentration of enzyme in that tissue. For removal of some of the impurities it was necessary to use CHAPS detergent and to add a second cycle of forskolin affinity chromatography. An SDS–polyacrylamide gel of the pure product revealed the presence of a M_r 115,000 peptide (Fig. 1B), but some preparations also contained variable amounts of α_s proteins (M_r 42,000 and 48,000), although the enzyme was isolated in its nonactivated form.[7,11] Compared with crude solubilized enzyme, the stimulation of the purified cyclase by forskolin, by activated G_s, and especially by Ca^{2+}/calmodulin was diminished although most of the enzyme could bind to calmodulin-Sepharose (Table III). We ascribe this lower stimulation to the presence of residual $G_{s\alpha}$ in the preparations. As with the purified activated myocardial enzyme, β,γ subunits appeared to be absent from the purified brain enzyme. Interestingly, the readdition of β,γ subunits generally improved the stimulation, especially that by calmodulin/Ca^{2+}.[11] The use of monoclonal antibodies has indicated that brain tissues contain more than one form of adenylyl cyclase.[12] In bovine brain cortex, a 150-kDa enzyme is present in somewhat lower amounts than the 115-kDa species. The larger enzyme, however, was lost

[12] S. Mollner and T. Pfeuffer, *Eur. J. Biochem.* **171,** 265 (1988).

during the procedure described here. Preliminary studies have indicated that coupling to activated G_s significantly improves its affinity for forskolin matrices.[12]

The availability of a simple and reproducible method for the purification of the catalytic unit of adenylyl cyclase makes, for the first time, all of the presently known protein components of the signal transmission chain available. Indeed, reconstitution of pure components into artificial lipids, conducted in this laboratory, has revealed that receptor, G_s protein, and catalyst are sufficient to generate a hormonally stimulated adenylyl cyclase.[13] Affinity chromatography on immobilized forskolin has now been successfully applied by several other groups for the purification of the calmodulin-sensitive enzyme from rat and bovine brain[14–16] and of the enzyme from frog erythrocytes.[17]

Acknowledgments

We thank Mrs. B. Weber and Mr. M. Brenner for expert technical assistance. The generous supply of forskolin from Drs. Metzger, Schöne, and Schorr (Hoechst Company, Frankfurt) is gratefully acknowledged. This work was supported by grants from the Deutsche Forschungsgemeinschaft (Sonderforschungsbereich 176) and by the Fonds der Chemischen Industrie.

[13] D. Feder, M. J. Im, H. W. Klein, M. Hekman, A. Holzhöfer, C. Dees, A. Levitzki, E. J. M. Helmreich, and T. Pfeuffer, *EMBO J.* **5**, 1509 (1986).
[14] M. D. Smigel, *J. Biol. Chem.* **20**, 1976 (1986).
[15] F. Coussen, M. Guermah, J. d'Alayer, A. Monneron, J. Haiech, and J.-C. Cavadore, *FEBS Lett.* **206**, 213 (1986).
[16] G. B. Rosenberg and D. R. Storm, *J. Biol. Chem.* **262**, 7623 (1987).
[17] I. Yoshimasa, D. R. Sibley, M. Bouvier, R. J. Lefkowitz, and M. G. Caron, *Nature (London)* **327**, 67 (1987).

[8] Calmodulin-Mediated Adenylyl Cyclase from Equine Sperm

By WILLIAM A. TOSCANO, JR., and MARY K. GROSS

Introduction

Understanding the detailed molecular basis of the regulation of sperm adenylyl cyclase [ATP pyrophosphate-lyase (cyclizing), EC 4.6.1.1, adenylate cyclase] is important because cyclic AMP (cAMP) is implicated in a number of sperm processes. These include (1) sperm maturation during passage through the epididymis; (2) regulation of motility and metabolism;

(3) the capacitation reaction in the female reproductive tract; and (4) the acrosome reaction just before fertilization.[1] There is, however, a gap between these physiological observations and an understanding of these processes at the molecular level because efforts to purify sufficient quantities of spermatozoal adenylyl cyclase for physicochemical studies have largely been unsuccessful, primarily because no suitable affinity column is available for the enzyme from sperm.

Sperm are unique cells in which to study adenylyl cyclase and its regulation, because adenylyl cyclase from spermatozoa is not stimulated by common mediators of the enzyme. Despite many attempts, spermatozoal adenylyl cyclase has never been shown to be responsive to hormones, guanine nucleotides, cholera toxin, or forskolin.[1,2] Even though differences exist between the enzyme found in sperm and that in somatic cells, both types of adenylyl cyclase are modulated by Ca^{2+} and Mn^{2+}.[3-5]

It is now generally accepted that Ca^{2+} plays a pivotal role in regulating sperm motility and the acrosome reaction.[1] It has also been known for some time that extracellular levels of calcium will increase cAMP levels in sperm.[5] In guinea pig sperm, Ca^{2+} has been shown to activate adenylyl cyclase by a process that is dependent on bicarbonate ion.[5] It has been suggested that both Ca^{2+} and cAMP regulate sperm motility,[6-8] presumably by regulation of a cAMP-dependent protein kinase.[9] Because of the large body of evidence that Ca^{2+} modulates adenylyl cyclase activity in spermatozoa, it has been generally assumed that the enzyme should be mediated by calmodulin (CaM). Even though CaM is present in sperm at high concentrations,[4,10,11] a definitive role for CaM regulation of adenylyl cyclase in mammalian sperm has been difficult to establish in spite of many elegant attempts. For example, guinea pig sperm adenylyl cyclase is inhibited by phenothiazine,[4] however, no reversal of inhibition could be demonstrated by the addition of exogenous CaM. Definitive involvement of CaM in regulating adenylyl cyclase in sperm depends on the restoration of enzymatic activity by the addition of exogenous CaM after inhibition of enzyme activity by anti-CaM drugs.

[1] D. L. Garbers and G. S. Kopf, *Adv. Cyclic Nucleotide Res.* **13,** 251 (1980).
[2] C. Y. Cheng and B. Boettcher, *Biochem. Biophys. Res. Commun.* **91,** 1 (1979).
[3] T. Braun and R. F. Dods, *Proc. Natl. Acad. Sci. U.S.A.* **72,** 1097 (1975).
[4] R. V. Hyne and D. L. Garbers, *Biol. Reprod.* **21,** 1135 (1979).
[5] D. L. Garbers, D. J. Tubb, and R. V. Hyne, *J. Biol. Chem.* **257,** 8980 (1982).
[6] D. D. Hoskins and E. R. Casillas, *Adv. Sex Horm. Res.* **1,** 283 (1975).
[7] C. Y. Cheng and B. Boettcher, *Arch. Androl.* **7,** 313 (1981).
[8] J. R. Tash and A. R. Means, *Biol. Reprod.* **28,** 75 (1982).
[9] J. A. Horowitz, H. Toeg, and G. A. Orr, *J. Biol. Chem.* **259,** 832 (1984).
[10] G. S. Kopf and D. L. Garbers, *Biol. Reprod.* **22,** 1118 (1980).
[11] G. S. Kopf and V. D. Vacquier, *J. Biol. Chem.* **259,** 7590 (1984).

The question of whether mammalian sperm adenylyl cyclase is mediated by CaM was examined in our laboratory using both a pharmacological approach, namely, inhibition of the enzyme by anti-CaM drugs followed by reactivation of the enzyme with CaM and reconstitution of CaM mediation of adenylyl cyclase after removal of CaM from the enzyme.[12] This approach has been useful in elucidating a role for CaM stimulation of a number of target enzymes in various tissues and cells.[13-16]

Materials and Methods

Phosphocreatine di-Tris salt (P-4635), creatine phosphokinase (1170 U/mg; C-3755), cAMP sodium salt (A-9501), bovine serum albumin, essentially fatty acid free (A-7511), imidazole (I-0125), papaverine (P-3510), isopropylidine adenosine (I-4002), phenylmethylsulfonyl fluoride (PMSF) (P-7626), $LaCl_3 \cdot 7H_2O$ (L-6640), and neutral alumina WN-3 (A-9003) were purchased from Sigma (St. Louis, MO). ATP (106-356), calmidazolium (R-24-571), and Tris base (604-207) were from Boehringer Mannheim (Indianapolis, IN). $H_3{}^{32}PO_4$ (64013) and [2,8-^{3}H]cAMP (24011) were from ICN (Costa Mesa, CA). [2,8-^{3}H]cAMP is purified by chromatography on DEAE-Sephadex A-25 (Pharmacia, Piscataway, NJ) before use in the adenylyl cyclase assay.[17] [α-^{32}P]ATP is synthesized from isopropylidine adenosine and $H_3{}^{32}PO_4$ using published procedures[18] and is purified by chromatography on DEAE-Sephadex A-25 (Pharmacia). Dowex columns used in adenylyl cyclase assays are prepared by pouring 1 ml of a slurry of Dowex AG 50W-X4 (200–400 mesh) in water into an Econo-column (Evergreen Scientific, Los Angeles, CA, 208-3366-050). The columns are regenerated before use in the assay by sequential washing with 4 ml each of HCl (1 *N*), HCl, water, NaOH (0.5 *N*), water, HCl, and 3 successive water washes. When Mg^{2+} is used as a cation, regeneration begins at the NaOH step. Alumina columns are prepared by pouring 0.5 g of dry alumina into an Econo-column. These columns are regenerated just prior to their use in the assay by washing twice with 4 ml of 0.1 *M* imidazole, pH

[12] M. K. Gross, D. G. Toscano, and W. A. Toscano, Jr., *J. Biol. Chem.* **262,** 8672 (1987).
[13] B. Weiss and R. M. Levin, *Adv. Cyclic Nucleotide Res.* **9,** 285 (1978).
[14] B. Weiss, W. Prozialeck, M. Cimino, M. S. Barnette, and T. L. Wallace, *Ann. N.Y. Acad. Sci.* **356,** 319 (1980).
[15] B. D. Roufogalis, *in* "Calcium and Cell Physiology" (D. Marme, ed.), p. 148. Springer-Verlag, Berlin and New York, 1985.
[16] J. A. Ryan and W. A. Toscano, Jr., *Arch. Biochem. Biophys.* **241,** 403 (1985).
[17] R. Iyengar and L. Birnbaumer, *in* "Laboratory Methods Manual for Hormone Action and Molecular Endocrinology" (W. T. Schrader and B. W. O'Malley eds.), 6th Ed., p. 9. Houston Biological Associates, Houston, Texas, 1981.
[18] R. H. Symons, this series, Vol. 29, p. 102.

7.3 (room temperature). These columns must often be resuspended on regeneration.

CaM-sensitive adenylyl cyclase from *Bordetella pertussis* was a generous gift from Dr. D. R. Storm, University of Washington. All other reagents were of the highest grade commercially available and were used without further purification.

Adenylyl Cyclase Assays

Adenylyl cyclase activity is routinely measured in a final volume of 50 μl with purified [α-^{32}P]ATP as the substrate and [^{3}H]cAMP to monitor recovery. (Also see [1] in this volume.) The assay mixture contains 1 m*M* [α-^{32}P]ATP (120–400 cpm/pmol), 10 m*M* $MnCl_2$ (or 5 m*M* $MgCl_2$), 2 m*M* [2,8-^{3}H]cAMP (200,000 cpm/ml) to monitor recovery, 1 m*M* EDTA, 1 m*M* 2-mercaptoethanol, 100 μM papaverine, 1 μg/ml bovine serum albumin, and an ATP-regenerating system consisting of 20 m*M* creatine phosphate and 120 units creatine phosphokinase/ml in a 25 m*M* Tris-HCl buffer, pH 7.5. We typically determine the concentration of ATP and cAMP spectrophotometrically at 259 nm, using ε values of 15.4 and 14.6 mM^{-1} cm^{-1} for ATP and cAMP, respectively. The concentrations of cAMP in the assays are always twice that of ATP. Assays are usually initiated by the addition of protein, and incubation is carried out at 37° for 10 min. The reaction is stopped with 100 μl of a stopping solution made up of 2% SDS/45 m*M* ATP/13 m*M* cAMP. The mixture is heated at 100° for 3 min in a dry bath, and the cyclic [^{32}P]AMP formed is recovered by a modification of the two-step elution method described by Salomon *et al.*[19] Briefly, the volume of each assay tube is made up to 1 ml with 0.85 ml of water and poured onto an equilibrated Dowex AG 50W-X4 column. The flow-through is discarded into the radioactive waste. The Dowex columns are washed with 3.0 ml of water, and the eluate is discarded into the radioactive waste. The rack of columns containing Dowex AG 50W-X4 are placed directly over preequilibrated alumina columns and eluted with 2.5 ml of water. The eluant is collected directly on the alumina columns. After the column is drained, cyclic [^{32}P]AMP is eluted directly into scintillation vials containing a high water capacity scintillation fluid with 2 ml of 0.1 *M* imidazole, pH 7.3 (room temperature). The yield of cyclic [^{32}P]AMP recovered in the assay is typically over 50%. The elution volumes are obtained by assessing the elution profile of ATP, ADP, AMP, and cAMP at each step by thin-layer chromatography using cellulose (Eastman Kodak, Rochester, NY, 13254) as a support.[20] It is important to standardize the columns periodi-

[19] Y. Salomon, C. Londos, and M. Rodbell, *Anal. Biochem.* **58,** 541 (1974).

[20] K. Randerath, "Thin Layer Chromatography," 2nd Ed., p. 223. Academic Press, New York, 1968.

cally. Unless otherwise noted, assays are always performed in triplicate, and the standard error of the mean is typically less than 5%.

Protein concentration is estimated by a modified Lowry assay as described by Peterson,[21] with bovine serum albumin as a standard. We have found this protocol to be generally applicable to assay of protein even under conditions where substances such as 2-mercaptoethanol or detergents are present.

Calculations

Routine calculations and statistical analyses are performed using a Tandy Model TRS 80-4 computer.[22,23]

Purification of Calmodulin

The protocol used in our laboratory for purification of CaM is one originally suggested to us by Dr. David C. LaPorte (University of Minnesota), and it has been applied to the purification of CaM from many tissues.[24] This procedure is convenient, inexpensive, and readily adaptable to large-scale preparation of highly purified CaM. Bovine brain is obtained from a local abattoir, the medulla is discarded, and the cortex is stored frozen for up to 2 years. About 2 kg of cerebral cortex is thawed and homogenized in 2000 ml of buffer A (25 m*M* sodium phosphate, 1 m*M* EDTA, 100 m*M* NaCl, 1 m*M* PMSF, pH 6.5) in a Waring blendor with 30-sec bursts at low, medium, and high speeds. After centrifugation at 12,000 *g* for 30 min, the supernatant is filtered through multiple layers of glass wool wrapped in cheesecloth. Two subsequent homogenizations are performed on the pellets from the above centrifugation step with the same buffer in a ratio of 1 volume to 1 g of pellet.

The combined supernatants, adjusted to pH 6.5 with 1 *N* NaOH and to a conductivity of 6 mS (millisiemens) with NaCl, are added to 2000 ml of precycled DE-52 cellulose anion-exchange resin (Whatman, Clifton, NJ) previously equilibrated with Buffer A, and stirred at 4° for 1.5 hr. The resulting slurry is filtered on a coarse glass filter and washed with buffer A containing 50 m*M* NaCl until the conductivity of the effluent is equivalent to that of the wash buffer. The resin is poured into a glass column (5

[21] G. L. Peterson, this series, Vol. 91, p. 95.

[22] R. J. Tallarida and R. B. Murray, "Manual of Pharmacologic Calculations with Computer Programs," 2nd Ed., p. 218. Springer-Verlag, Berlin and New York, 1987.

[23] Y. Salomon, *Adv. Cyclic Nucleotide Res.* **10,** 35 (1979).

[24] M. D. Coyne, P. Cornelius, N. Venditti, D. G. Toscano, M. K. Gross, and W. A. Toscano, Jr., *Arch. Biochem. Biophys.* **236,** 629 (1985).

× 100 cm) containing 300 ml of precycled DE-52 cellulose on the bottom. The column is eluted in the ascending direction (200 ml/hr) with a linear gradient of 6 liters each of 100 and 400 m*M* NaCl in buffer A. Twenty-milliliter fractions are collected and monitored for conductivity and a 17,000 molecular weight band by polyacrylamide gel electrophoresis in the presence of sodium dodecyl sulfate (SDS). Fractions enriched in CaM elute at NaCl concentrations of approximately 200–300 m*M*. The fractions enriched in CaM are combined and precipitated by the addition of solid ammonium sulfate to 0.90 saturation. After ammonium sulfate addition and stirring for an additional 60 min, the solution is acidified to pH 4.0 by the addition of 6 *N* HCl, followed by stirring for an additional 30 min, and centrifugation for 30 min at 20,000 *g* at 4°. The resulting pellet is resuspended in a minimal volume of double-distilled water and dialyzed against 1000 ml of buffer B (20 m*M* Tris-HCl/200 m*M* NaCl/50 μM $CaCl_2$, pH 7.5) for 16 hr and 3 changes of buffer. The dialyzate (150 ml) is applied to a 5 × 100 cm column packed with Sephadex G-150 (Pharmacia) and equilibrated with buffer B containing 150 m*M* NaCl and is eluted in the ascending direction (45 ml/hr) with the same buffer. The fractions enriched in CaM containing a single band on polyacrylamide gel electrophoresis in the presence of SDS are combined and concentrated with an Amicon (Danvers, MA) UM2 membrane filter. The concentrated solution of CaM is stored in the elution buffer at −70°, and is stable for several years.

Preparation of Calmodulin-Sepharose

CaM-Sepharose resin is prepared with CNBr-activated Sepharose 4B (Pharmacia). Briefly, 30 g of dry resin is swollen for 15 min in 1 m*M* HCl and rinsed with 6 liters of the same solution on a sintered glass filter. The gel is resuspended in 150 ml of coupling buffer (0.1 *M* $NaHCO_3$, pH 8.3, 0.5 *M* NaCl) containing approximately 40 mg of CaM purified from bovine brain and incubated overnight at 4° in a roller bottle on a rocker platform. After the binding incubation, the gel is filtered, resuspended in 1 *M* ethanolamine, pH 8, and incubated overnight at 4° to block remaining active groups on the resin. The resin is washed successively with coupling buffer, acetate buffer (0.1 *M*, pH 4, plus 0.5 *M* NaCl), and coupling buffer, then stored in this buffer at 4°.

High-Performance Liquid Chromatography of Equine Sperm Extracts

HPLC separations were performed at the HPLC Core Facility at the Laboratory of Human Reproduction and Reproductive Biology, Harvard Medical School. The HPLC system used was a Bio-Rad (Richmond, CA) Liquid Chromatograph equipped with a variable wavelength ultraviolet detector. Elution gradients were programmed on an Apple computer. All chromatography is performed at room temperature.

For gel filtration separations, samples of equine sperm proteins are concentrated in a minimum volume of 50 m*M* MOPS, pH 7/150 m*M* NaCl/250 m*M* sucrose with an Amicon Centricon 30 and applied to a Bio-Rad Bio-Sil TSK 250 column (7.5 × 600 mm) equilibrated with the same buffer. The column is eluted isocratically with this buffer at a rate of 0.8 ml/min.

For anion-exchange separations, samples of equine sperm proteins are diluted into 50 m*M* Tris-HCl, pH 7.4/25 m*M* sucrose/50 m*M* NaCl (buffer C) and applied to a Bio-Rad TSK DEAE-5-PW column (7.5 × 75 mm) equilibrated with the same buffer. The column is eluted at 0.8 ml/min with the following time gradient: 0–5 min, isocratic elution with buffer C; 5–7 min, linear gradient to 11.4% buffer D (50 m*M* Tris-HCl, pH 7.4/25 m*M* sucrose/400 m*M* NaCl); 7–12 min, isocratic elution with 11.4% w/v Buffer D; 12–14 min, linear gradient to 14.3% w/v buffer D; 14–24 min, isocratic elution with 14.3% w/v buffer D; 24–26 min, linear gradient to 17.6% buffer D; 26–36 min, isocratic elution with 17.6% w/v buffer D; 36–40 min, linear gradient to 21.5% w/v buffer D; 40–45 min, isocratic elution with 21.5% buffer D; 45–65 min, linear gradient to 100% buffer D.

Preparation and Properties of Equine Sperm Adenylyl Cyclase

Spermatozoa are separated from seminal fluid by centrifugation of a fresh ejaculate at 1500 *g* for 15 min at 4°. The remaining soft pellet is resuspended by gentle vortexing in one-twentieth the original volume in buffer E (50 m*M* HEPES, pH 8.0/150 m*M* NaCl/250 m*M* sucrose/0.5 m*M* EDTA/1 m*M* 2-mercaptoethanol/1 m*M* PMSF) and quickly frozen in a dry ice–ethanol bath. The sperm suspension is thawed and centrifuged at 150,000 *g* for 1 hr. The supernatant fraction (operationally called supernatant I) is removed and placed on ice. The pellet is suspended in buffer E containing 500 m*M* NaCl and again centrifuged at 150,000 *g* (supernatant II); the process is repeated a third time to obtain supernatant III. The results from a typical preparation are summarized in Table I. Approximately 90% of the adenylyl cyclase activity is released as a "soluble" fraction, whereas about 70% of the total protein remains in the pellet. It is not known whether adenylyl cyclase obtained in this manner is truly soluble or is loosely associated with the membrane as a peripheral membrane protein. It has been reported by various workers that adenylyl cyclase from mature spermatozoa is a membrane-bound enzyme[1]; therefore, the ease of solubilization of the enzyme was somewhat suprising to us. We had at first employed these washing steps as a prelude to solubilization in Lubrol PX. There is, however, ample precedent in the literature

TABLE I
FRACTIONATION OF EQUINE SPERM ADENYLYL CYCLASE

Step	Volume (ml)	Protein		Activity[a]		Yield (%)
		mg/ml	Total mg	Specific[b]	Total	
Homogenate	6.0	19	114	860	98,040	100
Supernatant I	8.4	1.7	14.3	1650	23,595	24
Supernatant II	6.0	0.82	4.9	9030	44,247	45
Supernatant III	6.0	0.72	4.3	3930	16,899	17
Final pellet	6.0	12.5	75	140	10,500	11

[a] Activity was measured using MnATP as the substrate.
[b] Specific activity is expressed as pmol cyclic AMP formed/10 min/mg protein.

for salt extraction of peripheral membrane proteins. These proteins are indeed membrane proteins, but they also exhibit properties of soluble proteins when removed from the membrane. Well-known examples include cytochrome *c*, pyruvate oxidase from *Escherichia coli,* and myelin A1 basic protein.[25,26]

It has recently been reported that conditions similar to those that we use to extract adenylyl cyclase from equine sperm will release a serine protease from sperm.[27] This was troublesome initially, but most of the proteolytic action in our extracts was attenuated by including a serine protease inhibitor in the extraction buffers. When supernatants I, II, and III are prepared in the absence of protease inhibitors, we observed approximately 30% of the activity in the first supernatant, and the same levels in the other two supernatants were observed. These results agree with published studies where the action of proteolysis on CaM-sensitive systems has been examined. Limited proteolysis of CaM-modulated enzymes results in a stimulated activity that is no longer sensitive to the actions of either CaM or anti-CaM drugs, whereas prolonged proteolysis results in the loss of enzymatic activity.[28] This phenomenon is not uncommon for enzymes that possess CaM binding sites and are subsequently proteolyzed. Examples include phosphodiesterase, calcium/CaM ATPase, myo-

[25] R. L. Blake, Ph.D. Thesis, University of Illinois, Urbana, Illinois (1977).
[26] R. B. Gennis and A. Jonas, *Annu. Rev. Biophys. Bioeng.* **6,** 195 (1977).
[27] R. A. Johnson, K. H. Jakobs, and G. Schultz, *J. Biol. Chem.* **260,** 114 (1985).
[28] Y. M. Lin and W. Y. Cheung, *in* "Calcium and Cell Function" (W. Y. Cheung, ed.), Vol. 1, p. 79. Academic Press, New York, 1980.

TABLE II
ACTION OF MODULATORS OF ADENYLYL CYCLASE ON EQUINE SPERM ADENYLYL CYCLASE

Addition[a]	cAMP formed[b] (pmol/10 min/mg)	Effect (-fold)
None (MnATP, 10 m*M*)	2315 ± 22	1.0[c]
GPP(NH)P, 50 μ*M*	2259 ± 24	1.0
Forskolin, 50 μ*M*	2132 ± 80	0.9
NaF, 15 m*M*	2250 ± 30	1.0
None (MgATP, 5 m*M*)	46 ± 2	1.0[d]
GPP(NH)P, 50 μ*M*	48 ± 4	1.0
Forskolin, 50 μ*M*	38 ± 3	0.8
NaF, 15 m*M*	50 ± 3	1.1

[a] Effector molecules were added to the assay at the final concentrations shown.
[b] Assays were performed in triplicate. The values shown are means ± SE.
[c] Effect (-fold) relative to MnATP control.
[d] Effect (-fold) relative to MgATP control.

sin light chain kinase, NAD^+ kinase, and adenylyl cyclase from cerebral cortex.[29–36]

The response of equine sperm adenylyl cyclase solubilized as described above to typical regulators of somatic cell adenylyl cyclase was examined, and, as shown in Table II, this solubilized adenylyl cyclase behaves similarly to other testicular forms of the enzyme.[37] The preferred substrate for equine sperm adenylyl cyclase is MnATP. As summarized in Table II, adenylyl cyclase from equine spermatozoa exhibited regulatory properties similar to other mammalian spermatozoal adenylyl cyclases, namely, manganese dependence and lack of sensitivity to GTP or forskolin.[1,38] In

[29] W. Y. Cheung, *Science* **207,** 19 (1980).
[30] K. Geitzen, I. Scdorf, and H. Bader, *Biochem. J.* **207,** 541 (1982).
[31] O. Orellana, C. C. Allende, and J. E. Allende, *Biochem. Int.* **3,** 663 (1981).
[32] C. Klee, T. H. Crouch, and P. G. Richman, *Annu. Rev. Biochem.* **49,** 489 (1980).
[33] U. Niggli, E. S. Adunyah, and E. Carfoli, *J. Biol. Chem.* **256,** 8588 (1981).
[34] M. P. Walsh, R. Dabrowska, S. Hinkins, and D. J. Hartshorne, *Biochemistry* **21,** 1919 (1982).
[35] L. Meijer and R. Guerrier, *Biochim. Biophys. Acta* **702,** 143 (1982).
[36] C. H. Keller, D. C. LaPorte, D. R. Storm, W. A. Toscano, Jr., and K. R. Westcott, *Ann. N.Y. Acad. Sci.* **356,** 205 (1980).
[37] K. L. Olgiati, D. G. Toscano, W. M. Atkins, and W. A. Toscano, Jr., *Arch. Biochem. Biophys.* **231,** 411 (1984).
[38] M. K. Gross, Ph.D. Thesis, Harvard University, Cambridge, Massachusetts (1988).

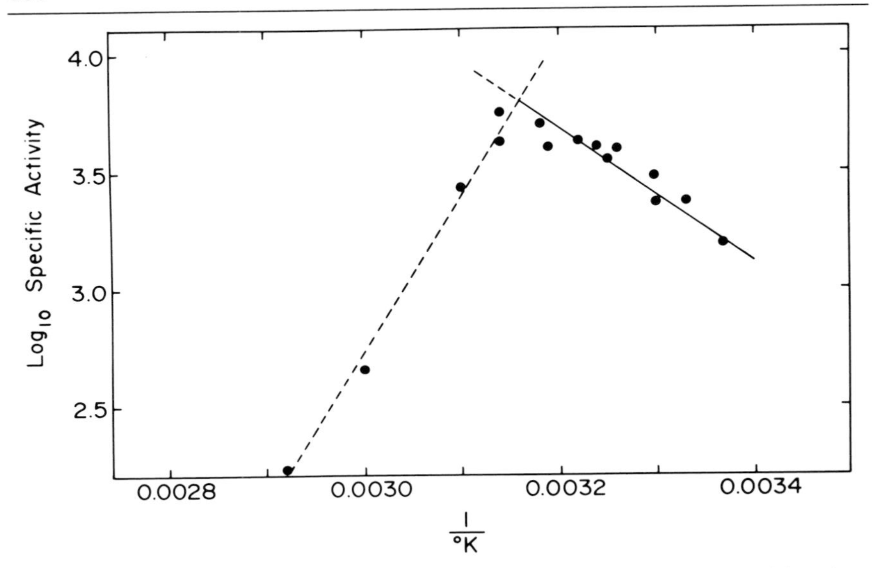

FIG. 1. Arrhenius plot of the temperature dependence of equine sperm adenylyl cyclase. Adenylyl cyclase activity was measured during 10-min enzyme assays at the temperatures indicated with 1 m*M* [α-^{32}P]ATP and 10 m*M* $MnCl_2$ as the substrate. The values shown were calculated from triplicate assays (S.E.M. $\leq$ 5%).

addition, this form of adenylyl cyclase displayed high thermal stability. Equine sperm adenylyl cyclase showed an activity maximum at 45° (Fig. 1) and retained approximately 50% of its catalytic activity following an incubation for 45 min at 50° and assay at 37° for 10 min. The activity of equine sperm adenylyl cyclase was highly sensitive to pH. Enzyme activity was optimal at slightly alkaline pH; the maximum adenylyl cyclase activity was observed at pH 8.0. Kinetic studies of equine sperm adenylyl cyclase yielded a K_m for MnATP of 6 m*M* and a V_{max} of 12 nmol cyclic AMP formed min/mg protein.[39]

Evidence That Equine Sperm Adenylyl Cyclase is Mediated by Calmodulin

There is extensive evidence supporting the hypothesis that cooperative regulatory interactions occur between calcium and cAMP in mature mammalian spermatozoa.[1,8] Demonstration of Ca^{2+}–CaM modulation of a mammalian sperm adenylyl cyclase, then, provides a biochemical mechanism to explain the functional association between calcium and cAMP in sperm. To establish unambiguously whether adenylyl cyclase from equine

[39] L. R. Forte, D. B. Bylund, and W. L. Zahler, *Mol. Pharmacol.* **24,** 42 (1983).

spermatozoa is modulated by CaM, we used anti-CaM drugs to inhibit the enzyme and then restored stimulation of the enzyme by the addition of CaM isolated from bovine brain. In addition, we removed CaM from the enzyme using lanthanum ion and isolated CaM-free adenylyl cyclase using CaM-Sepharose affinity chromatography. Adenylyl cyclase isolated in this manner was activated by the addition of exogenous CaM to the assay.

The CaM content of the soluble equine sperm extracts is estimated by the use of the *Bordetella pertussis* assay,[40] which measures CaM levels on the basis of functional activation of the bacterial adenylyl cyclase. CaM represented approximately 5% (4.5 mg CaM per 82 mg protein) of the protein present in the soluble equine sperm supernatant fractions.

Inhibition of Equine Sperm Adenylyl Cyclase by Anti-CaM Agents and Restoration of Activity by Authentic CaM

Calmidazolium, a highly specific CaM antagonist,[41] elicited concentration-dependent inhibition of equine sperm adenylyl cyclase activity (Fig. 2, inset). Even though we examined the actions of other anti-CaM drugs,[12] calmidazolium was the most potent inhibitor of enzyme activity in our preparations, correlating with the established relative potency and specificity of these drugs as anti-CaM agents.[41]

Calmidazolium inhibition of equine sperm adenylyl cyclase was biphasic; 60% of the total enzyme activity responded to relatively low drug concentrations. At the limits of solubility of calmidazolium (200 μM), about 30% of the initial enzyme activity remained unaffected. The apparent IC_{50} for the overall inhibition curve was 50 μM compared to a value of 15 μM when just the first phase of the calmidazolium inhibition curve was considered. The addition of purified CaM from bovine brain restored full enzymatic activity to equine sperm adenylyl cyclase that had been maximally inhibited by calmidazolium (Fig. 2). The restoration of enzyme activity was dependent on the concentration of CaM added back to the calmidazolium-inhibited enzyme; the EC_{50} of CaM to elicit this response was 15 μM, and enzyme activity was completely restored to precalmidazolium levels by the addition of 50 μM CaM.

Chromatographic Resolution of CaM-Depleted Equine Sperm Adenylyl Cyclase by Use of Lanthanum Dissociation

Even though equine sperm adenylyl cyclase gave evidence of being modulated by CaM because of its inhibition by various anti-CaM agents, the enzyme did not initially bind to a CaM-Sepharose affinity column

[40] D. V. Greenlee, T. J. Andraesen, and D. R. Storm, *Biochemistry* **21,** 2759 (1982).

[41] H. Van Belle, *Adv. Cyclic Nucleotide Protein Phosphorylation Res.* **17,** 557 (1984).

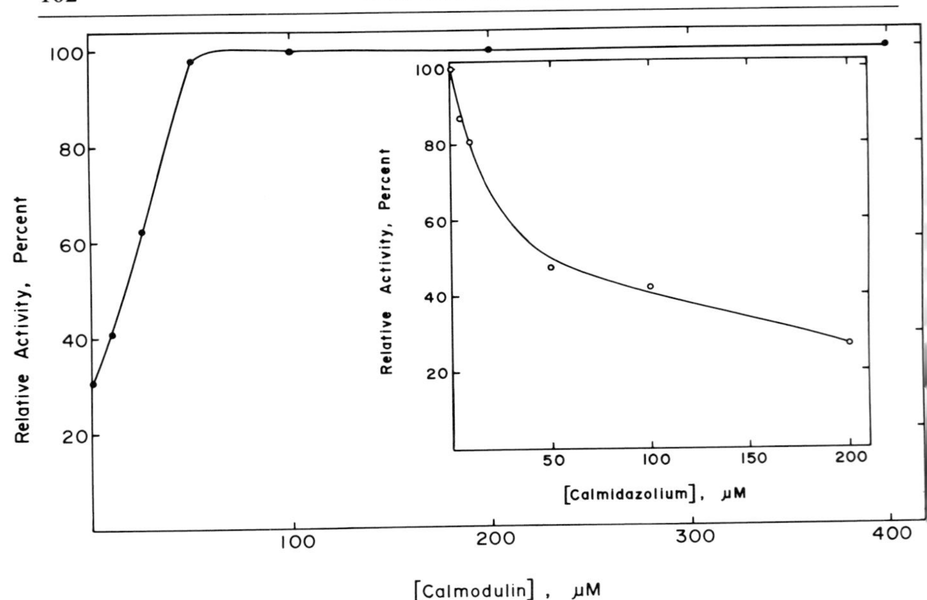

FIG. 2. Inhibition of equine sperm adenylyl cyclase by calmidazolium and restoration of enzyme activity with calmodulin. Equine sperm adenylyl cyclase was incubated at 4° for 15 min with 200 μM calmidazolium. After the initial incubation, CaM was added to the mixture at the final concentrations shown, and the incubation was continued for an additional 15 min at 4°, prior to measurement of enzyme activity. (Inset) Calmidazolium was added to the enzyme assay at the final concentrations shown. Assay conditions were the same as in Fig. 1 except the temperature was 37°. (Reprinted with permission from The American Society for Biochemistry and Molecular Biology.[12])

under standard column conditions.[16] This observation suggested either that the solubilized equine sperm adenylyl cyclase was not sensitive to CaM or that it was tightly complexed with CaM. To distinguish between these alternatives, the sensitivity of equine sperm adenylyl cyclase to La^{3+} was examined. Lanthanum ion has been reported to dissociate CaM from tightly bound CaM–enzyme systems,[42] and it is a potent inhibitor of CaM-dependent adenylyl cyclase from cerebral cortex.[43] The molecular basis for the action of lanthanum is to displace Ca^{2+} from CaM, resulting in a conformational change that is unfavorable for CaM binding to target enzymes.[44–46] Equine sperm adenylyl cyclase activity was inhibited by

[42] J. E. Schultz and S. Klumpp, *FEMS Microbiol. Lett.* **13,** 303 (1982).

[43] J. A. Nathanson, R. Freedman, and B. J. Hoffer, *Nature* (*London*) **261,** 331 (1976).

[44] J. S. Mills and J. D. Johnson, *J. Biol. Chem.* **260,** 15100 (1985).

[45] J. M. Buccigross and D. J. Nelson, *Biochem. Biophys. Res. Commun.* **138,** 1243 (1986).

[46] J. M. Buccigross and D. J. Nelson, *J. Inorg. Biochem.* **33,** 139 (1988).

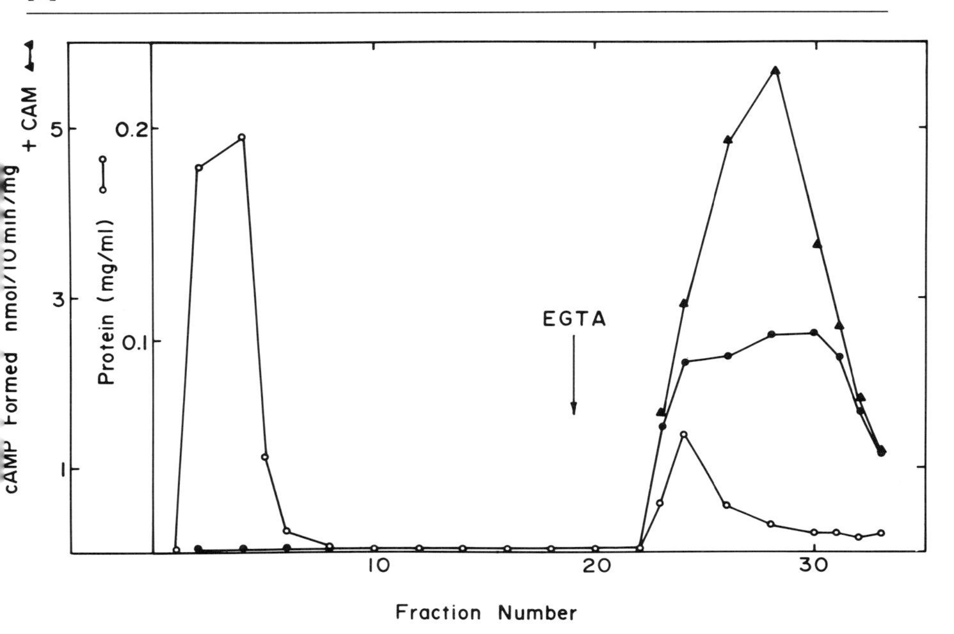

FIG. 3. CaM-Sepharose chromatography of La^{3+}-inhibited equine sperm adenylyl cyclase. Equine sperm adenylyl cyclase was incubated with 3 mM lanthanum at 4° for 15 min and applied to a CaM-Sepharose column (1 × 10 cm) that had been preequilibrated with 50 mM Tris-HCl, pH 7.5, 3 mM $CaCl_2$, 150 mM NaCl, 1 mM 2-mercaptoethanol, 250 mM sucrose. The column was washed with the same buffer containing 150 mM NaCl but no Ca^{2+} to remove nonspecifically bound proteins. Following the salt wash, CaM-binding proteins were eluted with the above buffer containing 1 mM EGTA. Assay conditions were the same as in Fig. 2 except that 1.5 mM $CaCl_2$ was added when CaM was present in the assay. (Reprinted with permission of The American Society for Biochemistry and Molecular Biology.[12])

La^{3+} in a concentration-dependent manner with an apparent IC_{50} value of 1 mM (approximately one-third the Ca^{2+} concentration).[12] Based on this observation, it was reasoned that La^{3+} pretreatment of equine sperm extracts would dissociate native CaM from adenylyl cyclase and allow CaM-Sepharose to compete for binding to the enzyme in the presence of calcium. Thus, equine sperm adenylyl cyclase was incubated with La^{3+} under conditions that showed maximum inhibition and that were reported to dissociate CaM from target enzymes.[41] The mixture was subsequently applied to a CaM-Sepharose column equilibrated with Ca^{2+} (Fig. 3). Following a low-salt wash to remove nonspecifically bound proteins, the column was eluted with EGTA. Equine sperm adenylyl cyclase that was specifically eluted with EGTA showed a 2-fold stimulation of enzymatic activity when assayed in the presence of exogenous CaM.

Several criteria have been proposed to establish CaM regulation of a target enzyme in impure systems.[47] These criteria include (1) inhibition of enzyme activity by appropriate doses of anti-CaM agents, (2) correlation between the potency of structurally unrelated anti-CaM agents as enzyme inhibitors and their potency as CaM inhibitors, and (3) restoration of activity to the inhibited enzyme by the addition of exogenous CaM. Based on these criteria, equine sperm adenylyl cyclase appears to be mediated by Ca^{2+}–CaM.

High-Performance Liquid Chromatography of Adenylyl Cyclase in Extracts from Equine Spermatozoa

The biphasic response of equine sperm adenylyl cyclase to CaM antagonists (see above) suggests that more than one enzyme species is present in soluble extracts of equine sperm proteins. Analytical HPLC has the potential to resolve biological molecules that differ only slightly in their biochemical properties; therefore, this technique was used to identify multiple forms of adenylyl cyclase in equine sperm protein isolates.

In order to determine whether the adenylyl cyclase activity in equine sperm extracts represented a homogeneous or heterogeneous enzyme population, soluble proteins from equine sperm were fractionated on analytical HPLC by the use of gel filtration or anion-exchange resins. As shown in Fig. 4, equine sperm adenylyl cyclase is eluted as a single peak of enzyme activity of approximately 32,000 molecular weight from a Bio-Sil TSK-250 gel-filtration column. These results were reproducible with several different equine sperm protein preparations. In contrast, multiple peaks of adenylyl cyclase activity were resolved from equine sperm extracts by anion-exchange HPLC on a TSK DEAE-5-PW column (see Figs. 5 and 6). The anion-exchange elution profile for adenylyl cyclase activity was somewhat variable between individual preparations of equine sperm protein extracts. These differences can be primarily attributed to donor-specific variation rather than isolation artifacts, because samples of sperm from the same horse that were processed separately produced similar enzyme elution profiles. Also, more variability was seen in supernatant I extracts than in supernatant II extracts. The enzyme elution profile of a given equine sperm protein extract was highly reproducible and not due to chromatographic artifact.

In separate experiments, the calmidazolium sensitivity of the various HPLC-resolved adenylyl cyclase peaks from several different equine sperm extracts was examined. In all cases, the fractionated enzyme peaks

[47] B. D. Roufogalis, *Calcium Cell Funct.* **3,** 129 (1982).

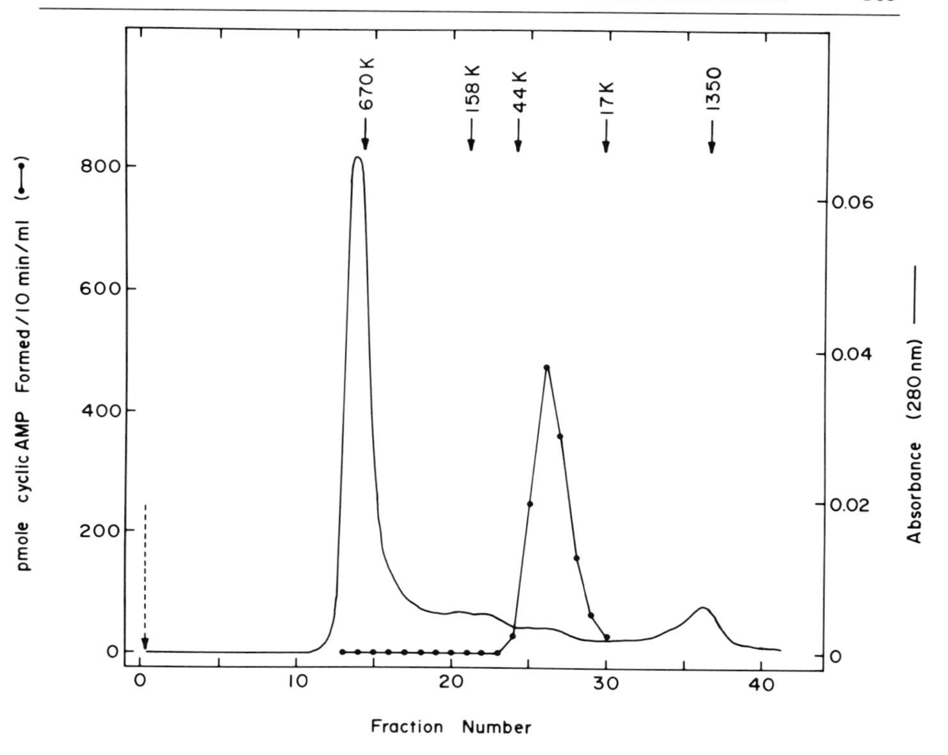

FIG. 4. Gel-filtration HPLC of soluble extracts from equine sperm. A sample of equine sperm proteins (0.8 mg) was applied in 800 μl to a 7.5 × 600 mm Bio-Sil TSK-250 (Bio-Rad) column, equilibrated with 50 m*M* MOPS, pH 7, 150 m*M* NaCl, 250 m*M* sucrose (room temperature), and eluted in the same buffer. Adenylyl cyclase activity of the fractions was measured for 10 min at 37° with Mn[α-^{32}P]ATP as substrate under standard assay conditions. Data represent means of duplicate assays with an S.E.M. of at most 5%. Molecular weight standards (Bio-Rad) were thyroglobulin (670K), γ-globulin (158K), ovalbumin (44K), myoglobin (17K), and vitamin B_{12} (1350).

were equally sensitive to calmidazolium inhibition; approximately 50% inhibition of enzyme activity was observed in the presence of 30 μM calmidazolium. Therefore, differences in CaM modulation did not appear to explain the observation of multiple sperm adenylyl cyclase forms.

Because the different forms of adenylyl cyclase separated by anion-exchange HPLC were apparently not distinguishable by their CaM sensitivity, other explanations for the existence of these multiple forms were investigated. Proteolytic processing and phosphorylation are two types of posttranslational modification that regulate the activity of numerous

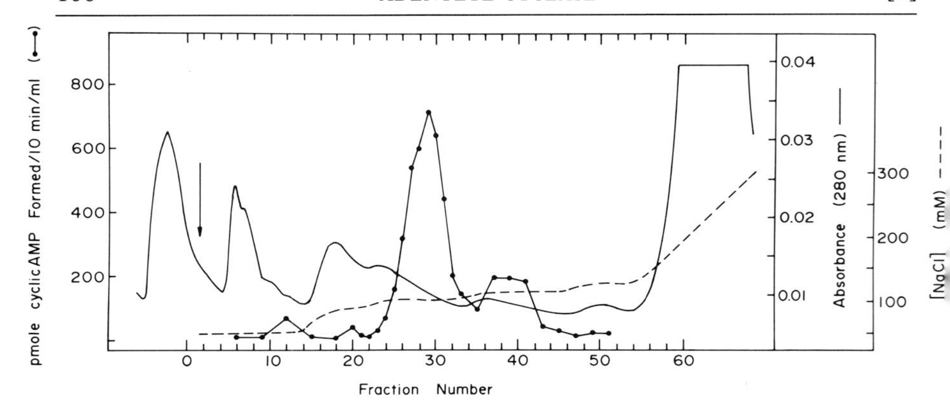

FIG. 5. Anion-exchange HPLC of soluble extracts from equine sperm. Supernatant II (1.2 mg) was applied to a 7.5 × 75 mm TSK DEAE-5-PW column, equilibrated with 50 mM Tris-HCl, pH 7.4 (room temperature), 25 mM sucrose, 50 mM NaCl, and eluted using the NaCl gradient shown. Adenylyl cyclase activity was measured at 37° as described earlier. The data shown are the means of duplicate assays with an S.E. of less than 5%. The arrow indicates the final sample injection and the start of the elution program. This chromatogram is the control for Figs. 7 and 8.

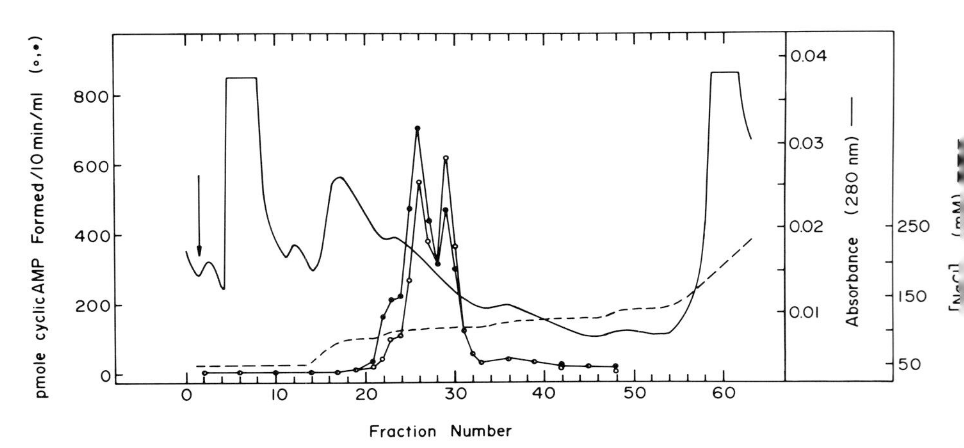

FIG. 6. Anion-exchange HPLC of soluble equine sperm extracts isolated in the presence of different protease inhibitors. Sperm proteins were extracted in the presence of either 1 mM PMSF (○) or 0.1 μM α_2-macroglobulin plus 0.1 μM soybean trypsin inhibitor (●). Conditions were the same as in Fig. 5, except that parallel HPLC separations were carried out on 1.02 mg of each extract.

enzyme systems.[48] To date there is no compelling evidence suggesting that adenylyl cyclase from sperm is mediated by phosphorylation; therefore, we focused only on the actions of proteases and their effect on both enzymatic activity and chromatographic behavior of equine sperm adenylyl cyclase to determine whether any of the observed sperm adenylyl cyclase forms were interconvertible by proteolysis.

Action of Proteases on Activity and Chromatographic Behavior of Adenylyl Cyclase from Equine Spermatozoa

No size heterogeneity of the sperm adenylyl cyclase was detected by gel-filtration HPLC. It was possible, however, that minor proteolytic degradation could account for some of the multiple adenylyl cyclase forms separated by anion-exchange HPLC. It was previously reported that extraction conditions similar to those used to isolate the equine sperm adenylyl cyclase in soluble form will solubilize a bovine sperm protease that has adenylyl cyclase-activating properties.[27] Because this trypsinlike protease was reported to be insensitive to PMSF but very sensitive to either α_2-macroglobulin or soybean trypsin inhibitor, extracts of soluble equine sperm proteins were prepared in parallel with either 1 m*M* PMSF (standard extraction conditions) or 0.1 μM α_2-macroglobulin plus 0.1 μM soybean trypsin inhibitor included in all extraction buffers. The concentrations of the latter protease inhibitors were selected based on the reported IC_{50} of 30 n*M* for either α_2-macroglobulin or soybean trypsin inhibitor alone to inhibit the purified sperm protease.[27] As shown in Fig. 6, the presence of either PMSF or α_2-macroglobulin plus soybean trypsin inhibitor during the extraction yielded soluble extracts of sperm adenylyl cyclase that displayed virtually identical elution profiles on anion-exchange HPLC. Furthermore, residual protease activity measured in these extracts, with flourescein isothiocyanate-labeled casein as a substrate,[49] was minimal and equivalent between samples processed in the presence of either PMSF or α_2-macroglobulin. Protease activity in the supernatant II fraction of the extract processed with PMSF was equivalent to 22 n*M* trypsin whereas protease activity in the parallel sample prepared with α_2-macroglobulin plus soybean trypsin inhibitor was equivalent to 23 n*M* trypsin.

In order to examine whether any of the HPLC-resolved adenylyl cyclase peaks were interconvertible by limited proteolysis, replicate samples of an equine sperm extract were left untreated or were exposed to exogenous protease prior to anion-exchange HPLC. The elution profile of the untreated control sperm extract is shown in Fig. 5. This extract exhibited

[48] F. Wold, *Annu. Rev. Biochem.* **50,** 783 (1981).
[49] S. S. Twining, *Anal. Biochem.* **143,** 30 (1984).

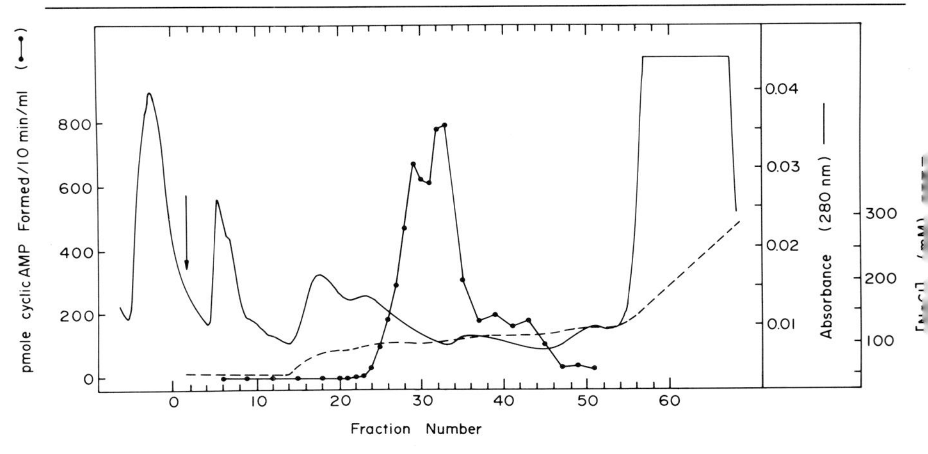

FIG. 7. Anion-exchange HPLC of soluble equine sperm proteins subjected to limited chymotrypsin proteolysis. Conditions were the same as in Fig. 6 except that, prior to column application, 1.19 mg of the supernatant II extract was pretreated with chymotrypsin (0.5 U/mg sperm protein) for 5 min at 30°. Proteolysis was terminated with 0.7 U of α_2-macroglobulin/mg sperm protein.

two broad peaks of adenylyl cyclase activity; a major peak representing 70% of the enzyme eluted between 90 and 100 m*M* NaCl (peak I), and a minor peak representing 30% of the enzyme eluted between 100 and 112.5 m*M* NaCl (peak II). As shown in Fig. 7, limited proteolysis of this sperm extract for 5 min with chymotrypsin activated total adenylyl cyclase activity 1.4-fold; 2-fold activation of adenylyl cyclase was observed in another equine sperm extract treated for a slightly shorter time (4 min) with chymotrypsin (specific activities were 3569 ± 445 and 6974 ± 144 pmol cyclic AMP formed/10 min/mg protein for the control and chymotrypsin-activated sperm adenylyl cyclase, respectively). Chymotrypsin proteolysis also generated an intermediate enzyme peak eluting between peaks I and II (Fig. 7, fraction 33) and a small peak eluting after peak II (Fig. 7, fraction 43). More extended proteolysis of this equine sperm extract with chymotrypsin for 10 min decreased total adenylyl cyclase activity to 80% of the untreated control extract activity; enzyme activity was decreased proportionately throughout the enzyme elution profile (data not shown). Regardless of the length of proteolysis with chymotrypsin, the ratio of adenylyl cyclase activity eluting at 90–100 m*M* NaCl to that eluting at 100–112.5 m*M* NaCl remained the same as controls, suggesting that peak I and II were not interconvertible by proteolysis. Likewise, treatment of this equine sperm extract with an equivalent amount of trypsin for 10 min decreased total adenylyl cyclase activity to 50% of control. Although

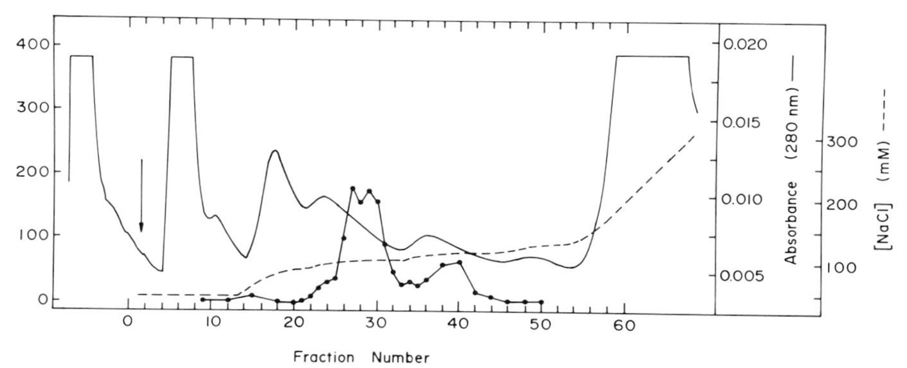

FIG. 8. Anion-exchange HPLC of soluble equine sperm proteins subjected to extended trypsin proteolysis. Conditions were the same as in Fig. 6 except that, prior to column application, 0.59 mg of the supernatant II extract was pretreated with trypsin (0.5 U/mg sperm protein) for 10 min at 30°. Proteolysis was terminated with 0.7 U of α_2-macroglobulin/mg sperm protein.

trypsin treatment did not appear to generate any new adenylyl cyclase peaks in the HPLC elution profile, enzyme activity was decreased proportionately in both peak I and peak II, so that the ratio of enzyme in peak I versus peak II was the same as the untreated control (Fig. 8).

Limited proteolysis of calmodulin-regulated target enzymes typically results in enhanced enzyme activity that is no longer sensitive to the actions of either calmodulin or anticalmodulin drugs, whereas prolonged proteolysis is associated with the loss of enzyme activity.[28] Therefore, the actions of exogenous proteases on the calmidazolium sensitivity of sperm adenylyl cyclase were also characterized. The fraction of equine sperm adenylyl cyclase activity that was inhibited by 30 μM calmidazolium was surprisingly constant, regardless of the type of protease used or whether the conditions of protease treatment increased or decreased the basal adenylyl cyclase activity (data not shown).

A freeze–thaw cycle is sufficient to solubilize adenylyl cyclase from bovine sperm membranes,[50] and this procedure has been adopted in the extraction protocol for equine sperm adenylyl cyclase. Other investigators have used limited chymotrypsin proteolysis to activate and solubilize adenylyl cyclase from ram sperm membranes.[51–53] A soluble extract of

[50] C. A. Herman, W. L. Zahler, G. A. Doak, and R. J. Campbell, *Arch. Biochem. Biophys.* **177,** 622 (1976).

[51] D. Stengel and J. Hanoune, *J. Biol. Chem.* **256,** 5394 (1981).

[52] D. Stengel and J. Hanoune, *Ann. N.Y. Acad. Sci.* **438,** 18 (1984).

[53] D. Stengel, L. Guenet, and J. Hanoune, *J. Biol. Chem.* **257,** 10818 (1982).

equine sperm adenylyl cyclase prepared in this manner, however, exhibited an adenylyl cyclase specific activity 70-fold lower than a parallel sperm extract prepared by our procedures (107 ± 2 pmol cAMP formed/10 min/mg for the chymotrypsin-solubilized extract versus 7567 ± 123 pmol cAMP formed/10 min/mg for the standard preparation).

We have applied affinity chromatography on CaM-Sepharose to isolate a CaM-mediated adenylyl cyclase from equine sperm. Even though the methods described here are amenable to the ultimate purification of adenylyl cyclase from sperm to homogeneity, methods for large-scale preparation of the enzyme are still lacking. Future development in this area will include preparation of antibodies against spermatozoal adenylyl cyclase for use as immunoaffinity chromatography agents.

Acknowledgments

We thank Diane Toscano for technical assistance in carrying out the experiments and for the artwork used in this chapter. We thank Dr. Clark Millette for expert advice and discussion. We are grateful to Ruth Yuan of the HPLC Core Facility at the Laboratory of Human Reproduction and Reproductive Biology at the Harvard Medical School for performing the HPLC experiments. The facility is funded by Grant No. 5 P-30 HD 06645 from the National Institutes of Health.

[9] Isolation of Polyclonal Antibodies against Bovine Brain Calmodulin-Sensitive Adenylyl Cyclases

By GARY B. ROSENBERG, PING WANG, and DANIEL R. STORM

Introduction

The difficulty in purifying mammalian adenylyl cyclases (EC 4.6.1.1, adenylate cyclase) has stimulated interest in isolation of antibodies against the enzymes that could be used for purification, cloning, and characterization of different forms of the enzyme. Although the catalytic subunit of the enzyme can be identified on SDS gels[1] and purified to apparent homogeneity by preparative SDS gel electrophoresis, attempts at raising antiadenylyl cyclase antibodies against these polypeptides in rabbits or mice have not been successful in this laboratory. On the other hand,

[1] CaM, Calmodulin; SDS, sodium dodecyl sulfate; WGA, wheat germ agglutinin; DTT, dithiothreitol; PMSF, phenylmethylsulfonyl fluoride; GPP(NH)P, guanyl-5′-yl imidodiphosphate; CHAPS, 3-[(3-cholamidopropyl)dimethylammonio]-1-propane sulfonate; IgG, immunoglobulin G.

we have been able to raise polyclonal antibodies against brain adenylyl cyclases using highly purified, calmodulin (CaM)-sensitive adenylyl cyclase preparations which have not been denatured by SDS. In this chapter we describe the properties of a mouse polyclonal antiserum that was specific for the CaM-sensitive form of the enzyme from brain. This antiserum was used to characterize the distribution of CaM-sensitive adenylyl cyclases in various rat tisues.[2]

Isolation of Mouse Polyclonal Antibodies That Recognize Calmodulin-Sensitive but Not Calmodulin-Insensitive Adenylyl Cyclase

Antigen. The CaM-sensitive adenylyl cyclase used for antigen is purified from bovine brain by the method of Minocherhomjee *et al.*[3] This protein runs as a single polypeptide of apparent molecular weight 135,000 on SDS gels. Adenylyl cyclase eluted from forskolin-Sepharose (forskolin pool) is concentrated approximately 100-fold by ultrafiltration using a PM10 ultrafiltration membrane (Amicon, Danvers, MA). The concentrated protein is mixed 1 : 1 with Freund's complete adjuvant for initial injections or 1 : 1 with Freund's incomplete adjuvant for subsequent boosts.

Antibody Production and Purification. BALB/c mice are injected subcutaneously with 1–2 μg of the antigen emulsified in 100 μl of Freund's complete adjuvant. At 2-week intervals the mice are boosted with 1–2 μg of antigen in 100 μl of Freund's incomplete adjuvant. The mice are bled prior to injection to obtain preimmune serum and 2 weeks after each boost to obtain immune serum. Antisera obtained after the seventh and eighth boosts are used in the experiments described in this chapter.

SDS–Polyacrylamide Gel Electrophoresis. SDS–polyacrylamide gels are prepared by the method of Laemmli.[4] Molecular weight standards are ovalbumin (M_r 45,000), bovine serum albumin (M_r 66,200), phosphorylase *b* (M_r 92,500), β-galactosidase (M_r 116,250), and myosin (M_r 200,000). The gels are stained with Coomassie blue and autoradiographs obtained using Kodak X-Omat AR film.

Protein Iodinations. CaM-sensitive adenylyl cyclase is iodinated with Iodogen as described by the manufacturer. CaM-sensitive adenylyl cyclase, purified through CaM-Sepharose affinity chromatography, is concentrated to 0.25 mg/ml by ultrafiltration with an Amicon PM10 membrane. The concentrated enzyme (250 μl) is mixed with an equal volume of 50

[2] G. B. Rosenberg and D. R. Storm, *J. Biol. Chem.* **262,** 7623 (1987).
[3] A. M. Minocherhomjee, S. Selfe, N. J. Flower, and D. R. Storm, *Biochemistry* **26,** 4444 (1987).
[4] U. K. Laemmli, *Nature (London)* **227,** 680 (1970).

mM Tris-HCl, 250 mM sucrose, 5 mM $MgCl_2$, and 1 mM EDTA, pH 7.4, and 2.5 mCi of $Na^{125}I$ in a glass test tube which had been previously coated with 15 μg of Iodogen (Sigma Chemical, St. Louis, MO). The reaction is allowed to proceed for 20 min at room temperature. The protein is then separated from unreacted $Na^{125}I$ on a 2-ml column of Sephadex G-25.

Preparation and Solubilization of Rat Tissues. Tissues are excised from freshly sacrificed Sprague-Dawley rats, and 5 g of tissue is placed immediately in 20 ml ice-cold 20 mM glycylglycine, pH 7.2, 5 mM $MgCl_2$, 1 mM EDTA, 250 mM sucrose, 1 mM PMSF, 3 mM DTT (homogenization buffer). The tissue is homogenized for 30 sec at the maximum setting with a Polytron homogenizer. After centrifugation at 12,000 g for 30 min at 4°, the pellets are resuspended with 40 ml homogenization buffer and again homogenized for 30 sec. The suspension is centrifuged at 12,000 g for 1 hr and the pellets resuspended with 5 ml of 20 mM Tris-HCl, pH 7.4, 250 mM sucrose, 1 mM $MgCl_2$, 1 mM EDTA, 0.5% (w/v) Lubrol PX. The preparations are placed on a rotating wheel at 4° for 1 hr and then centrifuged for 1 hr at 100,000 g. The supernatants are aliquoted and stored at −80° until use. Protein concentrations are determined by the method of Peterson[5] with bovine serum albumin as a standard.

Immunoprecipitation of Calmodulin-Sensitive Adenylyl Cyclase. Adenylyl cyclase preparations are incubated overnight on ice with various IgG fractions in a final volume of 0.1 ml. For each immunoprecipitation, 100 μl of Pansorbin (Calbiochem, La Jolla, CA) (10% suspension) is washed twice with 50 mM Tris-HCl, 250 mM sucrose, 5 mM $MgCl_2$, and 1 mM EDTA, pH 7.4 (buffer X), resuspended with 100 μl of buffer X, and mixed with 50 μl of rabbit anti-mouse IgG (12.7 mg total IgG/ml). After 1 hr on ice, the Pansorbin is washed 3 times with buffer X, resuspended in buffer X to a 10% suspension, and then 0.1 ml added to each tube containing adenylyl cyclase and serum. After 1 hr on ice, 0.2 ml of buffer X is added to each tube and the suspension centrifuged for 2 min in a Beckman Microfuge B. The supernatants are removed for adenylyl cyclase assays and the pellets washed twice with 1.2 ml of buffer X prior to resuspension with 0.4 ml of buffer X. The resuspended pellets are also assayed for adenylyl cyclase activity.

Serum from a mouse immunized with purified CaM-sensitive adenylyl cyclase is tested for its ability to precipitate adenylyl cyclase activity. As reported in Fig. 1A, the antiserum immunoprecipitates greater than 95% of the total enzyme activity from a highly purified preparation of CaM-sensitive adenylyl cyclase. The CaM-sensitive adenylyl cyclase activity is not immunoprecipitated by preimmune serum. A large fraction of the

[5] G. L. Peterson, *Anal. Biochem.* **83,** 346 (1977).

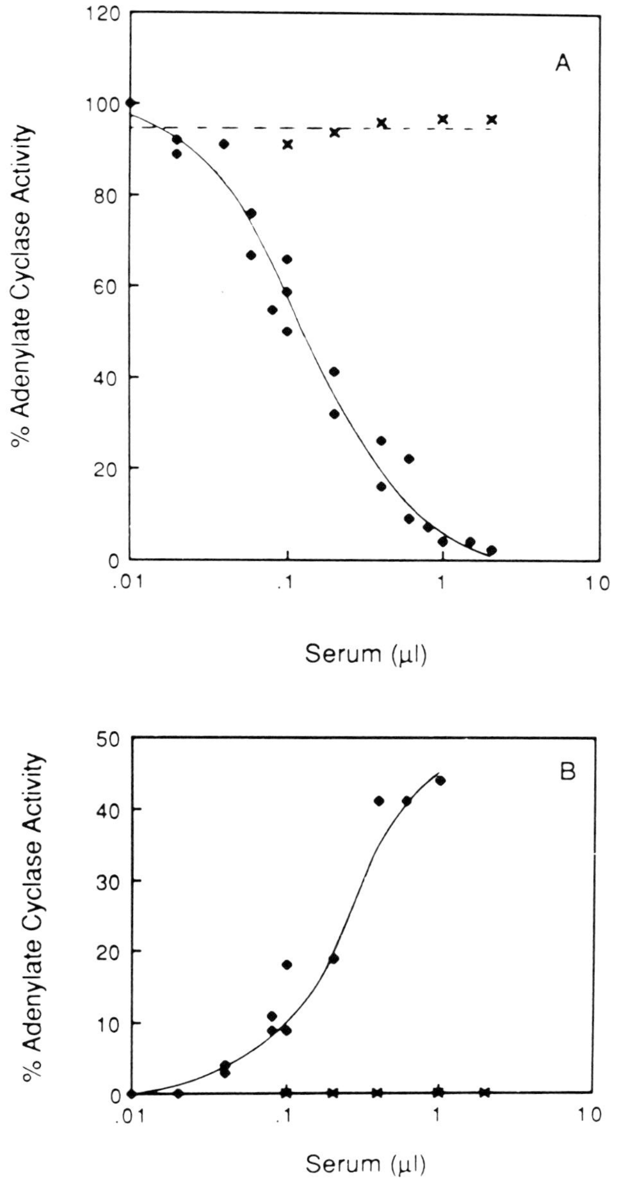

FIG. 1. Immunoprecipitation of CaM-sensitive adenylyl cyclase activity. The data presented are from three separate experiments. Adenylyl cyclase purified through CaM-Sepharose and forskolin-Sepharose was immunoprecipitated as described in the text, with either preimmune (x) or immune (◆) serum. After immunoprecipitation, adenylyl cyclase activity was determined in the supernatants (A) and the immunoprecipitates (B). The total starting adenylyl cyclase activity was 2.7–4.8 pmol cAMP/min. (Reprinted from Rosenberg and Storm[2] with permission from the American Society for Biochemistry and Molecular Biology.)

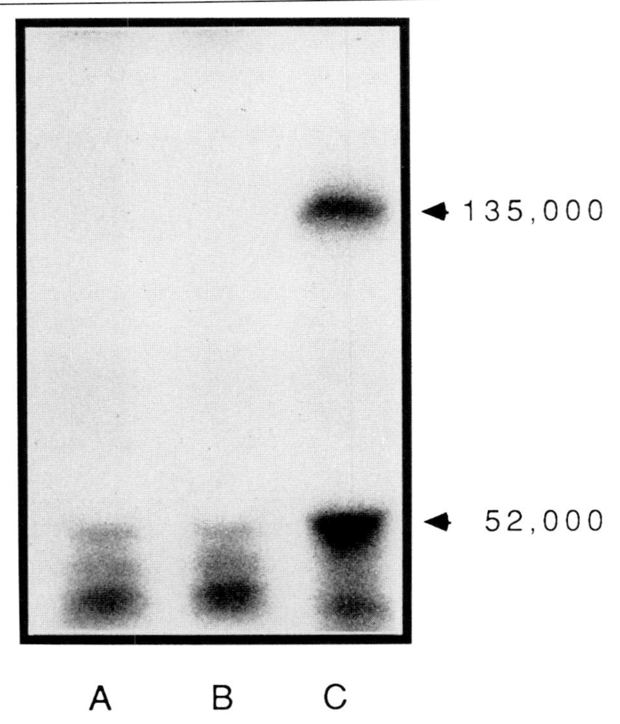

FIG. 2. Immunoprecipitation of ^{125}I-labeled CaM-sensitive adenylyl cyclase. CaM-sensitive adenylyl cyclase purified through CaM-Sepharose (F2) was iodinated as described in the text. The labeled adenylyl cyclase preparation was incubated with 0.2 μg of unlabeled adenylyl cyclase for 16 hr at 4° without serum (A), with 4 μl of preimmune serum (B), or with 4 μl of immune serum (C). Pansorbin–rabbit anti-mouse IgG was used to precipitate protein–antibody complexes as described in the text. The immunoprecipitates were then washed twice and resuspended in sample buffer.[4] After boiling for 2 min, the Pansorbin was removed by centrifugation and the supernatant applied to a 7.5% SDS–polyacrylamide gel. An autoradiogram from the gel is shown. (Reprinted from Rosenberg and Storm[2] with permission from the American Society for Biochemistry and Molecular Biology.)

adenylyl cyclase activity can be recovered in the immunoprecipitates (Fig. 1B), suggesting that the antiserum is not inhibitory. Incubation of the antiserum with adenylyl cyclase overnight at 4°, without immunoprecipitation and resuspension of pellets, has no effect on enzyme activity, confirming that the antibody is noninhibitory. The loss of activity in the immunoprecipitates is variable and reflects poor recovery of enzyme activity during resuspension and washing of the precipitates.

The purified catalytic subunit of CaM-sensitive adenylyl cyclase interacts directly with CaM and wheat germ agglutinin.[3] Therefore, the effect

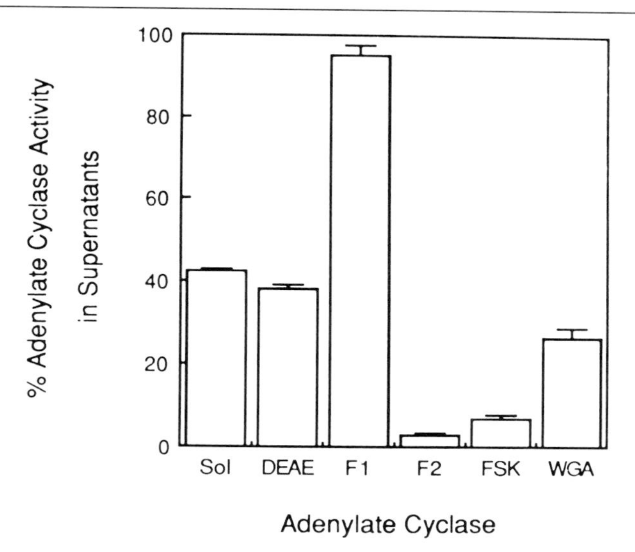

FIG. 3. Immunoprecipitation of various preparations of bovine brain adenylyl cyclase. The catalytic subunit of the CaM-sensitive adenylyl cyclase was purified to homogeneity as described in the text. A sample from each step through the purification was immunoprecipitated with 1 μl of either preimmune or immune serum. The values shown are the amounts of adenylyl cyclase activity remaining in the supernatants after precipitation with immune serum relative to the activity remaining after precipitation with preimmune serum. In no case did the preimmune serum precipitate a significant fraction of the activity. (Reprinted from Rosenberg and Storm[2] with permission from the American Society for Biochemistry and Molecular Biology.)

of these two proteins on immunoprecipitation of activity by immune serum has been examined. Neither CaM nor wheat germ agglutinin affects immunoprecipitation of adenylyl cyclase activity by immune serum, suggesting that the antibody recognition site(s) is distinct from the CaM- and wheat germ agglutinin-binding domains. Preincubation of the adenylyl cyclase preparation with the zwitterionic detergent CHAPS and NaCl, which have been shown to dissociate G_s from the catalytic subunit,[6] has no effect on the ability of the antiserum to precipitate enzyme activity.

Immunoprecipitation of ^{125}I-Labeled Calmodulin-Sensitive Adenylyl Cyclase. To identify the peptide(s) immunoprecipitated by immune serum, an ^{125}I-labeled preparation of CaM-sensitive adenylyl cyclase purified through CaM-Sepharose affinity chromatography is immunoprecipitated. After washing, the immunoprecipitated peptides are resolved by SDS–polyacrylamide gel electrophoresis and autoradiographed (Fig. 2).

[6] A. J. Bitonti, J. Moss, L. Hjelmeland, and M. Vaughan, *Biochemistry* **21,** 3650 (1982).

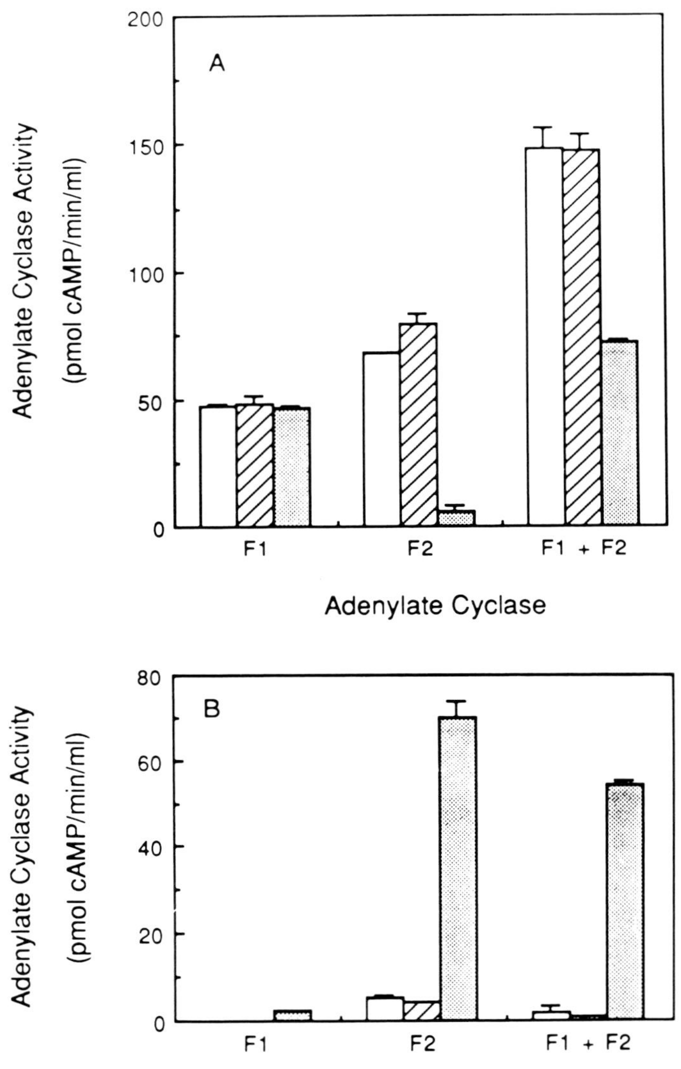

FIG. 4. Immunoprecipitation of adenylyl cyclase from a mixture of the CaM-sensitive and CaM-insensitive adenylyl cyclases. Immunoprecipitations were performed as described in the text without serum (□) or with 1 μl of preimmune (▨) or immune (▩) serum. Adenylyl cyclase activity was measured in both the supernatants (A) and the immunoprecipitates (B). F1 represents 47 μg of CaM-insensitive adenylyl cyclase; F2, 3 μg of CaM-sensitive adenylyl cyclase; F1 + F2, 47 μg of CaM-insensitive adenylyl cyclase and 3 μg of CaM-sensitive adenylyl cyclase. (Reprinted from Rosenberg and Storm[2] with permission from the American Society for Biochemistry and Molecular Biology.)

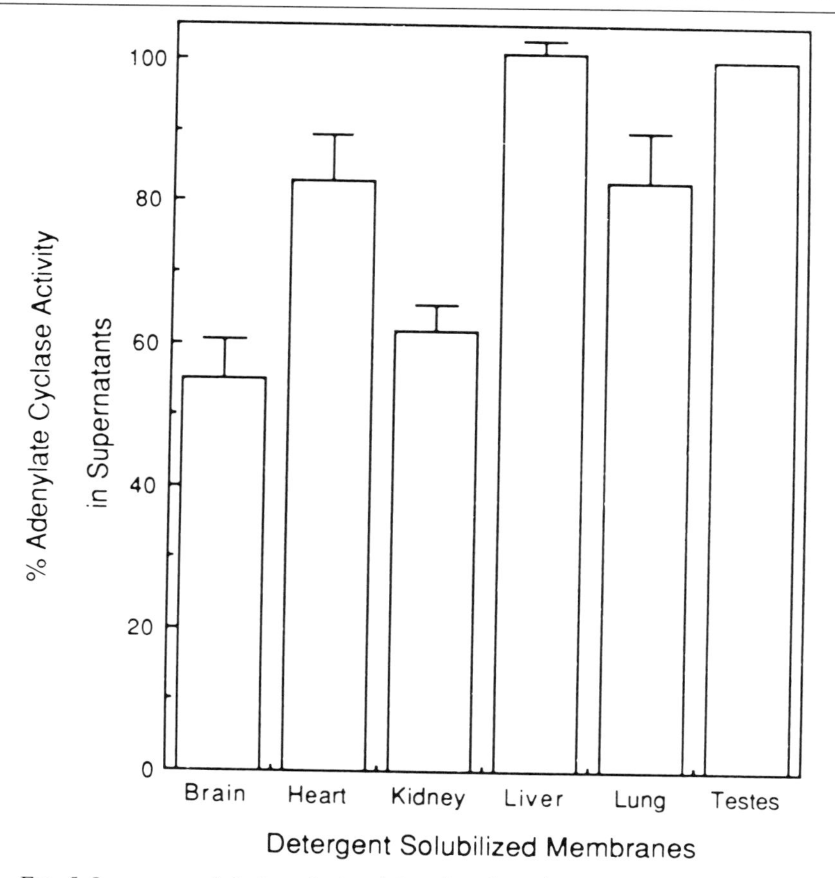

FIG. 5. Immunoprecipitation of adenylyl cyclase from detergent-solubilized membranes of various rat tissues. Detergent-solubilized membranes were prepared as described in the text. The values shown are the amounts of adenylyl cyclase activity remaining in the supernatants after precipitation with 2 μl of immune serum relative to the activity remaining after precipitation with 2 μl of preimmune serum. In no case did the preimmune serum precipitate a significant fraction of the activity. Total starting adenylyl cyclase activities (in pmol cAMP produced/min) were as follows: brain, 20.1; heart, 2,2; kidney, 2.3; liver, 2.4; lung, 2.3; testes, 12.6. (Reprinted from Rosenberg and Storm[2] with permission from the American Society for Biochemistry and Molecular Biology.)

An M_r 135,000 polypeptide is precipitated only with immune serum (lane C) and not with preimmune serum (lane B) or in controls without serum (lane A). Since the catalytic subunit of the CaM-sensitive adenylyl cyclase used as antigen has an M_r of 135,000, these data indicate that the antiserum interacts directly with the catalytic subunit of the enzyme. An unidentified

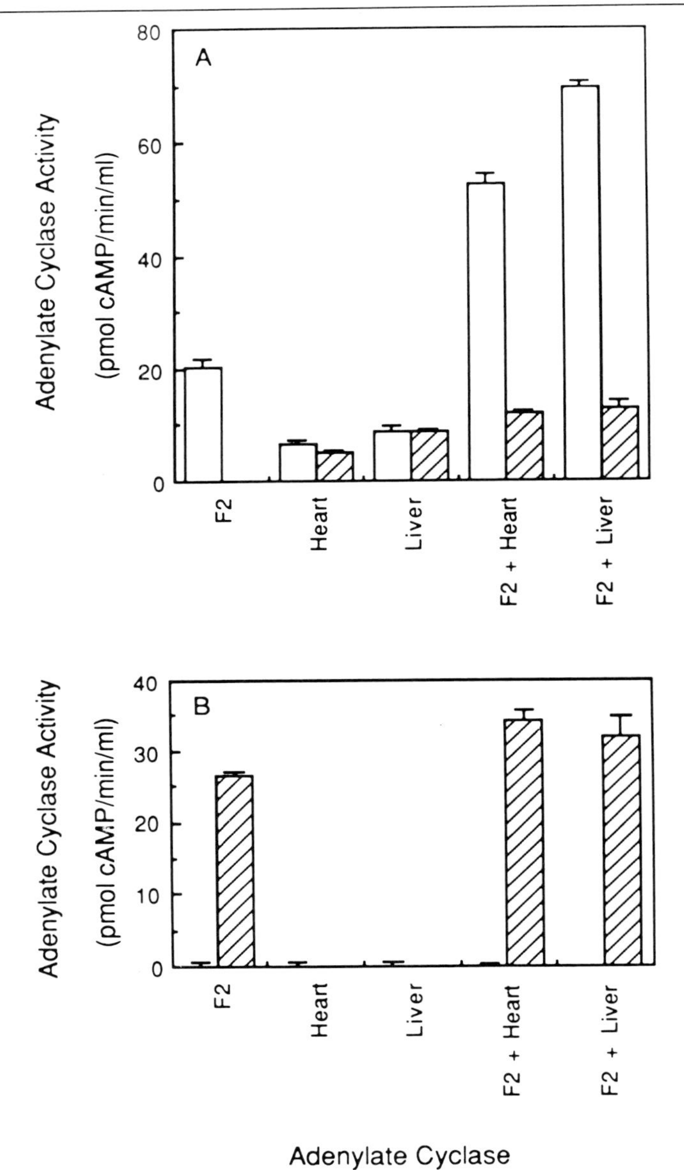

FIG. 6. Immunoprecipitation of adenylyl cyclase activity from a mix of bovine brain CaM-sensitive adenylyl cyclase with detergent-solubilized rat heart or liver membranes. Immunoprecipitations were performed as described in the text, with 1 μl of preimmune (□) or immune serum (▨). Adenylyl cyclase activity was measured in both the supernatants (A)

polypeptide of about M_r 52,000 is immunoprecipitated by immune serum; however, it is also present at lower levels in the preimmune control. The only polypeptide specifically precpitated by immune serum is the polypeptide of M_r 135,000. This antibody did not recognize denatured adenylyl cyclase, and Western blots using this polyclonal antibody show several nonspecifically labeled polypeptides but no interaction with the adenylyl cyclase catalytic subunit.

Immunoprecipitation of Partially Purified Calmodulin-Sensitive Adenylyl Cyclase. The ability of the antiserum to immunoprecipitate adenylyl cyclase activity at different stages of the purification has also been evaluated (Fig. 3). Approximately 60% of the total activity in detergent-solubilized membranes or the pooled enzyme from DEAE-Sephacel is immunoprecipitated. However, after separation of the CaM-insensitive adenylyl cyclase (F1) from the CaM-sensitive adenylyl cyclase (F2) by CaM-Sepharose affinity chromatography, the antiserum is able to immunoprecipitate enzyme activity only from the CaM-sensitive preparation of the enzyme.

To demonstrate that the antiserum interacts directly with the catalytic subunit of adenylyl cyclase, the ability of the antiserum to immunoprecipitate adenylyl cyclase activity from a homogeneous preparation of the catalytic subunit of CaM-sensitive adenylyl cyclase has been examined. The catalytic subunit of this enzyme is purified to homogeneity with CaM-Sepharose, forskolin-Sepharose, and WGA-Sepharose as described. As reported in Fig. 3, the antiserum is able to precipitate adenylyl cyclase activity from a pure preparation of the enzyme (WGA), indicating that the antibody interacts directly with the catalytic subunit of CaM-sensitive adenylyl cyclase purified from bovine brain. Even though 95% or more of the activity in CaM pool or the forskolin pool is immunoprecipitated by saturating antiserum, only 75% of the pure catalytic subunit from the WGA pool is immunoprecipitated by the antiserum. Although the reason for incomplete precipitation of WGA pool is not known, it may be the result of covalent modifications of the enzyme during the last purification step.

To ensure that there is nothing present in the CaM-insensitive adenylyl cyclase preparation (F1) that interferes with the immunoprecipitation, the immunoprecipitations are repeated with mixtures of the CaM-sensitive

and the immunoprecipitates (B). F2 represents 0.25 μg of CaM-sensitive adenylyl cyclase; heart, 160 μg detergent-solubilized rat heart membranes; liver, 660 μg of detergent-solubilized rat liver membranes; F2 + heart, 0.25 μg of CaM-sensitive adenylyl cyclase plus 160 μg detergent-solubilized rat heart membranes; F2 + liver, 0.25 μg of CaM-sensitive adenylyl cyclase plus 660 μg of detergent-solubilized rat liver membranes. (Reprinted from Rosenberg and Storm[2] with permission from the American Society for Biochemistry and Molecular Biology.)

and CaM-insensitive adenylyl cyclases (Fig. 4). The antiserum precipitates activity from a preparation of the CaM-sensitive form of the enzyme (F2) but not from a preparation of the CaM-insensitive adenylyl cyclase (F1). The antiserum precipitates the same amount of activity from a mixture of the two preparations (F1 + F2) as from the CaM-sensitive preparation alone. The total amount of adenylyl cyclase activity removed by the antiserum from the CaM-sensitive fraction of adenylyl cyclase is the same as that immunoprecipitated from a mixture of the two preparations. The adenylyl cyclase activity recovered in the immunoprecipitates corresponds well with the activity removed from the supernatants (Fig. 4B). These data indicate that the inability of the antiserum to immunoprecipitate CaM-insensitive adenylyl cyclase is not due to the presence of a factor that interferes with immunoprecipitation and that CaM-sensitive and CaM-insensitive adenylyl cyclase activities of bovine brain are immunologically distinct.

Distribution of Calmodulin-Sensitive Adenylyl Cyclase in Various Rat Tissues. The antiserum described above, which is obtained from the bovine enzyme, also immunoprecipitates adenylyl cyclase activity from detergent-solubilized membranes of several tissues from rat (Fig. 5). At saturating concentrations, the antiserum cross-reacts with significant fractions of rat brain and rat kidney adenylyl cyclase, but apparently does not cross-react with any adenylyl cyclase in either liver or testes. The antiserum also precipitates small but reproducible fractions of the adenylyl cyclase activity from rat heart and lung. The CaM sensitivity of the immunoprecipitates from these tissues can not be verified due to the presence of endogenous CaM in crude solubilized membrane preparations. To determine if there is any substance present in the rat heart or liver detergent-solubilized membranes that interferes with the immunoprecipitation, adenylyl cyclase activity is immunoprecipitated from mixture of purified bovine CaM-sensitive adenylyl cyclase and detergent-solubilized membranes from either heart or liver. The antiserum precipitates activity from a mix of the CaM-sensitive adenylyl cyclase (F2) and liver or heart extracts (Fig. 6). These results indicate that there is nothing in the detergent-solubilized extracts of heart or liver which inhibits immunoprecipitation of CaM-sensitive adenylyl cyclase.

Conclusions

It is possible to isolate polyclonal antiadenylyl cyclase antibodies from mice using highly purified CaM-sensitive adenylyl cyclases. The antibodies were obtained with native antigens, and they did not recognize the denatured catalytic subunit. Attempts at raising antibodies against the denatured catalytic subunit isolated from SDS gels were unsuccessful. The mouse polyclonal antibodies recognized the CaM-sensitive adenylyl cyclase but not the CaM-insensitive adenylyl cyclase. This was the first immunological evidence for two distinct forms of adenylyl cyclase in animal tissues. There was considerable variation in the amount of CaM-sensitive adenylyl cyclase found in various rat tissues, ranging from 60% of total adenylyl cyclase activity in brain to less than 5% in liver. These results are consistent with other data indicating that kidney[7] and heart[8] contain CaM-sensitive adenylyl cyclase activity whereas other tissues such as liver do not.

[7] S. Sulimovici, L. M. Pinkus, F. I. Susser, and M. S. Roginsky, *Arch. Biochem.* **234,** 434 (1984).
[8] M. P. Panchenko and V. A. Tkachuk, *FEBS Lett.* **174,** 50 (1984).

[10] Characteristics and Use of Monoclonal Antibodies to Various Forms of Adenylyl Cyclase

By STEFAN MOLLNER and THOMAS PFEUFFER

Introduction

The catalytic unit of adenylyl cyclase (EC 4.6.1.1, adenylate cyclase) has been purified to homogeneity from several sources thanks to the specificity of the hypotensive drug forskolin which was used as affinity ligand.[1] A Ca^{2+}-sensitive enzyme from brain and a Ca^{2+}-insensitive enzyme from heart have been obtained in homogeneous forms.[2–4] Both enzymes are coupled via G_s[5,6] to hormone receptors and appear to be glycoproteins. In contrast, the enzymes from lower eukaryotic organisms, like protozoa and slime molds, are not stimulated by hormones but are coupled

[1] T. Pfeuffer and H. Metzger, *FEBS Lett.* **146,** 369 (1982).
[2] E. Pfeuffer, R.-M. Dreher, H. Metzger, and T. Pfeuffer, *Proc. Natl. Acad. Sci. U.S.A.* **82,** 3086 (1985).
[3] E. Pfeuffer, S. Mollner, and T. Pfeuffer, *EMBO J.* **4,** 3675 (1985).
[4] M. D. Smigel, *J. Biol. Chem.* **201,** 1976 (1986).
[5] MOPS, 3-(*N*-Morpholino)propanesulfonic acid; CHAPS, 3-[(3-cholamidopropyl)dimethylammonio]propane 1-sulfonate; GPP(NH)P, guanylyl-5′-imidotriphosphate; GTPγS, guanosine 5′-(3-thiotriphosphate); G_s, stimulatory GTP-binding protein of adenylyl cyclase.
[6] D. Feder, M. J. Im, H. W. Klein, M. Hekman, A. Holzhöfer, C. Dees, A. Levitzki, E. J. Helmreich, and T. Pfeuffer, *EMBO J.* **5,** 1509 (1986).

to G proteins at least in some instances.[7] The adenylyl cyclase from mammalian sperm is not stimulated by hormones, guanine nucleotides, or even forskolin,[8] although the latter is supposed to act through the catalytic subunit itself.

An idea of how these different adenylyl cyclases may be related to each other could be obtained by the use of mono- or polyclonal antibodies directed against distinct epitopes of the protein. The isolation of the enzyme in a homogeneous form and in reasonable yield prompted us to prepare monoclonal antibodies against the 115K protein from bovine brain cortex.

Preparation of Pure Bovine Brain Adenylyl Cyclase

Buffers

Buffer A: 10 m*M* MOPS, pH 7.4, 2 m*M* EDTA, 2% (w/v) sucrose, 3 m*M* benzamidine, 10 μ*M* benzethonium chloride, 1 m*M* dithioerythritol, 0.2 m*M* phenylmethylsulfonyl fluoride, 5 μg/ml pepstatin, 10 μg/ml aprotinin, 10 μg/ml soybean trypsin inhibitor

Buffer B: 10 m*M* MOPS, pH 7.4, 1 m*M* $MgCl_2$, 1 m*M* EDTA, 1 m*M* benzamidine, 10 μ*M* benzethonium chloride, 1 μg/ml aprotinin, 0.2 m*M* phenylmethylsulfonylfluoride, 2% sucrose

Buffer C: Buffer B without sucrose but with 100 m*M* NaCl plus 1 m*M* Tween 60

Buffer D: 1 *M* glycine, 2 *M* NaCl, 1 m*M* $MgCl_2$, 3 m*M* benzamidine, 5 m*M* NaN_3, 10 μ*M* benzethonium chloride, 1 m*M* Tween 60, pH 8.2

The preparation of purified adenylyl cyclase is described in detail in [7], this volume. Purified enzyme preparations of two different properties are used: (1) the homogeneous enzyme, exhibiting a specific activity of 3–4 μmol/mg/min, and (2) the partially pure enzyme (0.5–1.1 μmol/mg/min) where the hydroxyapatite and the second affinity step are omitted. Briefly, for the latter preparation, the loaded forskolin-Affi-Gel 10 is washed with 25 column volumes of buffer C and then with 15 volumes of buffer C without aprotinin at a rate of 200 ml/hr. Adenylyl cyclase is released with 2 volumes of the same buffer containing 100 μ*M* forskolin. Both pure and partially pure enzymes are concentrated in a Centriflo CF50

[7] P. J. M. van Haastert, B. E. Snaar-Jagalska, and P. M. W. Janssens, *Eur. J. Biochem.* **162,** 251 (1987).

[8] D. Stengel, L. Guenet, and J. Hanoune, *J. Biol. Chem.* **257,** 10818 (1982).

filter cone (Amicon, Danvers, MA) to 0.1–0.2 and 1–2 mg protein/ml, respectively, prior to immunization.

Immunization Protocol and Hybridoma Production

BALB/c mice are immunized subcutaneously in the presence of Freund's complete adjuvant with 300 μg of partially purified adenylyl cyclase from bovine brain cortex. Booster injections are given after 4 and 8 weeks with the same amount of antigen but in the presence of Freund's incomplete adjuvant. The last immunization is done intraperitoneally with pure adenylyl cyclase, 30 μg antigen being applied 4, 3, and 2 days before fusion.[9] Spleen cells are fused with a nonsecreting NS0-1 myeloma cell line (M. R. Clark, B. Wright, and C. Milstein, unpublished and quoted in Ref. 10). The line was kindly donated by the Milstein laboratory. Fusion is performed according to standard methods.

ELISA Screening of Hybridoma Supernatants. Microtiter plates (Falcon 3912) are coated with 100 ng/well of pure bovine brain cortical adenylyl cyclase in 50 μl phosphate-buffered saline (PBS) for 16 hr at 4°. After washing twice with PBS (0.05% Tween 20), blocking is achieved by treatment with 10% fetal calf serum for 1 hr at 37°. Following incubation with 50 μl hybridoma supernatant for 1 hr at 37° and washing with PBS (0.05% Tween 20), 50 μl of 1 : 500 diluted horseradish peroxidase-coupled anti-mouse IgG [from goat, Sigma (St. Louis, MO) A-5278] is added and left for 1 hr at 37°. Following washing, 50 μl substrate solution [2,2-azinodi(3-ethylbenzthiazoline)sulfonic acid, NH_4^+ salt, 0.55 mg/ml] in 100 m*M* citrate, pH 4.5, containing 0.002% H_2O_2, is added. The absorption is measured at 410 nm with a Dynatech ELISA reader.

Preparation of Ascites Fluid. Ascites fluid is prepared by intraperitoneal injection of 10^7 cells into BALB/c mice which had been pretreated with 0.5 ml pristane.[10] Ascites fluid is centrifuged, diluted with 15 volumes of 20 m*M* Tris-HCl, pH 7.2, and immunoglobulin is isolated by chromatography on DEAE-Affi-Gel Blue (linear gradient of 0–120 m*M* NaCl in 20 m*M* Tris-HCl, pH 7.2). Classification and subclassification of monoclonal antibodies are achieved with a mouse immunoglobulin subtype identification kit (Bio-Rad, Richmond, CA). Monoclonal antibodies BBC-1 to BBC-4 are of the IgG_1 type. Additional characteristics are summarized in Table I.

[9] C. Stähli, T. Staehelin, and V. Miggiano, this series, Vol. 92, p. 26.
[10] G. Galfrè and C. Milstein, this series, Vol. 73, p. 26.

TABLE I
CHARACTERISTICS OF MONOCLONAL ANTIADENYLYL CYCLASE ANTIBODIES

Antibody	Subtype	ELISA[a]	Western blot	Inhibition[g]	Immunoprecipitation[g]	
					Native[d]	Denatured[e]
BBC-1	IgG_1	+	115K	–[b]	+	+
BBC-2	IgG_1	+	115K + 150K	–[b,c]	–	+[f]
BBC-3	IgG_1	+	115K	–[b]	+	+
BBC-4	IgG_1	+	115K + 150K	–[b,c]	(+)	+[f]

[a] Purified 115K enzyme was used as antigen.
[b] Purified 115K enzyme was tested for basal, forskolin-stimulated, G_s [GPP(NH)P]-, and calmodulin/Ca^{2+}-reconstituted activities.
[c] Membrane-bound 150K enzyme (rabbit myocardium) was tested for basal, forskolin-, and GTPγS-stimulated activities.
[d] Based on cyclase activity measurements and Western blot analysis.
[e] Based on Western blot analysis.
[f] Including 115K and 150K antigens.
[g] +, Strong reaction; (+), weak reaction; –, no reaction.

Adenylyl Cyclase Assay

Adenylyl cyclase activity is measured in the presence of 20 m*M* MOPS, pH 7.4, 10 m*M* creatine phosphate, 50 μg/ml creatine kinase, 5 m*M* $MgCl_2$, 1 m*M* cAMP, 0.1 m*M* [^{32}P]ATP (350–3500 Bq/nmol). Incubation is for 20 min at 30°. Protein is estimated according to Lowry *et al.*[11] or by the method of Schultz *et al.*[12] (See also [1], this volume.)

SDS–Polyacrylamide Gel Electrophoresis and Immunoblotting

Proteins, separated on 5–15% polyacrylamide gradient gels according to Laemmli,[13] are transferred to nitrocellulose according to Towbin *et al.*[14] (but the transfer buffer was 25 m*M* Tris base, 192 m*M* glycine, 20% v/v methanol, 0.025% SDS). Blocking of nitrocellulose is performed essentially according to Naaby-Hansen *et al.*[15] by incubation with 2% Tween 20 in PBS for 5 min and washing with 0.05% Tween-20 in PBS. Antigens

[11] O. H. Lowry, N. J. Rosebrough, A. L. Farr, and R. J. Randall, *J. Biol. Chem.* **193,** 265 (1951).
[12] R. M. Schultz, J. D. Bleil, and P. M. Wassarman, *Anal. Biochem.* **91,** 354 (1978).
[13] U. K. Laemmli, *Nature* (*London*) **227,** 680 (1970).
[14] H. Towbin, T. Staehelin, and J. H. Gordon, *Proc. Natl. Acad. Sci. U.S.A.* **76,** 4350 (1979).
[15] S. Naaby-Hansen, A. O. F. Lihme, T. C. Bøg-Hansen, and O. J. Bjerrum, "Lectins," Vol. 4. p. 24. de Gruyter, Berlin and New York, 1985.

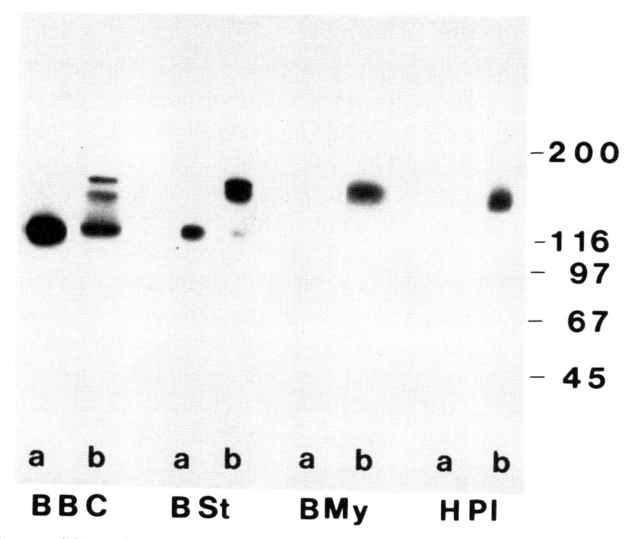

FIG. 1. Recognition of adenylyl cyclases from different species and tissues by monoclonal antibodies BBC-1 and BBC-2. Membranes from bovine brain cortex (240 μg), bovine brain striatum (180 μg), human platelets (130 μg), and solubilized bovine myocardial membranes (410 μg) were subjected to electrophoresis on 5–15% SDS polyacrylamide gels. Separated proteins were blotted onto nitrocellulose and probed with radioiodinated antibodies BBC-1 (lanes a) and BBC-2 (lanes b). To avoid interference by high concentrations of myosin present in myocardial membranes, myocardial membranes were treated with 10 m*M* Lubrol PX and the solubilized equivalent applied. BBC, Bovine brain cortex; BSt, bovine striatum; BMy, bovine myocardium; HPl, human platelets. Numbers refer to M_r values ($\times 10^{-3}$) of radioiodinated standard proteins.

are detected with ^{125}I-labeled IgGs (purified as above, specific activity 1.85–7.4 MBq/mg, 10^3–10^4 Bq per lane). Following incubation for 2 hr at 22° and washing with 0.05% Tween-20 in PBS, blots are exposed to Kodak XAR 5 film with an intensifying screen (Siemens Titan 2 HS). Myosin (200K), β-galactosidase (116K), phosphorylase (97K), bovine serum albumin (67K), and ovalbumin (45K) serve as standard proteins. Standards, adenylyl cyclase, and IgGs are iodinated according to Greenwood *et al.*[16]

The results of Western blots of adenylyl cyclase from different species and tissues are shown in Fig. 1. While monoclonal antibody BBC-1 (and BBC-3) shows strict tissue specificity (brain) and species specificity (bovine, not shown) antibodies BBC-2 and BBC-4 also cross-react with the enzymes from peripheral sources (myocardial membranes, platelets, etc.), thus matching a more conserved region of the catalytic subunit.

[16] F. C. Greenwood, W. M. Hunter, and J. S. Glover, *Biochem. J.* **89,** 114 (1963).

Immunoprecipitation of Adenylyl Cyclase by Monoclonal Antibodies

Immunoprecipitation of Native Antigens. Lubrol PX (Sigma, St. Louis, MO) solubilized adenylyl cyclase (see this volume [7]) from bovine brain cortex or striatum (80 μg each) or purified bovine adenylyl cyclase (20 ng) in 20 μl is incubated with 20 μl (0.2–0.5 μg) of IgG prepared from BBC-1 and BBC-3 ascites, respectively, for 2 hr at 4°. After this period 1.2 μg anti-mouse IgG (Sigma M-8890) is added and further incubated for 90 min at 4°, followed by addition of 60 μl of a 10% suspension of *Staphylococcus aureus* cells (Cowan I strain, Pansorbin from Calbiochem, San Diego, CA) serving as the source of protein A. Following agitation in a tumbling mixer for 30 min at 4° and centrifugation at 4000 *g*, the supernatant is saved and the pellet is washed 4 times with 1 ml buffer B plus 6 m*M* CHAPS and 1000 m*M* NaCl and twice with 1 ml buffer C. The pellet is then resuspended in 130 μl buffer C, and 10-μl aliquots are withdrawn for measurement of adenylyl cyclase activity. For audioradiography the procedure is conducted with ^{125}I-labeled adenylyl cyclase. Immunoabsorbed proteins are released by addition of 130 μl of 50 m*M* diethylamine buffer, pH 11.5, containing 0.75% deoxycholate. After 30 min of shaking the mixture is centrifuged at 9000 *g*, and the supernatant is neutralized by addition of 1/10 volume of 1 *M* NaH_2PO_4.

Direct immunoprecipitation of adenylyl cyclase in the native state, by omitting the anti-mouse IgG, could be achieved in the presence of a high salt/high pH buffer, which is necessary to bind IgG_1 (to which BBC-1 to BBC-4 belong) to protein A. Pansorbin, 125 μl in buffer D, is incubated with 20 μl ascites fluid or purified IgG (60 μg) for 90 min at 22° under gentle agitation. Following washing of the Pansorbin 3 times with 1.5 ml of buffer D, 60 μl of solubilized adenylyl cyclase (4 mg/ml in buffer D) is added, and the mixture is shaken for 60 min at 4°. After removal of unbound antigen by centrifugation and washing with 1.5 ml of buffer D (3 times), the cells are resuspended in 130 μl of buffer C, and 10-μl aliquots are removed for estimation of adenylyl cyclase activity.

With both methods about 70% of the Ca^{2+}/CaM-stimulated cyclase activity could be absorbed from solubilized bovine brain cortical membranes, while more than 50% of the activity could be recovered in the washed pellet.

Immunoprecipitation of Denatured Antigens. Immunoprecipitation of denatured antigens has been performed with all four antibodies BBC-1 to BBC-4 in the presence of a detergent mix (0.5% SDS, 1% deoxycholate, 1% (w/v) Nonidet P-40).[17,18] As described before, Pansorbin is incubated

[17] B. M. Sefton, K. Beemon, and T. Hunter, *J. Virol.* **28,** 947 (1978).

[18] H. Böttinger and R. Simmoteit, unpublished.

directly with the ascites fluid, but bound IgG is chemically cross-linked to protein A by adopting the method of Schneider *et al.*[19] The denaturing conditions allow for precipitation of both the 115K and 150K antigens (see Table I). By the use of pure radioiodinated $C\alpha_s$ complex from rabbit myocardial membranes as tracer,[2] more than 70% of the radioactivity within the 150K band could be specifically precipitated from crude SDS-solubilized membranes. This method has been successfully applied in this laboratory to prepare samples for Western blot analysis of tissues containing either low amounts of antigen or weakly reacting antigen. For some tissues the method is used for removal of other proteins prior to Western blot analysis, proteins which would otherwise obscure the specific antigen due to their higher abundance (e.g., C6 glioma cells, NG 108/15 neuroblastoma cells, turkey erythrocytes, bovine lung). This has previously been achieved by only partial purification of solubilized adenylyl cyclase by chromatography on forskolin-agarose.[20]

Separation of Ca^{2+}-Sensitive and Ca^{2+}-Insensitive Adenylyl Cyclase by Calmodulin-Sepharose

Crude solubilized bovine striatal adenylyl cyclase (or another Ca^{2+}/CaM-sensitive form) is depleted of endogenous calmodulin according to Westcott *et al.*[21] Two milliliters of depleted cyclase preparation in buffer B (plus 200 m*M* KCl and 2 m*M* $CaCl_2$) is treated with 500 μl calmodulin-Sepharose (Pharmacia, Piscataway, NJ) in the same buffer for 1 hr in a tumbling mixer. Following removal of unbound proteins, the resin is washed with 2 ml buffer B containing 200 m*M* KCl and 1 m*M* $CaCl_2$. The resin is shaken for 10 min and washed with another 6 ml of this buffer. Calmodulin-Sepharose-bound proteins are released by shaking the resin with 2 ml buffer B containing 200 m*M* KCl plus 2 m*M* EGTA. All steps are performed at 4°.

Discussion

We have raised several monoclonal antibodies against the pure 115K bovine brain adenylyl cyclase, four of which were characterized in more detail. Antibodies BBC-1 and BBC-3 were strictly specific for the Ca^{2+}-sensitive 115K enzyme from bovine brain cortex, while BBC-2 and BBC-4

[19] C. Schneider, R. A. Newman, D. A. Sutherland, U. Asser, and M. F. Greaves, *J. Biol. Chem.* **257,** 10766 (1982).

[20] S. Mollner and T. Pfeuffer, *Eur. J. Biochem.* **171,** 265 (1988).

[21] K. R. Westcott, D. C. La Porte, and D. R. Storm, *Proc. Natl. Acad. Sci. U.S.A.* **76,** 204 (1979).

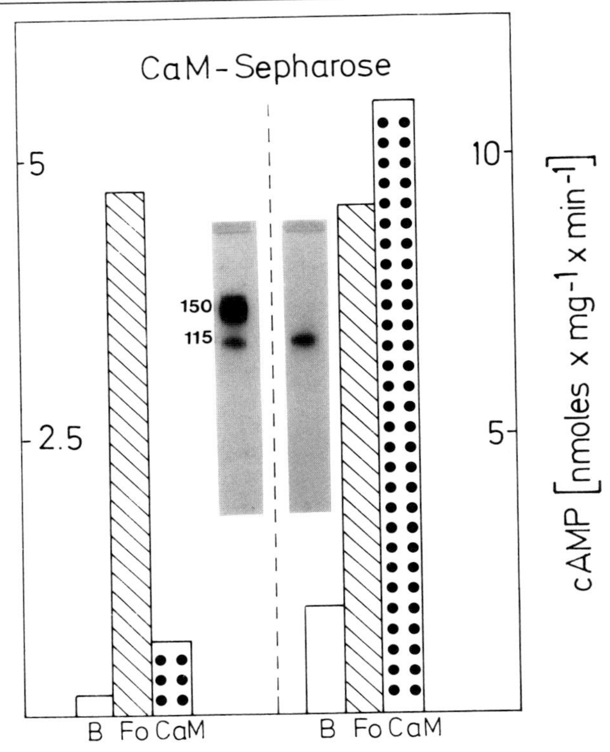

FIG. 2. Affinity chromatography on calmodulin-Sepharose of bovine striatal adenylyl cyclase. Crude extract from bovine striatal membranes depleted of calmodulin (left, 105 μg/lane) and calmodulin-Sepharose eluate (right, 9 μg/lane) were electrophoresed on 5–15% polyacrylamide gels. Immunoblots (insert) were performed with iodinated monoclonal antibody BBC-2. Bars indicate specific adenylyl cyclase activities in the absence (B, basal) and presence of forskolin (Fo) (100 μ*M*) or calmodulin (CaM) (30 μg/ml plus 50 μ*M* $CaCl_2$). (Reprinted from Ref. 20 with the permission of Springer-Verlag.)

cross-reacted additionally with the Ca^{2+}-insensitive enzymes from several tissues and species and therefore matched a more conserved region of the catalytic subunit. While BBC-1 and BBC-3 also reacted with the native antigen, BBC-2 and BBC-4 did so weakly, if at all. They also suggested lower affinity in the ELISA[20] but showed clear preference for the denatured antigen. This was demonstrated by their ability to precipitate the latter (via immobilized protein A) only in the presence of SDS and deoxycholate.[18] From Fig. 1 it is also evident that brain tissues (cortex and striatum) contain at least two antigens, recognized by antibodies BBC-2 and BBC-4. Besides the 115K and 150K proteins a sharp band at 160K was observed, the origin of which is unclear. In contrast to the other two proteins, the 160K species was less readily retained by either forskolin-agarose or immobilized wheat germ agglutinin.

It has been suggested that brain tissues contain a mixture of calmodulin-sensitive and calmodulin-insensitive cyclases.[21] The experiment in Fig. 2 unequivocally shows this to be the case. It is evident that only the smaller 115K form of the enzyme from bovine striatum was retained by the calmodulin-Sepharose column and therefore represents the Ca^{2+}-sensitive form of the enzyme. This is corroborated by the functional analysis of crude and retained cyclase: following purification on calmodulin-Sepharose, there was a marked increase of Ca^{2+}-stimulated cAMP production as compared with stimulation by forskolin. In bovine cortical membranes the intensities of the 115K and the 150K bands were reciprocal. This is in good correlation with the much higher stimulation by Ca^{2+}/calmodulin of the cortical adenylyl cyclase.

Recently two papers appeared describing the production of polyclonal antibodies against brain adenylyl cyclase purified by the use of the forskolin-Sepharose method applied in this laboratory. Rosenberg *et al.*[22] raised antibodies against the calmodulin-sensitive bovine brain enzyme in mice which only recognized this form and not the calmodulin-insensitive (more ubiquitous) form of the enzyme. Monneron *et al.*[23] obtained polyclonal antibodies against the calmodulin-sensitive enzymes from *Bordetella pertussis* and rat brain. Interestingly, both antibodies were able to react with both homologous as well as heterologous antigens, pointing to close similarity of at least part of these bacterial and mammalian enzymes.

Acknowledgments

We thank Mrs. B. Weber, Mrs. C. Harmening, Mrs. E. Pfeuffer, and Mr. B. Gerull for expert technical assistance. We also thank U. Heinz for doing some of the ELISA with purified adenylyl cyclase. We further appreciate that Dr. H. Böttinger and Dipl. Biol. R. Simmoteit provided us with data prior to publication. This work was supported by grants from the Deutsche Forschungsgemeinschaft (Sonderforschungsbereich 176) and by the Fonds der Chemischen Industrie.

[22] G. B. Rosenberg and D. R. Storm, *J. Biol. Chem.* **262,** 7623 (1987).

[23] A. Monneron, Ladant, J. d'Alayer, J. Bellalou, O. Bârzu, and A. Ullmann, *Biochemistry* **27,** 536 (1988).

[11] Purification of Soluble Form of Adenylyl Cyclase from Testes

By THEODOR BRAUN

Introduction[1]

It is now well established that the testis germ cells contain a soluble adenylyl cyclase[2-5] (EC 4.6.1.1, adenylate cyclase) designated by some as class III of adenylyl cyclase.[6,7] The functional and physical properties of the soluble germ cell adenylyl cyclase are fundamentally different from membrane-bound adenylyl cyclases derived from somatic cells. It is active only in the presence of manganese and uses only MnATP as substrate. In contrast to adenylyl cyclases of somatic cells, the germ cell enzyme is not stimulated by fluoride, forskolin, GTP and nonhydrolyzable GTP analogs, and gonadotropic hormones. Hence, it appears that the soluble form of adenylyl cyclase from testis germ cells is devoid of regulatory components which confer fluoride, forskolin, and hormonal responsiveness on the catalytic entity of the somatic cell membrane-bound adenylyl cyclase system. All findings thus far available indicate that the soluble testis adenylyl cyclase is unique and different from somatic cell adenylyl cyclase systems. The procedure for the purification of the soluble enzyme as described below differs markedly from the procedures used to isolate and purify adenylyl cyclases derived from somatic cells. In this chapter, the method of purification of the soluble form of adenylyl cyclase is presented.

Assay Method

Principle. The assay is based on the formation of ^{32}P-labeled cyclic AMP (cAMP) by the adenylyl cyclase from [α-^{32}P]ATP and subsequent separation and quantitation of the cAMP product. The reaction requires the presence of at least stoichiometric amounts of manganese, because the utilizable substrate complex *in vitro* is MnATP.

Procedure. The assay system contains, in a final volume of 50 μl, 40

[1] This work was supported by National Institutes of Health Grant HD-10006.

[2] T. Braun and R. F. Dods, *Proc. Natl. Acad. Sci. U.S.A.* **72,** 1097 (1975).

[3] E. J. Neer, *J. Biol. Chem.* **253,** 5808 (1978).

[4] A. R. Kornblihtt, M. M. Flavia, and H. N. Torres, *Biochemistry* **29,** 1262 (1981).

[5] J. O. Gordeladze and V. Hanson, *Mol. Cell. Endocrinol.* **23,** 125 (1981).

[6] J. E. Neer, *Adv. Cyclic Nucleotide Res.* **9,** 69 (1978).

[7] B. M. Sanborn, J. J. Heindel, and G. A. Robinson, *Annu. Rev. Physiol.* **42,** 37 (1980).

mM Tris-HCl buffer, pH 7.6, 5 mM $MgCl_2$, 0.2 mM [α-^{32}P]ATP, 1 mg/ml bovine serum albumin (BSA), 3% (v/v) glycerol, 0.12 μg gentamicin, 0.4 mM dithiothreitol (DTT), 0.2 mM EDTA, 5 μM bacitracin, 0.3 μM pepstatin A, 10 μM 2-phenyl ketone and L-lysine ketone, and, depending on the purity of the enzyme preparation, 0.13–10 μg of protein. Up to step 4, the assay system contains 1 mM creatine phosphate, 200 μg/ml creatine kinase, and 0.5 mM unlabeled cyclic AMP. After Step 4, the enzyme is assayed without the ATP-regenerating system and excess of cAMP, for the enzyme preparations at this purification stage do not exhibit either ATPase or phosphodiesterase activity. After incubation for 5–10 min at 37°, the reaction is terminated by the addition of 5 μl of a solution containing 50 mM of ATP, ADP, and cAMP and 200 mM EDTA. The incubation vial contents are concentrated by evaporation to dryness and resuspension in 10 μl of water. The [^{32}P]cAMP formed is separated from other radioactive substances by chromatography on fluorescent silica gel thin-layer sheets (Kodak, Rochester, NY, #6060) developed in an organic solvent system [chloroform/methanol/water, 40 : 20 : 3 (v/v/v)].[8] Before being used, the silica gel thin-layer sheets are dried in an oven at about 100° for at least 24 hr. (See [1], this volume, for alternative methods for the assay of adenylyl cyclase activities.)

Definition of Unit and Specific Activity. One unit is defined as the amount of enzyme in 1 ml which catalyzes the conversion of 1 μmol of ATP to cAMP in 1 min under the assay conditions described. Specific activity is expressed as units per milligram of protein. Protein concentrations in fractions from Steps 1 to 5 are determined by the Coomassie dye-binding assay of Bradford[9] and in the fractions from Step 6 by the bicinchoninic acid protein assay[10] modified into a micromethod (Pierce, Rockford, IL, Booklet No. 23235), using BSA standards.

Isolation and Preparation of Soluble Enzyme Fraction

Testes from 510 Holtzman Sprague-Dawley rats (retired breeders, 350–700 g body weight) are collected in ice-cold (ice bath) 0.9% saline solution. The saline solution is decanted, and the testes are rinsed twice with ice-cold 5 mM Tris-HCl buffer, pH 7.2, and processed fresh, avoiding freezing and thawing the tissue. The tunica albuginea is stripped off, visible blood vessels are dissected free, and a 20% homogenate is then prepared

[8] G. Flouret and O. Hechter, *Anal. Biochem.* **58,** 276 (1974).

[9] M. M. Bradford, *Anal. Biochem.* **72,** 248 (1976).

[10] P. K. Smith, R. I. Krohn, G. T. Hermanson, A. K. Mallia, F. H. Gartner, M. D. Provenzano, E. K. Fujimoto, N. M. Goeke, B. J. Olson, and D. C. Klenk, *Anal. Biochem.* **150,** 76 (1985).

from 1950 g of tissue. Aliquots of 20 g of tissue are suspended in 4 volumes of ice-cold 5 mM Tris-HCl buffer, pH 7.2, containing 1 mM EDTA, 3 mM $MgCl_2$, 2 mM DTT, 0.25 mM phenylmethylsulfonyl fluoride (PMSF), 25 μM bacitracin, 50 μM 2-phenyl ketone, 50 μM L-lysine ketone, 1.5 μM pepstatin A, and 12.5 μg/ml gentamicin and disrupted with a Waring blendor for 5 sec. Hypotonic (5 mM Tris-HCl) buffer is used for homogenization since apparently by lysing the cells it enables almost quantitative release of the soluble form of adenylyl cyclase without the necessity of using excessive or repetitive disruptive forces.[3] The homogenate, shortly after preparation (usually within 2 hr), is centrifuged at 20,000 g for 20 min. The supenatant is decanted and stored at 4°, and the pellet is discarded. The supernatant thus obtained is pooled and spun again at 28,000 g for 3.5 hr. The supernatant is decanted and filtered through a double layer of cotton gauze to remove lipids. The supernatant contains the soluble adenylyl cyclase and is used as the starting material for purification.

Purification

All procedures are carried out at 0° to 5°. Unless otherwise indicated all buffers contain 1 mM EDTA, 3 mM $MgCl_2$, 2 mM DTT, 0.25 mM PMSF, 12.5 μl/ml gentamicin, 1.5 μM pepstatin A, 25 μM bacitracin, 50 μM 2-phenyl ketone and L-lysine ketone, and 15% (v/v) glycerol. Concentration of combined enzymatic fractions is carried out in Amicon (Danvers, MA) ultrafiltration chambers with magnetic stirring, with nonionic polymer disks (PM30, MW cutoff 30,000) under nitrogen pressure. Dialysis with stirring is performed with standard cellulose tubing (MW cutoff 12,000–14,000).

Step 1: Ion-Exchange Chromatography on DEAE-Cellulose. DEAE-cellulose is packed by gravity in two columns (50 × 4.2 cm) and is equilibrated with 50 mM Tris-HCl buffer, pH 8.0, containing 7.5% (v/v) glycerol. The supernatant is diluted with an equal volume of 50 mM Tris-HCl buffer, pH 8.0, containing 15% (v/v) glycerol. The pH of the diluted supernatant is adjusted to 8.0; 7.8 liters of the supernatant, containing 33,345 mg of protein, is applied to each column at a flow rate of 125 ml/hr. The columns are washed with 1400 ml of 70 mM Tris-HCl buffer, pH 8.0, and then eluted with 2100 ml (about 3 column bed volumes) of 80 mM Tris-HCl buffer, pH 7.0. Fractions of about 200 ml are collected at a flow rate of 100 ml/hr. Fractions from both columns having enzyme with increased specific activity are pooled and concentrated.

Step 2: Gel Filtration Chromatography on Sephacryl S-200. The Tris-HCl concentration of the enzyme preparation from Step 1 is adjusted to 100 mM and the pH to 7.6. Aliquots of about 80 ml containing 2200 mg of protein are applied to a Sephacryl S-200 column (5 × 100 cm; 2000 ml column bed volume) previously equilibrated with 100 m*M* Tris-HCl buffer, pH 7.6. Elution is carried out with the same buffer at a flow rate of 100 ml/hr. The void effluent (about 650 ml) is collected into a beaker; then, the column is attached to a fraction collector, and 25-ml fractions are collected. The enzyme activity appears after the elution of a reddish contaminant at the 0.50 elution volume fraction and has its peak at the 0.53–0.55 elution volume fractions. Fractions with high enzyme activity, usually contained in 440–560 ml of elution fluid from four chromatographic runs, are pooled and concentrated to 100 ml with Amicon PM30 membranes, as above.

Step 3: Ion-Exchange Chromatography on DEAE-BioGel A. The concentrated enzyme preparation from the Sephacryl S-200 column is dialyzed against a 5-fold excess of 5 m*M* Tris-HCl buffer, pH 7.6, for 48 hr with 5 changes of the dialysis fluid. The dialyzed preparation (100 ml, containing 1010 mg protein) is then applied at a flow rate of 20 ml/hr to a DEAE-BioGel A column (2.6 × 20 cm; 42-ml column bed volume) equilibrated previously with 5 m*M* Tris-HCl buffer, pH 7.6. Elution is carried out successively with 40 ml of 20 m*M*, and 120 ml of 40 m*M* Tris-HCl buffer, pH 7.6, then with 120 ml each of 40, 60, and 100 m*M* Tris-HCl buffer, pH 7.0. The enzyme is eluted from the column as a broad, single peak with 40–60 m*M* Tris-HCl buffer, pH 7.0, in 175 ml of the elution fluid (Fig. 1).[11] The eluate is concentrated to about 20 ml and is dialyzed overnight against a 25-fold excess of 100 m*M* Tris-HCl buffer, pH 7.6.

Step 4: Second Gel-Filtration Chromatography on Sephacryl S-200. The dialyzed, concentrated enzyme from Step 3 is applied to the Sephacryl S-200 column (5 × 100 cm) previously equilibrated with 100 m*M* Tris-HCl buffer, pH 7.6. Elution is performed with the same buffer at a flow rate of 15 ml/hr, collecting 15-ml fractions. The enzyme is eluted as a distinct symmetrical peak at 0.50–0.56 elution volume fraction in 130 ml of the elution fluid. The peak fractions are pooled and concentrated.

Step 5: Affinity Chromatography on ATP Immobilized to Sepharose.

[11] As an alternative purification procedure, the DEAE-BioGel A can be replaced with preparative HPLC anion-exchange chromatography employing a Waters protein PAK DEAE PW (2.15 × 21 cm) column. The enzyme preparation containing 560 mg of protein in 500 ml of 60 m*M* Tris-HCl buffer, pH 8.0, is applied to the column, previously equilibrated with the same buffer, through a manifold valve attachment at a flow rate of 5 ml/hr. Elution of the enzyme is accomplished with 10 column volumes of a linear gradient formed from 60 m*M*, pH 8.0, to 120 m*M*, pH 7.0, Tris-HCl buffer, at a flow rate of 4 ml/hr, collecting 15-ml fractions. The purification effectiveness of the preparative HPLC anion-exchange column is similar to that of the DEAE-BioGel A open column. HPLC chromatography is less tedious, however, and less time consuming, for it can be repeated without great delay after regenerating the column.

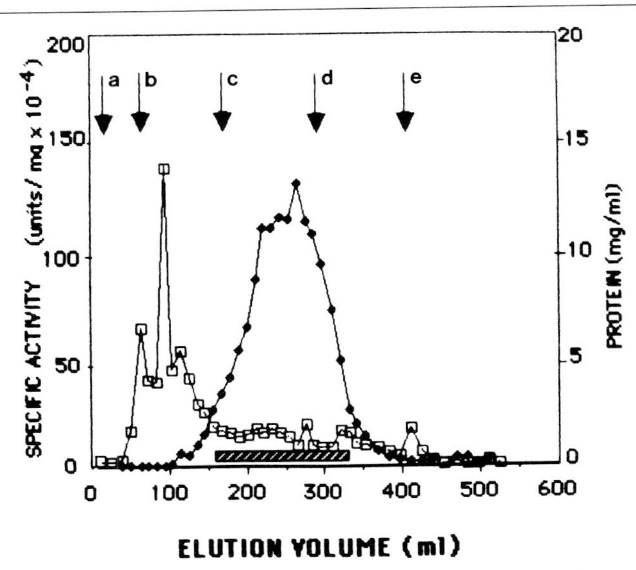

FIG. 1. Elution profile of the soluble adenylyl cyclase on DEAE-BioGel A ion-exchange column chromatography. (a–e, arrows) indicate the order of elution with 20 and 40 m*M* Tris-HCl buffer, pH 8.0, and with 40, 60, and 100 m*M* of Tris-HCl buffer, pH 7.0, respectively. The horizontal bar indicates the fractions that were pooled. ■, Activity; □, protein.

The enzyme solution from Step 4 is dialyzed against a 10-fold excess of 5 m*M* Tris-HCl buffer, pH 7.4, containing all the components as listed above, except $MgCl_2$. $MgCl_2$ is omitted from the buffer since it has been established that divalent cations (Mg^{2+}, Mn^{2+}) prevent binding of the enzyme to the ATP ligand. The preparation is dialyzed for 48 hr with 5 changes of the dialysis fluid. The dialyzed enzymatic preparation (27 ml, containing 60 mg of protein) is applied to a column (1.5 × 6.8 cm, 12-ml column bed volume) packed with ATP-Sepharose beads type 4 (Pharmacia, Piscataway, NJ), containing 4.5 μmol of ATP/ml of beads. The column is washed with 10 ml of 5 m*M* Tris-HCl buffer, pH 7.4, and the enzyme is eluted with 35 ml of 5 m*M* Tris-HCl buffer, pH 8.2, containing 2 m*M* ATP, collecting 5-ml fractions at a flow rate of 15 ml/hr. Peak activities are pooled, concentrated, and dialyzed for 48 hr to remove free ATP against a 25-fold excess of 5 m*M* Tris-HCl buffer, pH 7.4, with 5 changes of the dialysis fluid.

The dialyzed enzyme solution is applied again to the ATP-Sepharose column previously washed and equilibrated with the same buffer. The enzyme is eluted sequentially with 10 ml each of 5 mM Tris-HCl buffer, pH 7.4, pH 7.6, and pH 7.8, followed by elution with 25 ml of the same

TABLE I
PURIFICATION OF SOLUBLE FORM OF ADENYLYL CYCLASE

Step	Volume (ml)	Activity (units/ml × 10^{-4})	Protein (mg/ml)	Specific activity (units/mg × 10^4)	Yield (%)	Purification (-fold)
itial extract	15,600	1.7	4.27	0.04	100	—
DEAE-cellulose	315	76.7	27.8	2.76	91	6.9
First Sephacryl S-200	100	186	10.1	18.41	70	46
DEAE-BioGel A	20	936	11.4	82.1	70	205
Second Sephacryl S-200	27	590	2.23	246.6	60	661
ATP-Sepharose						
(a) ATP elution	10.5	1172	1.25	938	46	2345
(b) pH gradient elution	2.0	5080	0.75	6773	38	16,933
(c) Rechromatography	1.3	4720	0.33	14,303	23	35,757
HPLC anion exchange	2.0	2264	0.026	87,077	17	217,692

buffer, pH 8.2, collecting 5-ml fractions at a flow rate of 15 ml/hr. Peak activity fractions are pooled, the pH adjusted to 7.4, and the solution concentrated to about 10 ml. The sample is rechromatographed with the same column[12] previously washed and equilibrated with 5m*M* Tris-HCl buffer, pH 7.4. Elution is carried out with a discontinuous pH gradient as described above. Peak activity fractions are pooled, and the concenration of the Tris-HCl buffer is adjusted to 60 m*M* and pH to 8.0.

Step 6: HPLC Anion-Exchange Chromatography. The enzyme preparation from Step 5 (1.3 ml, containing 0.45 mg of protein) is injected through a 2-ml loop onto a Waters (Milford, MA) DEAE 5 PW (0.75 × 7.5 cm) column previously equilibrated with 60 m*M* Tris-HCl buffer, pH 8.0. The column is washed with 15 ml of the same buffer at a flow rate of 1 ml/min and then eluted at a flow rate of 1 ml/min for 45 min, with a linear gradient formed from Tris-HCl buffer, 60 m*M*, pH 8.0, and 120 m*M*, pH 7.0, by an automated gradient controller (Waters Model 680). The enzyme is eluted as a discrete peak between 95 and 105 m*M* Tris-HCl buffer concentration in 2 ml of the buffer. A summary of the purification procedure is given in Table I.

[12] The ATP-Sepharose beads can be reused several times (usually for 6–8 runs). When the amount of activity adsorbed to the packed beads falls below 90% of the activity present in the applied sample, it indicates that progressive release of the ATP ligand from the resin has begun. When this release occurs, use of the beads should be discontinued. Using ATP-Sepharose beads to purify crude preparations proved to be impractical, for the ATP ligand is hydrolyzed and washed out rapidly.

Properties

Yield and Purity. The overall yield of activity is 17%; however, the quantity of enzymatic protein recovered is extremely small and amounts to about $0.7 \times 10^{-4}\%$ of the total protein present in the initial extract. The final preparation appears about 90% homogeneous on silver-stained polyacrylamide gels following electrophoresis in the presence of sodium dodecyl sulfate (SDS) and reducing reagents. Moreover, a single protein band appears on Western blot analysis with polyclonal antibodies raised against a partially purified enzyme preparation.[13] The apparent molecular weight of the enzyme is estimated at 52,000 based on data from SDS-gel electrophoresis and Western blot analysis. This is in close agreement with the apparent molecular weight of approximately 50,000 estimated from gel filtration of the native enzyme.

Turnover Number. The turnover number, as calculated from the rate of reaction at a saturating ATP concentration (4 m*M*) in the presence of 24 m*M* Mn^{2+} and 300 μg BSA, is 2040.

Substrate Specificity. The enzyme is highly specific for ATP. Neither dATP nor GTP is utilized as substrate. The K_m for ATP is 0.95 m*M*.

pH Optimum. The optimum pH in Tris-HCl buffer at 37° is broad and lies between pH 7.0 and 9.0. The reaction rate falls off progressively below pH 7.0 and above pH 9.0.

Activators and Inhibitors. The rate of enzyme activity increases with increasing Mn^{2+} concentration. The K_m for Mn^{2+} is 2.6 m*M*, and maximal activation occurs at 18–20 mm. The enzyme is stimulated by BSA in a dose-dependent fashion in the range of 0.1 to 5 mg/ml BSA. The enzyme is inhibited by adenosine,[14] by catechol, and by some catechol derivatives.[15]

Stability. The purified enzyme is stable for about 4 weeks at 4° in the presence of protease enzyme inhibitors, gentamicin, and 15% (v/v) glycerol. Freezing–thawing leads to irreversible loss of activity.

[13] Prepared by Dr. Erwin Goldberg, Department of Molecular and Cell Biology and Biochemistry, Northwestern University.

[14] J. M. Onoda, T. Braun, and S. M. Wrenn, Jr., *Biochem. Pharmacol.* **36,** 1907 (1987).

[15] T. Braun, *Proc. Soc. Exp. Biol. Med.* **194,** 58 (1990).

[12] Purification and Assay of Cell-Invasive Form of Calmodulin-Sensitive Adenylyl Cyclase from *Bordetella pertussis*

By H. ROBERT MASURE, MAURA G. DONOVAN, and DANIEL R. STORM

Introduction

Bordetella pertussis is a gram-negative aerobic bacterium that is the causative agent for the childhood disease whooping cough. Several virulence factors produced by the bacterium have been identified and characterized.[1,2] One of these virulence factors is an extracellular adenylyl cyclase (EC 4.6.1.1, adenylate cyclase) that is stimulated by the Ca^{2+}-dependent regulator calmodulin (CaM).[3] Above 1 nM Ca^{2+} this stimulation is Ca^{2+} independent.[4,5] Furthermore, preparations of the adenylyl cyclase, when incubated with eukaryotic cells, have been shown to elevate intracellular cAMP levels.[6–8]

The enzyme is synthesized as a large 215-kDa precursor. This precursor is transported to the outer membrane of the bacterium, and a 45-kDa catalytic subunit is processed and released into the culture supernatant.[9] Preparations of the adenylyl cyclase, isolated from culture supernatants, are able to elevate cAMP levels in several cell types.[8] The catalytic subunit of the invasive form of the enzyme has a molecular weight of 45,000. In addition, another separable assisting subunit required for cell invasion is present in the culture supernatant.[10]

This chapter describes procedures for growth of the bacteria and the

[1] A. A. Weiss and E. L. Hewlett, *Annu. Rev. Microbiol.* **40,** 661 (1977).

[2] A. C. Wardlaw and R. Parton, "Pathogenesis and Immunity in *Pertussis*." Wiley, Chichester, New York, 1988.

[3] BSA, Bovine serum albumin; CaM, calmodulin; CHAPS, 3-[(3-cholamidopropyl)dimethylaminio]-1-propane sulfonate; DMEM, Dulbecco's modified Eagle's medium (high glucose); DTT, dithiothreitol; EDTA, ethylenediaminetetraacetic acid; EGTA, [ethyleneglycol bis(oxyethylenenitrilo)]tetraacetic acid; HEPES, *N*-2-hydroxyethylpiperazine-*N*′-2-ethanesulfonic acid; PBS, phosphate-buffered saline; Tween 20, polyoxyethylene sorbitan monolaurate; SDS, sodium dodecyl sulfate.

[4] D. V. Greenlee, T. J. Andreasen, and D. R. Storm, *Biochemistry* **21,** 2759 (1982).

[5] M. C. Kilhoffer, G. H. Cook, and J. Wolff, *Eur. J. Biochem.* **133,** 11 (1983).

[6] D. L. Confer and J. W. Eaton, *Science* **217,** 948 (1982).

[7] E. Hanski and Z. Farfel, *J. Biol. Chem.* **260,** 5526 (1985).

[8] R. L. Shattuck and D. R. Storm, *Biochemistry* **24,** 6323 (1985).

[9] H. R. Masure and D. R. Storm, *Biochemistry* **28,** 438 (1989).

[10] M. G. Donovan, H. R. Masure, and D. R. Storm, unpublished observations (1988).

isolation of an invasive preparation of the adenylyl cyclase from culture supernatants. We also describe a procedure for the assay of adenylyl cyclase activity and a biological assay for the invasive properties of the enzyme. Finally, we outline procedures for the detection of different forms of the adenylyl cyclase by SDS–polyacrylamide gel electrophoresis and the purification to apparent homogeneity of the 45-kDa catalytic subunit by this technique.

Adenylyl Cyclase Assay

Adenylyl cyclase is assayed by the general method of Salomon *et al.*[11] Modifications have been made to this method to accommodate the high intrinsic activity of the *B. pertussis* adenylyl cyclase as compared to the CaM-sensitive adenylyl cyclase from bovine brain[12,13] (see also [1], this volume).

Reagents

[α-^{32}P]ATP as a substrate
[^{3}H]cAMP to monitor product recovery
Bovine serum albumin (BSA)
Calmodulin, prepared from bovine brain by the method of Masure *et al.*[14] ATP, cAMP
Dowex AG 50W-X4, 200–400 mesh (Bio-Rad, Richmond, CA)
Alumina-Neutral WN-3 (Sigma, St. Louis, MO)

Procedure. Assays contain, in a final volume of 250 μl, 20 m*M* Tris-HCl, pH 7.5, 1 m*M* [α-^{32}P]ATP (10 cpm/pmol), 2 m*M* cAMP (400 cpm/μl [^{3}H]cAMP), 5 m*M* $MgCl_2$, 1 m*M* EDTA, 1 m*M* 2-mercaptoethanol, and 0.1% (w/v) BSA. The incubation temperature is 30°, and the length of incubation is usually 10 min. When assaying CaM-stimulated adenylyl cyclase activity, 2.4 μM CaM is included in the assay mixture. The adenylyl cyclase assay for the enzyme from *B. pertussis* does not contain phosphodiesterase inhibitors since there is no phosphodiesterase activity in the enzyme preparations.

[11] Y. Salomon, C. Londos, and M. Rodbell, *Anal. Biochem.* **58,** 541 (1974).
[12] R. L. Shattuck, D. J. Oldenburg, and D. R. Storm, *Biochemistry* **24,** 6356 (1985).
[13] P. Wang and D. R. Storm, this volume [6].
[14] H. R. Masure, J. F. Head, and H. M. Tice, *Biochem. J.* **218,** 691 (1984).

Growth of Bacteria

Bordetella pertussis is a slow growing aerobic bacterium that is successfully cultivated in a complex liquid medium as described by Cohn and Wheeler,[15] in a defined medium as described by Stainer and Scholte,[16] or on a solid medium such as Bordet–Gengou blood agar.[17] An invasive form of the enzyme is isolated from the culture supernatants of bacteria grown in the defined medium.[8]

Preparation of Stainer–Scholte Medium

1. All the ingredients for a 10 times concentrated stock solution of Stainer–Scholte medium are dissolved in distilled water (0.13 *M* Tris-HCl, pH 7.6, 0.56 *M* sodium glutamate, 20 m*M* L-proline, 0.43 *M* NaCl, 40 m*M* KH_2PO_4, 50 m*M* $MgCl_2$, 1.36 m*M* $CaCl_2$) except the Tris base.
2. The Tris base is dissolved in a small volume of distilled water, adjusted to pH 7.9, and added to the stock solution after the other ingredients have dissolved. The stock solution is adjusted to a final pH of 7.6.
3. The stock solution is diluted with distilled water, and 1.5 liters of the 1 times solution is aliquoted into 2.5-liter Fernbach flasks. The flasks are stuffed with cotton plugs, autoclaved, and stored for use.
4. A 1 : 100 volume of a freshly made, filter-sterilized supplement solution (Table I) is added to the medium immediately before inoculating with bacterial cultures.

Preparation of Supplement for Stainer–Scholte Medium

1. Dissolve the L-cystine in 2 *N* HCl.
2. Add distilled water and the remaining components except the 2 *M* Tris base. Stir until dissolved.
3. Adjust the solution with 2 *M* Tris base to pH 7.4.
4. Filter-sterilize with a 0.22-μm filter (Millipore, Bedford, MA).
5. Add to Stainer–Scholte medium immediately prior to inoculation with bacterial cultures.

Preparation of Starter Cultures and Growth Conditions

A single colony of *B. pertussis* (Tohama phase I) is isolated from Bordet–Gengou blood agar plates and grown in either Cohn–Wheeler or Stainer–Scholte medium. The bacterial cultures (1 to 2 liters) are grown

[15] S. M. Cohn and M. W. Wheeler, *Am. J. Public Health* **36,** 371 (1946).

[16] D. Stainer and M. Scholte, *J. Gen. Microbiol.* **63,** 211 (1971).

[17] J. Munoz and J. Bergman, "*Bordetella pertussis*: Immunological and Other Biological Activities" (A. C. Wardlaw and R. Parton, eds.), p. 193. Dekker, New York, 1977.

TABLE I
SUPPLEMENTS FOR STAINER–SCHOLTE MEDIUM

Component	Amount
L-Cystine	0.40%[a]
$FeSO_4 \cdot 7H_2O$	0.10%
L-Ascorbic acid	0.20%
Nicotinic acid	0.04%
Glutathione[b]	1.00%
2 *N* HCl	16.70%[c]
2 *M* Tris base	16.70%[d]
Water	66.70%

[a] Solid reagents are added weight to volume.
[b] Reduced form.
[c] Liquid reagents are added volume to volume.
[d] The volume of 2 *M* Tris base added is an approximate value. This reagent is added to adjust the solution to a final pH of 7.4.

to an OD_{650} of 1.0 to 1.5, divided into 30-ml aliquots, quick frozen in a dry ice–ethanol bath, and stored at $-80°$. These starter cultures are thawed, added to 600 ml of Stainer–Scholte medium, and grown for approximately 24 hr (OD_{650}, 1.0 to 1.2). Thirty to seventy milliliters of this culture is added to 1.5 liters of supplemented Stainer–Scholte medium and grown to an OD_{650} of 0.5 (16 to 18 hr). Flasks are incubated at 37° and shaken at 250 cycles/min. Typically, 18 liters of bacteria are grown for a single preparation.

Tohama phase I bacteria produce the highest levels of CaM-sensitive adenylyl cyclase activity in the culture supernatant when grown to an OD_{650} of 0.5. We have observed variable production of extracellular adenylyl cyclase with different strains of *B. pertussis*. Therefore, it is advisable to characterize the production of the extracellular adenylyl cyclase with each strain of *B. pertussis*. In addition to strain variations, conditions such as temperature and aeration may also play a role in the production and release of the enzyme into the culture supernatant.

Isolation of Invasive Form of Adenylyl Cyclase

Anion-Exchange Chromatography

An invasive preparation of the calmodulin-sensitive adenylyl cyclase is isolated from bacterial culture supernatants by anion-exchange chromatography with QAE-Sephadex resin (Pharmacia). This procedure gives a

60-fold purification of the invasive form of the adenylyl cyclase.[12] Other anion-exchange resins such as DEAE-Sephadex or DE-52 cellulose do not adsorb adenylyl cyclase activity.

Solutions and Materials

QAE-20 buffer: 20 m*M* Tris-HCl, pH 7.8, 5 m*M* $MgCl_2$, and 20 m*M* NaCl

QAE-40 buffer: 20 m*M* Tris-HCl, pH 7.8, 5 m*M* $MgCl_2$, and 40 m*M* NaCl

QAE-1000 buffer: 20 m*M* Tris-HCl, pH 7.8, 5 m*M* $MgCl_2$, and 1 m*M* NaCl

QAE-Sephadex resin: QAE anion-exchange resin (300 ml) is preequilibrated in QAE-20 buffer, and the resin is washed with QAE-20 buffer until a pH of 7.8 is obtained

Procedure

1. Bacterial cultures (18 liters) grown to an OD_{650} of 0.5 are spun for 1 hr at 4200 rpm in a preparative centrifuge (Beckman, J-6B).
2. The culture supernatant is decanted and adjusted to 2 m*M* $CaCl_2$.
3. Preequilibrated QAE-Sephadex anion exchange resin is added to the culture supernatant and allowed to mix for 30 min at 4°.
4. The resin is separated from the culture supernatant by filtration on a Büchner funnel. Unbound material is washed from the resin with 4 liters of QAE-40 buffer.
5. The resin is poured into a 5 × 20 cm column. Adenylyl cyclase activity is eluted from the resin with a 2-liter (total volume) NaCl gradient with QAE-40 and QAE-1000 buffers. The gradient is percolated through the resin at 1 ml/min, and 8-ml fractions are collected.
6. The conductivity of every eighth column fraction is measured and every fourth sample between conductivities of 5 and 14 mS (millisiemens) are assayed for calmodulin-sensitive adenylyl cyclase activity.

Two peaks of adenylyl cyclase activity are obtained with QAE anion-exchange chromatography as described above (Fig. 1). Peak 1 (Pk1) elutes from the column at a conductivity of 7 mS, and Peak 2 (Pk2) elutes from the column at a conductivity of 10 mS. While both Pk1 and Pk2 adenylyl cyclase preparations contain the 45-kDa catalytic subunit, only the Pk1 preparation contains the ability to elevate cAMP levels in eukaryotic cells.[18] This is due to a distinct, separable assisting subunit present in Pk1 that facilitates cell invasion.[10]

[18] H. R. Masure, M. G. Donovan, D. J. Oldenburg, and D. R. Storm, *J. Biol. Chem.* **263,** 6933 (1988).

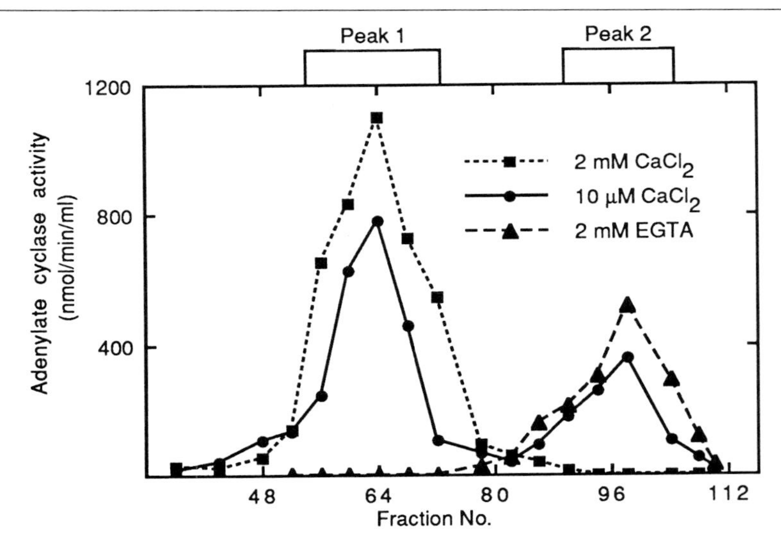

FIG. 1. Anion-exchange chromatography of the culture supernatant from *B. pertussis*. Eighteen liters of culture supernatant was adjusted to a final concentration of 2 m*M* $CaCl_2$ (■, ●) or 2 m*M* EGTA (▲), applied to a QAE-Sephadex column, washed with 4 liters of QAE-40 buffer, and eluted with a 40 m*M* to 1 *M* linear NaCl linear gradient (QAE-40, QAE-1000 buffer) that contained either 2 m*M* (■) $CaCl_2$, no added $CaCl_2$ (●), or 2 m*M* EGTA (▲). The free Ca^{2+} in the gradient which contained no added $CaCl_2$ was 10 μM. Fractions were collected and assayed for adenylyl cyclase activity in the presence of CaM.

The addition of exogenous Ca^{2+} to the gradient buffers affects elution from the QAE-Sephadex column.[18] Pk1 and Pk2 are obtained if there is no Ca^{2+} added to the QAE-Sephadex elution buffers. The contaminating Ca^{2+} in these buffers is 10 μM. Only Pk1 is obtained if the QAE-Sephadex elution buffers are adjusted to 2 m*M* $CaCl_2$. Conversely, only Pk2 is obtained if the freshly isolated culture supernatant and the QAE-Sephadex elution buffers are adjusted to 2 m*M* EGTA. This property is probably the result of the direct interaction of Ca^{2+} with the 45-kDa in catalytic subunit and a subsequent association with the assisting subunit present in Pk1.[18]

Peak fractions of adenylyl cyclase activity are pooled, concentrated approximately 10-fold by ultrafiltration (Amicon PM10 membrane), and stored at $-80°$ for further use.

Fetuin-Sepharose Affinity Chromatography

Pertussis toxin, another virulence factor produced by *B. pertussis*, catalyzes the ADP-ribosylation of the eukaryotic regulator protein, G_i. Under certain conditions, this results in activation of the eukaryotic adenylyl cyclase and an elevation in cAMP levels.[19-21] Pertussis toxin is also present as a contaminant in *B. pertussis* adenylyl cyclase preparations.[12] Fetuin is a serum sialoglycoprotein that binds pertussis toxin and has been used in its purification.[22,23] Therefore, invasive adenylyl cyclase preparations can be passed over fetuin-Sepharose to remove contaminating pertussis toxin.

Procedure

1. Fetuin (200 mg; Sigma) is coupled to 25 mg of cyanogen bromide-activated Sepharose (Pharmacia) according to the manufacturer's recommended procedure.
2. The column is equilibrated with 20 m*M* Tris-HCl, pH 7.8, 5 m*M* $MgCl_2$, and 150 m*M* NaCl.
3. Concentrated adenylyl cyclase from QAE-Sephadex chromatography is applied to the column. Unbound material is washed from the column with 3 column volumes of a solution containing 20 m*M* Tris-HCl, pH 7.8, 5 m*M* $MgCl_2$, 150 m*M* NaCl, and 3.5-ml fractions are collected. Fractions are assayed for calmodulin-sensitive adenylyl cyclase activity, and the peak samples are pooled and used for cell invasion studies.
4. The fetuin-Sepharose resin can be regenerated by washing alternately with 5 to 10 column volumes of high and low pH buffers (high pH buffer: 0.1 *M* $NaHCO_3$, pH 8.0, 0.5 *M* NaCl; low pH buffer: 0.1 *M* $NaCH_3CO_2$, pH 4.0, 0.5 *M* NaCl).

Assay for Cell-Invasive Properties of Adenylyl Cyclase

Several studies have reported that various preparations of the *B. pertussis* adenylyl cyclase will increase cAMP levels in animal cells.[6-8] In particular, it has been shown that the Pk1 preparation, described above, is able to increase intracellular cAMP levels in human erythrocytes and N1E-115 mouse neuroblastoma cells.[8] We describe procedures for assaying the elevation of cAMP in these two cell types. Human erythrocytes provide an excellent model for studying cell entry of the *B. pertussis* adenylyl cyclase for several reasons: they are relatively stable, they have no detectable endogenous adenylyl cyclase activity, and they are not

[19] T. Katada and M. Ui, *J. Biol. Chem.* **256,** 8310 (1981).

[20] T. Katada and M. Ui, *Proc. Natl. Acad. Sci. U.S.A.* **79,** 3129 (1982).

[21] T. Katada and M. Ui, *J. Biol. Chem.* **257,** 7210 (1982).

[22] M. Yajima, K. Hosada, Y. Kanbayashi, T. Nakamura, K. Nogimori, Y. Mizushima, Y. Nakase, and M. Ui, *J. Biochem.* (*Tokyo*) **83,** 305 (1978).

[23] R. Sekura, F. Fish, C. Manclark, B. Meade, and Y. Zhang, *J. Biol. Chem.* **258,** 14647 (1983).

thought to undergo receptor-mediated endocytosis.[24] Neuroblastoma cells are a stable, well-established, easily cultivated cell line. Other cell lines such as Chinese hamster ovary cells and Vero cells have also been used in this assay.[10]

Neuroblastoma Cell Growth. N1E-115 cells (passages 12–28) are grown at 37° in DMEM (Gibco, Grand Island, NY) supplemented with 5% fetal calf serum, without antibiotics, in an atmosphere of 10% CO_2/90% (v/v) humidified air. The cells are grown to 80–100% confluency in 60-mm plastic tissue culture dishes prior to the start of each experiment. Cells are subcultured weekly, and the culture medium is changed on days 3, 5, and daily thereafter. The cells are usually used for experiments on day 6 following subculture.

Isolation of Human Erythrocytes. Whole blood, collected in 0.15% EDTA, is centrifuged at 200 *g* for 5 min, and the upper layer is discarded. The erythrocytes are washed twice in PBS containing 5 m*M* theophylline. The cells are then diluted to a hematocrit value of 50%, layered onto Ficoll-Paque (Pharmacia), and centrifuged at 200 *g* for 25 min. The lymphocyte layer is removed, after which the packed cells are washed 2 times in PBS and again diluted to a 50% hematocrit value prior to the start of the experiment. Purified erythrocytes, when examined by Wright staining for contaminating white blood cells, contain less than 50 lymphocytes/10^6 erythrocytes.[8]

Determination of Intracellular cAMP Levels in Neuroblastoma Cells. Cell cultures are washed with serum-free DMEM and then preincubated for 20 min at 37° with serum-free DMEM buffered with 20 m*M* HEPES-HCl (pH 7.4) and supplemented with 5 m*M* theophylline. After preincubation, cells are treated with preparations of the adenylyl cyclase and incubated for 20 min at 37°. Typically a minimum of 100 nmol/min of enzyme activity is applied per plate of cells. The total volume of concentrated material applied to the cells varies from 50 to 1000 μl depending on the adenylyl cyclase activity. The enzyme-containing solution is removed, and the cells are washed twice with PBS. The cells are lysed with cold 5% (w/v) trichloroacetic acid, and the isolated supernatants are assayed for cAMP by the method described by Gilman.[25] The protein pellets are solubilized in 2 ml of 1 *N* NaOH, and the amount of protein is determined according to the method of Lowry *et al.*[26]

Determination of Intracellular cAMP Levels in Human Erythrocytes.

[24] P. Stahl and A. C. Schwartz, *J. Clin. Invest.* **77,** 657 (1986).

[25] A. G. Gilman, *Proc. Natl. Acad. Sci. U.S.A.* **67,** 305 (1970).

[26] O. H. Lowry, N. J. Rosebrough, A. L. Farr, and R. J. Randall, *J. Biol. Chem.* **193,** 265 (1951).

Isolated human erythrocytes are incubated for 15 min at 37° with preparations of the adenylyl cyclase (100 nmol/min of enzyme activity per 1 ml of diluted erythrocytes). The erythrocytes are centrifuged through Ficoll-Paque and washed twice with PBS. Packed cells are lysed with 5% trichloroacetic acid and assayed for cAMP.[25]

Assay for cAMP

Materials and Solutions

3′,5′-cAMP-dependent protein kinase from bovine heart (Sigma)
0.5 *M* Sodium acetate, pH 4.0
20 mg/ml BSA
0.2 *M* Potassium phosphate (monobasic), pH 6.0
2 μ*M* cAMP
[^{3}H]cAMP (specific activity ~28 Ci/mmol)
Assayed cocktail for 150 samples: 3 ml of 0.5 *M* sodium acetate, pH 4.0; 0.75 ml water; 5.5 μl [^{3}H]cAMP; 0.75 ml 20 mg/ml BSA

Assay Mixtures

140–160 μl unknown sample
30 μl assay cocktail
10–30 μl 3′,5′-cAMP-dependent protein kinase (the actual amount of protein kinase is chosen by determining the amount that binds approximately 50% of the [^{3}H]cAMP; this is done by a preliminary binding experiment)

Preparation of Samples. Add approximately 15,000 cpm [^{3}H]cAMP to the lysed cells. These samples are centrifuged, and the supernatants are applied to 2-ml Dowex columns (Bio-Rad; AG 50W-X4, 200–400 mesh). The columns are eluted with water, and column recovery is determined. Aliquots of the eluted samples are then used in the cAMP assay.

The assay components are mixed at 4° and allowed to incubate for at least 2 hr. The samples are diluted to 1.2 ml with cold 20 m*M* potassium phosphate (monobasic), pH 6, and then applied to a prewashed nitrocellulose filter in 4 ml of this buffer. The filter retains the protein kinase and allows the separation of unbound cAMP. The filters are rinsed with 8 ml of the potassium phosphate buffer and then dissolved in scintillation vials with 2-methoxyethanol. After 1 hr, scintillation fluid is added to the vials and the ^{3}H counts are measured. A standard curve is generated by mixing varying amounts of unlabeled cAMP (from 1 to 100 pmol) with the assay cocktail and protein kinase.

Analysis of Data. The total (labeled + unlabeled) cAMP concentrations (pmol/tube) in the standard samples are plotted versus counts per

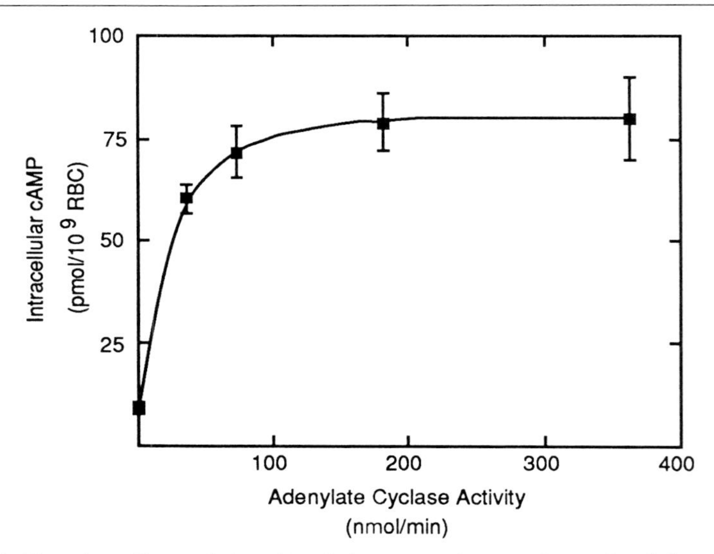

FIG. 2. Elevation of intracellular cAMP in human erythrocytes by a calmodulin-sensitive adenylyl cyclase preparation from *B. pertussis*. Human erythrocytes were incubated at 37° for 15 min with increasing concentrations of the Pk1 adenylyl cyclase preparation that was passed over fetuin-Sepharose to remove contaminating pertussis toxin. Samples containing 5×10^9 cells were centrifuged through Ficoll-Paque and washed with PBS. Packed cells were lysed and the supernatant assayed for cAMP. The results are presented as the average of duplicate points.

minute (cpm) bound to the membrane on a log–log scale. The unknowns are calculated by interpolation. In addition, the results are corrected for loss of cAMP on the Dowex columns.

Figure 2 shows a dose–response curve for increases in intracellular cAMP in human erythrocytes. Pk1 adenylyl cyclase purified through fetuin-Sepharose causes a substantial increase in cAMP levels, and this increase has been found to be dose dependent. Maximal increases in cAMP levels occur at adenylyl cyclase activities of approximately 100 nmol/min/10^9 cells. Figure 3 shows a time course for increases in intracellular cAMP in N1E-115 mouse neuroblastoma cells. Maximum levels of intracellular cAMP are reached within 30 min, and there appears to be a short lag time of approximately 5 min. A similar time course has been observed for increases in intracellular cAMP levels in human erythrocytes.[8]

Two controls are often included in these experiments to ensure that the levels of cAMP which are measured are, indeed, intracellular. The Pk2 preparation of adenylyl cyclase has catalytic activity but has been shown to be noninvasive.[18] The addition of exogenous CaM (5 μM) to the Pk1

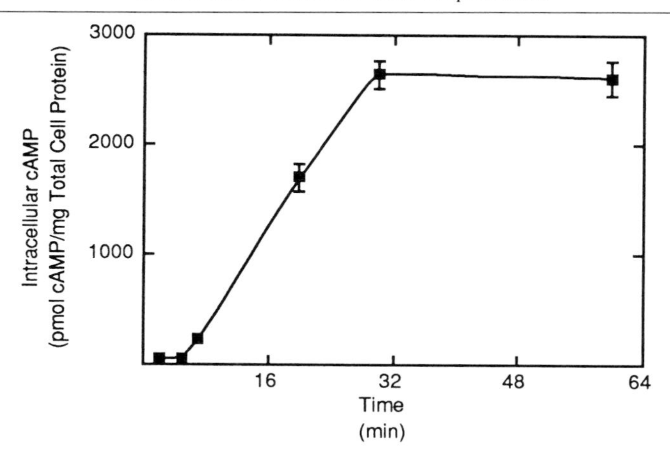

FIG. 3. Elevation of intracellular cAMP in N1E-115 mouse neuroblastoma cells by a calmodulin-sensitive adenylyl cyclase preparation from *B. pertussis*. Cells, grown to confluency, were washed with serum-free DMEM and then preincubated for 20 min at 37° with 5 m*M* theophylline in serum-free DMEM. After preincubation, cells were incubated with DMEM and 5 m*M* theophylline that contained fetuin-purified Pk1 adenylyl cyclase (100 nmol/min/plate of cells) at 37° for the indicated periods of time. Cells were then washed with PBS and lysed, and the supernatant was assayed for cAMP. The results are presented as the average of duplicate points.

adenylyl cyclase inhibits the formation of intracellular cAMP catalyzed by the enzyme, indicating that the enzyme–CaM complex does not enter animal cells.[8]

Identification and Purification of Catalytic Subunit by Preparative SDS–Polyacrylamide Gel Electrophoresis

Calmodulin-sensitive adenylyl cyclase activity can be recovered from preparations of *B. pertussis* adenylyl cyclase subjected to SDS–polyacrylamide gel electrophoresis.[27] This implies that the enzyme either retains some degree of secondary structure in the presence of SDS or regains structural integrity in the presence of calmodulin or a zwitterionic detergent such as Tween 20 or CHAPS.[18] This technique has been exploited to identify different forms of the adenylyl cyclase[9] and to purify microgram quantities of the catalytic subunit to apparent homogeneity.[18]

[27] R. H. Kessin and J. Franke, *J. Bacteriol.* **166,** 290 (1986).

Identification of Adenylyl Cyclase Activity from SDS–Polyacrylamide Gels

Preparations of the *B. pertussis* adenylyl cyclase are electrophoresed in an SDS–polyacrylamide gel. The gel is sliced, and each slice is soaked in a buffered low-salt solution. Samples from these solutions are then assayed for enzyme activity in the presence of 2.4 μM calmodulin and 0.08% (w/v) CHAPS. A migration profile is created based on a plot of enzyme activity versus distance migrated in the gel. Comparing this profile with the relative migration of molecular weight standards gives an apparent molecular weight for the adenylyl cyclase activity.

Procedure

1. Pour a discontinuous, SDS-10% (w/v) polyacrylamide slab gel according to procedures outlined by Laemmli.[28]

2. Adjust a sample containing adenylyl cyclase activity with an appropriate volume of a 5 times concentrated sample buffer solution [62.5 m*M* Tris-HCl, pH 6.8, 15 m*M* DTT, 10% (w/v) glycerol, and 1% (w/v) SDS]. As little as 0.2 nmol/min of total enzyme activity has been successfully applied to a gel to obtain detectable enzyme activity. High concentrations of salt (greater than 50 m*M*) in samples will cause the aberrant migration of enzyme activity. Therefore, it is imperative to remove or reduce the salt concentration prior to electrophoresis, preferably by dialysis into a low salt buffer. We routinely pressure dialyze and concentrate samples with an Amicon microultrafiltration system, equilibrating the sample in a low-salt buffer (12.5 m*M* Tris-HCl, pH 6.8).

3. Electrophorese the sample into the gel along with samples containing appropriate molecular weight standards. Gels are run in 2–4 hr. A water-cooled Bio-Rad Protean 12-cm gel apparatus is used to minimize gel heating and thus enzyme degradation.

4. Gel lanes containing enzyme activity are cut into 0.4- to 0.5-cm slices and placed into 250 μl of a low-salt buffer (20 m*M* Tris-HCl, pH 7.8, 40 m*M* NaCl, 2 m*M* $MgCl_2$). Gel lanes containing molecular weight standards are stained with Coomassie blue and are used to determine the apparent molecular weight of different forms of the enzyme.

5. The gel slice samples are frozen for a minimum of 12 hr at $-20°$, thawed, and aliquots of the soaking solution are assayed for adenylyl cyclase activity in the presence of 2.4 μM calmodulin and 0.08% CHAPS. Freeze–thawing the samples increases the amount of enzyme that diffuses out of the polyacrylamide matrix.

[28] U. Laemmli, *Nature (London)* **227,** 680 (1970).

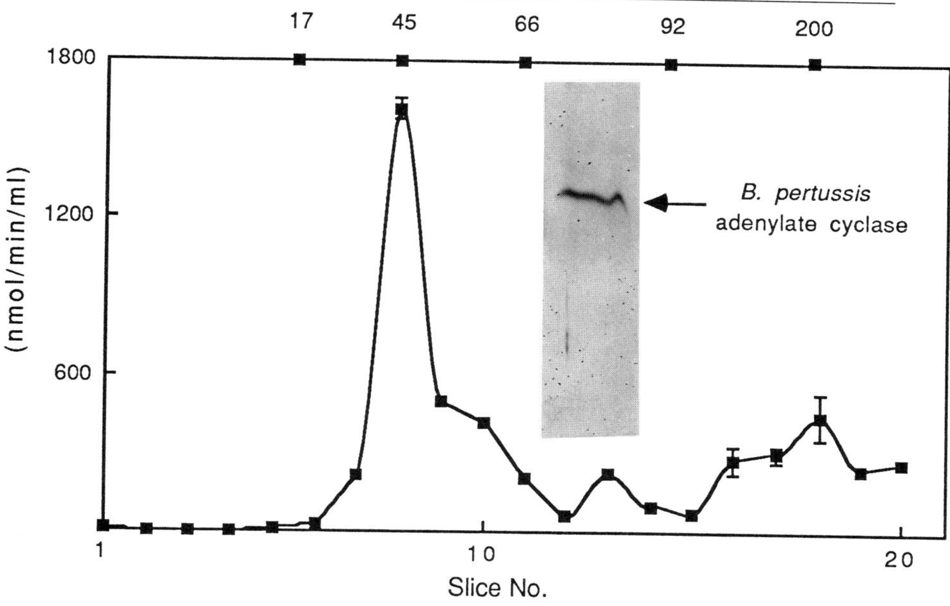

FIG. 4. Purification of the catalytic subunit of the *B. pertussis* adenylyl cyclase. Three hundred microliters of concentrated Pk1 adenylyl cyclase (330 nmol/min total activity) was equilibrated with sample buffer (defined in the text) and electrophoresed into an SDS-7% polyacrylamide gel. The gel was cut into 0.45-cm slices, and each slice was placed into 250 μl of a solution containing 20 m*M* Tris-HCl, pH 7.5, 40 m*M* NaCl, and 5 m*M* $MgCl_2$. The samples were then frozen for 12 hr at $-20°$, thawed, and assayed for a denylyl cyclase activity in the presence of 2.4 μM CaM and 0.08% CHAPS. The migration of molecular weight standards ($\times 10^{-3}$) is indicated on the top axis. The inset shows a silver-stained SDS-10% polyacrylamide gel of 50 μl of the peak adenylyl cyclase fraction recovered from the 7% polyacrylamide gel. [Reprinted from H. R. Masure, M. G. Donovan, D. J. Oldenburg, and D. R. Storm, *J. Biol. Chem.* **263,** 6933 (1988), with permission from the American Society for Biochemistry and Molecular Biology.]

Figure 4 shows the migration profile of Pk1 material applied to a SDS-7% polyacrylamide gel. The majority of activity migrated with an apparent molecular mass of 45 kDa. Smaller peaks of enzyme activity are observed at 215 and 85 kDa. Reapplication of the 45-kDa material to another SDS–polyacrylamide gel reveals a single polypeptide. Up to 1 μg of purified enzyme has been obtained by this procedure with a specific activity of 333 μmol/min/ml.

Purification by Electrophoretic Elution of Catalytic Subunit from Preparative SDS–Polyacrylamide Gel

Larger quantities of the homogeneous, catalytically active enzyme have been easily purified from SDS–polyacrylamide gels by electrophoretic elution. Yields of the purified adenylyl cyclase have approached 10 μg with this technique.

Materials and Solutions

Solutions for discontinuous polyacrylamide gel electrophoresis as described by Laemmli[28]
Gel soaking solution: 12.5 m*M* Tris-HCl, pH 6.8, 5 m*M* $MgCl_2$
Slab gel electrophoresis apparatus
Concentrated adenylyl cyclase preparation; up to 500 nmol/min of total enzyme activity of a Pk1 preparation has been successfully applied to a single lane of a preparative gel, but, again, it is imperative that the sample applied to the gel be in a low-salt solution

Procedure

1. Pour a preparative discontinuous SDS–polyacrylamide gel (12 × 15 × 1.5 cm) with 2.5 × 2 cm gel lanes. Gels with polyacrylamide concentrations between 5 and 7% (w/v) are used to separate proteins with a molecular weight range between 200,000 and 50,000. Proteins with molecular weights 50,000 and lower are applied to gels with polyacrylamide concentrations between 7 and 12.5%.

2. Adjust a concentrated adenylyl cyclase sample with an appropriate volume of 5 times sample buffer as described above and electrophoresis the sample into the SDS–polyacrylamide gel. Corun appropriate molecular weight standards.

3. Cut the enzyme-containing gel lane into 0.4- to 0.5-cm slices; place each slice into 400 μl of gel soaking solution (12.5 m*M* Tris-HCl, pH 6.8, 5 m*M* $MgCl_2$). Freeze the samples at $-20°$ overnight and assay a small aliquot of the gel soaking solution (10 μl) for CaM-sensitive adenylyl cyclase activity.

4. Adjust the gel slice and soaking solution that contains the peak of adenylyl cyclase activity to 1–5% (w/v) glycerol and 0.001% bromphenol blue.

5. Pour a second preparative discontinuous polyacrylamide gel (12 × 15 × 1.5 cm with 2 × 2 cm gel lanes) without any SDS. This gel will be used as a platform to electrophoretically elute the enzyme from the gel slice.

6. The gel slice containing the peak of adenylyl cyclase activity is suspended in an empty gel well, 1 cm from the bottom (Fig. 5). The gel soaking solution is layered above and below the suspended gel slice.

7. Running buffer containing no SDS is placed on top of the gel, and current (20 mA) is applied to the gel for approximately 15 min, the protein

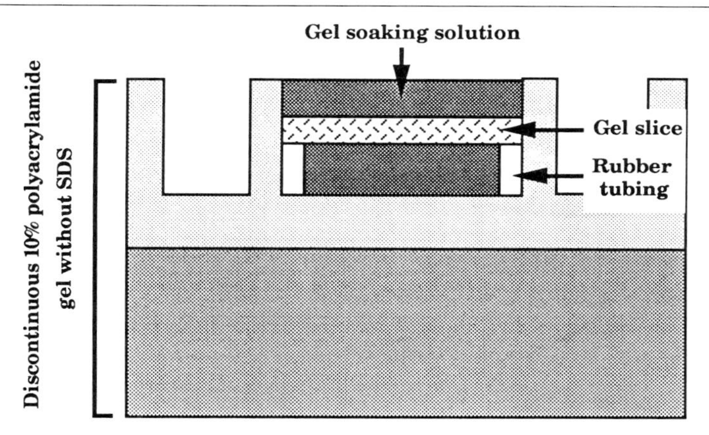

FIG. 5. Electrophoretic elution of the catalytic subunit of the *B. pertussis* adenylyl cyclase from an SDS–polyacrylamide preparative gel. A concentrated Pk1 adenylyl cyclase preparation is applied to an SDS–polyacrylamide preparative gel. The gel is sliced and each slice assayed for adenylyl cyclase activity as described in the text. The slice which contains the peak adenylyl cyclase activity (45 kDa) is suspended in the well of a second preparative polyacrylamide gel as depicted. A gel soaking solution is layered above and below the gel slice. Current (20 mA) is applied to the gel for approximately 20 min until a sharp band containing bromphenol blue and the eluted protein is observed in the solution below the gel. This material is carefully removed with a syringe. Excess glycerol and SDS are removed from the solution by dialysis.

thus being eluted from the gel. A sharp line containing bromphenol blue and the protein can be observed in the solution below the gel slice.

8. This solution below the gel slice is carefully removed with a syringe.
9. The glycerol, bromphenol blue, and excess SDS are removed by pressure dialysis with the Amicon microultrafiltration system.

We estimate that 70 to 80% of the purified enzyme is recovered by this procedure. The 45-kDa catalytic subunit present in Pk1 material, purified in this manner, is apparently homogeneous by SDS–polyacrylamide gel electrophoresis. It has catalytic activity in the presence and absence of calmodulin, and it has common immunogenic determinants with the native 45-kDa subunit purified from culture supernatants and the 215-kDa precursor isolated from the whole bacteria.[9,18] Other proteins have also been successfully eluted from polyacrylamide gels with this technique. The purity of the protein obtained with this technique relies on the ability of gel electrophoresis to resolve a mixture of polypeptides. This caveat should be carefully considered when using this technique.

Summary

An invasive form of the CaM-sensitive adenylyl cyclase from *Bordetella pertussis* can be isolated from bacterial culture supernatants. This isolation is achieved through the use of QAE-Sephadex anion-exchange chromatography. It has been demonstrated that the addition of exogenous Ca^{2+} to the anion-exchange gradient buffers will affect elution from the column and will thereby affect the isolation of invasive adenylyl cyclase.[18] This is probably due to a Ca^{2+}-dependent interaction of the catalytic subunit with another component in the culture supernatant. Two peaks of adenylyl cyclase activity are obtained. The Pk1 adenylyl cyclase preparation is able to cause significant increases in intracellular cAMP levels in animal cells. This increase occurs rapidly and in a dose-dependent manner in both N1E-115 mouse neuroblastoma cells and human erythrocytes. The Pk2 adenylyl cyclase has catalytic activity but is not cell invasive.[18] This material can serve, therefore, as a control to ensure that the cAMP which is measured is, indeed, intracellular. A second control is to add exogenous CaM to the Pk1 adenylyl cyclase preparation. The 45-kDa catalytic subunit–CaM complex is not cell invasive.[8] Although the mechanism for membrane translocation of the adenylyl cyclase is unknown, there is evidence that the adenylyl cyclase enters animal cells by a mechanism distinct from receptor-mediated endocytosis.[10,29,30]

Calmodulin-sensitive adenylyl cyclase activity can be recovered from preparations of the adenylyl cyclase that have been subjected to SDS–polyacrylamide gel electrophoresis. This property of the enzyme has enabled purification of the catalytic subunit to apparent homogeneity. The purified catalytic subunit from culture supernatants has a predicted molecular weight of 45,000. This polypeptide interacts directly with Ca^{2+}, and this interaction may be important for its invasion into animal cells.[18] Finally, the technique for purifying the catalytic subunit by SDS–polyacrylamide gel electrophoresis may prove useful in studying the interaction of the adenylyl cyclase with other components produced by the bacteria, as well as the interaction of the enzyme with eukaryotic target cells.

[29] F. Gentile, A. Raptis, L. G. Knipling, and J. Wolff, *Eur. J. Biochem.* **175,** 447 (1988).
[30] V. M. Gordon, S. H. Leppla, and E. L. Hewlett, *Infect. Immun.* **56,** 1066 (1988).

[13] Purification and Characterization of Adenylyl Cyclase from *Bacillus anthracis*

By STEPHEN H. LEPPLA

Introduction

Bacillus anthracis secretes three proteins which are collectively known as anthrax toxin.[1-8] The protective antigen (PA), lethal factor (LF), and edema factor (EF) proteins individually have no known toxic activities. Simultaneous injection of PA and LF ("lethal toxin") causes death of rats, while PA and EF ("edema toxin") produces edema in skin.[5] Recent studies have shown that PA (M_r 83,000) binds to receptors on sensitive cells and is cleaved by a cell surface protease at or near Arg^{167}, which is in a cluster of four basic amino acids, Arg^{164}-Lys^{165}-Lys^{166}-Arg^{167}. The N-terminal 20-kDa fragment is released, exposing a high-affinity site on the receptor-bound 63-kDa C-terminal fragment to which either LF or EF binds.[9,10] Proof that this cleavage is required for toxicity comes from the demonstration that deletion of residues 163–168 produces a mutant PA which cannot be cleaved and is completely nontoxic.[10] The complex of the 63-kDa PA fragment and LF or EF is internalized to an acidic vesicular compartment,

[1] The author's chapter in a previous volume in this series[2] describes procedures for the preparation and assay of anthrax toxin, of which the *B. anthracis* adenylyl cyclase is one component. Many of the procedures described here are taken from that report. However, some background information and certain procedures not specific to the adenylyl cyclase toxin component are not repeated here, so the reader may wish to consult the earlier chapter.

[2] S. H. Leppla, this series, Vol. 165, p. 103.

[3] H. Smith and J. Keppie, *Nature (London)* **173,** 689 (1954).

[4] C. B. Thorne, D. M. Molnar, and R. E. Strange, *J. Bacteriol.* **79,** 450 (1960).

[5] J. L. Stanley and H. Smith, *J. Gen. Microbiol.* **26,** 49 (1961).

[6] P. Hambleton, J. A. Carman, and J. Melling, *Vaccine* **2,** 125 (1984).

[7] S. H. Leppla, B. E. Ivins, and J. W. Ezzell, Jr., *in* "Microbiology—1985" (L. Leive, P. F. Bonventre, J. A. Morello, S. Schlessinger, S. D. Silver, and H. C. Wu, eds.), p. 63. American Society for Microbiology, Washington, D.C., 1985.

[8] J. Stephen, *in* "Pharmacology of Bacterial Toxins" (F. Dorner and J. Drews, eds.), p. 381. Pergamon, Oxford, 1986.

[9] S. H. Leppla, A. M. Friedlander, and E. Cora, *in* "Bacterial Protein Toxins" (F. Fehrenbach, J. E. Alouf, P. Falmagne, W. Goebel, J. Jeljaszewicz, D. Jurgen, and R. Rappouli, eds.), p. 111. Fischer, New York, 1988.

[10] Y. Singh, V. K. Chaudhary, and S. H. Leppla, *J. Biol. Chem.* **264,** 19103 (1989).

from which the LF or EF is apparently transferred to the cytosol.[11,12] The penetration of LF and EF across vesicular membranes may depend on the demonstrated ability of the 63-kDa PA fragment to form channels in membranes.[13]

EF is a bacterial adenylyl cyclase that combines in the cytosol with its required eukaryotic protein cofactor, calmodulin, and causes large and unregulated increases in intracellular cAMP concentrations.[14,15] The only other known calmodulin-dependent prokaryotic adenylyl cyclase, that of *Bordetella pertussis*,[16,17] also enters cells and increases cAMP concentrations. These two toxic proteins have been called "invasive adenylyl cyclases." Genetic evidence shows that the *B. pertussis* adenylyl cyclase contributes to virulence,[18] and this is probably also true of the *B. anthracis* adenylyl cyclase, although proof has not yet been provided. LF is assumed to be an enzyme that acts in the cytosol to damage cells, but no enzymatic activity has been identified.

The genes for PA,[19,20] LF,[21,22] and EF[23–28] have been cloned and sequenced. Each protein is expressed as a precursor containing a signal peptide, which is removed during secretion. The mature PA and LF proteins have deduced masses of 82.7 and 90.2 kDa. The EF gene contains

[11] A. M. Friedlander, *J. Biol. Chem.* **261,** 7123 (1986).

[12] R. Bhatnagar, Y. Singh, S. H. Leppla, and A. M. Friedlander, *Infect. Immun.* **57,** 2107 (1989).

[13] R. O. Blaustein, T. M. Koehler, R. J. Collier, and A. Finkelstein, *Proc. Natl. Acad. Sci. U.S.A.* **86,** 2209 (1989).

[14] S. H. Leppla, *Proc. Natl. Acad. Sci. U.S.A.* **79,** 3162 (1982).

[15] S. H. Leppla, *in* "Advances in Cyclic Nucleotide and Protein Phosphorylation Research, Volume 17" (P. Greengard, ed.), p. 189. Raven, New York, 1984.

[16] E. Hanski, *Trends Biochem. Sci.* **14,** 459 (1989).

[17] H. R. Masure, M. G. Donovan, and D. R. Storm, this volume [12].

[18] A. A. Weiss, E. L. Hewlett, G. A. Myers, and S. Falkow, *J. Infect. Dis.* **150,** 219 (1984).

[19] M. H. Vodkin and S. H. Leppla, *Cell (Cambridge, Mass.)* **34,** 693 (1983).

[20] S. L. Welkos, J. R. Lowe, F. Eden-McCutchan, M. Vodkin, S. H. Leppla, and J. J. Schmidt, *Gene* **69,** 287 (1988).

[21] D. L. Robertson and S. H. Leppla, *Gene* **44,** 71 (1986).

[22] T. S. Bragg and D. L. Robertson, *Gene* **81,** 45 (1989).

[23] M. T. Tippetts and D. L. Robertson, *J. Bacteriol.* **170,** 2263 (1988).

[24] M. Mock, E. Labruyere, P. Glaser, A. Danchin, and A. Ullmann, *Gene* **64,** 277 (1988).

[25] D. L. Robertson, M. T. Tippetts, and S. H. Leppla, *Gene* **73,** 363 (1988).

[26] V. Escuyer, E. Duflot, O. Sezer, A. Danchin, and M. Mock, *Gene* **71,** 293 (1988).

[27] D. L. Robertson, *Biochem. Biophys. Res. Commun.* **157,** 1027 (1988).

[28] The published DNA sequences for EF differ by several nucleotides in four regions, two of which are in the mature protein; in every case, the sequence in Escuyer *et al.*[26] is considered to be correct (Don Robertson, personal communication, 1989). Errors in only one of the four regions caused incorrect assignment of amino acids in the mature protein in the sequence in Robertson *et al.*,[25] these being residues 477–480.

an open reading frame of 800 codons, notable only for the lack of cysteines. N-terminal sequencing of the purified protein[25] showed that the mature protein begins at codon 34.[29] The mature protein therefore contains 767 residues and has a calculated mass of 88,808 daltons, close to the value of 89,000 estimated by electrophoresis.[7] The p*I* calculated from the sequence is 6.8, somewhat higher than the values of 5.9 and 6.4 determined by Quinn *et al.*[30] and Little,[31] respectively.

The deduced amino acid sequences show that EF and LF have substantial homology within the first 300 amino acids.[22] Thus, the N-terminal domains probably contain the sites which bind to PA. Amino acids 300–800 of EF contain the catalytic domain, as suggested by the recognition that residues 314–321 of the mature protein[25,26] perfectly match the consensus sequence, GXXXXGKS (where X is any amino acid), of ATP binding sites.[32] Assignment of the catalytic domain of EF to residues 300–800 also follows from the sequence homology to the *B. pertussis* adenylyl cyclase[33] of three short sequences in this region; one of these includes the consensus ATP site,[25,26] and the other two may constitute portions of the calmodulin binding site.[34] Consistent with the sequence homology, it was shown that the *B. anthracis* cyclase has immunological cross-reactivity to the *B. pertussis* adenylyl cyclase.[24]

The sequences of the *B. anthracis* and *B. pertussis* adenylyl cyclases do not show homology to the adenylyl cyclases of *Escherichia coli*[35] or *Saccharomyces cerevisiae*.[36] However, it was found that the *B. pertussis* adenylyl cyclase has immunological cross-reactivity with rat brain calmodulin-dependent adenylyl cyclase,[37] suggesting that the bacterial enzymes may have sequence homology to the eukaryotic cyclases; this might be confined to the calmodulin domains or extend to the catalytic domain as well. Therefore, when the sequence of a bovine brain adenylyl cyclase became available recently,[38] it was of interest to search for sequence homologies; none were found.

[29] The assumption[26] that signal peptide cleavage would occur after residue 29 was incorrect.

[30] C. P. Quinn, C. C. Shone, P. C. Turnbull, and J. Melling, *Biochem. J.* **252,** 753 (1988).

[31] S. F. Little, personal communication (1986).

[32] C. F. Higgins, I. D. Hiles, G. P. C. Salmond, *et al.*, *Nature (London)* **323,** 448 (1986).

[33] P. Glaser, D. Ladant, O. Sezer, F. Pichot, A. Ullmann, and A. Danchin, *Mol. Microbiol.* **2,** 19 (1988).

[34] D. Ladant, *J. Biol. Chem.* **263,** 2612 (1988).

[35] H. Aiba, K. Mori, M. Tanaka, T. Oci, A. Roy, and A. Danchin, *Nucleic Acids Res.* **12,** 9427 (1984).

[36] T. Kataoka, D. Broek, and M. Wigler, *Cell (Cambridge, Mass.)* **43,** 493 (1985).

[37] A. Monneron, D. Ladant, J. d'Alayer, J. Bellalou, O. Barzu, and A. Ullmann, *Biochemistry* **27,** 536 (1988).

[38] J. Krupinski, F. Coussen, H. A. Bakalyar, W. J. Tang, P. G. Feinstein, K. Orth, C. Slaughter, R. R. Reed, and A. G. Gilman, *Science* **244,** 1558 (1989).

For *B. anthracis* to be fully virulent, it must produce two materials, the anthrax toxin complex and a poly-D-glutamic acid capsule.[39] Virulent strains possess two large plasmids. Plasmid pXO1 [184.5 kilobase pairs (kbp)] codes for all three toxin components[19,40,41] while pXO2 (95.3 kbp) codes for the polyglutamic acid capsule.[42–44] Methods are available to eliminate either or both plasmids. Of particular value are strains possessing only pXO1, since these are at least 10^5-fold less virulent than strains that produce both toxin and capsule. The most widely used toxinogenic ($pXO1^+$), noncapsulated ($pXO2^-$) strain is that designated Sterne, after its originator.[45] Suspensions of Sterne spores are employed as a vaccine for livestock, and they have also been used as a vaccine in man.[6] The Sterne strain is effective as a vaccine because of its production of PA, which is the only *B. anthracis* antigen known to be involved in inducing immunity to infection.[46,47] Sterne-type strains are also preferred for production of toxin components (see below).

Safety Considerations

The anthrax toxins are not of extremely high potency,[48,49] and they are not known to be absorbed from the digestive tract. Therefore, the normal precautions used in handling toxic chemicals provide adequate protection. An added degree of safety applies when the toxin components have been separated, since they have no known action individually.

The principal hazard is possible infection by virulent strains of *B. anthracis*. For investigators who wish to grow only smaller volumes of the avirulent Sterne-type strains, normal precautions for handling potentially pathogenic organisms are sufficient. Many institutions require registration

[39] C. B. Thorne, *Ann. N.Y. Acad. Sci.* **88,** 1024 (1960).

[40] P. Mikesell, B. E. Ivins, J. D. Ristroph, and T. M. Dreier, *Infect. Immun.* **39,** 371 (1983).

[41] C. B. Thorne, *in* "Microbiology—1985" (L. Lieve, P. F. Bonventre, J. A. Morello, S. Schlessinger, S. D. Silver, and H. C. Wu, eds.), p. 56. American Society for Microbiology, Washington, D.C., 1985.

[42] I. Uchida, T. Sekizaki, K. Hashimoto, and N. Terakado, *J. Gen. Microbiol.* **131,** 363 (1985).

[43] B. D. Green, L. Battisti, T. M. Koehler, C. B. Thorne, and B. E. Ivins, *Infect. Immun.* **49,** 291 (1985).

[44] S. Makino, I. Uchida, N. Terakado, C. Sasakawa, and M. Yoshikawa, *J. Bacteriol.* **171,** 722 (1989).

[45] M. Sterne, *Onderstepoort J. Vet. Sci. Anim. Ind.* **8,** 279 (1937).

[46] B. E. Ivins and S. L. Welkos, *Eur. J. Epidemiol.* **4,** 12 (1988).

[47] B. E. Ivins, S. L. Welkos, G. B. Knudson, and S. F. Little, *Infect. Immun.* **58,** 303 (1990).

[48] M. Gill, *Microbiol. Rev.* **46,** 86 (1982).

[49] J. W. Ezzell, B. E. Ivins, and S. H. Leppla, *Infect. Immun.* **45,** 761 (1984).

of such organisms and submission of a safety plan. Investigators uncertain of the safety aspects should review the more extensive discussion in the author's earlier chapter.[2]

Bacterial Strain Selection

For production of toxin and adenylyl cyclase, most laboratories will prefer to use the Sterne strain. All isolates of *B. anthracis* produce approximately the same amounts of the toxin components, so other strains could be considered, provided they were proven to be free of pXO2. However, investigators who have not worked with *B. anthracis* should refer to the cautionary comments in the previous work[2] on the possible difficulties in proving the absence of pXO2.

The procedures described in this chapter were optimized for the Sterne strain, use of which is recommended. The Sterne strain is available from several laboratories, but is not currently deposited with the ATCC (American Type Culture Collection, Rockville, MD). Another suitable strain is a spontaneous, rifampicin-resistant Sterne mutant designated *SRI-1*, which has been found to produce 50 to 75% more toxin than Sterne. This strain appears defective in septum formation and grows in long filaments. This strain is not a good choice if a method generating high shear forces, such as tangential flow filtration, is used for removal of bacteria.

Strains are stored as either spore suspensions or frozen vegetative cells and are revived on blood agar or other appropriate media. In this laboratory, aliquots of vegetative cells grown in RMM medium (described below) and stored at −70° are thawed and spread on solid RMM medium lacking bicarbonate and grown 24 hr at 32° to prepare a fermentor inoculum. Growth at temperatures above 37° should be avoided since plasmid curing may occur.[40,41] If virulent *B. anthracis* are in use in the laboratory, aliquots of the fermentor inoculum should be grown in parallel on serum- or bicarbonate-containing medium and incubated in a CO_2 atmosphere to detect any capsulated contaminants.

Culture Medium and Growth Conditions

The features of a medium which appear important for toxin production are as follows: (1) inclusion of $NaHCO_3$, which aids pH control, specifically increases gene transcription,[50] and may also permeabilize the bacteria[51]; (2) maintenance of the pH above 7 (achieved here by $NaHCO_3$ and

[50] J. M. Bartkus and S. H. Leppla, *Infect. Immun.* **57,** 2295 (1989).
[51] M. Puziss and M. B. Howard, *J. Bacteriol.* **85,** 237 (1963).

Tris), which serves to limit the action of proteolytic enzymes[52]; and (3) growth under essentially anaerobic conditions. Besides enhancing yields, anaerobic growth is advantageous because it places fewer demands on the fermentation equipment and decreases the potential for contamination of the laboratory.

A number of investigators have developed media that promote production of PA[53,54] or toxin.[4,55,56] A successful synthetic medium, "R" medium,[56] was derived from a semisynthetic formulation[55] by replacement of casamino acids with an equivalent L-amino acid mixture (except that alanine was omitted to limit sporulation). Further media development work[57] was done to optimize yields of LF and EF in addition to PA and to facilitate product recovery. In these trials, yields of all three toxin components increased or decreased in parallel. The modified medium developed through these trials was designated RM.[2] Recently, further adjustments have been made to the medium[58] that appear to improve growth and increase the reliability of the method. The recipe for the adjusted medium, RMM, is provided below. RMM differs from RM by having lower concentrations of Tris (50 m*M* in RMM versus 75 m*M* in RM) and $NaHCO_3$ [0.4% (w/v) in RMM versus 0.8% (w/v) in RM] and a higher concentration of $KHPO_4$ (17 m*M* in RMM versus 3.4 m*M* in RM). While these modifications improve production in the fermentor, the lower concentrations of buffering ions may not maintain a sufficiently high pH in culture conditions where pH is not automatically controlled.

RMM medium contains the following ingredients at the indicated final concentrations (mg/liter), with all amino acids being of the L configuration: tryptophan (35), glycine (65), tyrosine (144), lysine hydrochloride (230), valine (173), leucine (230), isoleucine (170), threonine (120), methionine (73), aspartic acid (184), sodium glutamate (612), proline (43), histidine hydrochloride (55), arginine hydrochloride (125), phenylalanine (125), serine (235), NaCl (2920), KCl (3700), adenine sulfate (2.1), uracil (1.4), thiamin hydrochloride (1.0), cysteine (25), KH_2PO_4 (2300), 2-amino-2-(hydroxymethyl)-1,3-propandiol [Tris] (6040), glucose (5000), $CaCl_2 \cdot 2H_2O$ (7.4), $MgSO_4 \cdot 7H_2O$ (9.8), $MnCl_2 \cdot H_2O$ (1.0), and $NaHCO_3$ (4000). The medium is prepared with good quality distilled or deionized

[52] R. E. Strange and C. B. Thorne, *J. Bacteriol.* **76,** 192 (1958).

[53] L. C. Puziss, J. W. Manning, J. W. Lynch, E. Barclay, I. Abelow, and G. G. Wright, *Appl. Microbiol.* **11,** 330 (1963).

[54] C. B. Thorne and F. C. Belton, *J. Gen. Microbiol.* **17,** 505 (1957).

[55] B. W. Haines, F. Klein, and R. E. Lincoln, *J. Bacteriol.* **89,** 74 (1965).

[56] J. D. Ristroph and B. E. Ivins, *Infect. Immun.* **39,** 483 (1983).

[57] S. H. Leppla, unpublished studies (1984).

[58] S. H. Leppla, unpublished studies (1988).

water. If extremely high quality water is used, addition of iron and trace metals may be required to support growth.

To prepare medium for a 50-liter fermentor culture, the first 21 ingredients in the above list (ending at thiamin) are added as solids to 40 liters of distilled water and sterilized in the fermentor vessel. The remaining 8 ingredients are individually dissolved in sterile water and sequentially pumped through a disposable capsule filter into the vessel, followed by water as needed to reach 50 liters. The medium in the vessel is then titrated to pH 8.0. For RMM medium to be used in flasks, it is more convenient to group the ingredients into several stock solutions (for examples, see Haines *et al.*[55]); these are separately filter-sterilized and added to the flasks. Stocks of $NaHCO_3$ should be made fresh and filtered by pressure to avoid loss of CO_2, and media to which this stock has been added should be kept in tightly closed flasks. Failure of *B. anthracis* to grow in aged medium is usually due to alkaline conditions resulting from CO_2 loss.

One medium modification that consistently increases yields of all three toxin components by 50–100% is addition of horse serum to 3–5%.[4,49] Adding serum precludes purification (except by immunoadsorption, see below), but the increased yields are useful when the goal is small-scale production of immunochemically or enzymatically active toxin. In cases where growth conditions are not optimal, adenylyl cyclase activity may be detectable only in serum-supplemented cultures. Other proteins or putative protective agents and dialyzed horse serum seem less effective than horse serum. The basis of this effect is not known. Another additive that merits testing is cyclodextrin; in preliminary trials, addition of 0.5 g/liter improved yields.[59] Cyclodextrin may act in the same way as the charcoal which was added to many earlier media formulations.

To grow a 50-liter fermentor culture, the bacteria on five RMM agar plates are suspended in 25 ml RMM medium, giving an A_{540} of 8; this suspension is added to the vessel. The culture is stirred at 150 rpm, regulated at 35°, controlled at pH 8.0 by addition of 1 *M* NaOH, and no aeration is used. If dissolved oxygen is measured, it is found that this falls to 0 once significant growth of the culture has occurred. The culture grows to an A_{540} of 2–2.5 by 18–24 hr, with logarithmic growth evident from the rate of NaOH consumption. Approximately 1–3 mol of NaOH is consumed. Fermentor cultures are harvested promptly after growth has ceased.

For growth in flasks, the containers are half-filled with medium, inoculated, tightly capped, and shaken at a speed just sufficient to maintain the bacteria in suspension. Cultures grown in RMM medium in flasks or in

[59] C. P. Quinn, personal communication (1989).

fermentors without pH control will show a fall in pH to 6.8–7.2. Growth in flasks often does not exceed an A_{540} of 1.5. Toxin yields in such cultures are usually comparable to those in pH-controlled fermentor cultures.

Recovery of Toxin from Culture Supernatants

In cultures grown as described above, PA, LF, and EF are present at approximately 20, 5, and 1 μg/ml, respectively, and collectively constitute more than one-half of the extracellular protein. Since separating the components from each other and from impurities is not difficult (see below), the principal challenge lies in recovering the dilute proteins from the culture supernatant. In the author's experience, the most effective method found employs "salting out" the proteins[60,61] by ammonium sulfate onto cross-linked agarose beads. This process depends on the fact that a concentration of ammonium sulfate somewhat below that needed to precipitate a particular protein will cause the protein to adsorb to surfaces having hydrophobic character. Various unsubstituted or substituted celluloses and agaroses have been used successfully.[60] This method is effective even when protein concentrations are too low to produce a protein precipitate that could be collected by centrifugation.

The method to be described is used for 50-liter fermentor cultures but may be adapted to smaller volumes. When growth has ceased, 1,10-phenanthroline hydrochloride is added to give a final concentration of 50 μ*M*; EDTA is added to 2 m*M*. These chelators inhibit a *B. anthracis* metalloprotease,[52,57] which is similar to that produced by *Bacillus cereus*.[62,63] Phenylmethylsulfonyl fluoride is added to 0.1 m*M*, although it has not been proven that this enhances yields. Mercaptoethanol is added to 2 m*M*. The culture is chilled if the fermentor has this capability and is pumped via a stainless steel coil immersed in an ice bath into a continuous-flow centrifuge (Sorvall TZ-28 rotor, or a comparable system that does not generate an aerosol). The TZ-28 rotor is operated at 12,000 rpm, and a flow rate of 400 ml/min is maintained with a peristaltic pump. The supernatant is moved to a cold room, and all subsequent steps are performed at 4°. The supernatant is sterilized by filtration by using a Pellicon tangential flow ultrafiltration unit containing 10 ft^2 of Durapore 0.45-μm membrane (Millipore Corp., Bedford, MA). Filtration rates exceeding 500 ml/min can be obtained in this unit. The membrane is treated with bovine

[60] F. von der Haar, *Biochem. Biophys. Res. Commun.* **70,** 1009 (1976).

[61] R. K. Scopes, "Protein Purification: Principles and Practice," p. 141. Springer-Verlag, New York, 1982.

[62] B. Holmquist, *Biochemistry* **16,** 4591 (1977).

[63] See also this series, Vol. 19, p. 569.

serum albumin before its first use in order to prevent adsorptive losses. The membranes may be used to process a number of 50-liter cultures, provided that each use is followed by washing with 0.1 *M* NaOH and water.

The proteins are concentrated from the sterile supernatant by the "salting out" adsorption process described above.[60] Approximately 1 liter of cross-linked agarose beads [Sepharose CL-4B, Pharmacia, (Piscataway, NJ) or a similar resin] is added to the supernatant, followed by the slow addition of 25 kg $NH_4(SO_4)_2$. The suspension is stirred gently until the salt dissolves (2–3 hr), and then the agarose beads are allowed to settle. Every effort is made to reach this point on the same day the culture is harvested. If successful, it is convenient to let the resin settle overnight. The supernatant is pumped off, passing it through a porous plastic funnel (Bel-Art Plastics, Pequannock, NJ), if necessary, to collect any resin that has not settled. The agarose is then placed in a 14-cm-diameter filter funnel or column and eluted at 25 ml/min with 2 liters of 50 m*M* Tris, 1 m*M* EDTA, 2 m*M* 2-mercaptoethanol, pH 8.0. The fractions containing over 95% of the protein (coeluting with the pigment) are pooled and precipitated by slow addition of solid $NH_4(SO_4)_2$ to 75% saturation. After 2–24 hr the precipitated toxin is collected by centrifugation, redissolved in 100 ml of 10 m*M* Tris, 50 μM 1,10-phenanthroline, 2 m*M* 2-mercaptoethanol, pH 8.0, and dialyzed against the same buffer. At the author's facility, it was found convenient and feasible at this stage to filter-sterilize the toxin and remove it from the BL3 laboratory. Subsequent purifications are done at the BL1 containment level. Successful preparations contain 2–4 g protein, as determined by UV absorption, assuming an $E_{1\%}$ value (280 nm, 1 cm) of 10. The phenanthroline in the sample absorbs at 280 nm, but this does not interfere if dialyzate is used to blank the spectrophotometer.

Alternate methods for bacterial growth and recovery of the toxin components have been described.[13,30] Cultures grown in static cultures in a semisynthetic medium are filter sterilized, diluted 5-fold to decrease the ionic strength, and batch adsorbed to DEAE-cellulose.[30] After chromatographic purifications, the yield of each component was 0.3–1.0 mg/liter. In the author's experience, batch adsorptions to DEAE, although successful with *Pseudomonas* exotoxin A,[64] did not give complete adsorption of the anthrax toxin proteins, especially PA. Ultrafiltration has been used successfully to concentrate supernatants of Sterne cultures for purification of PA[13]; effective inhibition of proteases will be needed to assure good recoveries with this method.

Immunoadsorbent chromatography has also been used to recover and

[64] S. H. Leppla, *Infect. Immun.* **14,** 1077 (1976).

purify the toxin components.[65–67] The proteins obtained from Sterne supernatants by immunoadsorbent methods, however, were degraded, a problem that might be solved by use of appropriate protease inhibitors. A mutationally altered form of PA expressed in *Bacillus subtilis* was successfully and conveniently purified on a small scale using an immobilized monoclonal antibody to PA.[10]

For recovery and purification of EF, it was expected that calmodulin-agarose would be useful, because this step has been effective for purification of the *B. pertussis* adenylyl cyclase.[34] However, Quinn *et al.*[30] found that EF did not bind to this resin under the several conditions tested. Perhaps particular concentrations of Ca^{2+} are needed to obtain binding.

Alternate Expression Systems for *Bacillus anthracis* Adenylyl Cyclase

Production of EF occurs at lower levels than PA or LF in *B. anthracis* in most media, perhaps reflecting a lower transcriptional efficiency of the EF gene.[25] Transcription of all three genes may require a trans-acting positive control element, which has been proposed to explain the induction of PA mRNA by bicarbonate.[50] The PA protein is expressed well from its own promoter in *B. subtilis* and is secreted,[68] allowing relatively easy recovery and purification,[10] so it may be useful to place the EF gene behind the PA promoter and secrete EF from *B. subtilis*.

It is probable that *E. coli* expression systems could also prove effective. However, expression of *E. coli* adenylyl cyclase at high levels in *E. coli* is lethal to the bacterium.[69] Expression of EF may also be lethal if intracellular proteolysis produces fragments that are active in the absence of calmodulin. Some evidence was obtained for existence of active fragments of the *B. pertussis* cyclase in *E. coli*.[33] If expression of EF in *E. coli* is lethal, then a tightly controlled expression system such as that developed to express *E. coli* adenylyl cyclase and other lethal gene products might be needed.[69] The facts that neither EF nor the *B. pertussis* adenylyl cyclase contain cysteine and that the latter enzyme can be renatured from urea or

[65] E. J. Machuga, G. J. Calton, and J. W. Burnett, *Toxicon* **24,** 187 (1986).
[66] D. K. Larson, G. J. Calton, S. F. Little, S. H. Leppla, and J. W. Burnett, *Toxicon* **26,** 913 (1988).
[67] D. K. Larson, G. J. Calton, J. W. Burnett, S. F. Little, and S. H. Leppla, *Enzyme Microb. Technol.* **10,** 14 (1988).
[68] B. E. Ivins and S. L. Welkos, *Infect. Immun.* **54,** 537 (1986).
[69] P. Reddy, A. Peterkofsky, and K. McKenney, *Nucleic Acids Res.* **17,** 10473 (1989).

SDS[34,70] suggest that EF might successfully be renatured from inclusion bodies.

Chromatographic Separation of Toxin Components

A number of chromatography methods are available to separate the toxin components. Details are given here for processing the amount of toxin obtained from a 50-liter fermentor, by using sequential chromatography on hydroxyapatite and DEAE-Sepharose. This order is preferred over the inverse, since EF was found to elute in several distinct peaks when crude toxin was run on DEAE, probably reflecting EF binding to PA fragments like the 63-kDa one described above. In the protocols described below, all operations are performed at 4°, and dialysis times should not exceed 16 hr, except for the final product.

The dialyzed crude toxin (2–4 g) is pumped onto a 2.6 × 38 cm (200 ml) column of hydroxyapatite (Fast Flow type, Calbiochem, San Diego, CA) previously equilibrated to 5 m*M* potassium phosphate, 0.1 *M* NaCl, 50 μM 1,10-phenanthroline, 2 m*M* 2-mercaptoethanol, pH 7.0 (buffer A). The column is washed with at least 150 ml buffer A at 50 ml/hr and then eluted with a gradient of 500 ml each of 0.0 and 0.5 *M* potassium phosphate, pH 7.0, both in buffer A. Fractions of 10 ml are collected in tubes containing 0.1 ml of 100 m*M* EDTA. The components elute in the order PA > LF > EF, and each is evident as a peak of UV-absorbing material. The EF peak is small compared with the others but is easily identified because it is the last significant peak. PA and LF are concentrated from pooled fractions by ammonium sulfate precipitation at 75% saturation and dialysis against 10 m*M* Tris, 25 m*M* NaCl, 50 μM 1,10-phenanthroline, 2 m*M* 2-mercaptoethanol, 1% glycerol, pH 8.0 (buffer B). The EF is concentrated by ultrafiltration to about 25 ml and is dialyzed against buffer B.

Final purification of each component is achieved by chromatography on DEAE-Sepharose CL-4B (Pharmacia), with a column containing about 1 ml resin per 5–10 mg input protein. As an example, 700 mg PA is purified on a 1.6 × 50 cm (100 ml) column, with a gradient of 750 ml each of 0 and 0.25 *M* NaCl in buffer B. For LF and EF, the high-salt buffers should contain 0.40 and 0.25 *M* NaCl, respectively. The protein in the pooled fractions is dialyzed against 5 m*M* HEPES, 50 m*M* NaCl, pH 7.5, filter-sterilized with low protein binding filters (Millex-GV, Millipore), quick frozen in small aliquots, and stored at −70°.

[70] D. Ladant, C. Brezin, J.-M. Alonso, I. Crenon, and N. Guiso, *J. Biol. Chem.* **261,** 16264 (1986).

Other chromatographic systems have been used successfully to purify the toxin components. Quinn *et al.*[30] used a fast protein liquid chromatography (FPLC) apparatus for sequential anion exchange on Mono Q, gel filtration on Superose 12, and hydrophobic interaction chromatography on Phenyl-Sepharose. Addition of Triton X-100 to buffers was considered necessary to obtain elution of EF in narrow peaks and with adequate yields. EF did not bind to several ATP-agaroses and could not be eluted from Blue-Sepharose.

Purifications using the preferred methods described above yield, from 50-liter fermentors, approximately 400 mg PA, 75 mg LF, and 20 mg EF. The LF and EF proteins appear as single species on SDS gels and analytical ion-exchange HPLC (Mono Q resin, Pharmacia); these components appear homogeneous. In contrast, two types of heterogeneity in PA have been observed. A variable fraction (usually <10%) of the PA contains a cryptic polypeptide cleavage; electrophoresis under denaturing conditions reveals fragments of 37 and 47 kDa. The site which is cleaved by the endogenous *B. anthracis* protease is also highly susceptible to specific cleavage by other proteases, including chymotrypsin.[2,58] N-terminal sequencing of the purified 47-kDa fragment showed that this fragment is C-terminal and that chymotrypsin cleavage occurred at residue 314, next to a pair of phenylalanine residues.

The other type of heterogeneity in PA consists of differences in net charge. During the final chromatography on DEAE, most preparations show one or two partially separated, trailing species. These species are indistinguishable on SDS gels, but migrate as 2–4 evenly spaced bands after isoelectric focusing or nondenaturing, continuous slab gel electrophoresis. Careful selection of fractions, or chromatography on Mono Q with shallow gradients (Tris buffer, pH 8.0, 0.25–0.35 *M* NaCl), yields distinct, homogeneous preparations. The species with the least negative charge (eluting first from DEAE and Mono Q) has been shown to have significantly greater potency (with LF) in the macrophage lysis assay described below. The nature of the alteration that introduces additional negative charge and decreases potency is not known. A similar type of heterogeneity has been observed in *Pseudomonas* exotoxin A.[71]

Toxin Assays

Chemical Assays. For laboratories beginning purification of the toxin and not yet having component-specific antisera, it may be most convenient to perform direct chemical measurement of the toxin components by rapid

[71] L. T. Callahan III, D. Martinez, S. Marburg, R. L. Tolman, and D. R. Galloway, *Infect. Immun.* **43,** 1019 (1984).

electrophoresis or chromatography methods. The large size (83–89 kDa) and the high concentrations of the proteins in culture supernatants makes it relatively simple to locate the bands or peaks of PA and LF. While the three toxin components are similar in size, they can be distinguished on 8 or 10% polyacrylamide SDS slab gels if small amounts of protein (0.1 μg/band) are loaded. The three proteins are well separated during ion-exchange high-performance liquid chromatography (Mono Q), eluting in the order PA > EF > LF when the column is developed with a NaCl gradient at pH 8.[2,30]

Immunochemical Assays. Purified PA and LF induce high-titer polyclonal antisera, while EF seems less immunogenic and has not yielded antisera useful in gel-diffusion systems. Typically, antisera to EF produce only weak precipitan bands when diffused in agar against pure EF.[30] Antisera to PA may be obtained by immunization with the licensed human vaccine (Michigan Department of Public Health), which contains principally this toxin component. Immunization of guinea pigs or rats with the live spore veterinary vaccine (Anvax, Jensen-Salsbery Laboratories, St. Louis, MO) induces good titers to PA and lower titers to LF and EF.[46,72] For routine detection of PA and LF in column eluates, gel diffusion in agar is preferred because antigens can be detected over a wide range of concentrations. For quantitative measurement of PA, radial immunodiffusion employing specific goat antiserum has been most useful. Monoclonal antibodies to all three toxin components have been developed[73,74] and can be expected to replace polyclonal sera in some assays. Owing to the low potency of the available sera, EF is usually assayed enzymatically (see below).

Adenylyl Cyclase Assay. Of the three toxin proteins, only EF is known at this time to have enzymatic activity. The EF adenylyl cyclase has enzymatic properties[2,14,15] resembling those of the *B. pertussis* adenylyl cyclase.[16,33,75–77] EF has a high enzymatic activity, V_{max} 1.2 mmol cAMP/min/mg protein,[15,30] similar to values of 1.6 mmol cAMP/min/mg protein found for purified *B. pertussis* adenylyl cyclase.[33,78] The K_m for ATP is

[72] B. E. Ivins, J. W. Ezzell, Jr., J. Jemski, K. W. Hedlund, J. D. Ristroph, and S. H. Leppla, *Infect. Immun.* **52,** 454 (1986).

[73] S. F. Little, S. H. Leppla, and E. Cora, *Infect. Immun.* **56,** 1807 (1988).

[74] S. F. Little, unpublished studies (1989).

[75] E. Hewlett and J. Wolff, *J. Bacteriol.* **127,** 890 (1976).

[76] J. Wolff, G. H. Cook, A. R. Goldhammer, and S. A. Berkowitz, *Proc. Natl. Acad. Sci. U.S.A.* **77,** 3840 (1980).

[77] D. L. Confer and J. W. Eaton, *Science* **217,** 948 (1982).

[78] A. Rogel, Z. Farfel, S. Goldschmidt, J. Shiloach, and E. Hanski, *J. Biol. Chem.* **263,** 13310 (1988).

0.16 mM in the presence of Mg^{2+}. EF differs from the *B. pertussis* cyclase in that the dependence on calmodulin is absolute; no calmodulin-independent forms have been detected. In the presence of calcium, 2 nM calmodulin gives half-maximal enzyme activity. Calmodulin still is able to activate EF when all calcium is chelated, but approximately 5 μM is needed to get half-maximal activity.[15] Enzyme activity is inhibited by Ca^{2+} concentrations exceeding 1 mM in the presence of 5 mM Mg^{2+}, but not in the presence of 5 mM Mn^{2+}. The previous claim[2] that phosphate inhibits the enzyme is now considered to be incorrect.[79]

The high activity of the enzyme makes it easy to assay, and any of the methods developed for the more difficult task of measuring eukaryotic adenylyl cyclases may be used (see [1], this volume). The assay described below is modified from that of Salomon[80] by dilution of [^{32}P]ATP to lower specific activity, omission of a phosphodiesterase inhibitor, omission of the [^{3}H]cAMP added to determine chromatographic recoveries, and optional use of [^{3}H]- or [^{14}C]ATP as a substitute for [^{32}P]ATP. This method has higher sensitivity than is needed, but it is preferred because the Dowex and alumina columns are reusable. Manganese ion is used in order to make the enzyme activity independent of Ca^{2+} concentration. In assays designed to study the Ca^{2+} requirement, consideration must be given to the probable inability of chelators rapidly to extract Ca^{2+} from a preformed complex of Ca^{2+}–calmodulin–cyclase.

Reagents

Assay buffer (5×), pH 7.5: 100 mM HEPES, 25 mM $MnCl_2$, 2.5 mM $CaCl_2$, 2.5 mM EDTA, 0.25 mM cAMP, 2.5 mM dithiothreitol, 0.5 mg/ml bovine serum albumin (calmodulin-free[76])

[^{32}P]ATP (10×): 20 μCi/ml, 5 mM ATP; if assays are done infrequently, it may be preferred to use [^{3}H]- or [^{14}C]ATP. These provide adequate sensitivity and long half-life, but require counting in scintillation fluid

Calmodulin, bovine (10×): 50 μg/ml

Protocol. The reaction mixtures contain 0.020 ml assay buffer (5×), 0.010 ml calmodulin (10×), EF sample (2–10 ng) and water totalling 0.060 ml, and 0.010 ml [^{32}P]ATP (10×). Mixtures are preincubated 20 min before addition of ATP, to allow association of EF, calmodulin, and calcium. Controls should include reactions (1) with excess EGTA to remove calcium, (2) omitting calmodulin, and (3) with excess EF (10 μg/ml) to cause complete ATP conversion. The latter allows calculation of chromato-

[79] Roger Johnson, personal communication (1989).

[80] Y. Salomon, *Adv. Cyclic Nucleotide Res.* **10,** 35 (1979).

graphic recovery. After 60 min at 23°, 0.10 ml of stopping solution (2% sodium dodecyl sulfate, 45 m*M* ATP, 1.3 m*M* cAMP) is added, the samples are heated 5 min at 95–100°, and 1.0 ml of water is added. The samples are poured into disposable plastic chromatography columns having integral 10-ml reservoirs (Bio-Rad, Richmond, CA) and packed with 1.0 ml Dowex AG 50W-X4 (Bio-Rad), followed by two portions of 3.0 ml water, taking care to let the fluid run completely into the resin after each addition. The Dowex columns are then placed over columns of the same design containing 1.0 ml neutral alumina WN-3 (Sigma, St. Louis, MO), and 7 ml water is added to elute the cAMP from the Dowex and transfer it to the alumina. After the 7.0 ml has drained through the alumina columns, these are placed above scintillation vials, and the cAMP is eluted with 7.0 ml 0.1 M imidazole hydrochloride, pH 7.0. The [^{32}P]cAMP is measured by Cerenkov counting. If [^{3}H]- or [^{14}C]ATP is used as substrate, an aliquot is transferred to another vial containing an aqueous scintillation fluid. When the Dowex and alumina columns are first set up, ATP and cAMP standards should be run and fractions collected to verify elution positions. The results obtained may show that the [^{3}H]- or [^{14}C]cAMP can be collected in a smaller volume so as to facilitate scintillation counting.

Toxicity Assays. Anthrax toxin was originally defined as an agent causing edema in skin[81] (now known to reflect the combined action of PA and EF), and subsequently was found to contain a material lethal for guinea pigs and rats[5] (LF, when combined with PA). The skin edema assay is laborious and rather variable, and it has been replaced in this laboratory by the Chinese hamster ovary (CHO) cell elongation assay[15] (see below). The rat lethality assay can provide a rapid and relatively accurate measure of potency for PA and LF,[49] but in the author's laboratory it has been replaced by the macrophage cytotoxicity assay.[11] These toxicity assays can be performed directly on *B. anthracis* culture supernatants, which do not contain other substances having measurable toxicity, although appropriate controls should always be included. All bioassays of unfractionated anthrax toxin must take into account the competitive action of LF and EF[5,14,49]; these components will inhibit the toxicity of the heterologous component unless diluted to concentrations below 0.1 μg/ml. Also, cell culture methods should take into account the protective activity of amines,[11,82,83] including ammonia, which can accumulate in tissue culture medium.

[81] H. Smith, J. Keppie, and J. L. Stanley, *Br. J. Exp. Pathol.* **36,** 460 (1955).
[82] V. M. Gordon, S. H. Leppla, and E. L. Hewlett, *Infect. Immun.* **56,** 1066 (1988).
[83] V. M. Gordon, W. W. Young, Jr., S. M. Lechler, S. H. Leppla, and E. L. Hewlett, *J. Biol. Chem.* **264,** 14792 (1989).

Cell Culture Assay of Bacillus anthracis Adenylyl Cyclase. The internalization of EF and LF by eukaryotic cells is a receptor-mediated process,[11,82,83] whereas uptake of the *B. pertussis* adenylyl cyclase appears to involve direct penetration of the membrane, mediated by the hemolytic domain at the C terminus.[16,83] Nearly all cultured primary and continuous cell lines tested respond to edema toxin (PA + EF) with an increase in intracellular cAMP,[7,14,15,83] which can be measured by radioimmunoassay.[84] CHO cells are one of several cell types that also respond to elevated cAMP by changing shape. If CHO cells are plated thinly and used before they reach confluence, the shape change is more easily detected. Dose–response curves can be done with serial dilutions of PA or EF in the presence of a fixed concentration of the complementary component, usually 1 μg/ml. The morphological change is evident in about 60 min, and should be observed at intervals, since it is reversible at low EF concentrations. Kinetics experiments[14] showed that internalized EF is unstable, and since the cells are not killed they eventually deplete low extracellular concentrations of EF and recover. Dose–response curves usually show saturation of each component at or below 1 μg/ml. When radioimmunoassay is used to detect cellular cAMP, controls must be included to show that the cAMP is intracellular, since recent work showed that ATP released from cells can be a substrate for the added, extracellular adenylyl cyclase.[85]

Summary and Future Developments

Study of anthrax toxin has been essential to increasing our understanding of the virulence of *B. anthracis* and to improving the design of vaccines. In addition, the anthrax toxins have been found to be useful tools in cell biology. The adenylyl cyclase toxin (PA + EF) provides another tool to analyze the consequences of increased intracellular concentrations of cAMP. For this purpose it may be preferred over cholera toxin because the action of EF is rapidly reversed after toxin removal.[14,82] Identification of the PA receptor and the protease that nicks PA will provide new tools to study the normal functions of these membrane proteins. Expanded use of the anthrax toxins as pharmacological tools is likely after the molecular basis of LF action is determined and improved methods not involving growth of *B. anthracis* are developed for production of the toxin proteins.

[84] G. Brooker, J. F. Harper, W. L. Terasaki, and R. D. Moylan, *Adv. Cyclic Nucleotide Res.* **10,** 1 (1979).

[85] F. Gentile, R. Anastassios, L. G. Knipling, and J. Wolff, *Biochim. Biophys. Acta* **971,** 63 (1988).

Section II

Guanine Nucleotide-Dependent Regulatory Proteins

A. Purification and Characterization of G Proteins
Articles 14 through 21

B. Labeling and Quantitating of G Proteins
Articles 22 through 28

C. Reconstitution
Article 29

[14] Preparation and Application of GTP-Agarose Matrices

By ELKE PFEUFFER and THOMAS PFEUFFER

Introduction

The recent discovery of a still growing family of GTP-binding proteins serving as cellular mediators of signal transmission has raised the need for their purification and characterization. Since the affinity constants for GTP and these proteins are in the 10^7 to 10^9 M^{-1} range, affinity techniques could represent an important step within a purification protocol. An affinity column with periodate-oxidized GTP coupled through hydrazide agarose[1] was successfully used for purification of bacterial dihydroneopterin-triphosphate synthetase[2] and soluble guanylyl cyclase from sea urchin sperm,[3] but attempts to purify adenylyl cyclase-associated G proteins[4] analogously were unsuccessful.[5] Regarding the successful application of protein-reactive GTP analogs we made use of the same strategy for the construction of an immobilized GTP derivative.[6] Synthesis of γ-substituted phosphate derivatives of GTP provided potently activating and binding ligands of the adenylyl cyclase-associated G protein.[7] Furthermore, these analogs were resistant to the action of nucleoside triphosphatases, which are usually in vast excess over cyclase components in biological membranes.

Preparation of GTP-Sepharose

Guanosine triphosphate tetrasodium salt (GTP · Na_4) (33 μmol) and *N*-ethyl-*N*′-(3-dimethylaminopropyl)carbodiimide hydrochloride (20 μmol)

[1] R. Lamed and C. Villar-Palasi, *Curr. Top. Cell. Regul.* **3,** 195 (1971).
[2] R. J. Jackson, R. M. Wolcott, and T. Shista, *Biochem. Biophys. Res. Commun.* **51,** 428 (1973).
[3] D. L. Garbers, *J. Biol. Chem.* **251,** 4071 (1976).
[4] MOPS, 3-(*N*-Morpholino)propane sulfonate; PBS, phosphate-buffered saline; 4-aminoanilido-GTP, P^3-(4-aminoanilido)-P^1-guanosine triphosphate; 4-azidoanilido-GTP, P^3-(4-azidoanilido)-P^1-guanosine triphosphate; GPP(NH)P, guanosine 5′-(β,γ-imido)triphosphate; GTPγS, guanosine 5′-(γ-thio)triphosphate; G proteins, GTP-binding proteins; C, catalytic component of adenylyl cyclase; G_s, guanine nucleotide-binding regulatory component responsible for stimulation of adenylyl cyclase; G_i, guanine nucleotide-binding regulatory component responsible for inhibition of adenylyl cyclase; SDS, sodium dodecyl sulfate.
[5] A. M. Spiegel, R. W. Downs, and G. D. Aurbach, *J. Cyclic Nucleotide Res.* **5,** 3 (1979).
[6] T. Pfeuffer, *J. Biol. Chem.* **252,** 7224 (1977).
[7] T. Pfeuffer and F. Eckstein, *FEBS Lett.* **67,** 354 (1976).

dissolved in 1 ml of distilled water are added to a solution of 60 μmol of *p*-phenylenediamine in 1 ml of peroxide-free dioxane and left for 3 hr at 22° in the dark. The reaction is monitored by thin-layer chromatography on polyethyleneimine cellulose (Merck, Darmstadt) with 0.75 *M* LiCl as solvent (R_f values: GTP, 0.1; GDP, 0.3; 4-aminoanilido-GTP, 0.32). When a trace of [γ-^{32}P]GTP is included in the reaction mixture, over 90% of the radioactivity appears in the product, further indicating that it is a triphosphate analog. Owing to the aromatic NH_2 moiety the spot of 4-aminoanilido-GTP becomes colored on exposure to air and light.

The reddish brown reaction mixture is poured under stirring into 50 ml ice-cold ethanol, and the precipitate is collected by centrifugation at 10,000 *g* at 4°. The precipitate is washed with ice-cold ethanol until no color remains in the supernatant. The 4-aminoanilido-GTP, which is obtained at greater than 80% final yield, is dissolved in 4 ml of water, mixed with 4 ml of packed carboxypropylamino-Sepharose CL-4B, and the slurry adjusted to pH 4.7. Following the addition of 200 μmol *N*-ethyl-*N*′-(3-dimethylaminopropyl)carbodiimide hydrochloride, the mixture is stirred for 12 hr at 22° in the dark while the pH is kept at 4.7. The resin is filtered and washed with 1 *M* NaCl and distilled water until the absorption (252 nm) of the wash fluid returns to zero. The GTP-Sepharose CL-4B is stored in 50% ethanol at −20°.

Preparation of Carboxypropylamino-Sepharose

Carboxypropylamino-Sepharose CL-4B is prepared by coupling 4 mmol γ-aminobutyric acid in 100 ml of 0.1 *M* Na_2CO_3, pH 10.0, to 50 ml of settled Sepharose CL-4B (Pharmacia, Uppsala, Sweden), which had been previously activated with 10 g CNBr.[8] After 16 hr at 4° the resin is washed with 500 ml of 0.1 *M* Na_2CO_3, pH 10, containing 1 *M* NaCl and then with 500 ml of distilled water. Instead of activation by CNBr, the method of Bethell *et al.*[9] with carbonyldiimidazole may also be applied.

A GTP-agarose derivative with a 13-atom spacer can be prepared by reacting 4-aminoanilido-GTP (25 μmol) with 4 ml settled Affi-Gel 10 (Bio-Rad, Richmond, CA) in 0.1 *M* MOPS, pH 7.4, at 22° for 16 hr. These different forms of GTP-Sepharose exhibit nearly identical results (Fig. 1). Bound GTP is estimated spectroscopically at 256 nm after hydrolysis with 1 *M* HCl for 2 hr at 37°.

[8] P. Cuatrecasas, *J. Biol. Chem.* **245,** 3059 (1970).

[9] G. S. Bethell, J. S. Ayers, W. S. Hancock, and M. T. W. Hearn, *J. Biol. Chem.* **254,** 2572 (1979).

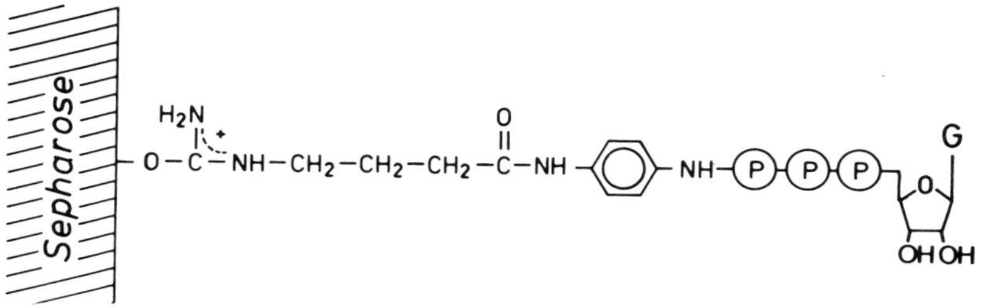

FIG. 1. GTP-Sepharose. G, Guanine.

Preparation and Solubilization of Avian Erythrocyte Membranes

Avian (pigeon or turkey) erythrocyte membranes are prepared according to Øye and Sutherland[10] with slight modifications.[11] The blood of decapitated animals is collected in heparinized glass beakers. After repeated washes with 10 m*M* Tris buffer, pH 7.4, 0.145 *M* NaCl, 1 m*M* EDTA, centrifugation at 5,000 *g*, and careful removal of the buffy coat, the packed blood cells are lysed for 5 min in 15 volumes of 20 mOsm phosphate buffer, pH 7.4, 1 m*M* dithiothreitol, 10 μM phenylmethylsulfonyl fluoride. After lysis, the same volume of 0.3 *M* NaCl is added to prevent damage of the nuclei. Following centrifugation, the ghosts are washed with 20 mOsm phosphate, pH 7.4, 0.15 *M* NaCl, 1 m*M* EDTA, 1 m*M* dithiothreitol, 10 μM phenylmethylsulfonyl fluoride, 3 m*M* $MgSO_4$ (buffer A), until the supernatant is only faintly pink. The ghosts are suspended in 5 volumes of buffer A and homogenized with a Buehler blender at setting 80 for 8 sec. Centrifugation at 5000 *g* for 30 min yields a two-layered pellet. The upper layer is removed, repeatedly washed, and centrifuged at 500 *g* to remove nuclei or cells. Membranes in buffer A are adjusted to a protein concentration of 20 mg/ml and kept at $-80°$.

For solubilization of adenylyl cyclase activity, erythrocyte membranes (10 mg/ml) are treated with 20 m*M* Lubrol PX in 10 m*M* MOPS, pH 7.4, 1 m*M* EDTA, 1 m*M* MgCl, 2 m*M* dithiothreitol, 0.25 *M* sucrose buffer for 30 min at 0°. After centrifugation at 20,000 *g* for 20 min at 4°, 80–90% of the adenylyl cyclase activity is recovered in the supernatant. The soluble adenylyl cyclase preparation is stored at $-80°$ without loss of activity for at least 6 months. Adenylyl cyclase activities are measured at 37° in 20 m*M* MOPS buffer, pH 7.4, containing 10 m*M* creatine phosphate, 20 μg/ml creatine kinase, 3 m*M* $MgCl_2$, 3 m*M* theophylline, 0.1 m*M* [α-^{32}P]ATP (0.6–2.5 Bq/pmol). (See also [1], this volume.)

[10] I. Øye and E. W. Sutherland, *Biochim. Biophys. Acta* **127,** 347 (1966).

[11] G. Puchwein, T. Pfeuffer, and E. J. M. Helmreich, *J. Biol. Chem.* **249,** 3232 (1974).

Affinity Chromatography of Solubilized Avian Erythrocyte Membrane Proteins on GTP-Sepharose

The following chromatography is conducted as a batch procedure, although the column version is also successfully applied. Solubilized pigeon erythrocyte membranes (5 ml) are added to 2.5 ml of packed GTP-Sepharose in 300 m*M* NaCl, such that the final concentration is 100 m*M*, and transferred to a 10-ml plastic syringe equipped with a nylon net. The syringe is shaken end-over-end for 60 min at 22°. This is the time necessary for maximal reduction of GPP(NH)P-stimulated adenylyl cyclase activity. Thereafter the resin is filtered off and the flow-through saved. The GTP-Sepharose is washed 4 times, 5 ml each time, with 10 m*M* MOPS, pH 7.4, 1 m*M* $MgCl_2$, 1 m*M* EDTA, 1 m*M* Lubrol PX (buffer B).

For release of bound proteins 5 ml of buffer B containing 0.15 m*M* GPP(NH)P (or 0.15 m*M* GTP when AlF_4^--stimulated activity is measured) is added, the resin shaken for 30 min at 22°, and the eluate separated from the resin. The GTP-binding protein fraction is stored at −80°, with no loss of activity.

For accurate quantitation of stimulation by GPP(NH)P, control (C) and nonretained proteins (NR) are incubated for 30 min at 22° with 0.1 m*M* GPP(NH)P in order to fully activate the cyclase prior to the assay. The protein fraction retained on GTP-Sepharose (R, 25 μl) receives either 25 μl of buffer B or 25 μl of the preactivated, nonretained proteins (NR + R). Control and nonretained protein samples (25 μl) receive 25 μl of buffer B only. $[AlF_4]^-$-stimulated adenylyl cyclase is tested without preincubation, but 0.1 m*M* GTP is included in the assay mixture. Virtually no reconstitutive activity is observed when retained or nonretained fractions are treated with trypsin or *N*-ethylmaleimide prior to recombination.

SDS–polyacrylamide gel electrophoresis is conducted according to Laemmli.[12]

Discussion

The affinity technique described here was originally designed for isolation of the GTP-dependent adenylyl cyclase holoenzyme. Although not fully understood, this procedure caused (reversible) resolution. Nevertheless, this finding could demonstrate, for the first time, that the GTP-binding site was a dissociable "subunit" of an oligomeric complex.

Figure 2 shows that stimulation by both GPP(NH)P and $[AlF_4]^-$ is diminished following treatment of solubilized pigeon erythrocyte adenylyl

[12] U. K. Laemmli, *Nature* (*London*) **227,** 680 (1970).

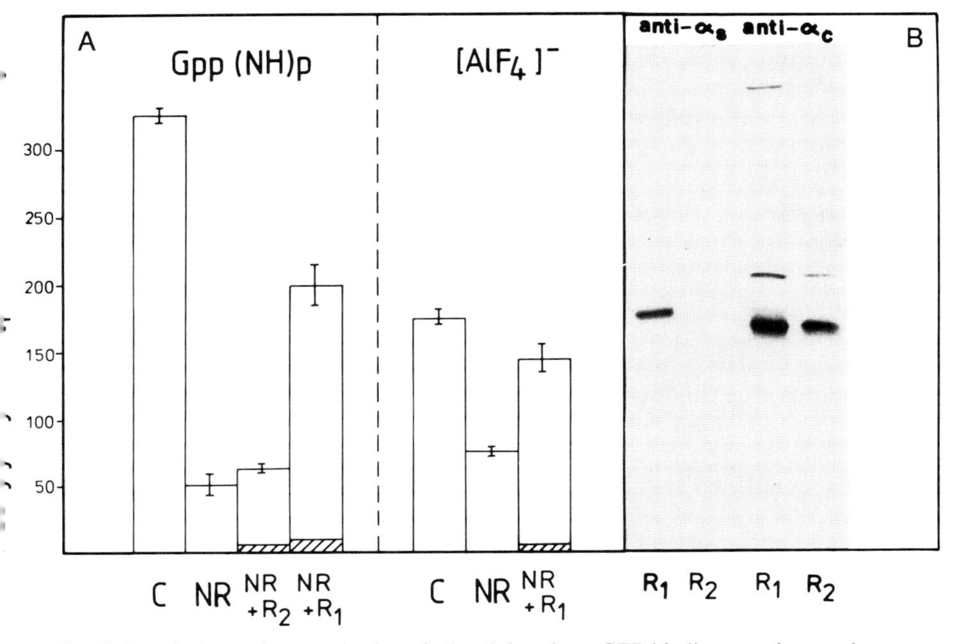

FIG. 2. Resolution and reconstitution of adenylyl cyclase–GTP-binding protein complex. (A) Solubilized pigeon erythrocyte adenylyl cyclase from GMP/isoproterenol-treated or nontreated membranes was incubated with GTP-Sepharose and retained (R_1 or R_2, respectively) and nonretained (NR) fractions were assayed for GPP(NH)P-stimulated (left) or $[AlF_4]^-$-stimulated (right) adenylyl cyclase activity, alone or following recombination (NR + R_1 or R_2). Nonfractionated starting material (C, control) was likewise tested. R_1 is the retained fraction from the GTP-Sepharose loaded with solubilized adenylyl cyclase from membranes pretreated with GMP/isoproterenol, while R_2 is the same fraction from membranes not treated with GMP/isoproterenol. The shaded areas represent the cyclase activity of R_1 and R_2 alone. The NR fraction was derived from membranes treated with GMP/isoproterenol. The detailed procedure is described in the text. (B) Western blot analysis of $G_{s\alpha}$ and $G_{i\alpha}$ proteins released from GTP-Sepharose. Solubilized adenylyl cyclase from turkey erythrocyte membranes treated with or without GMP/isoproterenol was chromatographed on GTP-Sepharose as described for the pigeon erythrocyte enzyme. Retained proteins (20 μg in 50 μl) from GMP/isoproterenol-treated membranes (R_1) or nontreated membranes (R_2) were run on 11% polyacrylamide gels and blotted onto nitrocellulose. Blocking of nitrocellulose was conducted by incubation with 2% Tween 20 (Serva, Heidelberg) in phosphate-buffered saline (PBS). Incubation with anti-α antisera (1 : 1000 diluted) was at 4° for 2 hr in PBS containing 0.05% Tween 20. Following 5 washes with PBS/Tween, nitrocellulose strips were incubated with horseradish peroxidase-labeled anti-rabbit IgG (Sigma 8275, 1 : 750) in PBS/Tween for 60 min. After washing with PBS/Tween, antigens were visualized in PBS/Tween containing 0.5 mg/ml 3,3′-diaminobenzidine tetrahydrochloride (Sigma) and 0.015% H_2O_2. Antibodies against sequences from α subunits of G proteins were prepared by Dr. D. Palm, Würzburg. The following peptides of the α_s sequence were synthesized and coupled to bovine serum albumin: 46–59 [α-common, cf. M. S. Mumby, R. A. Kahn, D. R. Manning, and A. G. Gilman, *Proc. Natl. Acad. Sci. U.S.A.* **83,** 265 (1986)] and 379–394 (C terminus specific for α_s). Note that the α-common antiserum (α_c) considerably underestimates α_s compared to the α_i-like peptide with M_r 39,000.

cyclase with GTP-Sepharose, yet to different degrees. Pretreatment of membranes with GMP/isoproterenol before solubilization was necessary in order to achieve reduction of enzymatic activity in the pass-through. It is conceivable that this treatment, the necessity of which seems to be unique to the avian erythrocyte system, results in clearance of the nucleotide binding site from barely exchangeable GDP. The GMP-liganded or empty site now easily allows for binding or exchange of free or immobilized nucleotide. On the other hand, loss of $[AlF_4]^-$ activation was only observed when GTP was present in the adenylyl cyclase assay, obviously because only GDP-, but not GMP-liganded or empty $G_{s\alpha}$, is complexed by $[AlF_4]^-$.[6,13] Figure 2A shows that both nucleotide and $[AlF_4]^-$ stimulation was partially restored on recombination of retained and nonretained fractions, the functional components of which must be proteins, because reconstitution did not occur following treatment with trypsin or *N*-ethylmaleimide.[6] The results with pigeon erythrocyte cyclase could be essentially confirmed with the turkey enzyme by Spiegel *et al.*[5] and Nielsen *et al.*[13]

The Western blot in Fig. 2B shows that both G_s and a G_i-like protein were retained by the GTP-Sepharose following chromatography of solubilized turkey erythrocyte membranes. Figure 2B further underscores the necessity for GMP/isoproterenol pretreatment of membranes, since in its absence no G_s and less G_i-type G protein was bound, as demonstrated by Western blot analysis with anti-G antibodies raised against defined amino acid sequences. At least for G_s it is established that this G protein binds to the affinity support as a α,β,γ heterotrimer, since there is a difference in the sedimentation rate whether the protein is released by a guanosine triphosphate or diphosphate analog.[14] GTP-Sepharose has also been successfully applied for the isolation of ADP-ribosylated G_s (catalyzed by cholera toxin), which could at last verify that the GTP-binding component of adenylyl cyclase was the cellular target of the toxin.[15]

The presented protocol describes the synthesis of an immobilized nucleoside triphosphate derivative as a simple two-step procedure. Although the general applicability of this affinity resin remains to be proved, at least the GTP-binding proteins of the α,β,γ heterotrimer type seem to be good candidates. This group of regulatory proteins, which seems to participate in a variety of transmembrane signaling processes, has a GTP binding site in common which readily tolerates a bulky hydrophobic group at its γ-

[13] T. B. Nielsen, R. W. Downs, and A. M. Spiegel, *Biochem. J.* **190,** 439 (1980).

[14] T. Pfeuffer, *FEBS Lett.* **101,** 85 (1979).

[15] D. Cassel and T. Pfeuffer, *Proc. Natl. Acad. Sci. U.S.A.* **75,** 2669 (1978).

phosphate. On the other hand, this substituent protects the modified triphosphate against hydrolysis by nucleotidases present in crude preparations that would quickly cause its destruction.

It is assumed that GTP-binding proteins are present in membranes at a 10- to 100-fold (or more) excess over receptors and effectors, such that their isolation has indeed been achieved by use of classic (although multistep) purification procedures.[16] However the affinity chromatographic approach may be useful for small-scale isolation of G proteins, dictated by limiting amounts of biological material or by the application of expensive radioisotopes. Furthermore, the presented affinity technique may be indicated for the isolation of those GTP-binding proteins which cannot be isolated by the general G-protein purification protocol developed by Gilman's group.[16] Indeed, GTP-Sepharose was also able to retain a group of low M_r GTP-binding proteins (M_r 23,000–26,000) recently identified in avian erythrocyte membranes by means of the affinity label 4-azidoanilido[^{32}P]GTP.[6,17]

[16] P. Sternweis, M. D. Smigel, and A. G. Gilman, *J. Biol. Chem.* **256,** 11517 (1981).
[17] R. Thomas and T. Pfeuffer, this volume [24].

[15] Purification of G Proteins

By JUAN CODINA, DONNA J. CARTY, LUTZ BIRNBAUMER, and RAVI IYENGAR

G proteins are heterotrimeric signal transducers that are widely distributed. They consist of distinct α subunits that have the guanine nucleotide binding site and β,γ subunits that are very similar for all G proteins. Though only a few G proteins have been isolated by protein purification, the cDNAs for many G-protein α subunits have been isolated. Given the multiplicity of G proteins, the major thrust has been not only to purify the G proteins from other unrelated contaminants, but also to resolve one G protein from another so that they can be individually analyzed for functions. The procedures described here have been used in our laboratories for the purification of G proteins from human erythrocytes, bovine brain, and rat liver. The basic protocol used is similar to the one published in

METHODS IN ENZYMOLOGY, VOL. 195

1984.[1] However, the addition of fast protein liquid chromatography (FPLC) chromatography as the last purification step allows the resolution of closely related G proteins. This has allowed us to isolate all three forms of G_i, as well as two forms of the major brain G protein, G_o.

Materials

All reagents used during the purification are analytical grade or better. Ethylene glycol is purchased from Fisher (Springfield, NJ). The detergents, cholate and Lubrol PX, have to be purified before they can be used. Cholic acid is purified by recrystallization. Approximately 160 g of cholic acid is dissolved in 600 ml of boiling 95% (v/v) ethanol. The cholic acid is precipitated by adding the hot ethanol solution to 4 liters of water. This procedure is repeated 6 times over 4 days. The final precipitate is dried in a 37° oven. The cholic acid, starting as a pale yellow granular powder, will become a white, flaky crystalline powder. Typically a 10% (v/v) stock solution of cholic acid is made by adding NaOH to a suspension of cholic acid in water at room temperature until a pH of 7.5 is reached. Lubrol PX is dissolved in distilled water to yield a 10% solution. One liter of this solution is passed through 30 ml of mixed-bed resin (AG-501-X8, Biorad, Rockville Centre, NY) 3 times prior to use. The final 10% Lubrol solution is adjusted to pH 8.0 with 10 *N* NaOH.

Assays for G Proteins

G_s Assay. G_s, the G protein responsible for stimulation of adenylyl cyclase, is assayed by its capacity to stimulate S49 *cyc*$^-$ cell membrane adenylyl cyclase in the presence of a nonhydrolyzable analog of GTP, such as GPP(NH)P or GTPγS. Generally 10 μg of crude detergent-extract protein will contain measurable amounts of G_s.

The G_s-containing extract (see below) is mixed with crude S49 *cyc*$^-$ membranes (10–20 μg protein) in a final volume of 20 μl and held on ice for 15 min. This mixture is then added to 30 μl of adenylyl cyclase assay mixture such that the final concentrations of the various agents are as follows: sodium HEPES (25 m*M*), 1 m*M* EDTA, 1 m*M* [^{3}H]cAMP (~10,000 cpm), 20 m*M* $MgCl_2$, 50 μM GTPγS or GPP(NH)P, and 0.1 m*M* [α-^{32}P]ATP (5000 cpm/pmol); an ATP-regenerating system is included, consisting of 20 m*M* creatine phosphate, 0.2 mg/ml creatine phosphokinase, and 0.02 mg/ml myokinase. The reaction is allowed to proceed for 30 min at 32°.

[1] J. Codina, J. D. Hildebrandt, R. D. Sekura, M. Birnbaumer, J. Bryan, C. R. Manclark, R. Iyengar, and L. Birnbaumer, *J. Biol. Chem.* **259,** 5871 (1984).

The [^{32}P]/cAMP formed is quantified by the method of Salomon *et al.*[2] (see [1], this volume).

It is crucial that the detergent concentrations be low enough that the adenylyl cyclase activity is not inhibited. An acceptable detergent concentration in the final assay is 0.05–0.1% Lubrol or 0.05% cholate. It is also essential that the G_s assay is in the linear range (i.e., proportional to the amounts of added G_s). To be certain, it is useful to perform the assay with at least two dilutions of G_s that are 5-fold apart.

Pertussis Toxin-Catalyzed [^{32}P]/ADP-Ribose Labeling of G Proteins. The pertussis toxin assay is useful for G proteins such as the various G_i and G_o proteins that have a pertussis toxin-sensitive ADP-ribosylation site, usually a cysteine in the fourth position from the carboxy terminus.[3] This is a very sensitive assay, but it identifies classes of G proteins rather than individual G proteins. (See also [2], this volume.)

Briefly, 0.5–2 μg of G-protein-containing extract is incubated with 25 m*M* sodium HEPES, 1 m*M* EDTA, 100 μM GDP, 1 m*M* ATP, 1 m*M* dithiothreitol (DTT), 0.5% Lubrol, 10 m*M* thymidine, 10 μM [^{32}P]NAD (5×10^6 cpm), and 100 ng/ml activated pertussis toxin for 30 min at 32° in a final volume of 50 μl. Pertussis toxin is activated by incubation with 20 m*M* DTT at a concentration of 10 mg/ml for 20 min at 32°. After the incubation, the sample is diluted to 500 μl with −20° acetone. Two micrograms of bovine serum albumin (BSA) is added as carrier. The acetone precipitation is very useful for the removal of the Lubrol PX, which contributes to anomalous running of the sodium dodecyl sulfate (SDS)–polyacrylamide gel. The proteins are allowed to precipitate for 30 min at 0° and then collected by centrifugation and washed twice with 0.5 ml of 10% (w/v) trichloroacetic acid. The final pellet is washed with cold ether and then dissolved in SDS–PAGE sample buffer, for electrophoresis according to the procedure of Laemmli.[3] It is important that the proteins not be boiled in the sample buffer since they will aggregate. To ensure that all the protein will enter the resolving region of the gel, we incubate the protein for 30 min at 32° in sample buffer.

[^{35}S]GTPγS Binding Assay. The [^{35}S]GTPγS binding assay measures GTP-binding sites and is the least specific, and hence the most risky, of the G-protein assays. It is a useful assay in systems such as the brain where G proteins are particularly abundant, but in other systems, such as the liver, the assay is not useful during the early stages of G-protein purification, since the majority of the GTP-binding sites are not G-protein-

[2] Y. Salomon, C. Londos, and M. Rodbell, *Anal. Biochem.* **58,** 541 (1974).
[3] U. K. Laemmli, *Nature (London)* **227,** 680 (1970).

related sites.[4] However, in the later stages of purification, especially during the resolution of the individual G proteins, this is a very useful assay to detect separation rapidly with minimal expenditure of proteins.

The assay can be conducted with 10–20 ng of G protein per assay tube. The G protein is combined with 25 m*M* sodium HEPES, 1 m*M* EDTA, 20 m*M* 2-mercaptoethanol, 25 m*M* $MgCl_2$, 100 m*M* NaCl, 0.1% Lubrol, and 50 n*M* [^{35}S]GTPγS (~500,000 cpm) in a final volume of 50 μl. Nonspecific binding is measured in the presence of 10 m*M* GTPγS or 10 m*M* GTP. Unlabeled GTPγS and GTP generally give equivalent backgrounds, and since GTP is far less expensive than GTPγS, it is at least economically advisable to use GTP. The binding reaction is allowed to proceed for 30 min at 32°. The reaction is stopped by the addition of 2 ml of ice-cold 25 m*M* Tris-HCl, 25 m*M* $MgCl_2$, and 100 m*M* NaCl. The samples are filtered through nitrocellulose filters (Millipore, Bedford, MA), after which the filters are rapidly washed twice (2 × 10 ml) with ice-cold stopping solution. The filters are dried and counted in 3 ml of liquid scintillation fluid in the liquid scintillation counter.

Quantitative Immunoblotting with Sequence-Specific Antisera. This is the most unequivocal assay for the various G proteins that do not have easy functional assays. This category unfortunately includes all nonretinal G proteins except G_s. However, even when the specific antisera are available, this is a very expensive assay, since it generally involves large expenditure of proteins (100 ng–2 μg of pure proteins). This assay is dealt with in detail in [26], this volume.

Preparation of Membranes

For the purification of G proteins it is generally sufficient to prepare crude particulate preparations (10,000–40,000 *g*) fractions. The red cells from about 3 liters of blood are washed by centrifugation with four 1-liter batches of 5 m*M* potassium phosphate containing 100 m*M* NaCl, to remove the white buffy coats. This procedure is repeated 3–4 times until all of the white cells are removed. The washed cells (~1.5 liters) are lysed in 40 liters of potassium phosphate buffer and collected by continuous flow centrifugation at 35,000 *g*, at a flow rate of 100–130 ml/min. The membranes are repeatedly washed with 5 m*M* potassium phosphate buffer at 16,500 *g* for 30 min, until light pink membranes are obtained. The membranes are finally washed with 10 m*M* sodium HEPES, pH 8.0, and stored at −70° until used.

[4] Y. Salomon and M. Rodbell, *J. Biol Chem.* **254,** 7245 (1975).

Bovine brain membranes are prepared essentially according to the procedure of Sternweis and Robishaw.[5] S49 *cyc*$^-$ cell membranes that are used for G_s assays are prepared by the method of Ross *et al.*,[6] except that Mg^{2+}-free buffers are used throughout.

Extraction of Membranes

It is generally advisable to wash the membranes thoroughly prior to extraction into detergent. Typically 6–8 g of membrane protein/liter is washed 2 times with 5 liters of 20 m*M* sodium HEPES, 200 m*M* NaCl, and 20 m*M* 2-mercaptoethanol, pH 8.0. The membranes are then preextracted by suspension in 20 m*M* sodium HEPES, 200 m*M* NaCl, 20 m*M* 2-mercaptoethanol, and 0.1% cholate, pH 8.0. This mixture is stirred for 30 min in the cold and then centrifuged at 70,000 g for 1 hr. The pellet is resuspended in 10 m*M* sodium HEPES, 20 m*M* 2-mercaptoethanol, and 1 m*M* EDTA, in a final volume of about 1300 ml. The 10% cholate solution is then added so that the final concentration is 1% cholate. For human erythrocyte preparations, $MgCl_2$ is also added if needed to obtain a final concentration of 10 m*M*. The protein suspension is stirred in the cold for 1 hr. The extract is collected by centrifugation at 100,000 *g* for 90 min.

It is necessary to remove the supernatant carefully so as not to include any particulates. We generally siphon as much as possible into a beaker and collect the rest separately using a plastic Pasteur pipette. If $MgCl_2$ is used during the extraction, EDTA is added at this stage to a final concentration of 11 m*M*. The extract is made 30% with respect to ethylene glycol and loaded onto the first DEAE-Sephacel column.

All solutions after this stage are filtered through 0.45-μm filters to avoid clogging the columns. From this stage on, all chromatographies, dilutions, etc., are carried out in buffer A [10 m*M* sodium HEPES, 1 m*M* EDTA, 20 m*M* 2-mercaptoethanol, and 30% (v/v) ethylene glycol, pH 8.0 (final)].

Purification Procedure

First DEAE-Sephacel Chromatography

The DEAE-Sephacel should be preequilibrated with buffer A containing 0.9% cholate. This is done by passing through at least 2, preferably 3,

[5] P. C. Sternweis and J. Robishaw, *J. Biol. Chem.* **259,** 13806 (1984).

[6] E. M. Ross, M. E. Maguire, T. M. Sturgill, R. L. Biltonin, and A. G. Gilman, *J. Biol. Chem.* **252,** 3568 (1977).

column volumes of the cholate-containing buffer A. Generally a 1-liter column is used, and 3–4 g of extracted protein in 1.5 to 2 liters is loaded onto the column at 1 ml/min. After the sample is loaded, the column is washed with 1 liter of 0.9% cholate in buffer A. The proteins are eluted with a linear gradient (2 × 1 liter) of 0–0.5 *M* NaCl in 0.9% cholate-containing buffer A.

At this stage, the G proteins are not separated significantly from each other. In both human erythrocytes and brain preparations, G_s and G_i are overlapping. Two hundred fractions of 10 ml are collected, and every third fraction is assayed. The peak tubes are identified and collected. It is necessary to cut the peak sharply so as to avoid high molecular weight contaminants that are quite difficult to get rid of in later chromatographic steps. We collect no more than 20 tubes (~200 ml) representing only the top half of the peak.

The pooled sample is dialyzed twice against 5 liters each time of 0.9% cholate in buffer A overnight. The dialyzed pool is then concentrated to 40–50 ml over an Amicon (Danvers, MA) PM/30 filter by ultrafiltration.

Gel Filtration

The concentrated material in buffer A containing 0.9% cholate is applied to a 1-liter Ultrogel AcA 34 (Pharmacia-LKB, Piscataway, NJ) column (5 × 60 cm) equilibrated with 0.9% cholate and 100 m*M* NaCl in buffer A. The column is then eluted with 1 liter of the equilibration buffer. Two hundred fractions of 5 ml each are collected. For human erythrocyte preparations, G_s activity eluted as two distinct peaks which are labeled A and B and processed separately. For brain preparations, generally only a single G-protein peak is seen, though double peaks are sometimes seen. Typical yields at this stage are 20–30% of the activity.

Heptylamine Chromatography

Heptylamine chromatography is crucial in resolution of G proteins. The gel filtration peak is diluted to reduce the concentration of cholate to 0.4%. The approximately 500 ml column (5 × 30 cm) is equilibrated with buffer A containing 0.4% cholate and 100 m*M* NaCl. The flow rate is maintained at 0.5 ml/min. After the column is loaded, it is washed with 1 liter of buffer A containing 0.4% cholate and 100 m*M* NaCl. This is followed by a wash with 1 liter of buffer A containing 0.4% (w/v) cholate and 500 m*M* NaCl. The G proteins are eluted with a double-reciprocal gradient (2 × 1 liter) of 0.4 → 4% cholate and 500 → 0 m*M* NaCl at a flow rate of 1 ml/min. Four hundred 5-ml fractions are collected. The G_i proteins elute earlier than the G_s protein. The G proteins elute at 2–2.5% cholate in our laboratories. Since the proteins are unstable in such a high concentration of cholate, the sample should be diluted at least 5-fold into buffer A without cholate but containing 0.1% Lubrol. If the chromatographies are run correctly and the peaks cut sharply, by this stage the G proteins should be substantially free of other contaminants (~80–95% pure).

Before reuse, the heptylamine-Sepharose is regenerated by washing with 5 column volumes of 0.1% SDS and 3 *M* urea in deionized and filtered water. This procedure is done by unpacking the gel from the column and washing the gel in a sintered glass funnel at room temperature. After regeneration, 500 ml of the gel is washed 10 times with 2 liters of filtered water at room temperature. This extensive washing at room temperature with filtered water is crucial to remove all traces of the urea and SDS used during the regeneration. The gel is then equilibrated with 5 volumes of buffer A containing 0.4% cholate and 100 m*M* NaCl.

Second DEAE Ion-Exchange Chromatography

The heptylamine-Sepharose peaks are diluted into buffer A containing 0.5% Lubrol PX such that the final concentration of cholate is no greater than 0.3%. This generally entails a 7- to 10-fold dilution. The proteins are then loaded onto a 20-ml DEAE-Toyopearl column (Pierce Chemicals, Rockford, IL) that is packed in a 50-ml plastic syringe. It is crucial that, after this stage, all the materials, including the columns, be made of plastic to minimize protein losses by adsorption to glass surfaces.

The DEAE-Toyopearl is equilibrated in buffer A containing 0.1% Lubrol PX, after which the diluted sample is loaded at a rate of about 0.5 ml/min. The column is washed with 10 volumes of buffer A containing 0.1% Lubrol PX to accomplish the complete removal of cholate. The column is then eluted with a linear gradient (2 × 30 ml) of 0–300 m*M* NaCl. Sixty 1-ml fractions are collected and analyzed by both activity measurements and SDS–PAGE. The SDS–PAGE analysis is important to determine where the high molecular weight contaminants are, in order to pool the peaks of desired proteins while minimizing these contaminants. Generally, brain preparations are so enriched in G proteins that the high molecular weight contaminants are not a significant (>5%) proportion of the total protein. In contrast, the erythrocyte and liver preparations contain much more (10–20%) contaminating protein. In a typical G-protein preparation at least two second DEAE ion-exchange chromatographies are run since the G_s is processed separately from the G_i. For brain preparations, we do not separate G_o from G_i since excellent separation can be achieved during the FPLC chromatography. The G-protein-containing fractions from the second DEAE chromatography are pooled. At this stage

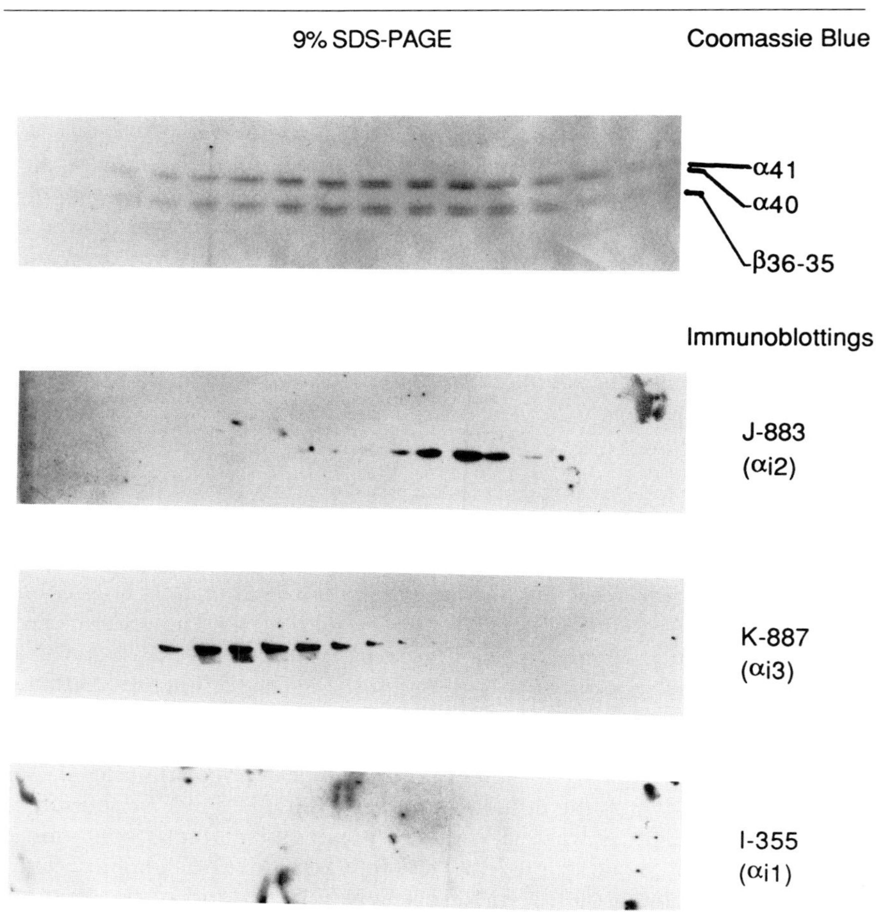

FIG. 1. Resolution of human erythrocyte G_{i2} and G_{i3} by a 100–300 m*M* NaCl gradient on a Mono Q column using FPLC chromatography. Alternate fractions were resolved on 9% SDS–polyacrylamide gels by electrophoresis and visualized either by Coomassie blue staining or by immunoblotting with a_i-subtype-specific antisera.

we have 5–10 mg protein in about 10–12 ml. The proteins can be diluted in buffer A containing 0.1% Lubrol such that the NaCl concentration is 100 m*M* and stored at −70°.

FPLC Chromatography

Resolution of G proteins by chromatography over Mono Q columns using FPLC systems was introduced by Katada *et al.*[7] and is very useful

[7] T. Katada, M. Oinuma, K. Kusakabe, and M. Ui, *FEBS Lett.* **213,** 353 (1987).

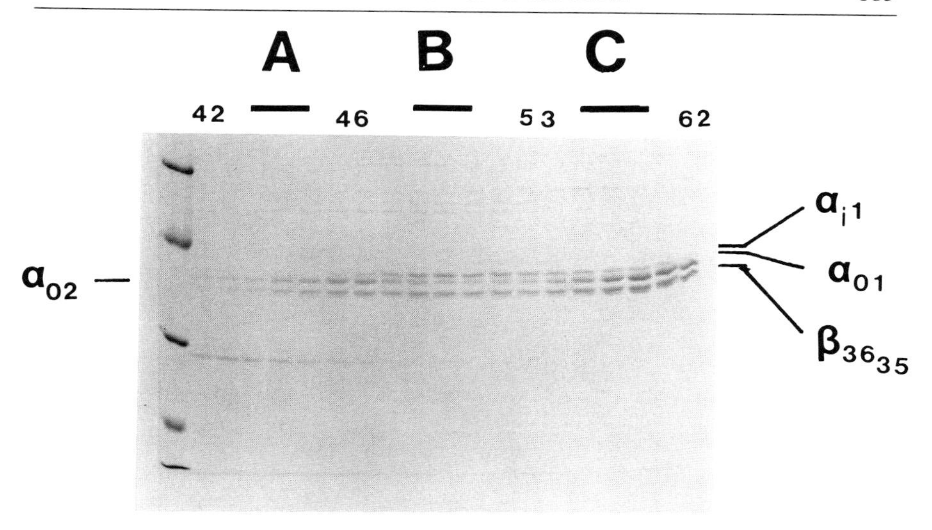

FIG. 2. Resolution of two forms of G_o and G_i from a bovine brain G-protein mixture by a 100–300 m*M* NaCl gradient on a Mono Q column using FPLC chromatography. Fraction numbers used to constitute pools A, B, and C are shown. The identity of the G proteins were determined by immunoblotting analysis (see Fig. 3).

in the resolution of the various G proteins. The pooled samples obtained from the DEAE-Toyopearl chromatography are diluted in a buffer containing 25 m*M* sodium HEPES, 1 m*M* EDTA, 20 m*M* 2-mercaptoethanol, 0.1% Lubrol PX, 11 m*M* CHAPS, and 100 m*M* NaCl, pH 8.0 (buffer B) and loaded onto a Mono Q column at a flow rate of 1 ml/min with a 50 ml Superloop (Pharmacia). The Mono Q column is equilibrated with at least 10 volumes (~10 ml) of buffer B prior to loading. It is crucial that only 5–6 mg protein be loaded on the Mono Q column; overloading generally results in poor resolution. After loading, the column is washed with 10 ml of buffer B. The proteins are then eluted by a linear NaCl gradient of buffer B to buffer C, which is identical to buffer B except that the NaCl concentration is 1 *M*. The gradient is run from 0 to 30% buffer B (100–300 m*M* NaCl) in 60 min at a flow rate of 0.8 ml/min. Ninety to one hundred fractions of approximately 0.5 ml are collected. The fractions are assayed generally by SDS–PAGE followed by staining or by immunoblotting. Separations of human erythrocyte G_{i2} and G_{i3} on a FPLC column are shown in Fig. 1.

As can be seen in Fig. 1, full separation of G_{i2} and G_{i3} is not possible, in spite of a programmable gradient former. Extremely shallow gradients resulted in smearing of the proteins along the length of the gradient. It was better to use a steeper gradient and to rechromatograph the pooled peak to obtain a pure protein. The same salt gradient could also resolve two

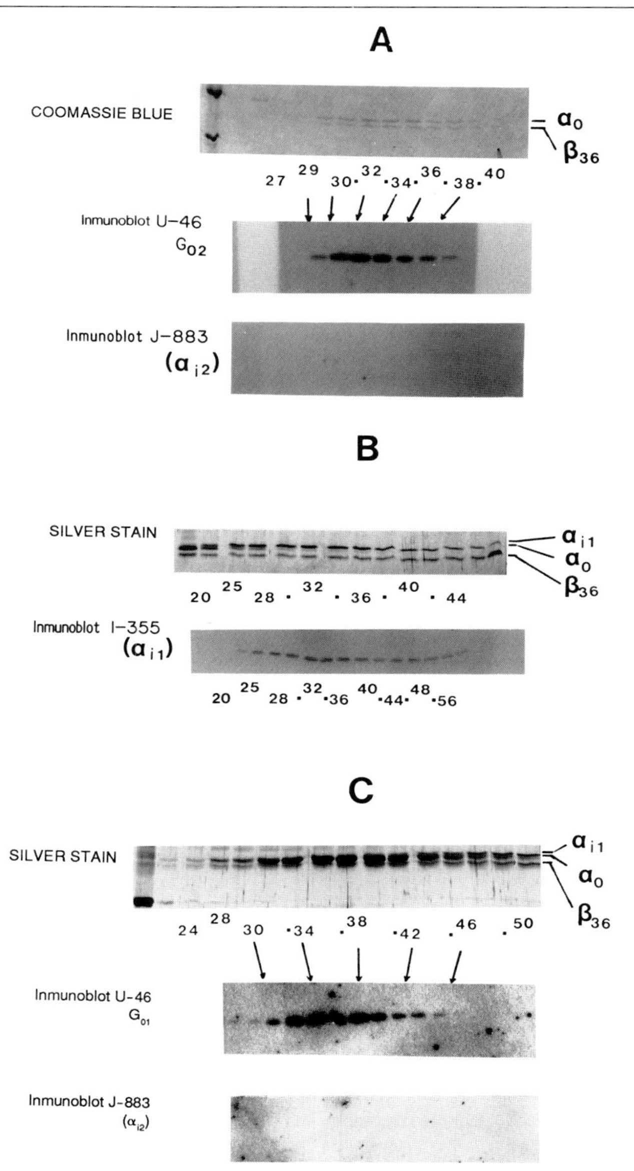

FIG. 3. FPLC resolution and immunoblotting of the multiple forms of the bovine brain G proteins. Peaks A, B, and C from the chromatography in Fig. 2 were rechromatographed individually on Mono Q columns using 0–3% cholate gradients. The column fractions were then electrophoresed on 9% SDS–polyacrylamide gels and stained with silver or blotted with the indicated antisera. (A) Individual fractions from 27 to 40 are shown. The upper band in

forms of G_o and G_{i1} from bovine brain G-protein mixtures as shown in Figs. 2 and 3. the three pools a, b, and c were constituted from the fractions. Each of these pools was rechromatographed, and the indicated fractions were blotted with the α-subunit-specific antiserum. This is shown in Fig. 3. It can be seen that both peaks A and C are composed mostly of G_o while peak B is composed mostly of G_{i1}. Such resolution led to the identification of two distinct forms of G_o.

Like the G proteins from human erythrocyte preparations, the brain G proteins can also be separated from each other by rechromatography. Each FPLC chromatography takes only 2–3 hr, and hence it is possible to run up to two chromatographies during 1 day. If, however, immunoblot analysis is required, then we always add ethylene glycol to the Mono Q column fractions and store them in the cold room (0°–2°) during the immunoblot analysis. While we have not determined in a rigorous fashion that CHAPS is essential for the FPLC chromatography, it appears that the yields as measured by G_s activity are better in the presence of CHAPS. Further, CHAPS does not have any negative effects on the resolution. Therefore we have always had CHAPS present during the FPLC chromatography.

Storage of Proteins

After the final chromatographies, the relevant fractions are pooled, diluted 20-fold with buffer A containing 0.1% Lubrol to reduce NaCl concentration to 10 m*M*, and loaded onto a 100–200 μl DEAE-Sephacel column. After the sample is loaded, the column is washed with 10 ml of buffer A with 0.1% Lubrol and finally eluted with buffer A with 0.1% Lubrol and 200 m*M* NaCl in 100-μl aliquots. The eluate is diluted 2-fold with buffer A containing 0.1% Lubrol PX to reduce the NaCl concentration to 100 m*M*. Protein concentrations are estimated by Coomassie blue staining of protein resolved by SDS-gel electrophoresis, with BSA (0.1–1 μg) as a standard. Details of these measurements are given in Chapter 26. Aliquots (10–30 μl) are stored at −70°. For the final storage, the protein concentration should be in the 100 μg/ml range. We have stored proteins at concentrations as low as 30 μg/ml. Repeated freezing and thawing of G proteins is not recommended. We have found that freezing and thawing 5 times results in a 75% loss of GTPγS binding activity for G_o.

---

fractions 30–33 was found to be α_{i1} by immunoblotting with antiserum I-355. (B) Alternate fractions from 28 to 48 are shown. The lower band in fractions 20–25 was identified as α_o by immunoblotting with the antiserum U-46 (not shown). (C) Alternate fractions from 30 to 50 are shown. The upper band in fractions 46–50 was identified as α_{i1} by immunoblotting with antiserum I-355. For details about the cholate gradient, see the text.

Alternative Sources of Erythrocytes

Given the current concerns about viral contamination of human blood we have recently started to utilize porcine blood as the source of erythrocytes. The recovery of G_s, G_{i2}, and G_{i3} from porcine erythrocytes is very similar to that obtained from human erythrocytes.

[16] Preparation of Guanine Nucleotide-Free G Proteins

By KENNETH M. FERGUSON and TSUTOMU HIGASHIJIMA

Introduction

Guanine nucleotide-binding regulatory proteins (G proteins) tightly bind guanine di- and triphosphates.[1] The affinity of these proteins for guanine nucleotides is sufficient to preserve the nucleotide-bound form during the common G-protein purification procedures.[2] This retention of the nucleotide complicates certain types of nucleotide-binding measurements. To facilitate such studies, methods were devised to measure the amount of nucleotide in the G-protein preparation and to remove the nucleotide without denaturing the protein.

Measurement of Bound Guanine Nucleotide

A spectrophotometric or a competitive-binding assay can be used to measure nucleotides released from G proteins. Spectrophotometry is convenient when a large amount of protein is available, whereas the competitive-binding assay is more sensitive and thus requires less protein.

Spectrophotometric Assay

The spectrophotometric assay requires 200-μl aliquots of at least 5 μM G protein. Buffer A is 20 mM sodium HEPES, pH 7.6, 1mM sodium EDTA, 1 mM dithiothreitol (DTT), and 0.1% Lubrol.

[1] A. G. Gilman, *Annu. Rev. Biochem.* **56,** 615 (1987).

[2] K. M. Ferguson, T. Higashijima, M. D. Smigel, and A. G. Gilman, *J. Biol. Chem.* **261,** 7393 (1986).

1. Dilute the G-protein samples in buffer A to provide 200-μl aliquots of at least 5 μM G protein. Prepare three additional 200-μl samples containing buffer A, 10 μM GTP in buffer A, or 10 μM GDP in buffer A. Boil the samples for 3 min, transfer each solution to a Centricon microconcentrator (PM30 membrane, Amicon, Danvers, MA), and collect the filtrate by centrifugation at 5000 rpm in an SS-34 rotor for 1 hr at 4°. Add an additional 200 μl of buffer A to each microconcentrator, collect it by centrifugation, and combine it with the original filtrate.

2. Measure the absorbance of the filtrates at 253 nm and determine the concentration of nucleotide. The extinction coefficient of either GDP or GTP is 13,700 M^{-1} cm^{-1} at 253 nm.[3]

3. (Optional) The guanine nucleotide composition of each filtrate can be determined by fast protein liquid chromatography (FPLC) ion-exchange chromatography. Prepare a 100-μl solution containing GMP, GDP, and GTP in buffer A, each at a concentration of 2.5 μM. Apply the nucleotide mixture (100 μl) to a Polyanion SI HR 5/5 (Pharmacia, Piscataway, NJ) anion-exchange column that has been equilibrated with 10 mM potassium phosphate (KP_i), pH 7.0. Wash the column with 2 ml of the equilibration buffer at a rate of 1 ml/min and elute the nucleotides with a gradient of KP_i, pH 7.0, consisting of the following concentration changes: 10 to 300 mM KP_i over 1 min, 300 mM KP_i for 2 min, 300 to 650 mM KP_i over 20 min, and 650 to 700 mM KP_i over 4 min. Monitor the column elute at 254 nm and note the time of elution of the three nucleotides. Repeat the procedure with each of the filtrates and compare the time of the elution of the ultraviolet-absorbing material with that of GMP, GDP, and GTP to identify the nucleotide in the filtrate.

Note: The chromatography described in Step 3 may not distinguish between a particular ribonucleotide and its deoxyribonucleotide form. A portion of the guanine nucleotide bound to purified *ras* proteins, which also bind guanine nucleotides tightly, was dGDP.[4] Thus, deoxyribonucleotides may be present in G-protein preparations.

Competitive-Binding Assay

A 50-μl aliquot of a 250 nM G-protein solution provides adequate material for the competitive-binding assay. Buffer B is 10 mM KP_i, pH 8.0, 1 mM $MgCl_2$, and 1 mM DTT.

[3] R. M. C. Dawson, D. C. Elliot, W. H. Elliot, and K. M. Jones (eds.), "Data for Biochemical Research," 2nd Ed., p. 174. Oxford Univ. Press, London, 1969.

[4] J. Tucker, G. Sczakiel, J. Feuerstein, J. John, R. S. Goody, and A. Wittinghofer, *EMBO J.* **5,** 1351 (1986).

1. Prepare a set of GDP-containing samples in buffer A (e.g., 7, 14, 35, 70, 140, 350, 700, 1400, and 7000 n*M* GDP). This set will provide a calibration curve for the assay.

2. Boil 50-μl aliquots of the G-protein-containing samples and of two GDP standards (the 35 and 350 n*M* standards) for 3 min. Dilute the protein-containing samples in buffer A, if needed, to bring the nucleotide concentration within the range of the calibration curve samples prepared in Step 1.

3. Mix duplicate 10-μl aliquots of the GDP samples prepared in Step 1 with 30 μl of buffer A containing 25 n*M* [^{35}S]GTPγS (250 cpm/fmol) in 10 $\times$ 75 mm plastic or siliconized glass test tubes. Prepare four additional samples by mixing 10 μl of buffer A with 30 μl of buffer A containing 25 n*M* [^{35}S]GTPγS (250 cpm/fmol). Reserve two of these additional samples; they will be used to assess the nonspecific binding of the nucleotide to the filter (see below). Keep the samples on ice until needed.

4. Mix duplicate 10-μl aliquots of the boiled samples prepared in Step 2 with 30 μl of buffer A containing 25 n*M* [^{35}S]GTPγS (250 cpm/fmol) in 10 $\times$ 75 mm plastic or siliconized glass test tubes. Keep them on ice until needed.

5. Dilute purified G protein (G_i or G_o, see Note 4 below) in buffer A to yield a 5 n*M* solution. Add 30 μl of the G protein to the first sample, mix, then incubate it at 30° (20° if G_o is used). Repeat for each sample at 1-min intervals, except for the two samples reserved for the measurement of nonspecific binding in Step 3 (see above). To these samples add 30 μl buffer A instead of the G-protein solution, mix, and incubate with the other samples.

6. After 1 hr, dilute each sample with 2 ml of buffer B and apply to a BA-85 nitrocellulose filter (Schleicher and Schuell, Keene, NH) mounted on a vacuum filtration unit. Add an additional 2 ml of buffer B to the assay tube and apply to the same filter. Wash each filter 4 times with 3 ml of buffer B and set aside to dry. Mix each of the dried filters with scintillation cocktail and measure the filter-bound radioactivity by scintillation counting.

7. The radioactivity bound to the filter when no G protein is included in the sample provides an estimate of the nonspecific binding of the nucleotide to the filter. The nonspecific binding should be less than 0.1% of the total radioactivity applied to the filter. Subtract the value for the nonspecific binding from each of the other samples. Construct a calibration curve by plotting the amount of radioactivity associated with the GDP-containing samples, prepared in Step 1 (see above), as a function of the GDP concentration. Use this calibration curve to determine the amount of "GDP equivalents" present in the protein-containing samples.

Notes: (1) Two of the GDP standards were boiled to determine the amount of degradation caused by the boiling. They should be unaffected. (2) [α-^{32}P]GDP or [α-^{32}P]GTP can be used instead of [^{35}S]GTPγS. (3) The affinity of G proteins for guanine nucleotides is affected by ionic strength and by various metal ions.[1,5-7] Therefore, the standard curve and protein samples must be in the same buffer. (4) This procedure has been tested with the G proteins G_i, purified from rabbit liver, and G_o, purified from bovine brain. When G_o is used, the incubation temperature should be changed to 20° because this protein is less stable at 30°. Presumably, other G proteins can be substituted.[8,9] (5) The term "GDP equivalents" (Step 7, see above) is used because this assay does not distinguish various types of guanine nucleotides.

Preparation of GDP-Free G Proteins

Ammonium sulfate increases the rate of GDP dissociation from G proteins. Chromatography in the presence of ammonium sulfate effectively removes the bound nucleotide.

1. Apply up to 100 μl of 1 μM G protein to a Sephadex G-25 column (0.9 × 25 cm) that has been equilibrated with buffer A containing 1 M $(NH_4)_2SO_4$ and 20% (v/v) glycerol. Run the column overnight at 1 ml/hr. Collect 0.5-ml fractions.
2. Locate the G-protein peak by mixing a 10-μl aliquot of each column fraction with 90 μl of buffer A containing 1 μM [^{35}S]GTPγS (10–20 cpm/fmol), incubating each sample for 1 hr at 30°, and measuring the bound nucleotide as described in Step 6 of the competitive-binding assay (see above).
3. Pool the G-protein-containing fractions and remove the $(NH_4)_2SO_4$ by chromatography on a Sephadex G-25 column (0.9 × 10 cm) equilibrated with buffer A containing 20% (v/v) glycerol. G_i can be stored at 4° for up to 1 week. G_o should be stored at −80°. It is stable for at least 1 month.

[5] T. Higashijima, K. M. Ferguson, P. C. Sternweis, E. M. Ross, M. D. Smigel, and A. G. Gilman, *J. Biol. Chem.* **262,** 752 (1987).

[6] T. Higashijima, K. M. Ferguson, M. D. Smigel, and A. G. Gilman, *J. Biol. Chem.* **262,** 757 (1987).

[7] T. Higashijima, K. M. Ferguson, P. C. Sternweis, M. D. Smigel, and A. G. Gilman, *J. Biol. Chem.* **262,** 762 (1987).

[8] P. C. Sternweis, J. K. Northup, M. D. Smigel, and A. G. Gilman, *J. Biol. Chem.* **256,** 11517 (1981).

[9] P. C. Sternweis and J. D. Robishaw, *J. Biol. Chem.* **259,** 13806 (1984).

Notes: (1) G-protein samples that are more concentrated than 1 μM may require a slower flow rate and a longer column to obtain complete separation of the protein from the nucleotide. (2) The recovery of active protein, as measured by nucleotide binding, varies with the type of G protein. Greater than 80% of the G_i, 50% of the G_o, and 30% of the $G_{o\alpha}$ applied to the column should be recovered. The amount of $G_{o\alpha}$ free of nucleotide should be about 80% of the total protein, while the amount of G_i and G_o should be greater than 90% free of nucleotide. (3) The 20% (v/v) glycerol included in the buffers increases the stability of the nucleotide-free G proteins. (4) The volume of the column specified in Step 3 can accommodate up to 1.5 ml of the $(NH_4)_2SO_4$-containing solution. The volume of the column should be increased when removing $(NH_4)_2SO_4$ from larger volumes. The total column volume should be 5 times the volume of the applied sample.

[17] Purification of Recombinant $G_{s\alpha}$

By MICHAEL P. GRAZIANO, MICHAEL FREISSMUTH, and ALFRED G. GILMAN

Introduction

Members of a family of guanine nucleotide-binding regulatory proteins (G proteins) serve to couple a large number of receptors to appropriate intracellular effectors.[1] G proteins exist as heterotrimers, composed of α, β, and γ subunits (designated in order of decreasing molecular weight). The α subunit, which is unique for each oligomer, contains a single high-affinity binding site for guanine nucleotides and possesses an intrinsic GTPase activity. To date, direct interactions of a G-protein subunit with an effector have been demonstrated only for the α subunit. Detailed reviews of the properties of individual members of the G protein family and their mechanisms of action have been presented elsewhere.[1,2]

The α subunit of G_s, the G protein responsible for stimulation of adenylyl cyclase, exists in either of two electrophoretically distinct classes that migrate on sodium dodecyl sulfate-polyacrylamide gels with apparent molecular weights of 45,000 or 52,000.[3] This heterogeneity is due to alterna-

[1] A. G. Gilman, *Annu. Rev. Biochem.* **56,** 615 (1987).
[2] P. J. Casey and A. G. Gilman, *J. Biol. Chem.* **263,** 2577 (1988).
[3] J. K. Northup, P. C. Sternweis, M. D. Smigel, L. S. Schleifer, E. M. Ross, and A. G. Gilman, *Proc. Natl. Acad. Sci. U.S.A.* **77,** 6516 (1980).

METHODS IN ENZYMOLOGY, VOL. 195

tive splicing of messenger RNA that is transcribed from a single gene. cDNA[4,5] and genomic[6] cloning studies indicate that the proteins in these two classes differ primarily by the inclusion or deletion of 15 amino acid residues that are encoded by exon III of the gene for $G_{s\alpha}$. Furthermore, each class apparently consists of two proteins that differ by the inclusion or deletion of a single serine residue that is encoded by nucleotides at the splice junction. The distribution of these various forms within a single tissue is variable. Some tissues contain only M_r 45,000 or 52,000 forms of $G_{s\alpha}$, although multiple forms of the protein coexist in most. Single cell types also contain multiple forms of $G_{s\alpha}$. Detailed study of individual $G_{s\alpha}$ subunits has been hampered by the fact that it is difficult to resolve them chromatographically. Expression in *Escherichia coli* permits unambiguous assessment of the activities of these various forms of $G_{s\alpha}$.

In this chapter we describe methods for the expression in *E. coli* and rapid purification of milligram quantities of two of the four forms of $G_{s\alpha}$. Recombinant $G_{s\alpha\text{-s}}$ ($rG_{s\alpha\text{-s}}$) and $rG_{s\alpha\text{-L}}$ encode 45- and 52-kDa forms of $G_{s\alpha}$, respectively. The two cDNAs employed are described in detail elsewhere.[4]

Expression of $G_{s\alpha}$ in *Escherichia coli*

Subcloning of the $G_{s\alpha}$ cDNA into a Vector for Expression in Escherichia coli. Methods employed in the construction of $G_{s\alpha}$ expression vectors are detailed in Maniatis *et al.*[7] A prokaryotic expression system developed by Tabor and Richardson has been utilized.[8] Characteristics of this two-plasmid system are outlined in Fig. 1. pT7-5 confers ampicillin resistance and contains a multicloning site downstream from a bacteriophage T7 RNA polymerase promoter. This plasmid was sequentially cleaved in its multicloning site with *Eco*RI and *Bam*HI and was purified by agarose gel electrophoresis. The following two oligonucleotides were synthesized, annealed, and ligated with the cleaved plasmid (Rbs, ribosomal binding sequence):

```
                Rbs              NcoI     SmaI
5'  AATTCTAAGGAGGTTTAACCATGGCCCGGGG          3'
3'         GATTCCTCCAAATTGGTACCGGGCCCCCTAG   5'
```

[4] J. D. Robishaw, M. D. Smigel, and A. G. Gilman, *J. Biol. Chem.* **261,** 9587 (1986).

[5] P. Bray, A. Carter, C. Simons, V. Guo, C. Puckett, J. Kamholtz, A. Spiegel, and M. Nirenberg, *Proc. Natl. Acad. Sci. U.S.A.* **83,** 8893 (1986).

[6] T. Kozasa, H. Itoh, T. Tsukamoto, and Y. Kaziro, *Proc. Natl. Acad. Sci. U.S.A.* **85,** 2081 (1988).

[7] T. Maniatis, E. Fritsch, and J. Sambrook, "Molecular Cloning." Cold Spring Harbor Laboratory, Cold Spring Harbor, New York, 1982.

[8] S. Tabor and C. C. Richardson, *Proc. Natl. Acad. Sci. U.S.A.* **82,** 1074 (1985).

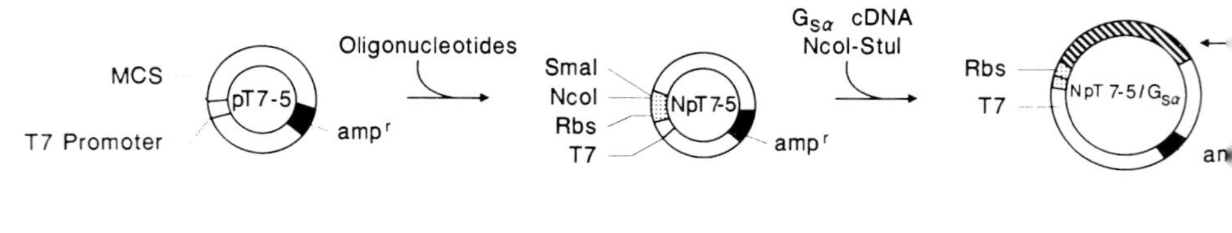

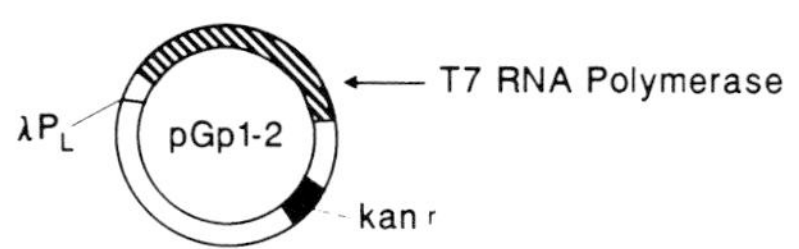

FIG. 1. Construction of plasmids employed for the expression of $rG_{s\alpha}$. For explanation, see text.

The modified plasmid (NpT7-5) was cleaved sequentially with *Nco*I and *Sma*I and ligated with the 1167-base pair *Nco*I–*Stu*I fragment of $G_{s\alpha\text{-}s}$ (1202 base pairs for $G_{s\alpha\text{-}L}$).[9] This construction, NpT7-5/$G_{s\alpha}$, places the initiation codon for $G_{s\alpha}$ 10 nucleotides downstream from a consensus ribosomal binding sequence (Rbs).[10] pGp1-2[8] confers kanamycin resistance and contains the cDNA for T7 RNA polymerase downstream from a bacteriophage λ P_L promoter. Transcription of the T7 RNA polymerase gene is repressed at 30° by the thermolabile cI857 repressor and is derepressed at 42°. *Escherichia coli* strain K38[11] was sequentially transformed with plasmids pGp1-2 and NpT7-5/$G_{s\alpha}$. Double transformants were selected at 30° on L broth plates containing 50 μg/ml kanamycin and ampicillin.

Culture and Lysis of Escherichia coli Expressing $G_{s\alpha}$. Routine preparation of $rG_{s\alpha}$ is performed as follows. Nine liters (1 × 9 liters in 2-liter Erlenmeyer flasks) of enriched medium[12] containing 50 μg/ml of kanamycin and ampicillin is inoculated with *E. coli* K38 harboring plasmids pGp1-2 and either NpT7-5/$G_{s\alpha\text{-}s}$ or NpT7-5/$G_{s\alpha\text{-}L}$. The flasks are maintained at 30° in a rotary air shaker at 300–350 rpm. No significant accumula-

[9] J. D. Robishaw, D. W. Russell, B. A. Harris, M. D. Smigel, and A. G. Gilman, *Proc. Natl. Acad. Sci. U.S.A.* **83,** 1251 (1986).

[10] J. Shine and L. Dalgarno, *Proc. Natl. Acad. Sci. U.S.A.* **71,** 1342 (1974).

[11] M. Russel and P. Model, *J. Bacteriol.* **159,** 1034 (1984).

[12] Solutions used in the expression and purification of recombinant $G_{s\alpha}$ are enriched medium: 2% Bacto-tryptone, 1% yeast extract, 0.2% glycerol, 0.5% NaCl; TEDP: 50 m*M* Tris-HCl, pH 8.0, 1 m*M* sodium EDTA, 1 m*M* dithiothreitol, 0.1 m*M* phenylmethylsulfonyl fluoride; TDPK: 50 m*M* Tris-HCl, pH 8.0, 1 m*M* dithiothreitol, 0.1 m*M* phenylmethylsulfonyl fluoride, 10 m*M* KH_2PO_4; Mono Q: 50 m*M* Tris-HCl, pH 8.0, 20 μ*M* sodium EDTA, 1 m*M* dithiothreitol; HPHT: 10 m*M* Tris-HCl, 1 m*M* dithiothreitol, 10 m*M* KH_2PO_4/K_2HPO_4, pH 8.0; HED: 50 m*M* sodium HEPES, 1 m*M* sodium EDTA, 1 m*M* dithiothreitol; HME: 20 m*M* sodium HEPES, 2 m*M* $MgCl_2$, 1 m*M* sodium EDTA.

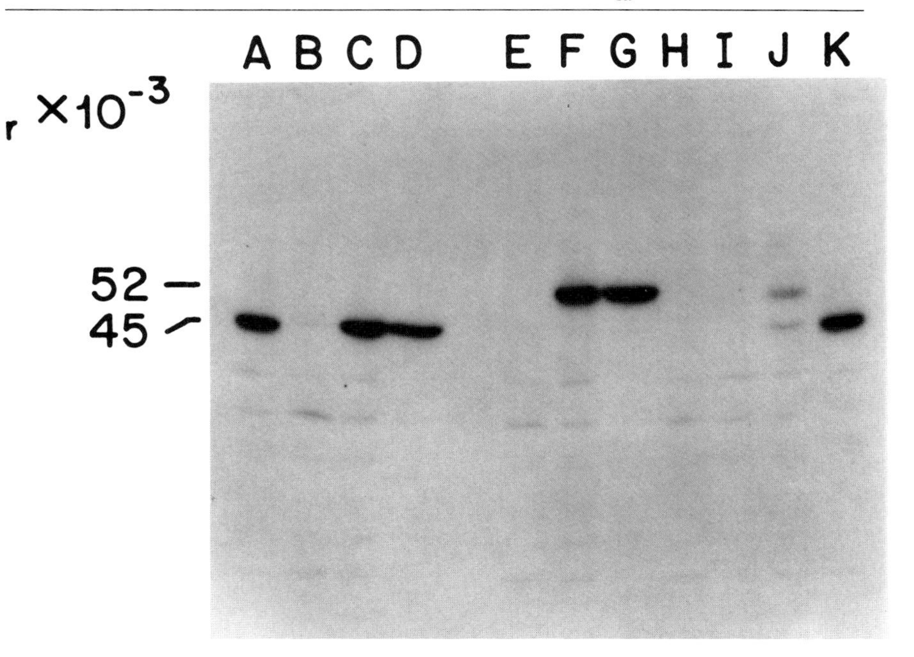

FIG. 2. Immunoblot analysis of K38/pGpl-2 harboring $G_{s\alpha}$ expression plasmids. Cultures of *E. coli* K38/pGpl-2 harboring plasmid NpT7-5/$G_{s\alpha\text{-S}}$ (lanes B, C, and D) or N_pT7-5/$G_{s\alpha\text{-L}}$ (lanes E, F, and G) or a similar plasmid with $G_{s\alpha\text{-L}}$ cDNA in the reverse orientation (lanes H and I) were grown with and without induction. After lysis, supernatants were prepared as described in the text. Aliquots of total cell pellets (lanes B, C, E, F, H, and I) representing 100 μl of culture or aliquots of 100,000 *g* supernatants prepared from 200 μl of culture (lanes D and G) were resolved by SDS–PAGE on an 11% polyacrylamide gel and immunoblotted as described [S. M. Mumby, R. A. Kahn, D. R. Manning, and A. G. Gilman, *Proc. Natl. Acad. Sci. U.S.A.* **83,** 265 (1986)]. Blots were processed with affinity-purified A-572, a $G_{s\alpha}$-specific antipeptide antibody (Mumby *et al., op. cit.*), and goat anti-rabbit ^{125}I-labeled F(ab′)$_2$ (2.5 × 10^5 cpm/ml). The blot was subjected to autoradiography for 3 days using Kodak XAR-5 film and an intensifying screen. Lanes B, E, and H, uninduced cultures; lanes C, D, F, G, and I, induced cultures; lanes A and K, identical to lane H with 100 ng of purified liver G_s added per lane; lane J, identical to lane H with 15 μg of S49 wild-type membranes added. Total protein loaded per lane was approximately 50–60 μg.

tion of r$G_{s\alpha}$ is observed under these conditions (Fig. 2, lanes B and E). At an OD_{600} of 1.2–1.4, r$G_{s\alpha}$ synthesis is induced by shifting the flasks to a 42° water bath. After 30 min at 42°, rifampicin (Sigma Chemical Company, St. Louis, MO, Cat. No. R-3501) is added to a final concentration of 0.1 mg/ml (5 ml of a 20 mg/ml solution in methanol per liter). Rifampicin inhibits *E. coli* RNA polymerase while having no significant effect on T7 RNA polymerase. Inhibition of transcription of *E. coli* genes theoretically increases the relative quantity of recombinant protein. Cultures are then maintained in the air shaker at 37° for 1 hr with shaking.

Cultures are decanted into chilled Beckman JA10 centrifuge bottles. All subsequent manipulations involving the purification of recombinant $G_{s\alpha}$ are performed at 4°, unless otherwise indicated. Cells are harvested by centrifugation at 8000 rpm for 5 min in a Beckman JA10 rotor. The supernatant is discarded, and the pellet is washed, without resuspension, with 100 ml of TEDP.[12] Following centrifugation as described above, the supernatant is discarded, and the pellet is suspended in the centrifuge bottle with 90 ml of TEDP. We find that suspension is most efficiently performed by first adding 10 ml of buffer to the pellet and resuspending it with a 35-ml syringe and a canula. A freshly prepared solution of 1 mg/ml chicken egg white lysozyme (Sigma, Cat. No. L-6876) in TEDP is added to the resuspended cells to a final concentration of 0.1 mg/ml. This constitutes a total protein to lysozyme ratio of 20–30 : 1. Higher ratios of total protein to lysozyme cause a decrease in the percentage of cells lysed. Following 30 min on ice, cell lysates are centrifuged at 30,000 *g* for 60 min in a Beckman JA10 rotor, and the supernatants are decanted into a chilled plastic beaker. Although variable, approximately 50% of the total $rG_{s\alpha}$ present in the cells is recovered in the supernatant (Fig. 2, lanes D and G). The short and the long forms of recombinant $G_{s\alpha}$ comigrate on SDS–polyacrylamide gels with the 45- and 52-kDa forms of $G_{s\alpha}$ found in murine S49 lymphoma membranes (Fig. 2, lane J).

Purification of Recombinant $G_{s\alpha}$

The purification scheme detailed below was developed with two major considerations: rapidity and the ability to scale up conveniently. The former consideration became paramount when we found that crude preparations of $rG_{s\alpha}$ are unstable with time. Experience indicates that the overall yield is improved dramatically if the initial steps are performed rapidly. The advantages of easy scale-up of the purification protocol are obvious.

$rG_{s\alpha}$ is quantified throughout its purification by two methods. The first is based on the ability of $rG_{s\alpha}$ to restore GTPγS-stimulated adenylyl cyclase activity to membranes from *cyc*⁻ S49 lymphocytes. The *cyc*⁻ S49 cell is deficient in $G_{s\alpha}$ and hence in hormone- and guanine nucleotide-stimulated adenylyl cyclase activity.[13] This assay is of particular advantage with crude fractions because it is completely specific for $G_{s\alpha}$. The second assay is based on the ability of $rG_{s\alpha}$ to bind guanine nucleotides such as GTPγS

[13] E. M. Ross, A. C. Howlett, K. M. Ferguson, and A. G. Gilman, *J. Biol. Chem.* **253,** 64[illegible] (1978).

TABLE I
PURIFICATION OF RECOMBINANT $G_{s\alpha-s}$[a]

raction	Volume (ml)	Protein (mg)	Adenylyl cyclase activity: nmol cAMP/min	Adenylyl cyclase activity: nmol cAMP/min/mg protein	GTP-γS binding activity: nmol	GTP-γS binding activity: nmol/mg protein
ble lysate	1200	2820	1040[b]	0.4	216[b]	0.07
ꓥE	350	1365	876[b]	0.6	136[b]	0.10
ɜel HTP	90	62.1	1218	19.6	130	2.09
ɪo Q	8	7.8	678	127	70	8.79
ɪT	1	1.6	252	157	30	18.10

Nine liters of *E. coli* strain K38 transformed with plasmids pGp1-2 and NpT7-5/$G_{s\alpha-s}$ was cultured, and recombinant $G_{s\alpha}$ was purified as described in the text. For the assay of adenylyl cyclase activity, aliquots of each fraction were incubated with 20 μM GTPγS, 5 mM $MgCl_2$, and 0.1% Lubrol for 45 min at 30°, diluted to an appropriate concentration with HME[12] containing 20 μM GTPγS and 0.1% Lubrol, and reconstituted with S49 *cyc*$^-$ membranes as described by P. C. Sternweis, J. K. Northup, M. D. Smigel, and A. G. Gilman, *J. Biol. Chem.* **256,** 11517 (1981). Assays of GTPγS binding were performed at a final concentration of 2 μM GTPγS [J. K. Northup, M. D. Smigel, and A. G. Gilman, *J. Biol. Chem.* **257,** 11416 (1982)].
The instability of $rG_{s\alpha}$ in these fractions makes accurate quantitation of the yield difficult. Estimates of $rG_{s\alpha}$ concentrations based on immunoblotting indicate that the overall yield of purified proteins is 10–20% of that present in the soluble lysate. From multiple preparations, the yield of $rG_{s\alpha}$ has ranged from 1.2 to 1.6 mg for $rG_{s\alpha\text{-}S}$ and from 0.8 to 0.9 mg for $rG_{s\alpha\text{-}L}$. Other guanine nucleotide-binding proteins are present in the lysate.

Although nucleotide binding is simple and rapid, it is of limited utility in the early stages of purification because of the presence of other guanine nucleotide-binding proteins (Table I; Fig. 3).

DEAE-Sephacel Chromatography. DEAE-Sephacel (Pharmacia, Piscataway, NJ) equilibrated with TEDP is added directly to the soluble lysate obtained from the lysozyme treatment (100 ml of packed resin; 1 ml packed resin per 25–30 mg protein). The resin is suspended by occasional stirring with a glass rod. After 15 min, the mixture is poured into a 10.5-cm Büchner funnel fitted with a Whatman #4 filter. The resin is washed on the filter with 1 liter of TEDP. (The flow rate can be improved with a mild vacuum; the resin should always be covered with buffer.) Protein adsorbed to the resin is eluted with three 100-ml washes of TEDP containing 300 mM NaCl. At this stage, $rG_{s\alpha}$ activity is stable to rapid freezing with liquid N_2 and rapid thawing at 30°. Approximately 40% of the applied protein is recovered in the eluate. Although the purification of $rG_{s\alpha}$ is minimal, this procedure achieves two major goals. It removes nucleic

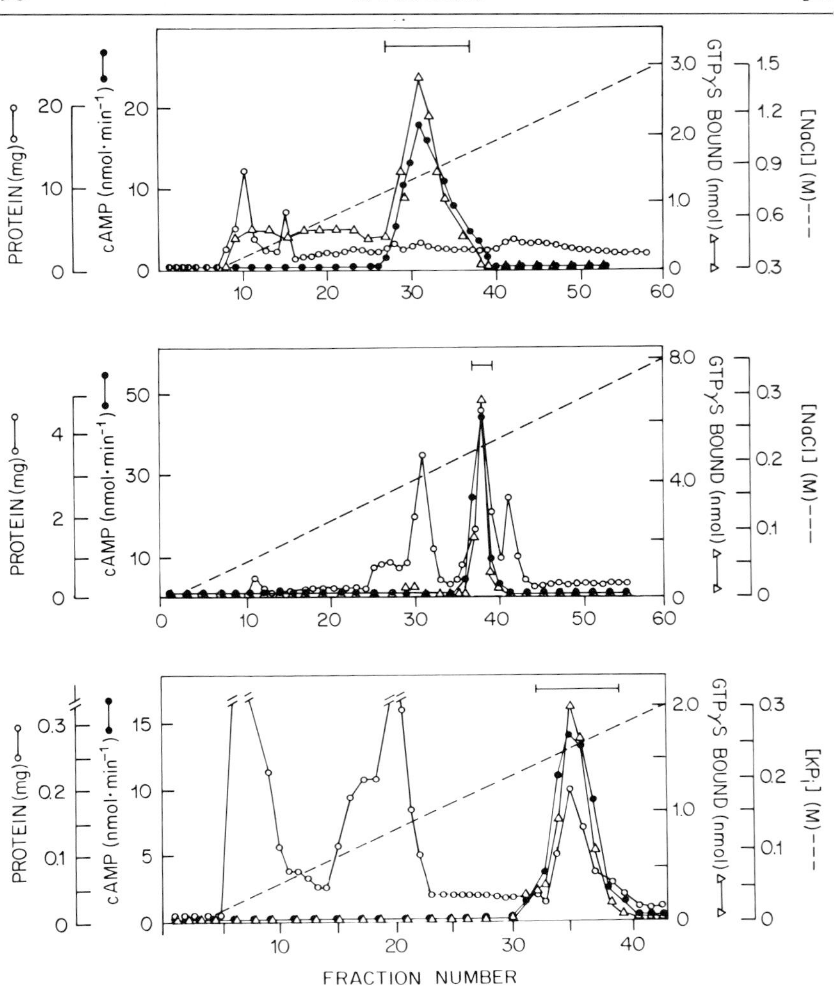

FIG. 3. Chromatography of recombinant $G_{s\alpha\text{-}s}$. Nine liters of *E. coli* K38 harborin plasmids pGp1-2 and NpT7-5/$G_{s\alpha\text{-}s}$ was grown at 30° to an OD_{600} of 1.2–1.4; bacteria wer then induced to express $rG_{s\alpha}$. After initial processing as described in the text, $rG_{s\alpha\text{-}s}$ wa purified from the DEAE-Sephacel eluate by sequential chromatography on BioGel HT (top), Mono Q HR 10/10 (middle), and Bio-Rad HPHT (bottom) columns. Regions of eac chromatogram denoted by the horizontal bar represent the fractions that were pooled fc further chromatography.

acids that bind to subsequent resins and inhibit flow. It also significantly reduces the volumes that must be manipulated; this becomes increasingly important as the preparation is scaled up. The resin is regenerated by extensive washing with TED containing 1 *M* NaCl and can be reused many times.

BioGel HTP Chromatography. The DEAE-Sephacel eluate is applied directly to a 100-ml column of hydroxyapatite (about 10–12 mg protein/ml resin) (BioGel HTP; Bio-Rad Laboratories, Richmond, CA) equilibrated with TDPK[12] containing 300 m*M* NaCl. The column is washed with 1 column volume of TDPK plus 300 m*M* NaCl, and adsorbed protein is eluted with a 500-ml linear gradient of NaCl (300–1500 m*M*) in TDPK. The M_r 45,000 form of $rG_{s\alpha}$ elutes as a single peak centered at 0.8 *M* NaCl (Fig. 3). The M_r 52,000 form of $rG_{s\alpha}$ elutes somewhat later in the gradient (at approximately 1 *M* NaCl). This step is particularly useful and is unusual in that it relies on NaCl to elute $rG_{s\alpha}$ from the hydroxyapatite column. Greater than 90% of the applied protein remains adsorbed to the column at NaCl concentrations that approach 1.5 *M*. Given this, we now frequently perform this step in a batchwise fashion with no discernible loss of purity in the final preparation. In the batch procedure, 100 ml of BioGel HTP is added directly to a 2-liter plastic beaker containing the DEAE-Sephacel eluate on ice. After 30 min with occasional stirring with a glass rod, the mixture is poured onto a Büchner funnel fitted with a Whatman #4 filter. Flow rates can be improved by the use of a gentle vacuum; however, excessive vacuum compresses the resin and impairs flow. The resin is washed with 150 ml of TDPK plus 300 m*M* NaCl, removed from the funnel, resuspended in 150 ml of the same buffer, reapplied to the filter-containing funnel, and washed with 150 ml of buffer. $rG_{s\alpha}$ is then eluted in a manner analogous to the wash procedure with 450 ml of TDPK containing 1.5 *M* NaCl. The resin is regenerated by washing with TDPK plus 500 m*M* KH_2PO_4 and can be reused several times.

Mono Q Fast Protein Liquid Chromatography. To permit $rG_{s\alpha}$ to adsorb to the Mono Q resin, the BioGel HTP eluate is desalted by ultrafiltration in an Amicon (Danvers, MA) stirred cell fitted with a PM30 membrane (76 mm diameter). The 450 ml eluate is concentrated approximately 10-fold (to 50 ml). The concentrated eluate is then diluted to 350 ml with Mono Q buffer[12] and is again concentrated to 50 ml. A second cycle of dilution and concentration results in a 50-ml sample with a NaCl concentration of less than 25 m*M*. The ultrafiltrate is applied to a Pharmacia Mono Q HR 10/10 FPLC column that has been equilibrated with Mono Q buffer. Under optimal conditions, a flow rate of 4 ml/min can be achieved. The column is washed with 1 volume of Mono Q buffer, and adsorbed protein is eluted with a linear gradient of NaCl (0–350 m*M*) in Mono Q buffer;

4-ml fractions are collected. $rG_{s\alpha}$ elutes as a single peak centered at 220 mM NaCl (Fig. 3). This step affords an approximate 7-fold purification of $rG_{s\alpha}$, with a recovery of greater than 60%.

BioGel HPHT Chromatography. $rG_{s\alpha}$ is purified to homogeneity by chromatography on a BioRad HPHT column (an HPLC hydroxyapatite column). Column fittings are modified to allow this column to be utilized with the FPLC system. Fractions from the Mono Q column comprising the peak of $rG_{s\alpha}$ activity (12–16 ml) are diluted 2-fold with HPHT buffer[12] and applied to the HPHT column equilibrated in the same buffer. The column is washed with 1 column volume of HPHT buffer, and adsorbed protein is eluted with a 30-ml linear gradient of potassium phosphate (0–290 mM; pH 8.0) in HPHT buffer; 0.75-ml fractions are collected. $rG_{s\alpha}$ elutes as a single peak at 240 mM potassium phosphate; recovery of activity is routinely in the range of 50%. $rG_{s\alpha}$ is unstable to freezing at this stage. Fractions comprising the peak of $rG_{s\alpha}$ (3–4 ml total volume) are pooled, and the buffer is exchanged in an Amicon Centricon 30 by sequential cycles of concentration and dilution into HED buffer.[12] Purified $rG_{s\alpha}$ (approximately 0.5–1.0 ml final volume; protein concentration >1 mg/ml; phosphate concentration <10 mM) is divided into aliquots, snap-frozen in liquid N_2, and stored at −80°. Under these storage conditions, $rG_{s\alpha}$ activity is stable for at least several months. Aliquots can be rapidly thawed (at 30°) and refrozen (in liquid N_2) several times with minimal loss of activity.

Characterization of $rG_{s\alpha}$. Discussion of the detailed characterization of purified $rG_{s\alpha}$ is beyond the scope of this chapter and is presented elsewhere.[14] A summary of those studies is presented here. Coomassie blue staining of SDS–polyacrylamide gels containing purified preparations of $rG_{s\alpha}$ reveals a single protein band that migrates with an apparent molecular weight of 45,000 ($rG_{s\alpha\text{-s}}$) or 52,000 ($rG_{s\alpha\text{-L}}$) (Fig. 4). Two-dimensional gel analysis of $rG_{s\alpha\text{-s}}$ reveals a single protein species with a pI of approximately 6.5. Amino-terminal amino acid sequence analysis demonstrates that the initiating methionine residue is cleaved from the recombinant proteins; the subsequent amino acid sequence is identical to that deduced from the cDNAs.

Both forms of the recombinant protein bind GTPγS stoichiometrically and hydrolyze GTP in a manner indistinguishable from that observed for G proteins purified from mammalian sources. The proteins are substrates for cholera toxin-catalyzed ADP-ribosylation, a reaction that requires the addition of G protein $\beta\gamma$ subunits. Both forms of recombinant $G_{s\alpha}$ reconstitute GTP-, isoproterenol plus GTP-, GTPγS-, and fluoride-stimulated adenylyl cyclase activity in S49 *cyc*⁻ membranes to maximal levels,

[14] M. P. Graziano, M. Freissmuth, and A. G. Gilman, *J. Biol. Chem.* **264,** 409 (1989).

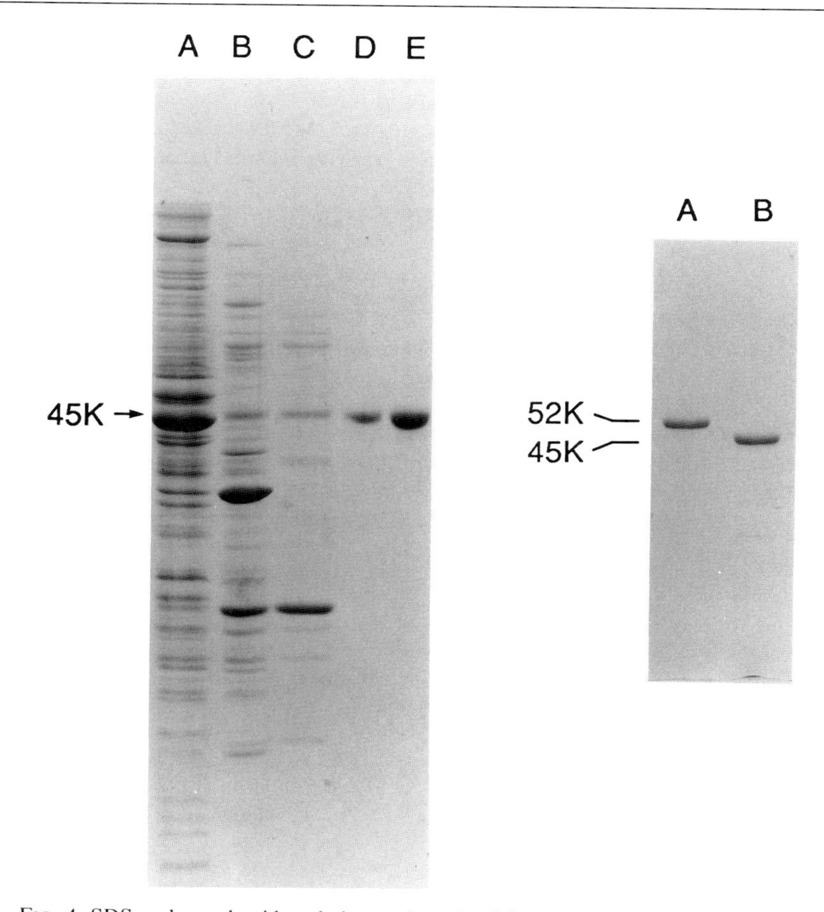

FIG. 4. SDS–polyacrylamide gel electrophoresis of fractions obtained from the purification of $rG_{s\alpha}$. (Left) Aliquots of fractions from the various stages in the purification of $rG_{s\alpha\text{-S}}$ were resolved by SDS–PAGE on an 11% polyacrylamide gel, and proteins were visualized by staining with Coomassie blue. Where applicable, fractions shown are aliquots of the pooled fractions identified in Fig. 3. Lane A, 25 μg of the DEAE-Sephacel eluate; lane B, 10 μg of the BioGel HTP pool; lane C, 5 μg of the Mono Q HR10/10 pool; lanes D and E, 2 and 6 μg, respectively, of the Bio-Rad HPHT pool. Apparent molecular weights were determined from the migration of protein standards resolved in the same gel. (Right) Eight micrograms each of $rG_{s\alpha\text{-L}}$ (lane A) or $rG_{s\alpha\text{-S}}$ (lane B) (Bio-Rad HPHT pool) was subjected to SDS–PAGE on an 11% polyacrylamide gel, and proteins were visualized by staining with Coomassie blue. Apparent molecular weights were determined from the migration of protein standards resolved on the same gel.

although their specific activity for this reaction is lower than that observed for G_s purified from mammalian sources. Experiments performed with purified bovine brain adenylyl cyclase indicate that the intrinsic capacity of the recombinant proteins to activate the enzyme is normal; however, their affinities for adenylyl cyclase are approximately 10-fold lower than that of liver G_s. Based on these data, we speculate that the recombinant proteins may lack a posttranslational modification that is necessary for high-affinity interaction of $G_{s\alpha}$ with adenylyl cyclase.

Scope of Method. The method described above can be scaled up readily. In order to facilitate this, cells have been grown in a fermentor. Under these growth conditions, the quantity of $rG_{s\alpha}$ recovered in the soluble lysate is approximately 5-fold higher than when the cells are grown in flasks. The cell paste can be frozen in liquid N_2, and rG_s activity is stable at $-80°$ for at least 2 months. Purification of $rG_{s\alpha}$ from approximately 35 liters of fermentation culture yielded 25 mg of homogeneous protein by direct scale-up (or repetition) of the procedures described above.

The method has also been applied to the purification of several mutant forms of $rG_{s\alpha}$. As is described in [18], this volume, the identical expression system can be used to synthesize other G protein α subunits in *E. coli*, and their purification has been achieved by a modification of the procedure employed for $rG_{s\alpha}$. Thus, this approach appears to be generally applicable for the expression and purification of milligram quantities of a variety of wild-type and mutant G-protein α subunits.

[18] Purification of Recombinant $G_{i\alpha}$ and $G_{o\alpha}$ Proteins from *Escherichia coli*

By MAURINE E. LINDER and ALFRED G. GILMAN

Introduction

The term G_i originally referred to the guanine nucleotide-binding regulatory protein (G protein) that is responsible for receptor-mediated inhibition of adenylyl cyclase. Today, this term is assigned to a subfamily of highly homologous G proteins that are substrates for pertussis toxin but are distinct from G_o, an abundant G protein in brain, and G_t, the signal-transducing protein of the retinal cyclic GMP phosphodiesterase system. The functions of the proteins in this group are not clearly defined, but they are activated by a large number of receptors and are implicated in the regulation of a variety of effector systems, including adenylyl cyclase, phospholipases, and ion channels.[1,2]

G_i was first purified from rabbit liver as a heterotrimer of α, β, and γ subunits.[3,4] The α subunit has an apparent molecular weight of 41,000 and can be ADP-ribosylated by pertussis toxin. The β and γ subunits, which are tightly associated with each other, are indistinguishable from those that are purified with $G_{s\alpha}$, the G protein that stimulates adenylyl cyclase. More recent data obtained with peptide antisera (see below) indicate that G_i purified from liver and elsewhere is heterogeneous in its composition and that more than one form of α_i is present.[5] The heterogeneity of G_i proteins became apparent with the cloning of their corresponding cDNAs. The first cDNA clones for G_i were detected with oligonucleotide probes based on partial amino acid sequences of G_i proteins purified from bovine and rat brain.[6,7] The amino acid sequence deduced from the bovine cDNA clone (subsequently termed $G_{i\alpha1}$) differed from that deduced from the rat clone ($G_{i\alpha2}$) by 11%. It was not clear at the time if this was a species difference or if the clones encoded distinct proteins. The situation was clarified when Jones and Reed[8] isolated both $G_{i\alpha1}$ and $G_{i\alpha2}$ clones from a rat olfactory epithelial library. In addition, they also cloned and sequenced a related cDNA, termed $G_{i\alpha3}$. $G_{i\alpha1}$ and $G_{i\alpha3}$ encode proteins that are 94% identical in their amino acid sequence. cDNAs for $G_{o\alpha}$ and $G_{s\alpha}$ were also detected in the rat olfactory epithelial library.

Antisera against peptides unique to each form of $G_{i\alpha}$ have been useful in determining the prevalence of each polypeptide in various cells and tissues and in discerning the composition of "purified" preparations of G_i. For example, 41,000- and 40,000-dalton proteins purified from bovine brain have been identified as $G_{i\alpha1}$ and $G_{i\alpha2}$, respectively.[5] Although the forms can be resolved chromatographically from one another (with some effort), it is particularly difficult to resolve these

[1] A. G. Gilman, *Annu. Rev. Biochem.* **56,** 615 (1987).

[2] P. J. Casey and A. G. Gilman, *J. Biol. Chem.* **263,** 2577 (1988).

[3] G. M. Bokoch, T. Katada, J. K. Northup, E. L. Hewlett, and A. G. Gilman, *J. Biol. Chem.* **258,** 2072 (1983).

[4] G. M. Bokoch, T. Katada, J. K. Northup, M. Ui, and A. G. Gilman, *J. Biol. Chem.* **259,** 3560 (1984).

[5] S. M. Mumby, I. Pang, A. G. Gilman, and P. C. Sternweis, *J. Biol. Chem.* **263,** 2020 (1988).

[6] T. Nukada, T. Tanabe, H. Takahashi, M. Noda, K. Haga, T. Haga, A. Ichiyama, K. Kangawa, M. Hiranaga, H. Matsuo, and S. Numa, *FEBS Lett.* **197,** 305 (1986).

[7] H. Itoh, T. Kozasa, S. Nagata, S. Nakamura, T. Katada, M. Ui, S. Iwai, E. Ohtsuka, H. Kawasaki, K. Suzuki, and Y. Kaziro, *Proc. Natl. Acad. Sci. U.S.A.* **83,** 3776 (1986).

[8] D. T. Jones and R. R. Reed, *J. Biol. Chem.* **262,** 14241 (1987).

proteins completely from $G_{o\alpha}$. Preparations of "G_i" from human erythrocytes can be shown to contain G_{i1}, G_{i2}, and G_{i3}.[9]

Given the difficulty in resolving the individual G_i proteins from each other (and, perhaps, from yet to be discovered G proteins), we have chosen to express the cDNA clones for $G_{o\alpha}$ and the $G_{i\alpha}$ proteins in *Escherichia coli* in order to purify each protein. Using a modification of the method developed by Graziano *et al.*[10] for $G_{s\alpha}$, we have succeeded in purifying recombinant (r) $G_{i\alpha 1}$, $G_{i\alpha 2}$, $G_{i\alpha 3}$, and $G_{o\alpha}$.

Expression

Rat olfactory epithelial cDNA clones that encode $G_{o\alpha}$ and all three forms of $G_{i\alpha}$ were generously provided by Reed.[8] The methods used to construct the plasmids for expression of G-protein α subunits are those described by Maniatis *et al.*[11]

The prokaryotic expression system of Tabor and Richardson[12] has been utilized; it is described in more detail by Graziano *et al.*[10] in this volume. Briefly, the cDNA to be expressed is subcloned into a plasmid, NpT7-5, behind a promoter for T7 RNA polymerase and a ribosomal binding site (Rbs). This construction is transformed into *E. coli* strain K38(pGp1-2), where pGp1-2 is a plasmid that encodes T7 RNA polymerase under the control of the λ P_L promoter and a temperature-sensitive repressor.

The cDNAs for $G_{o\alpha}$ and $G_{i\alpha 1}$ have *Nco*I sites at the initiation codon. Thus, they can be inserted directly into the NpT7-5 expression vector, maintaining the correct spacing between the initiating methionine codon and the ribosomal binding site engineered into the vector by Graziano *et al.*[10] The sequence of this region of the plasmid is shown below.

```
       EcoRI      Rbs                NcoI
      ______     _____              ______
5′  AATTCTA AGGA GGTTTAA CCATGG 3′
    TTAAGAT TCCT CCAAATT GGTACC
```

Fragments of the cDNAs are prepared with the restriction enzymes indicated in Table I. The cDNAs for $G_{o\alpha}$ and $G_{i\alpha 1}$ are digested completely with *Sma*I and *Xba*I, respectively, to generate the correct 3′ end of the fragment. However, since both of these cDNAs have an internal *Nco*I site in the coding sequence, as well as one at the codon for the initiating

[9] D. J. Carty and R. Iyengar, *FEBS Lett.* **262,** 101 (1990).

[10] M. P. Graziano, M. Freissmuth, and A. G. Gilman, this volume [17].

[11] T. Maniatis, E. Fritsch, and J. Sambrook, "Molecular Cloning." Cold Spring Harbor Laboratory, Cold Spring Harbor, New York, 1982.

[12] S. Tabor and C. C. Richardson, *Proc. Natl. Acad. Sci. U.S.A.* **82,** 1074 (1985).

TABLE I
RESTRICTION ENZYME SITES USED IN SUBCLONING cDNAs FOR $G_{o\alpha}$ AND $G_{i\alpha}$ INTO NpT7-5[a]

Protein	cDNA 5′	cDNA 3′	NpT 7-5 5′	NpT 7-5 3′
G_o	*Nco*I(p)[b]	*Dra*I	*Nco*I	*Sma*I
G_{i1}	*Nco*I(p)	*Xba*I	*Nco*I	*Xba*I
G_{i2}[c]	*Hae*II	*Xba*I	*Eco*RI	*Xba*I
G_{i3}[c]	*Hae*II(p)	*Hin*dIII	*Eco*RI	*Hin*dIII

[a] The cDNA clones and NpT7-5 were digested with appropriate restriction enzymes and ligated as described in the text. Partial digests of $G_{o\alpha}$, $G_{i\alpha 1}$, and $G_{i\alpha 3}$ were necessary to obtain full-length coding sequence.
[b] Partial digest.
[c] Complementary synthetic oligonucleotides were annealed and ligated with the restriction fragments as described in the text.

methionine residue, partial digests are done with *Nco*I to obtain full-length coding sequence. This protocol generates a 1.13-kilobase (kb)[13] fragment for $G_{i\alpha 1}$ and a 1.29-kb fragment for $G_{o\alpha}$. The fragments are gel purified and ligated with NpT7-5 that has been digested with the appropriate enzymes (see Table I). The ligated DNA is first transformed into *E. coli* HB101 cells and characterized by restriction enzyme mapping; this plasmid DNA is then transformed into K38 cells that harbor pGp1-2.

$G_{i\alpha 2}$ and $G_{i\alpha 3}$ cDNAs do not have convenient restriction sites for insertion directly into NpT7-5. To construct the expression plasmid for $G_{i\alpha 3}$, synthetic complementary oligonucleotides are synthesized and annealed. The sequences are shown below.

```
        EcoRI    Rbs              Met                 HaeII
5′ AATTCA AGGA GAT AT ACAT ATG GGCTGCACGTTGA GCGC 3′
        GT TCCT CTAT ATGTA TAC CCGACGTGCAACT
```

The sequence contains an *Eco*RI site, a consensus ribosomal binding site, and the 5′ portion of the coding sequence for $G_{i\alpha 3}$. It ends with a *Hae*II site compatible with the first *Hae*II site within the coding sequence of $G_{i\alpha 3}$. The cDNA for $G_{i\alpha 3}$ has a second *Hae*II site 102 nucleotides downstream. The $G_{i\alpha 3}$ restriction fragment is prepared by first digesting the

[13] kb, Kilobase; SDS, sodium dodecyl sulfate; GTPγS, guanosine 5′-(γ-thio)triphosphate; DTT, dithiothreitol.

DNA completely with *Hin*dIII to generate the appropriate 3′ end. A partial digest with *Hae*II then yields a full-length coding sequence of 1.71 kb. A three-part ligation reaction is done with NpT7-5, the annealed oligonucleotides, and the $G_{i\alpha3}$ restriction fragment. The ligated DNA is transformed into HB101 cells, characterized, and then transformed into K38(pGp1-2) cells.

The expression plasmid for $G_{i\alpha2}$ is constructed in the same way, except the oligonucleotides are synthesized with $G_{i\alpha2}$ sequence 3′ to the initiating methionine codon, as follows:

```
EcoRI     Rbs             Met                                      Hae
AATTCAAGGAGATATACATATGGGCTGCACCGTGAGCGCCGAGGACAAGGCGGCAGCCGAGCC
    GTTCCTCTATATGTATACCCGACGTGGCACTCGCGGCTCCTGTTCCGCCGTCGGCT
```

The synthetic oligonucleotides utilized to construct the $G_{i\alpha2}$ and $G_{i\alpha3}$ expression plasmids replace the corresponding existing (5′) sequences in NpT7-5 that are retained in the expression plasmids containing cDNAs for $G_{s\alpha}$, $G_{o\alpha}$, and $G_{i\alpha1}$. The spacing between the ribosomal binding site and the initiating methionine codon is retained, but the nucleotide sequence differs at five positions. The sequence that we placed between the ribosomal binding site and the codon for the initiating methionine residue is A–T rich in the $G_{i\alpha2}$ and $G_{i\alpha3}$ expression plasmids, since Dunn and Studier reported a correlation between A–T-rich sequences at this site and efficient initiation of translation of late T7 proteins.[14] However, we do not observe increased expression of $rG_{i\alpha2}$ or $rG_{i\alpha3}$ compared with that of $rG_{i\alpha1}$ and $rG_{o\alpha}$. In fact, the yield of purified $rG_{i\alpha2}$ or $rG_{i\alpha3}$ protein is consistently lower than that of $rG_{o\alpha}$ or $rG_{i\alpha1}$. Western blot analysis of soluble fractions of *E. coli* show equivalent accumulation of all four proteins, approximately 0.5–1 mg/liter of culture. (The amount of soluble protein is variable from experiment to experiment.) Thus, the difference in yield may be due more to the stability of the proteins than to the efficiency of initiation of translation from the two different ribosomal binding sequences.

Assays

ADP-Ribosylation. Pertussis toxin-catalyzed ADP-ribosylation is the primary assay used to monitor the purification of recombinant $G_{i\alpha}$ and $G_{o\alpha}$. Soluble lysates of *E. coli* cells that express these proteins are ADP-ribosylated by pertussis toxin with [^{32}P]NAD. The reactions are stopped with electrophoresis sample buffer containing sodium dodecyl sulfate

[14] J. J. Dunn and F. W. Studier, *J. Mol. Biol.* **166,** 477 (1983).

(SDS), and samples are subjected to electrophoresis on SDS–polyacrylamide gels. Autoradiography of the gels reveals a single major product with an apparent molecular weight of 41,000 in the case of $rG_{i\alpha1}$ or $rG_{i\alpha3}$, 40,000 for $rG_{i\alpha2}$, and 39,000 for $rG_{o\alpha}$. These bands are absent in lysates of K38 cells that lack the expression plasmids. Parallel samples that are precipitated with trichloroacetic acid (TCA), filtered, and counted yield similar results. Since no *E. coli* proteins are labeled significantly, recombinant $G_{o\alpha}$ or $G_{i\alpha}$ can be easily identified in column fractions with this precipitation and filtration technique.

ADP-ribosylation is carried out as described by Bokoch *et al.*,[4] with minor modifications. Samples to be assayed (15 μl) are added to 25 μl of reaction mixture such that final concentrations in the assay are as follows: 100 m*M* Tris-HCl (pH 8.0), 10 m*M* thymidine, 1 m*M* ATP, 0.1 m*M* GTP, 2.5 m*M* $MgCl_2$, 1 m*M* EDTA, 10 m*M* dithiothreitol (DTT), 0.5 m*M* dimyristoylphosphatidylcholine, 2.5 μ*M* [^{32}P]NAD (approximately 10,000 cpm/pmol), and 5 μg/ml pertussis toxin (List Biologicals, Campbell, CA). The reaction is supplemented with brain $\beta\gamma$ subunits, usually 5 pmol/assay. Brain $\beta\gamma$ was generously provided by P. Casey; the purification is described elsewhere in this volume.[15] The reactions are incubated at 30° for 1 hr and are terminated by the addition of 500 μl of a solution containing 2% SDS and 50 μ*M* NAD. Protein is precipitated by the addition of 500 μl of 30% TCA and collected on BA85 nitrocellulose filters (Schleicher and Schuell, Keene, NH). The filters are washed with a total volume of 16 ml of 6% TCA, dissolved in Liquiscint (National Diagnostics, Manville, NJ), and counted. (See also chapter [22].)

GTPγS Binding. Guanine nucleotide binding is assessed with a modification of the method described by Sternweis and Robishaw.[16] Samples are assayed at a final concentration of 50 m*M* sodium HEPES (pH 8), 1 m*M* EDTA, 1 m*M* DTT, 0.1% Lubrol PX, 10 m*M* $MgSO_4$, and 5 μ*M* GTPγS (2500 cpm/pmol). Binding is performed at 30° for $G_{i\alpha}$ and at 20° for $G_{o\alpha}$. The kinetics of nucleotide binding for $rG_{i\alpha}$ subunits is slow; thus, reaction mixtures are usually incubated for 2 hr. Incubations with $rG_{o\alpha}$ are for 30 min. Alterations in these conditions are indicated in the figure legends. The samples (usually 50 μl) are diluted with 2 ml of ice-cold filtration buffer (20 m*M* Tris-HCl, pH 8, 100 m*M* NaCl, 25 m*M* $MgCl_2$) and are filtered through BA85 nitrocellulose filters. The filters are washed with a total volume of 12 ml of the same buffer, dried, dissolved in Liquiscint, and counted.

Protein Assays. Protein concentrations are measured by staining with

[15] P. J. Casey, I.-H. Pang, and A. G. Gilman, this volume [27].

[16] P. C. Sternweis and J. D. Robishaw, *J. Biol. Chem.* **259**, 13806 (1984).

Amido black[17] or as described by Bradford,[18] with bovine serum albumin as a standard.

Electrophoresis. SDS–polyacrylamide gel electrophoresis of proteins is done according to the method of Laemmli.[19] SDS sample buffer is supplemented with 20 m*M* DTT.

Purification

Culture Conditions. Eight 1-liter cultures are grown overnight in enriched medium (2% tryptone, 1% yeast extract, 0.5% NaCl, 0.2% glycerol, and 50 m*M* KH_2PO_4, pH 7.2) with 50 μg/ml ampicillin and 50 μg/ml kanamycin at 30° with shaking. At an OD_{600} of 1.2, the cultures are shifted to 42° for 30 min to induce expression of the T7 RNA polymerase. Rifampicin is added to the cultures to a final concentration of 100 μg/ml, and the cells are grown for 60 min at 37° with vigorous shaking.

Preparation of Soluble Fraction. The cells are harvested in a Beckman JA10 rotor at 5000 rpm for 5 min at 4°. The pellets are washed in ice-cold TEDP (50 m*M* Tris-HCl, pH 8, 1 m*M* EDTA, 1 m*M* DTT, and 0.1 m*M* phenylmethylsulfonyl fluoride) and centrifuged for 5 min at 5000 rpm. The cells are resuspended in TEDP to a final volume of 1.2 liters. All subsequent steps are performed at 0–4°. The cells are treated with lysozyme at 0.1 mg/ml for 30 min on ice. The lysate is centrifuged for 1 hr at 14,000 rpm in a Beckman JA14 rotor, and the supernatant is collected.

DEAE-Sephacel. The soluble lysate is mixed with 100 ml of DEAE-Sephacel (Pharmacia, Piscataway, NJ) that has been equilibrated in TEDP. The resin is incubated with the extract for 20 min with occasional stirring and is then collected on a Whatman #4 filter in a Büchner funnel. The resin is washed with 1 liter of TEDP. Protein is eluted from the resin with three 100-ml volumes of TEDP containing 300 m*M* NaCl. This material can be frozen in liquid nitrogen and stored at −80° without significant loss of activity (assessed by ADP-ribosylation). The material should be stored in aliquots that can be thawed quickly at 30° and immediately placed on ice.

Heptylamine-Sepharose Chromatography. Heptylamine (C_7)-Sepharose was synthesized by the method of Shaltiel as modified by Northup *et al.*[20] The DEAE eluate is made 1.2 *M* in $(NH_4)_2SO_4$ by the addition of one-half volume (150 ml) of 3.6 *M* $(NH_4)_2SO_4$. GDP is added to a

[17] W. Schaffner and C. Weissmann, *Anal. Biochem.* **56,** 502 (1973).

[18] M. M. Bradford, *Anal. Biochem.* **72,** 248 (1976).

[19] U. K. Laemmli, *Nature* (*London*) **227,** 680 (1970).

[20] J. K. Northup, P. C. Sternweis, M. D. Smigel, L. S. Schleifer, E. M. Ross, and A. G. Gilman, *Proc. Natl. Acad. Sci. U.S.A.* **77,** 6516 (1980).

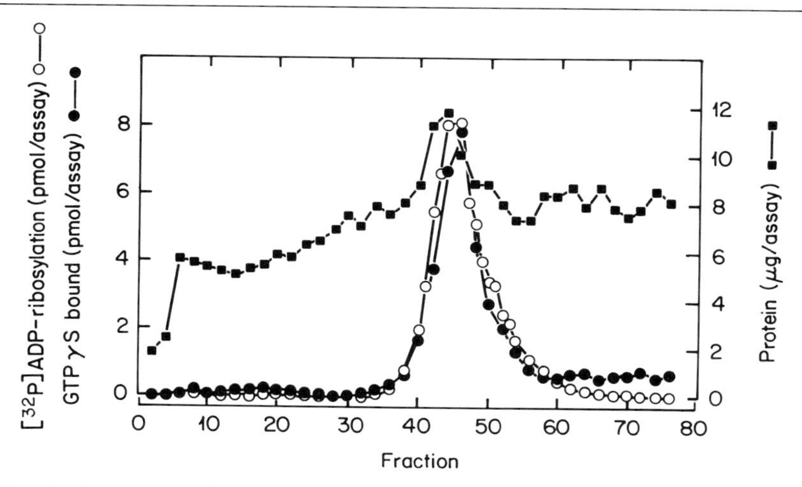

FIG. 1. Heptylamine-Sepharose chromatography of recombinant $G_{i\alpha 1}$. The DEAE eluate was chromatographed on a C_7-Sepharose column, as described in the text. The volumes assayed were as follows: ADP-ribosylation, 15 μl; GTPγS binding, 10 μl; and protein, 10 μl. ADP-ribosylation reactions were supplemented with 2.5 pmol $\beta\gamma$ subunits. GTPγS binding was performed at a final concentration of 2 μM GTPγS; assays were incubated for 1 hr at 30°. Values for protein, GTPγS binding, and ADP-ribosylation represent the means of duplicate determinations from a single representative preparation.

concentration of 25 μM. (GDP is included in the buffers during this stage of purification since high ionic strength facilitates dissociation of the nucleotide from G-protein α subunits.) The mixture is incubated on ice for 10 min and is centrifuged at 8500 rpm for 10 min in a Beckman JA14 rotor to remove any precipitated protein. (Less than 5% of the protein precipitates under these conditions.) The supernatant is applied to a 100-ml C_7-Sepharose column (2.6 × 19 cm) that is equilibrated in TEDP containing 25 μM GDP and 1.2 M ammonium sulfate. Greater than 90% of the total protein binds to the column. The column is first washed with 100 ml of TEDP, 25 μM GDP, 1.2 M ammonium sulfate, and protein is then eluted overnight with a 600-ml descending gradient of $(NH_4)_2SO_4$ (1.2 to 0 M) in TEDP, 25 μM GDP. Glycerol (35%) is added to increase the density of the dilution buffer (to stabilize gradient formation) and to slow the rate of dissociation of GDP from the G protein.[21] The elution profile from C_7-Sepharose for all four G-protein α subunits is similar. However, the position of elution is dependent on the batch of C_7-Sepharose. In the $rG_{i\alpha 1}$ preparation shown in Fig. 1, fractions 40–54 were pooled. In another experiment, with a

[21] K. M. Ferguson, T. Higashijima, M. D. Smigel, and A. G. Gilman, *J. Biol. Chem.* **261**, 7393 (1986).

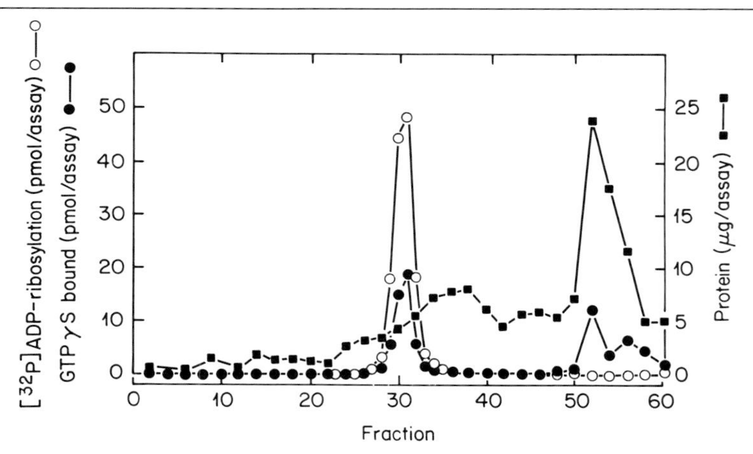

FIG. 2. Mono Q chromatography of recombinant $G_{i\alpha 1}$. The C_7 pool was chromatographed on a Mono Q HR 10/10 column as described in the text. The volumes assayed were as follows: ADP-ribosylation, 15 μl; GTPγS binding, 10 μl; and protein, 10 μl. Assay conditions were the same as those described in the legend to Fig. 1.

different C_7 resin, the peak of activity was late in the gradient, and an additional 50 ml of TEDP, 25 μ*M* GDP was washed through the column to elute the protein completely.

The fractions are assayed by ADP-ribosylation and peak fractions are pooled, desalted, and concentrated in an Amicon (Danvers, MA) pressure filtration device with a PM30 membrane. The pool is concentrated to 1/10 volume and then diluted 10-fold with Mono Q buffer (50 m*M* Tris-HCl, pH 8, 20 μ*M* EDTA, 1 m*M* DTT). After concentration again to 10 ml, the pool is diluted to 50 ml with Mono Q buffer. This 50-fold dilution is sufficient to permit binding of G-protein α subunits to the ion-exchange column used in the next step of purification.

Mono Q Chromatography. The C_7 pool is chromatographed on a Mono Q HR 10/10 anion-exchange system using a Pharmacia fast protein liquid chromatography (FPLC) system. The column is equilibrated with Mono Q buffer. Protein is eluted at 4 ml/min with a 240-ml gradient of increasing salt concentration from 0 to 300 m*M* NaCl. $rG_{i\alpha}$ and $rG_{o\alpha}$ proteins elute at approximately 150 m*M* NaCl in a volume of 8–16 ml (Fig. 2). Other GTPγS-binding proteins are resolved from $rG_{o\alpha}$ or $rG_{i\alpha}$ proteins on the Mono Q column.

The Mono Q pool (fractions 29–32 in the experiment shown in Fig. 2) is quite stable; the fractions can be held overnight at 4° or concentrated, frozen, and stored at −80°. If the fractions are to be held for more than a few hours, they should be supplemented with 50 μ*M* GDP.

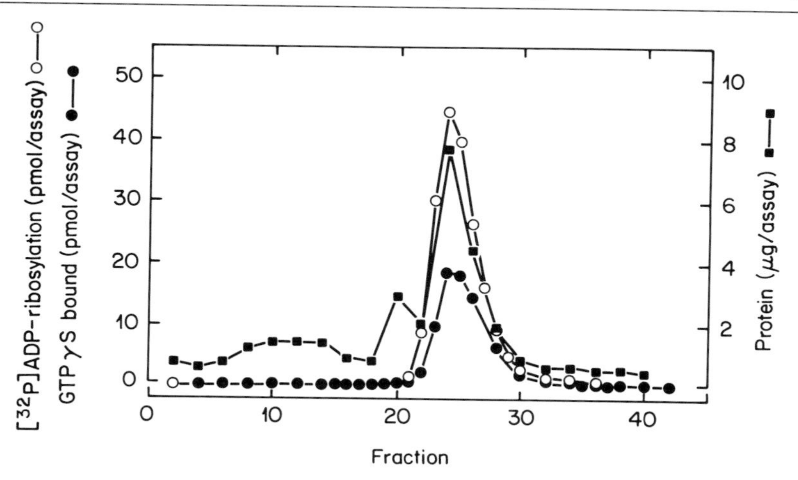

FIG. 3. High-performance hydroxyapatite chromatography of recombinant $G_{i\alpha 1}$. The Mono Q pool was chromatographed on hydroxyapatite as described in the text. Assay volumes were as follows: ADP-ribosylation, 15 μl; GTPγS binding, 5 μl; and protein, 10 μl. Assay conditions were the same as those described in the legend to Fig. 1.

HPHT Chromatography. The peak fractions from the Mono Q column are pooled and diluted 2-fold with HPHT buffer (10 m*M* Tris-HCl, pH 8, 10 m*M* potassium phosphate, pH 8, 1 m*M* DTT). The pool is applied to a Bio-Rad (Richmond, CA) HPLC hydroxyapatite column (7.8 × 100 mm) that has been adapted for the FPLC system. Protein is eluted with a 30-ml gradient from 10 to 200 m*M* potassium phosphate (pH 8) at 0.75 ml/min. $rG_{o\alpha}$ and $rG_{i\alpha}$ proteins elute at approximately 120 m*M* phosphate (Fig. 3).

The peak fractions (23–28 in Fig. 3) are pooled and concentrated in a Centricon PM30 (Amicon), diluted 10-fold with HED (50 m*M* sodium HEPES, pH 8.0, 1 m*M* EDTA, 1 m*M* DTT), and concentrated again. This cycle is repeated to decrease the phosphate concentration to less than 10 m*M*. Aliquots of the HPHT fraction are frozen and stored at −80° at a protein concentration of 1 mg/ml or greater. A summary of the purification of $rG_{i\alpha 1}$ is shown in Table II and Fig. 4.

At this stage of purification, the $rG_{i\alpha}$ proteins are greater than 90% pure as judged from polyacrylamide gels stained with either Coomassie blue or silver (see Fig. 5). However, an additional chromatographic step is required to purify $rG_{o\alpha}$ to homogeneity.

Phenyl-Superose Chromatography. The final step in the purification of $rG_{o\alpha}$ utilizes high-resolution hydrophobic interaction chromatography (Pharmacia Phenyl-Superose HR 5/5). The column is equilibrated in buffer

TABLE II
PURIFICATION OF RECOMBINANT $G_{i\alpha 1}$[a]

Step	Volume (ml)	Protein (mg)	GTPγS binding activity		ADP-Ribosylation	
			nmol	nmol/mg protein	nmol	nmol/mg protein
Lysozyme	1150	1219	184[b]	0.15	34.5[c]	0.03
DEAE	300	567	132[b]	0.23	57[c]	0.10
C_7 pool	47	87	115[b]	1.32	43	0.49
Mono Q pool	16	5.9	42	7.1	27	4.6
HPHT pool	0.4	2.1	27	12.9	22	10.5

[a] Eight liters of *E. coli* harboring pGp1-2 and NpT7-5/$G_{i\alpha 1}$ was cultured, and recombinant $G_{i\alpha 1}$ was purified as described in the text. Aliquots of the pools were assayed; the C_7 and HPHT pools were assayed after concentration and desalting. ADP-ribosylation assays were supplemented with 25 pmol of $\beta\gamma$ subunits.
[b] Other GTPγS-binding proteins are present in these fractions.
[c] Recombinant $G_{i\alpha 1}$ is unstable in these fractions, making quantitation difficult.

A [50 m*M* Tris-HCl, pH 8, 1 m*M* EDTA, 1 m*M* DTT, 1.2 *M* $(NH_4)_2SO_4$, 20% glycerol, 50 μM GDP]. One-half volume of 3.6 *M* $(NH_4)_2SO_4$ is added to the concentrated HPHT pool, and an equal volume of buffer A is also added. The material is applied to the Phenyl-Superose column. Protein is eluted with a 20-ml descending gradient of $(NH_4)_2SO_4$ (1.2 to 0 *M*) at a flow rate of 0.2 ml/min. Forty 0.5-ml fractions are collected, and $rG_{o\alpha}$ is recovered in fractions 23–28 [0.5 *M* $(NH_4)_2SO_4$]. $rG_{o\alpha}$ is resolved from the proteins that copurify on HPHT (see Fig. 5, first lane). The pooled fractions

TABLE III
PURIFICATION OF RECOMBINANT $G_{o\alpha}$ AND $G_{i\alpha}$ PROTEINS[a]

Preparation	Protein (mg)	GTPγS binding activity		ADP-Ribosylation	
		nmol	nmol/mg	nmol	nmol/mg
$G_{o\alpha}$	1.6	21	13.1	18	11.3
$G_{i\alpha 1}$	3.0	52.8	17.6	40.8	13.4
$G_{i\alpha 2}$	0.38	4.9	12.9	3.8	10.0
$G_{i\alpha 3}$	0.75	9.5	12.7	8.9	11.0

[a] Recombinant proteins were assayed for GTPγS binding and ADP-ribosylation as described in the text. GTPγS binding incubation conditions were as follows: $rG_{o\alpha}$, 20° for 45 min; $rG_{i\alpha 2}$, 30° for 75 min, $rG_{i\alpha 1}$ and $rG_{i\alpha 3}$, 30° for 2 hr. ADP-ribosylation reactions were supplemented with 15 pmol $\beta\gamma$ subunits.

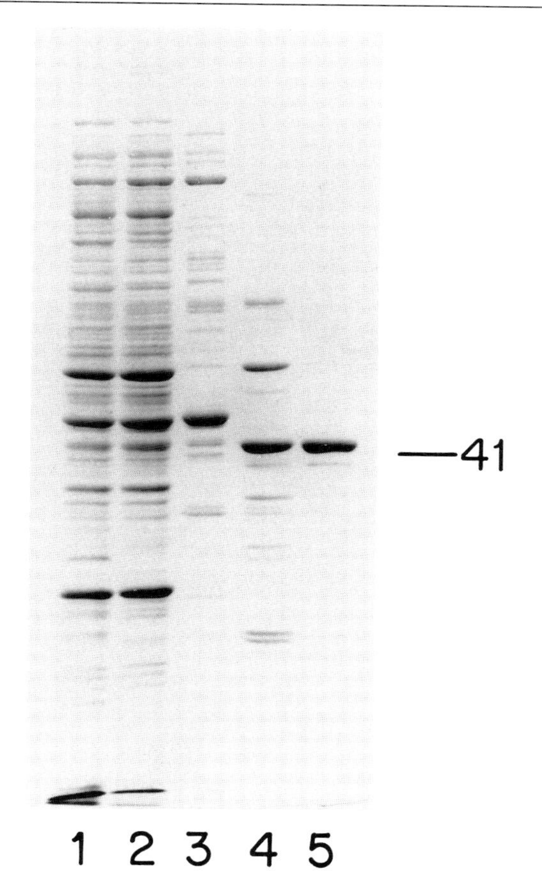

FIG. 4. SDS–polyacrylamide gel electrophoresis of fractions obtained during the purification of $rG_{i\alpha 1}$. Aliquots from the pools from each step in the purification were electrophoresed on a 9% SDS–polyacrylamide gel. The gel was stained with Coomassie blue. Lane 1, 20 μg of soluble lysate. Lane 2, 20 μg of DEAE fraction. Lane 3, 8 μg of C_7 pool. Lane 4, 4 μg of Mono Q pool. Lane 5, 1.6 μg of Bio-Rad HPHT pool. The apparent molecular weight ($M_r \times 10^{-3}$) is indicated at right.

are concentrated and desalted in a Centricon PM30. Four cycles of dilution with HED and concentration are done to reduce the concentrations of $(NH_4)_2SO_4$ to less than 1 mM, glycerol to less than 1%, and GDP to less than 1 μM.

Table III summarizes the purification of representative preparations of all four proteins. From 8 liters of culture, the yields have ranged from 1.3 to 1.6 mg for $rG_{o\alpha}$, 2.1–3 mg for $rG_{i\alpha 1}$, 300 μg to 1 mg for $rG_{i\alpha 2}$, and 200–800

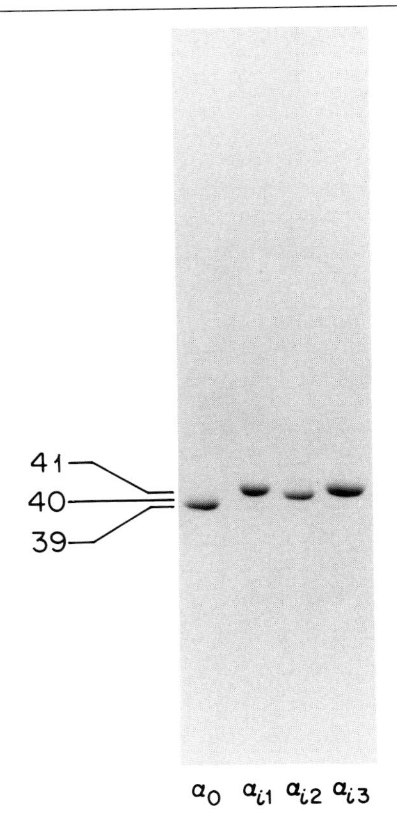

FIG. 5. SDS–polyacrylamide gel electrophoresis of purified $rG_{o\alpha}$, $rG_{i\alpha 1}$, $rG_{i\alpha 2}$, and $rG_{i\alpha 3}$. Each protein (1 μg) was electrophoresed on a 9% SDS–polyacrylamide gel. The gel was stained with Coomassie blue. The apparent molecular weights ($M_r \times 10^{-3}$) are indicated at left.

μg for $rG_{i\alpha 3}$. Although the amount of soluble protein that accumulates (estimated from Western blots) is similar for each, the fraction of protein in the soluble lysate that can be ADP-ribosylated is higher from cells expressing $rG_{o\alpha}$ or $rG_{i\alpha 1}$ than from those expressing $rG_{i\alpha 2}$ or $rG_{i\alpha 3}$. The reason for this is not known.

The stoichiometry of binding of GTPγS to $rG_{o\alpha}$ and $rG_{i\alpha}$ proteins ranges from 40 to 75%. The stoichiometry of ADP-ribosylation is similar. However, when ADP-ribosylation is assessed by shifts in electrophoretic mobility on SDS–polyacrylamide gels,[5] the stoichiometry approaches 100%. We suspect, therefore, that there may be a systematic error in protein determination and that a higher fraction of the purified protein is capable of binding guanine nucleotide.

The purified proteins have their amino termini intact. Amino-terminal amino acid sequencing was performed for six cycles on preparations of each purified protein.[22] In all cases, the first six amino acids faithfully represented the sequence deduced from the cDNA. The initiating membrane residue is cleaved from all four proteins.

In summary, we have succeeded in obtaining reasonable quantities of homogeneous recombinant $G_{o\alpha}$ and $G_{i\alpha}$ proteins by application of a purification procedure that can be completed in 2 days. It should be possible to increase the yields of purified protein by scaling up the procedures and by growth of larger quantities of bacteria in a fermenter. This approach has been successful with $rG_{s\alpha}$.[10]

Acknowledgments

We thank Leslie Perry and Todd Guthrie for skillful technical assistance. Work from the authors' laboratory was supported by United States Public Health Service Grant GM34497, American Cancer Society Grant BC555I, and the Raymond and Ellen Willie Chair of Molecular Neuropharmacology. We also acknowledge support from the Perot Family Foundation and The Lucille P. Markey Charitable Trust.

[22] Amino acid sequence analysis was performed by Kim Orth, Caroline Moomaw, and Clive Slaughter.

[19] Synthetic Peptide Antisera with Determined Specificity for G Protein α or β Subunits

By SUSANNE M. MUMBY and ALFRED G. GILMAN

Introduction

We have generated a number of antisera that react with G proteins by immunization of rabbits with either purified G-protein subunits or peptides (coupled to carrier protein) synthesized according to amino acid sequences of G proteins.[1] The peptide antisera have proved to be particularly useful, since one can usually determine their specificity by choice of peptides with sequences unique to a given subunit or common to a family of subunits. This approach was initiated when limited sequence information had been

[1] S. M. Mumby, R. A. Kahn, D. R. Manning, and A. G. Gilman, *Proc. Natl. Acad. Sci. U.S.A.* **83,** 265 (1986).

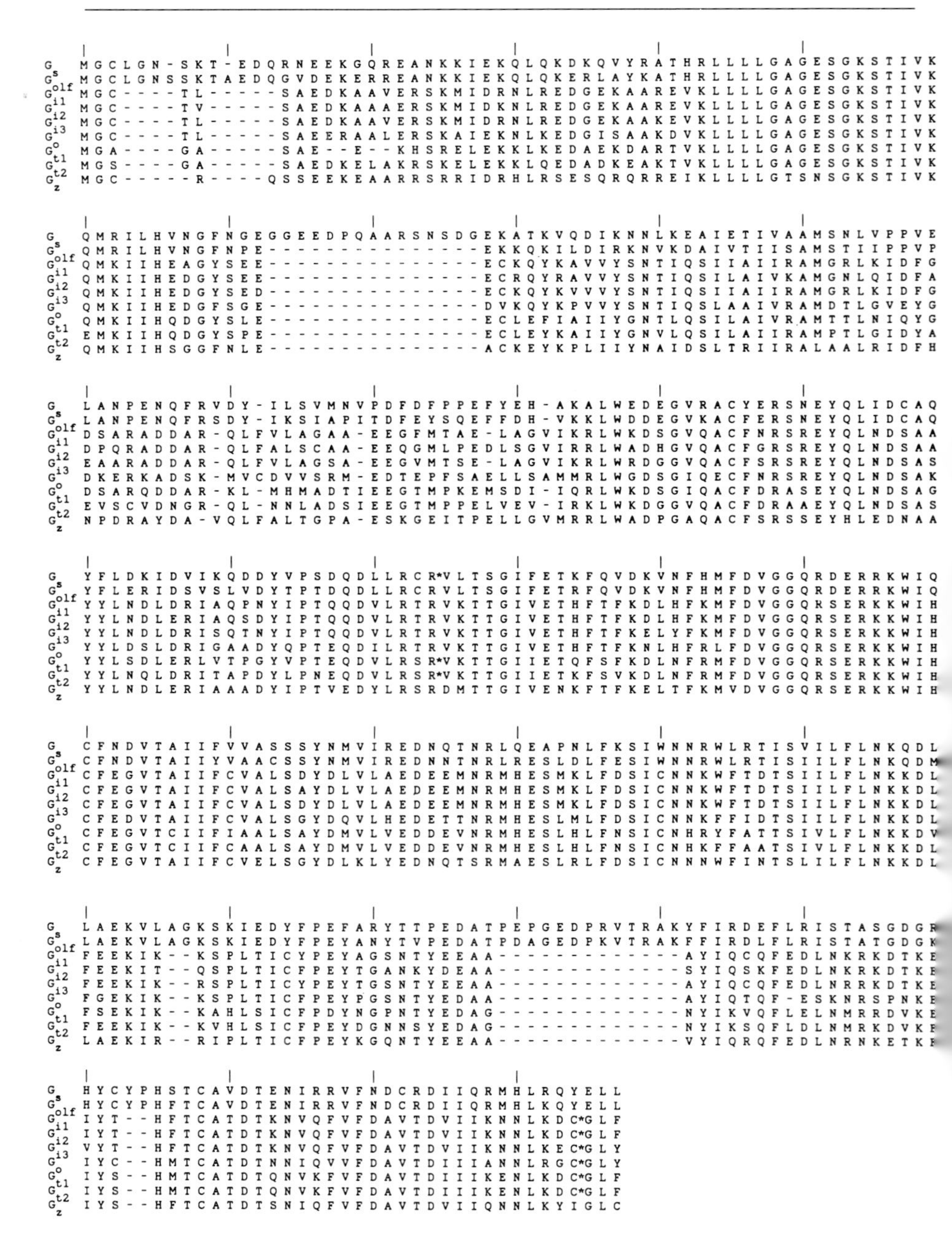

Gs M G C L G N - S K T - E D Q R N E E K G Q R E A N K K I E K Q L Q K D K Q V Y R A T H R L L L L G A G E S G K S T I V K
Golf M G C L G N S S K T A E D Q G V D E K E R R E A N K K I E K Q L Q K E R L A Y K A T H R L L L L G A G E S G K S T I V K
Gi1 M G C - - - - T L - - - - - S A E D K A A V E R S K M I D R N L R E D G E K A A R E V K L L L L G A G E S G K S T I V K
Gi2 M G C - - - - T V - - - - - S A E D K A A A E R S K M I D K N L R E D G E K A A R E V K L L L L G A G E S G K S T I V K
Gi3 M G C - - - - T L - - - - - S A E D K A A V E R S K M I D R N L R E D G E K A A K E V K L L L L G A G E S G K S T I V K
Go M G C - - - - T L - - - - - S A E E R A A L E R S K A I E K N L K E D G I S A A K D V K L L L L G A G E S G K S T I V K
Gt1 M G A - - - - G A - - - - - S A E - - E - - K H S R E L E K K L K E D A E K D A R T V K L L L L G A G E S G K S T I V K
Gt2 M G S - - - - G A - - - - - S A E D K E L A K R S K E L E K K L Q E D A D K E A K T V K L L L L G A G E S G K S T I V K
Gz M G C - - - - - R - - - - Q S S E E K E A A R R S R R I D R H L R S E S Q R Q R R E I K L L L L G T S N S G K S T I V K

Gs Q M R I L H V N G F N G E G G E E D P Q A A R S N S D G E K A T K V Q D I K N N L K E A I E T I V A A M S N L V P P V E
Golf Q M R I L H V N G F N P E - - - - - - - - - - - - - - - - E K K Q K I L D I R K N V K D A I V T I I S A M S T I I P P V P
Gi1 Q M K I I H E A G Y S E E - - - - - - - - - - - - - - - - E C K Q Y K A V V Y S N T I Q S I I A I I R A M G R L K I D F G
Gi2 Q M K I I H E D G Y S E E - - - - - - - - - - - - - - - - E C R Q Y R A V V Y S N T I Q S I L A I V K A M G N L Q I D F A
Gi3 Q M K I I H E D G Y S E D - - - - - - - - - - - - - - - - E C K Q Y K V V V Y S N T I Q S I I A I I R A M G R L K I D F G
Go Q M K I I H E D G F S G E - - - - - - - - - - - - - - - - D V K Q Y K P V V Y S N T I Q S L A A I V R A M D T L G V E Y G
Gt1 Q M K I I H Q D G Y S L E - - - - - - - - - - - - - - - - E C L E F I A I I Y G N T L Q S I L A I V R A M T T L N I Q Y G
Gt2 E M K I I H Q D G Y S P E - - - - - - - - - - - - - - - - E C L E Y K A I I Y G N V L Q S I L A I I R A M P T L G I D Y A
Gz Q M K I I H S G G F N L E - - - - - - - - - - - - - - - - A C K E Y K P L I I Y N A I D S L T R I I R A L A A L R I D F H

Gs L A N P E N Q F R V D Y - I L S V M N V P D F D F P P E F Y E H - A K A L W E D E G V R A C Y E R S N E Y Q L I D C A Q
Golf L A N P E N Q F R S D Y - I K S I A P I T D F E Y S Q E F F D H - V K K L W D D E G V K A C F E R S N E Y Q L I D C A Q
Gi1 D S A R A D D A R - Q L F V L A G A A - E E G F M T A E - L A G V I K R L W K D S G V Q A C F N R S R E Y Q L N D S A A
Gi2 D P Q R A D D A R - Q L F A L S C A A - E E Q G M L P E D L S G V I R R L W A D H G V Q A C F G R S R E Y Q L N D S A A
Gi3 E A A R A D D A R - Q L F V L A G S A - E E G V M T S E - L A G V I K R L W R D G G V Q A C F S R S R E Y Q L N D S A S
Go D K E R K A D S K - M V C D V V S R M - E D T E P F S A E L L S A M M R L W G D S G I Q E C F N R S R E Y Q L N D S A K
Gt1 D S A R Q D D A R - K L - M H M A D T I E E G T M P K E M S D I - I Q R L W K D S G I Q A C F D R A S E Y Q L N D S A G
Gt2 E V S C V D N G R - Q L - N N L A D S I E E G T M P P E L V E V - I R K L W K D G G V Q A C F D R A A E Y Q L N D S A S
Gz N P D R A Y D A - V Q L F A L T G P A - E S K G E I T P E L L G V M R R L W A D P G A Q A C F S R S S E Y H L E D N A A

Gs Y F L D K I D V I K Q D D Y V P S D Q D L L R C R*V L T S G I F E T K F Q V D K V N F H M F D V G G Q R D E R R K W I Q
Golf Y F L E R I D S V S L V D Y T P T D Q D L L R C R V L T S G I F E T R F Q V D K V N F H M F D V G G Q R D E R R K W I Q
Gi1 Y Y L N D L D R I A Q P N Y I P T Q Q D V L R T R V K T T G I V E T H F T F K D L H F K M F D V G G Q R S E R K K W I H
Gi2 Y Y L N D L E R I A Q S D Y I P T Q Q D V L R T R V K T T G I V E T H F T F K D L H F K M F D V G G Q R S E R K K W I H
Gi3 Y Y L N D L D R I S Q T N Y I P T Q Q D V L R T R V K T T G I V E T H F T F K E L Y F K M F D V G G Q R S E R K K W I H
Go Y Y L D S L D R I G A A D Y Q P T E Q D I L R T R V K T T G I V E T H F T F K N L H F R L F D V G G Q R S E R K K W I H
Gt1 Y Y L S D L E R L V T P G Y V P T E Q D V L R S R*V K T T G I I E T Q F S F K D L N F R M F D V G G Q R S E R K K W I H
Gt2 Y Y L N Q L D R I T A P D Y L P N E Q D V L R S R*V K T T G I I E T K F S V K D L N F R M F D V G G Q R S E R K K W I H
Gz Y Y L N D L E R I A A A D Y I P T V E D Y L R S R D M T T G I V E N K F T F K E L T F K M V D V G G Q R S E R K K W I H

Gs C F N D V T A I I F V V A S S S Y N M V I R E D N Q T N R L Q E A P N L F K S I W N N R W L R T I S V I L F L N K Q D L
Golf C F N D V T A I I Y V A A C S S Y N M V I R E D N N T N R L R E S L D L F E S I W N N R W L R T I S I I L F L N K Q D M
Gi1 C F E G V T A I I F C V A L S D Y D L V L A E D E E M N R M H E S M K L F D S I C N N K W F T D T S I I L F L N K K D L
Gi2 C F E G V T A I I F C V A L S A Y D L V L A E D E E M N R M H E S M K L F D S I C N N K W F T D T S I I L F L N K K D L
Gi3 C F E G V T A I I F C V A L S D Y D L V L A E D E E M N R M H E S M K L F D S I C N N K W F T D T S I I L F L N K K D L
Go C F E D V T A I I F C V A L S G Y D Q V L H E D E T T N R M H E S L M L F D S I C N N K F F I D T S I I L F L N K K D L
Gt1 C F E G V T C I I F I A A L S A Y D M V L V E D D E V N R M H E S L H L F N S I C N H R Y F A T T S I V L F L N K K D V
Gt2 C F E G V T C I I F C A A L S A Y D M V L V E D D E V N R M H E S L H L F N S I C N H K F F A A T S I V L F L N K K D L
Gz C F E G V T A I I F C V E L S G Y D L K L Y E D N Q T S R M A E S L R L F D S I C N N N W F I N T S L I L F L N K K D L

Gs L A E K V L A G K S K I E D Y F P E F A R Y T T P E D A T P E P G E D P R V T R A K Y F I R D E F L R I S T A S G D G R
Golf L A E K V L A G K S K I E D Y F P E Y A N Y T V P E D A T P D A G E D P K V T R A K F F I R D L F L R I S T A T G D G K
Gi1 F E E K I K - - K S P L T I C Y P E Y A G S N T Y E E A A - - - - - - - - - - - - - - - A Y I Q C Q F E D L N K R K D T K E
Gi2 F E E K I T - - Q S P L T I C F P E Y T G A N K Y D E A A - - - - - - - - - - - - - - - S Y I Q S K F E D L N K R K D T K E
Gi3 F E E K I K - - R S P L T I C Y P E Y T G S N T Y E E A A - - - - - - - - - - - - - - - A Y I Q C Q F E D L N R R K D T K E
Go F G E K I K - - K S P L T I C F P E Y P G S N T Y E D A A - - - - - - - - - - - - - - - A Y I Q T Q F - E S K N R S P N K E
Gt1 F S E K I K - - K A H L S I C F P D Y N G P N T Y E D A G - - - - - - - - - - - - - - - N Y I K V Q F L E L N M R R D V K E
Gt2 F E E K I K - - K V H L S I C F P E Y D G N N S Y E D A G - - - - - - - - - - - - - - - N Y I K S Q F L D L N M R K D V K E
Gz L A E K I R - - R I P L T I C F P E Y K G Q N T Y E E A A - - - - - - - - - - - - - - - V Y I Q R Q F E D L N R N K E T K E

Gs H Y C Y P H S T C A V D T E N I R R V F N D C R D I I Q R M H L R Q Y E L L
Golf H Y C Y P H F T C A V D T E N I R R V F N D C R D I I Q R M H L K Q Y E L L
Gi1 I Y T - - H F T C A T D T K N V Q F V F D A V T D V I I K N N L K D C*G L F
Gi2 I Y T - - H F T C A T D T K N V Q F V F D A V T D V I I K N N L K D C*G L F
Gi3 V Y T - - H F T C A T D T K N V Q F V F D A V T D V I I K N N L K E C*G L Y
Go I Y C - - H M T C A T D T N N I Q V V F D A V T D I I I A N N L R G C*G L Y
Gt1 I Y S - - H M T C A T D T Q N V K F V F D A V T D I I I K E N L K D C*G L F
Gt2 I Y S - - H M T C A T D T Q N V K F V F D A V T D I I I K E N L K D C*G L F
Gz I Y S - - H F T C A T D T S N I Q F V F D A V T D V I I Q N N L K Y I G L C

obtained from tryptic peptides of α_o ($G_{o\alpha}$) and α_t ($G_{t\alpha}$).[2] Use of the technique has expanded greatly following the recent spate of cDNA cloning and sequencing. Antisera to all known G-protein α and β subunits (at least from higher eukaryotes) have now been produced. Most of these have excellent specificity; some, intentionally, react indiscriminately with all α or β subunits. Application of this technique to γ subunits can be anticipated. In general, the antisera are useful for enzyme-linked immunosorbent assays (ELISA) and Western immunoblotting. A competitive ELISA has been developed to quantitate α_s.[3] Some G-protein-reactive peptide antisera are suitable for immunoprecipitation[4,5] or for studies of immuno-

[2] J. B. Hurley, M. I. Simon, D. B. Teplow, J. D. Robishaw, and A. G. Gilman, *Science* **226,** 860 (1984).

[3] L. A. Ransnas and P. A. Insel, *J. Biol. Chem.* **263,** 17239 (1988).

[4] J. E. Buss, S. M. Mumby, P. J. Casey, A. G. Gilman, and B. M. Sefton, *Proc. Natl. Acad. Sci. U.S.A.* **84,** 7493 (1987).

[5] P. O. Rothenberg and C. R. Kahn, *J. Biol. Chem.* **263,** 15546 (1988).

FIG 1. Amino acid sequences deduced from cDNAs that encode G-protein α subunits are listed. These include bovine G_s [J. D. Robishaw, D. W. Russell, B. A. Harris, M. D. Smigel, and A. G. Gilman, *Proc. Natl. Acad. Sci. U.S.A.* **83,** 1251 (1986)], rat G_{olf} [D. T. Jones and R. R. Reed, *Science* **244,** 790 (1989)], bovine G_{i1} [T. Nukada, T. Tanabe, H. Takahashi, M. Noda, K. Haga, T. Haga, A. Ichiyama, K. Kangawa, M. Hiranaga, H. Matsuo, and S. Numa, *FEBS Lett.* **197,** 305 (1986)], rat G_{i2} [H. Itoh, T. Kozasa, S. Nagata, S. Nakamura, T. Katada, M. Ui, S. Iwai, E. Ohtsuka, H. Kawasaki, K. Suzuki, and Y. Kaziro, *Proc. Natl. Acad. Sci. U.S.A.* **83,** 3776 (1986)], rat G_{i3} [D. T. Jones and R. R. Reed, *J. Biol. Chem.* **262,** 14241 (1987)], bovine G_o [K. P. Van Meurs, C. W. Angus, S. Lavu, H. Kung, S. K. Czarnecki, J. Moss, and M. Vaughan, *Proc. Natl. Acad. Sci. U.S.A.* **84,** 3107 (1987)], bovine G_{t1} [T. Tanabe, T. Nukada, Y. Nishikawa, K. Sugimoto, H. Suzuki, H. Takahashi, M. Noda, T. Haga, A. Ichiyama, K. Kangawa, N. Minamino, H. Matsuo, and S. Numa, *Nature (London)* **315,** 242 (1985)], bovine G_{t2} [M. A. Lochrie, J. B. Hurley, and M. I. Simon, *Science* **228,** 96 (1985)], and human G_z [H. K. W. Fong, K. K. Yoshimoto, P. Eversole-Cire, and M. I. Simon, *Proc. Natl. Acad. Sci. U.S.A.* **85,** 3066 (1988)]. Asterisks (*) indicate sites of ADP-ribosylation catalyzed by cholera or pertussis toxin. Additional sequences have been published for the α subunits of human G_{i1}, G_{i2}, and G_{i3} [H. Itoh, R. Toyama, T. Kozasa, T. Tsukamoto, M. Matsuoka, and Y. Kaziro, *J. Biol. Chem.* **263,** 6656 (1988)], human G_s [P. Bray, A. Carter, C. Simons, V. Guo, C. Puckett, J. Kamholtz, A. Speigel, and M. Nirenberg, *Proc. Natl. Acad. Sci. U.S.A.* **83,** 8893 (1986)], rat G_o (H. Itoh, T. Kozasa, S. Nagata, S. Nagata, S. Nakamura, T. Katada, M. Ui, S. Iwai, E. Ohtsuka, H. Kawasaki, K. Suzuki, and Y. Kaziro, *Proc. Natl. Acad. Sci. U.S.A.* **83,** 3776 (1986)], rat G_s [D. T. Jones and R. R. Reed, *J. Biol. Chem.* **262,** 14241 (1987)], and mouse G_{i2} and G_s [K. A. Sullivan, Y. Liao, A. Alborzi, B. Biederman, F. Chang, S. B. Masters, A. D. Levinson, and H. R. Bourne, *Proc. Natl. Acad. Sci. U.S.A.* **83,** 6687 (1986)].

cytochemical localization.[6–9] There is little published information about the effects of peptide antibodies on G-protein function. However, antibodies generated against the carboxy-terminal 10 amino acid residues of α_t inhibit the rhodopsin-stimulated GTPase activity of G_t in phospholipid vesicles.[10]

Choice of Synthetic Peptide Sequence

The first and most critical step in production of a synthetic peptide antiserum is the choice of the peptide to be utilized. The degree of specificity desired obviously dictates the regions of G-protein subunit structure to be considered. Figures and tables are presented to facilitate the decision. Figures 1 and 3 list many of the sequences currently available for α and β subunits, respectively. Figures 2 and 4 highlight regions of sequence similarity and divergence and show the location of sequences that have been utilized to produce synthetic peptide antisera. Details of peptide sequences and the immunological reactivity of peptide antisera that have been generated for α and β subunits are listed in Tables I and II, respectively. Understanding the implications of the data on immunological reactivity listed in Tables I and II requires knowledge of the members of the G-protein family and the cDNAs that encode them. A brief description is given below; more details can be found in the references listed in the tables and from review articles.[11–13]

α Subunits. The α subunits are the most diverse and confer specificity for interactions between G proteins and particular receptors and effectors (Figs. 1 and 2). They range in apparent molecular weight on sodium dodecyl sulfate (SDS)–polyacrylamide gels from 39,000 for α_o and α_t to 52,000 for the large form of α_s. The molecular weights deduced from their cDNAs range from 40,000 to 46,000. Two forms of α_s protein are distinguished on gels as species with apparent molecular weights of 45,000 and 52,000. They are derived from the same gene by alternative splicing

[6] R. R. Anholt, S. M. Mumby, D. A. Stoffers, P. R. Girard, J. F. Kuo, and S. H. Snyder, *Biochemistry* **26,** 788 (1987).

[7] C. L. Lerea, D. E. Somers, J. B. Hurley, I. B. Klock, and A. H. Bunt-Milan, *Science* **234,** 77 (1986).

[8] D. T. Jones and R. R. Reed, *Science* **244,** 790 (1989).

[9] K.-J. Chang, W. Pugh, S. G. Blanchard, J. McDermed, and J. P. Tam, *Proc. Natl. Acad. Sci. U.S.A.* **85,** 4929 (1988).

[10] R. A. Cerione, S. Kroll, R. Rajaram, C. Unson, P. Goldsmith, and A. M. Spiegel, *J. Biol. Chem.* **263,** 9345 (1988).

[11] A. G. Gilman, *Annu. Rev. Biochem.* **56,** 615 (1987).

[12] M. A. Lochrie and M. I. Simon, *Biochemistry* **27,** 4957 (1988).

[13] E. J. Neer and D. E. Clapham, *Nature (London)* **333,** 129 (1988).

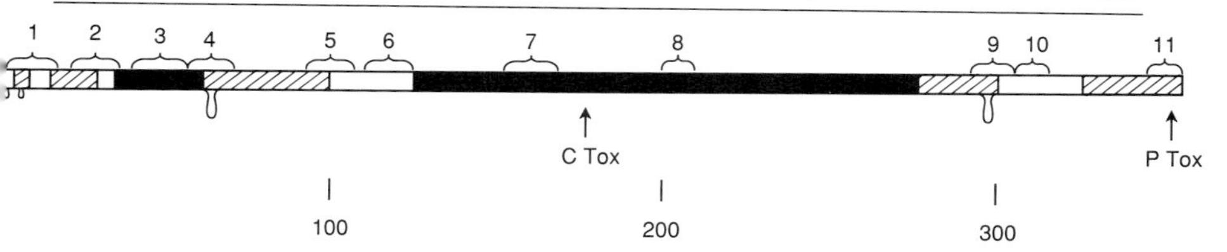

FIG. 2. Location of α-subunit sequences used for generation of synthetic peptide antisera. The horizontal line represents the primary amino acid sequence (left = amino terminus). The black areas represent regions with the highest degree of conservation of amino acid sequence among different α subunits, the white areas designate regions of greatest divergence, and the hatched areas are intermediate. The loops that extend below the line represent sequences that are present in α_s but not other α subunits. The numbered braces indicate regions that have been utilized to generate synthetic peptide antisera (see Table I). The arrows labeled C Tox and P Tox indicate the location of sites of ADP-ribosylation of α_s and α_t (arginine) catalyzed by cholera toxin and α_i, α_o, and α_t (cysteine) catalyzed by pertussis toxin. The numerals below the line represent numbering of the amino acid sequence of α_{i1}.

of mRNA.[14] In fact, there appear to be four different splice variants of α_s.[15] mRNAs for both the long and short forms of α_s may include or exclude a codon for a serine residue at the splice junction. (Antibodies capable of recognizing this variation have not been developed.) The larger forms of the protein include 15 (or 16) amino acid residues encoded by exon III of the gene for α_s.[16] An antiserum raised to this amino acid sequence reacts only with the 52-kDa form of α_s, whereas other α_s-specific antisera react with both forms of the protein (Table I).[1,11,14] α_{olf}, an olfactory-specific α subunit which is 88% identical to α_s, has been identified in olfactory sensory neurons by the use of synthetic peptide antisera. Two peptide antigens have been developed, based on the amino acid sequence deduced from an α_{olf} cDNA.[8] The antisera directed against peptides unique to α_{olf} react specifically with α_{olf} and not α_s. α_{olf} is presumed to function in odorant receptor-mediated signal transduction.

The term G_i was originally coined to designate the inhibitory G protein that regulates adenylyl cyclase. As a result of appreciation of the multiplicity of highly homologous forms of this protein, the term $G_{i\alpha}$ or α_i now denotes a group of three α subunits that are substrates for pertussis toxin and that differ significantly from α_o and α_t. All three forms of α_i cDNA have been isolated from a single rat olfactory library.[17] The α_{i1} and α_{i3}

[14] J. D. Robishaw, M. D. Smigel, and A. G. Gilman, *J. Biol. Chem.* **261,** 9587 (1986).

[15] P. Bray, A. Carter, C. Simons, V. Guo, C. Puckett, J. Kamholtz, A. Spiegel, and M. Nirenberg, *Proc. Natl. Acad. Sci. U.S.A.* **83,** 8893 (1986).

[16] T. Kozasa, H. Itoh, T. Tsukamoto, and Y. Kaziro, *Proc. Natl. Acad. Sci. U.S.A.* **85,** 2081 (1988).

[17] D. T. Jones and R. R. Reed, *J. Biol. Chem.* **262,** 14241 (1987).

TABLE I
G-PROTEIN α-SUBUNIT PEPTIDE ANTISERA

Designation	Code number[a]	Region[b]	Amino acid number	Sequence (one-letter code)	Reactivity[c]	
					Positive	Negative
α_{common}	A-569[d]	3	40–54	GAGESGKSTIVKQMK	$\alpha_o > \alpha_i, \alpha_t > \alpha_s$	α_z
	1398[e]	3	40–54	GAGESGKSTIVKQMK	$\alpha_o, \alpha_i, \alpha_t > \alpha_s$	
	AS 8[f]	3	40–54	GAGESGKSTIVKQMK	$\alpha_i > \alpha_o > \alpha_s$	
	GA/1[g]	3	40–51	GAGESGKSTIVK	α_i, α_o	
	P-960[h]	3	40–54	GTSNSGKSTIVKQMK	$\alpha_s, \alpha_i, \alpha_o, \alpha_t, \alpha_z$	
	8645[i]	8	200–211	FDVGGQRSERKK	α_o, α_i	
$\alpha_{s,long}$	561[j]	4	72–85	GEEDPQAARSNSDG	α_s (α_{52})	
α_s	584[k]	9	325–339	TPEPGEDPRVTRAKY	α_s (α_{45}, α_{52})	$\alpha_o, \alpha_i, \alpha_t$
	C-519[d]	2	28–42	KQLQKDKQVYRATHR	α_s (α_{45}, α_{52})	$\alpha_o, \alpha_i, \alpha_t$
$\alpha_{s,Arg}$	A-572[d]	2	28–42	KQLQRDRQVYRATHR	α_s (α_{45}, α_{52})	$\alpha_o, \alpha_i, \alpha_t$
α_{olf}	DJ5.2[l]	1	9–24	KTAEDQGVDEKERREA	α_{olf}	α_s
	DJ1[l]	6	118–132	IKSIAPITDFEYSQE	α_{olf}	α_s
α_i	E-786[m]	10	306–317	SKFEDLNKRKDT	α_{i1}, α_{i2}	$\alpha_o, \alpha_s, \alpha_t$
	AS 19[f]	2	22–34	NLREDGEKAAREV	α_i	α_o
α_{i1}	I-355[m]	6	111–124	AGAAEEGFMTAELA	α_{i1}	$\alpha_s, \alpha_{i2}, \alpha_{i3}, \alpha_t, \alpha_z$
	LD[n]	7	159–168	LDRIAQPNYI	α_{i1}	$\alpha_o, \alpha_{i2}, \alpha_{i3}$
α_{i2}	J-883[m]	6	112–126	CAAEEQGMLPEDLSG	α_{i2}	$\alpha_s, \alpha_{i1}, \alpha_{i3}, \alpha_o, \alpha_t, \alpha_z$
	LE[n]	7	160–169	LERIAQSDYI	α_{i2}	α_{i1}, α_{i3}
	1521[i]	7	160–169	LERIAQSDYI	α_{i2}	
	AS 64[f]	9	292–302	TGANKYDEAAS	α_{i2}	$\alpha_i, \alpha_o, \alpha_s$
α_{i3}	K-887[o]	6	111–125	AGSAEEGVMTSELAG	α_{i3}	$\alpha_s, \alpha_{i1}, \alpha_{i2}, \alpha_o, \alpha_t, \alpha_z$
	SQ[n]	7	159–168	LDRISQSNYI	α_{i3}	α_i, α_{i2}
α_o	U-46[d]	2	22–35	NLKEDGISAAKDVK	α_o	$\alpha_s, \alpha_i, \alpha_t, \alpha_z$
	AS 6[f]	2	22–35	NLKEDGISAAKDVK	α_o	α_s, α_i
	IM/1[g]	2	22–35	NLKEDGISAAKDVK	α_o	
	9120[p]	2	26–34	DGISAAKDV	α_o	$\alpha_s, \alpha_i, \alpha_t, \alpha_z$
	GC/1[g]	1	2–17	GCTLSAGERAALERSK	α_o	$\alpha_s, \alpha_i, \alpha_t$

	GO/1[g]	11	345–354	ANNLRGCGLY	α_o	
	2353[q]	5	94–105	EYGDKERKADSK	$\alpha_o > \alpha_t$	α_i, α_s
α_{t1}	T_αIB[r]	7	153–166	SDLERLVTPGYVPT	α_{t1}	α_{t2}
α_{t2}	T_αIIB[r]	7	159–170	LDRITAPDYLPN	α_{t2}	α_{t1}
$\alpha_t/\alpha_i/\alpha_o$	AS/7[s]	11	341–350	KENLKDCGLF	α_t, α_o, α_{i1}, $\alpha_{i2} > \alpha_{i3}$	
	R16,17[t]	11	347–355	NNLKDCGLF	α_t, $\alpha_i > \alpha_o$	
	8730[i]	11	346–355	KNNLKDCGLF	$\alpha_i > \alpha_o$	α_s
α_i/α_o	H-660[m]	1	3–17	CTVSAEDKAAAERSK	$\alpha_o > \alpha_{i1}$, α_{i2}	α_s, α_t
	anti-α_i[u]	1	3–17	CTVSAEDKAAAERSK	α_{i2}	α_s, α_{i1}, α_o
α_z	P-961[h]	1	3–18	CRQSSEEKEAARRSRR	α_z	α_i, α_o
	$G_{z\text{-}3}$[v]	1	3–18	CRQSSEEKEAARRSRR	α_z	α_i, α_o
	$G_{z\text{-}111}$[v]	6	111–125	TGPAESKGEITPELL	α_z	α_i, α_o

[a] Code number assigned to antiserum by investigator that produced the antiserum.

[b] Region numbers indicate location of the peptide sequence on the linear map of α subunit shown in Fig. 2.

[c] Reactivity of antisera with α subunits that have been tested either positive or negative by Western immunoblotting.

[d] S. M. Mumby, R. A. Kahn, D. R. Manning, and A. G. Gilman, *Proc. Natl. Acad. Sci. U.S.A.* **83,** 265 (1986).

[e] M. W. Wilde, K. E. Carlson, D. R. Manning, and S. H. Zigmond, *J. Biol. Chem.* **264,** 190 (1989).

[f] K.-D. Hinsch, W. Rosenthal, K. Spicher, T. Binder, H. Gausepohl, R. Frank, G. Schultz, and H. G. Joost, *FEBS Lett.* **238,** 191 (1988).

[g] P. Goldsmith, P. S. Backland, K. Rossiter, A. Carter, G. Milligan, C. G. Unson, and A. Spiegel, *Biochemistry* **27,** 7085 (1988).

[h] P. J. Casey and A. G. Gilman, unpublished data (1988).

[i] K. E. Carlson, L. F. Brass, and D. R. Manning, *J. Biol. Chem.* **264,** 13298 (1989).

[j] J. D. Robishaw, M. D. Smigel, and A. G. Gilman, *J. Biol. Chem.* **261,** 9587 (1986).

[k] A. G. Gilman, *Annu. Rev. Biochem.* **56,** 615 (1987).

[l] D. T. Jones and R. R. Reed, *Science* **244,** 790 (1989).

[m] S. M. Mumby, I. Pang, A. G. Gilman, and P. C. Sternweis, *J. Biol. Chem.* **263,** 2020 (1988).

[n] P. Goldsmith, K. Rossiter, A. Carter, W. Simonds, C. G. Unson, R. Vinitisky, and A. M. Spiegel, *J. Biol. Chem.* **263,** 6476 (1988).

[o] D. J. Carty, S. M. Mumby, E. Padrell, J. Codina, R. Graf, L. Birnbaumer, A. G. Gilman, and R. Iyengar, unpublished results (1989).

[p] K.-J. Chang, W. Pugh, S. G. Blanchard, J. McDermed, and J. P. Tam, *Proc. Natl. Acad. Sci. U.S.A.* **85,** 4929 (1988).

[q] D. R. Manning, unpublished results (1988).

[r] C. L. Lerea, D. E. Somers, J. B. Hurley, I. B. Klock, and A. H. Bunt-Milan, *Science* **234,** 77 (1986).

[s] P. Goldsmith, P. Gierschick, G. Milligan, C. G. Unson, R. Vinitsky, H. L. Malech, and A. M. Spiegel, *J. Biol. Chem.* **262,** 14683 (1987).

[t] G. M. Bokoch, K. Bickford, and B. P. Bohl, *J. Cell Biol.* **106,** 1927 (1988).

[u] J. Lang and T. Costa, *Biochem. Biophys. Res. Commun.* **148,** 838 (1987).

[v] H. K. W. Fong, M. I. Simon, P. J. Casey, and A. G. Gilman, unpublished results (1988).

cDNAs encode proteins that are 94% identical in amino acid sequence. The deduced α_{i1} and α_{i2} amino acid sequences are 88% identical. Peptide antisera have been developed that can distinguish between the three forms of α_i.[18–21] The results of Western blotting experiments indicate that the α_{i1} cDNA encodes the major 41-kDa α subunit from brain, the α_{i2} cDNA encodes a 40-kDA α subunit present in brain and many other tissues, and the α_{i3} cDNA encodes another 41-kDa α subunit present in liver, HL-60 cells, and elsewhere. The functions of these similar but distinct α subunits are under active investigation. All three forms of α_i protein have been detected immunologically in a single cell type, the erythrocyte, suggesting that the proteins may respond to different receptors or serve to regulate different effectors.[19] Unlike the three forms of α_i, the two forms of α_t that have been identified by cDNA cloning appear to be cell type-specific in their expression. Immunocytochemical evidence, obtained with peptide antisera, indicates that α_{t1} is expressed in rod photoreceptor cells, whereas α_{t2} is expressed in cone photoreceptor cells.[7] G_{t1} purified from rod outer segments couples rhodopsin to a cyclic GMP-specific phosphodiesterase. It is presumed that G_{t2} serves the same function.

G_o is an abundant G protein that was purified initially from bovine brain.[22] G_o is capable of coupling neuropeptide receptors to Ca^{2+} channels in dorsal root ganglion cells and elsewhere.[23–25] A single type of cDNA that encodes α_o has been isolated.[17,26,27] Recent biochemical evidence indicates that two forms of the protein may exist[21]; it is not known if the difference between the two is due to primary structure or to posttranslational modification.

Phospholipases are regulated in a guanine nucleotide-dependent fash-

[18] S. M. Mumby, I. Pang, A. G. Gilman, and P. C. Sternweis, *J. Biol. Chem.* **263,** 2020 (1988).

[19] D. J. Carty and R. Iyengar, *FEBS Lett.* **262,** 101 (1990).

[20] P. Goldsmith, K. Rossiter, A. Carter, W. Simonds, C. G. Unson, R. Vinitisky, and A. M. Spiegel, *J. Biol. Chem.* **263,** 6476 (1988).

[21] P. Goldsmith, P. S. Backland, K. Rossiter, A. Carter, G. Milligan, C. G. Unson, and A. Spiegel, *Biochemistry* **27,** 7085 (1988).

[22] P. C. Sternweis and J. D. Robishaw, *J. Biol. Chem.* **259,** 13806 (1984).

[23] J. Hescheler, W. Rosenthal, W. Trautwein, and G. Schultz, *Nature (London)* **325,** 445 (1987).

[24] D. A. Ewald, P. C. Sternweis, and R. J. Miller, *Proc. Natl. Acad. Sci. U.S.A.* **85,** 3633 (1988).

[25] R. M. Harris-Warrick, C. Hammond, D. Paupardin-Tritsch, V. Homburger, B. Rouot, J. Bockaert, and H. M. Gerschenfeld, *Neuron* **1,** 27 (1988).

[26] H. Itoh, T. Kozasa, S. Nagata, S. Nakamura, T. Katada, M. Ui, S. Iwai, E. Ohtsuka, H. Kawasaki, K. Suzuki, and Y. Kaziro, *Proc. Natl. Acad. Sci. U.S.A.* **83,** 3776 (1986).

[27] K. P. Van Meurs, C. W. Angus, S. Lavu, H. Kung, S. K. Czarnecki, J. Moss, and M. Vaughan, *Proc. Natl. Acad. Sci. U.S.A.* **84,** 3107 (1987).

ion,[28,29] but the G proteins responsible for these phenomena have not been identified. In many (but not all) tissues and cell types, receptor-mediated stimulation of phospholipase C is insensitive to pertussis toxin. Two research groups have recently isolated a cDNA that encodes a putative G-protein α subunit that lacks the consensus site for ADP-ribosylation by pertussis toxin[30,31] (Fig. 1). The protein, designated α_z by Simon's group and α_x by Kaziro and co-workers, may be a candidate for regulation of one or more phospholipases. A synthetic peptide antiserum directed against a sequence unique to α_z has been utilized in a Western-immunoblotting assay for the purification of α_z from extracts of bovine brain membranes.[32]

β Subunits. Two forms of the G-protein β subunits have been identified. When purified from most tissues, β appears as a doublet on SDS–polyacrylamide gels; the two bands have been designated β_{35} and β_{36}, based on their mobility. G_t is purified with only β_{36}, whereas preparations of G_s, G_i, G_o, and G_z include both β_{35} and β_{36}. Functional differences between the two forms of the protein are not known. Two distinct β-subunit cDNAs, termed β_1 and β_2, encode proteins whose amino acid sequences are 90% identical (Figs. 3 and 4; Table II). Results of Western blotting with subunit-specific peptide antisera indicate that the β_1 cDNA encodes the β_{36} protein and the β_2 cDNA encodes the β_{35} protein.[33,34]

Practical Considerations. Careful consideration of the amino acid sequences presented in Figs. 1 and 3 should aid in choosing a peptide that will produce antibodies of the desired specificity. A literature search for new sequence information should obviously be conducted as well. Although there is very little species variation in G-protein sequences, it is wise to search for amino acid substitutions before settling on a particular sequence. A single amino acid substitution may change the reactivity of a given antiserum dramatically.[33]

In general, the same peptide will elicit antibodies with very similar specificity profiles when injected into multiple rabbits. However, there are exceptions. Antiserum J-883 produced against an amino-terminal

[28] S. Cockcroft, *Trends Biochem. Sci.* **12,** 75 (1987).

[29] R. D. Burgoyne, T. R. Cheek, and A. J. O'Sullivan, *Trends Biol. Sci.* **12,** 332 (1987).

[30] H. K. W. Fong, K. Yoshimoto, P. Eversole-Cire, and M. I. Simon, *Proc. Natl. Acad. Sci. U.S.A.* **85,** 3066 (1988).

[31] M. Matsuoka, H. Itoh, T. Kozasa, and Y. Kaziro, *Proc. Natl. Acad. Sci. U.S.A.* **85,** 5384 (1988).

[32] P. J. Casey, H. Fong, M. Simon, and A. G. Gilman, *J. Biol. Chem.* **265,** 2383 (1990).

[33] B. Gao, S. M. Mumby, and A. G. Gilman, *J. Biol. Chem.* **262,** 17254 (1987).

[34] T. T. Amatruda, N. Gautum, H. K. W. Fong, J. K. Northup, and M. I. Simon, *J. Biol. Chem.* **263,** 5008 (1988).

```
                Asp                                     Lys                                         Ala     Ala         Ser             Asn Asn Ile
Met Ser Glu Leu Glu Gln Leu Arg Gln Glu Ala Glu Gln Leu Arg Asn Gln Ile Arg Asp Ala Arg Lys Ala Cys Gly Asp Ser Thr Leu Thr Gln Ile Thr Ala Gly Leu Asp Pro Val  40

Gly Arg Ile Gln Met Arg Thr Arg Arg Thr Leu Arg Gly His Leu Ala Lys Ile Tyr Ala Met His Trp Gly Thr Asp Ser Arg Leu Leu Val Ser Ala Ser Gln Asp Gly Lys Leu Ile  80

                                                                                                                        Tyr
Ile Trp Asp Ser Tyr Thr Thr Asn Lys Val His Ala Ile Pro Leu Arg Ser Ser Trp Val Met Thr Cys Ala Tyr Ala Pro Ser Gly Asn Phe Val Ala Cys Gly Gly Leu Asp Asn Ile 120

                Asn                                                     Ala                                                                     Val
Cys Ser Ile Tyr Ser Leu Lys Thr Arg Glu Gly Asn Val Arg Val Ser Arg Glu Leu Pro Gly His Thr Gly Tyr Leu Ser Cys Cys Arg Phe Leu Asp Asp Asn Gln Ile Ile Thr Ser 160

                                                                    Thr Thr     Thr         Thr                                         Asp Thr     Leu
Ser Gly Asp Thr Thr Cys Ala Leu Trp Asp Ile Glu Thr Gly Gln Gln Thr Val Gly Phe Ala Gly His Ser Gly Asp Val Met Ser Leu Ser Leu Ala Pro Asn Gly Arg Thr Phe Val 200
                                                                                                                                        Asp

                            Ala                         Glu Gly                     Thr                                 Ile Cys                 Asn
Ser Gly Ala Cys Asp Ala Ser Ile Lys Leu Trp Asp Val Arg Asp Ser Met Cys Arg Gln Thr Phe Ile Gly His Glu Ser Asp Ile Asn Ala Val Ala Phe Phe Pro Asn Gly Tyr Ala 240

    Ala                                                                                 Met Thr                                                     Ser         Lys
Phe Thr Thr Gly Ser Asp Asp Ala Thr Cys Arg Leu Phe Asp Leu Arg Ala Asp Gln Glu Leu Leu Met Tyr Ser His Asp Asn Ile Ile Cys Gly Ile Thr Ser Val Ala Phe Ser Arg 280

                                                            Val             Leu     Ala
Ser Gly Arg Leu Leu Leu Ala Gly Tyr Asp Asp Phe Asn Cys Asn Ile Trp Asp Ala Met Lys Gly Asp Arg Ala Gly Val Leu Ala Gly His Asp Asn Arg Val Ser Cys Leu Gly Val 320
                Val

Thr Asp Asp Gly Met Ala Val Ala Thr Gly Ser Trp Asp Ser Phe Leu Lys Ile Trp Asn                                                                                 340
```

FIG. 3. Amino acid sequence deduced from human β_2 cDNA [B. Gao, A. G. Gilman, and J. D. Robishaw, *Proc. Natl. Acad. Sci. U.S.A.* **84,** 6122 (1987)]. Amino acid differences deduced from bovine β_1 cDNA [H. K. W. Fong, J. B. Hurley, R. S. Hopkins, R. Miake-Lye, M. S. Johnson, R. F. Doolittle, and M. I. Simon, *Proc. Natl. Acad. Sci. U.S.A.* **83,** 2162 (1986)] are indicated above the complete β_2 sequence. The two differences in the bovine β_2 amino acid sequences are indicated below the complete human β_2 sequence.

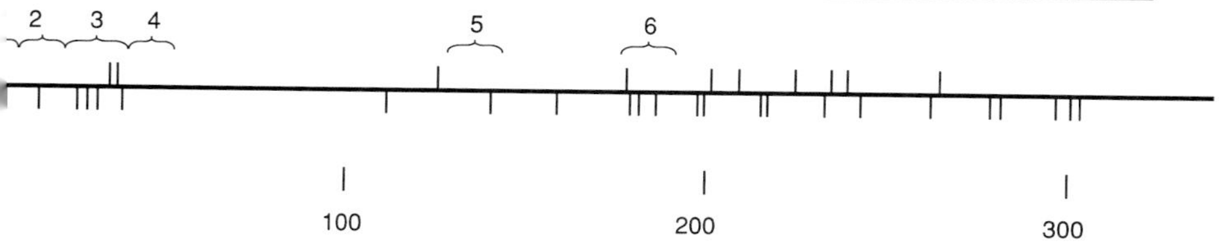

FIG. 4. Location of β-subunit sequences used for generation of antisera. β-subunit sequences deduced from the β_1 and β_2 cDNAs are compared schematically [B. Gao, S. M. Mumby, and A. G. Gilman, *J. Biol. Chem.* **262,** 17254 (1987)]. The horizontal line represents the primary sequence (left = amino terminus). Thirty-four residues differ between the two deduced sequences. Positions of the 10 nonconservative substitutions are indicated by the verticals above the line, while 24 conservative substitutions are indicated below the line. The numbered braces indicate regions utilized to synthesize peptides for generation of antisera (see Table II).

TABLE II
G-PROTEIN β-SUBUNIT PEPTIDE ANTISERA

Designation	Code number[a]	Region[b]	Amino acid number	Sequence (one-letter code)	Reactivity[c] Positive	Negative
β_{common}	MS/1[d]	1	1–10	MSELDQLRQE	$\beta_1(36)$, $\beta_2(35)$	
	βN1[e]	1–2	1–14	MSELDQLRQEAEQL	$\beta_1(36)$, $\beta_2(35)$	
	K-521[f]	2	8–23	RQEAEQLRNQIRDARK	$\beta_1(36)$, $\beta_2(35)$	
	8136[g]	4	38–52	DPVGRIQMRTRRTLR	$\beta_1(36)$, $\beta_2(35)$	
β_2	J-887[f]	3	25–38	CGDSTLTQITAGLD	$\beta_2(35)$	$\beta_1(36)$
	β2N[e]	3	25–39	CGDSTLTQITAGLDP	$\beta_2(35)$	$\beta_1(36)$
	BP[e]	6	177–191	TVGFAGHSGDVMSLS	$\beta_2(35)$	$\beta_1(36)$
	K-523[f]	5	130–145	EGNVRVSRELPGHTGY	$\beta_2(35)$	$\beta_1(36)$
	8129[g]	5	130–145	EGNVRVSRELPGHTGY	$\beta_2(35) > \beta_1(36)$	
β_1	U-49[f,h]	5	130–145	EGNVRVSRELAGHTGY	$\beta_1(36)$	$\beta_2(35)$
	8132[g]	5	130–145	EGNVRVSRELAGHTGY	$\beta_1(36) > \beta_2(35)$	

[a] Code number assigned to the antiserum by the investigator that produced the antiserum.
[b] Region numbers indicate location of the peptide sequence on the linear map of β subunit shown in Fig. 4.
[c] Reactivity of antisera with β subunits by Western immunoblotting. The number 35 or 36 in parentheses is the apparent molecular weight ($\times 10^{-3}$) of each form of β-subunit protein.
[d] P. Goldsmith, P. S. Backland, K. Rossiter, A. Carter, G. Milligan, C. G. Unson, and A. Spiegel, *Biochemistry* **27,** 7085 (1988).
[e] T. T. Amatruda, N. Gautum, H. K. W. Fong, J. K. Northup, and M. I. Simon, *J. Biol Chem.* **263,** 5008 (1988).
[f] B. Gao, S. M. Mumby, and A. G. Gilman, *J. Biol. Chem.* **262,** 17254 (1987).
[g] D. R. Manning, unpublished results (1988).
[h] S. M. Mumby, R. A. Kahn, D. R. Manning, and A. G. Gilman, *Proc. Natl. Acad. Sci. U.S.A.* **83,** 265 (1986).

peptide of α_{i2} reacts with α_o, α_{i1}, and α_{i2}; another antiserum (named anti-α_i) generated against the same peptide sequence reacts only with α_{i2} (Table I).

The length of peptides utilized successfully to produce G-protein-reactive antisera has ranged from 10 to 17 amino acid residues (Tables I and II). Of prime concern is the potential solubility of a peptide. The coupling of peptide to carrier protein (described below) requires solubility in aqueous buffer. Sequences rich in charged amino acid residues will obviously facilitate solubility and will often represent accessible (surface) regions of the corresponding protein. Inclusion of a phenylalanine, tryptophan, or tyrosine residue in the peptide sequence is useful for quantitating peptide in solution spectrophotometrically. Tyrosine residues are readily radiolabeled with ^{125}I. One may consider addition of such a residue (e.g., to the carboxy terminus of the peptide) if necessary. Coupling of peptide to carrier protein with the bifunctional cross-linking reagent *m*-maleimidobenzoyl-*N*-hydroxysuccinimide ester (MBS) requires the presence of a free sulfhydryl group, provided by a cysteine residue in the peptide. Cysteine is typically incorporated into the peptide as the last amino acid residue (to avoid synthetic problems), resulting in its placement at the amino terminus.

Coupling of Peptide to Carrier Protein

We have followed the protocol outlined by Green *et al.*[35] with success. Peptides are coupled to the carrier protein keyhole limpet hemocyanin (KLH) through an amino-terminal cysteine residue of the peptide with *m*-maleimidobenzoyl-*N*-hydroxysuccinimide ester as the coupling reagent (Pierce Chemical Co., Rockford, IL).[36] If one anticipates use of peptide antisera in the analysis of invertebrate systems, an alternative carrier protein (e.g., bovine serum albumin) may be preferable. Prior to coupling peptide to carrier protein, peptide solubility in aqueous buffer is determined and the presence of free sulfhydryl groups in the peptide is verified.

Peptide Solubility. Peptide solubility is tested in one of the following buffers, depending on the *pI* of the peptide: phosphate-buffered saline (pH 7.5), 0.1 *M* sodium borate buffer (pH 9.0), or 1.0 *M* sodium acetate buffer (pH 4.0). Solubility is determined at 5 mg/ml, since this is the concentration to be used in the coupling reaction. The efficiency of peptide dissolution

[35] N. Green, H. Alexander, A. Olson, S. Alexander, T. M. Shinnick, J. G. Sutcliffe, and R. A. Lerner, *Cell (Cambridge, Mass.)* **28,** 477 (1982).

[36] F. T. Liu, M. Zinnecker, T. Hamaska, and D. H. Katz, *Biochemistry* **18,** 690 (1979).

can be determined spectrophotometrically or with a fluorometric dye assay.[37]

Peptide Cysteine Content. The free sulfhydryl content of the peptide is determined to confirm the presence of a reactive cysteine residue. The protocol is derived from that developed by Ellman.[38] The following solutions are mixed gently to avoid undue aeration of the samples: 0.1 ml of 2-mercaptoethanol standard or peptide solution, 1 ml of deaerated 0.1 *M* sodium phosphate buffer (pH 8), and 0.1 ml of 4 mg/ml 5,5′-dithiobis(2-nitrobenzoic acid) in methanol. 2-Mercaptoethanol standards (25–400 μM) are diluted into deaerated phosphate-buffered saline from a 4 m*M* stock made fresh each day in water. Peptides are dissolved at 5 mg/ml immediately prior to assay. Dilutions of the peptide solution are made that should fit the 2-mercaptoethanol standard curve (100–500 μg/ml). The mixture is allowed to stand at room temperature for 15 min prior to determination of the absorbance at 420 nm. Typically, the content of free sulfhydryl groups calculated from the standard curve (absorbance versus 2-mercaptoethanol concentration) ranges between 0.3 and 1 mol per mole of peptide (assuming one cysteine residue).

Coupling Reaction. KLH is dissolved at 16 mg/ml in deaerated 10 m*M* sodium phosphate buffer (pH 7.2) (0.25 ml is required per peptide to be coupled). Warming to 37° may be required to effect dissolution. Undissolved KLH is removed by centrifugation. The KLH solution (0.25 ml per peptide) is gently mixed in a glass tube during the dropwise addition of 0.7 mg of MBS (28 μl of 25 mg/ml MBS in dimethylformamide) per peptide. High local concentrations of dimethylformamide can cause precipitation of KLH. Overzealous mixing results in undesirable aeration of the solution. The tube is purged with argon or nitrogen, capped, and incubated for 30 min at room temperature with intermittent manual shaking. Should precipitation occur, the precipitate is removed by centrifugation. The reaction product, KLH–MB, is separated from unreacted MBS by gel filtration. Sephadex G-25 (fine) is packed in a 1 × 30 cm column and equilibrated with deaerated 50 m*M* sodium phosphate buffer (pH 6) at 4°. Fractions (1–1.5 ml) are monitored by absorbance at 280 nm. The UV-absorbing fractions from the void volume are pooled, divided evenly by the number of peptides to be coupled, and deaerated in culture tubes inside a vacuum flask. Five milligrams of peptide, freshly dissolved in 1 ml of the chosen deaerated buffer, is added to an aliquot of the KLH–MB. The two solutions are mixed gently to avoid aeration. The pH of the mixture is adjusted to 7.0–7.5 by the addition of 1 *N* HCl or NaOH. The reaction

[37] J. R. Benson and P. E. Hare, *Proc. Natl. Acad. Sci. U.S.A.* **72,** 619 (1975).
[38] G. L. Ellman, *Arch. Biochem. Biophys.* **82,** 70 (1959).

tube is purged with argon or nitrogen, capped, and incubated at room temperature for 3 hr. The protein concentration of a portion of the reaction mixture is determined either during or following the incubation. Coupling efficiency can be monitored with radioactive peptide (^{125}I-labeled tyrosine) by gel filtration or dialysis of a sample of the conjugate against phosphate-buffered saline. The peptide-coupled KLH solution is stored frozen until use.

Injection and Bleeding of Rabbits

Initial Injection Schedule. A minimum of two rabbits (bled for preimmune serum) are injected with each peptide–KLH conjugate. Rabbits are immunized subcutaneously at four sites on the back on day 0 with 200 μg of peptide-coupled KLH, mixed homogeneously with Freund's complete adjuvant (volume ratio, 2 : 3). For ease of manipulation, the minimum volume of aqueous antigen should be 0.2 ml; dilution with sterile saline may be necessary. A booster injection of the antigen is administered subcutaneously with Freund's incomplete adjuvant on day 14. The final injection consists of 100 or 200 μg of peptide-coupled KLH adsorbed to alum, administered intraperitoneally at two sites in the lower abdomen on day 21.

Adsorption of Antigen to Nascent Alumina.[39] Adsorption occurs when alumina forms in the presence of protein. A trial floc is made as a guide to the amount of base that is required to produce the precipitate and to neutralize the resulting floc. Theoretically, 1 ml of 10% potassium alum requires 0.63 ml of 1.0 *N* NaOH. A mixture of 0.4 ml of 10% potassium alum ($AlKSO_4$), 0.4 ml of KLH at 2 mg/ml in phosphate-buffered saline and 0.2 ml of 1 *N* NaOH is a good starting point. Continue to add 1 *N* NaOH in 50-μl aliquots until precipitation occurs and the pH (determined with pH paper) is approximately 5. Reduce the volume of NaOH aliquot to 25 μl and continue additions until the pH is 6.8–7.0. The neutralized floc is allowed to stand at room temperature for 20 min, followed by separation of the supernatant and precipitated fractions in a clinical centrifuge. If KLH adsorption is efficient, the optical density of the supernatant at 280 nm should be less than 0.1.

When appropriate conditions for efficient adsorption of KLH are confirmed, the protocol is repeated using peptide-coupled KLH (1–2 mg/ml) and sterile solutions of $AlKSO_4$ and NaOH. The peptide-coupled KLH floc is centrifuged at low speed to avoid hard packing. The supernatant is removed, and its optical density (280 nm) is checked. The precipitate

[39] C. A. Williams and M. W. Chase, *Methods Immunol. Immunochem.* **1,** 201 (1967).

is washed with 10 ml of sterile phosphate-buffered saline, centrifuged, resuspended in 1 ml of the same buffer, and stored frozen until time of injection. The amount of peptide-coupled KLH adsorbed to the alumina precipitate is determined by the difference in the amount of starting material and the amount accounted for in the original supernatant fraction by UV absorption.

Bleeding and Serum Preparation. Rabbits are first bled 1 week after the intraperitoneal injection and then once every 2 weeks thereafter. Depending on the size of the rabbit, 15 to 30 ml of blood is taken by intravenous puncture of an ear. The blood is allowed to coagulate at room temperature for 1 or 2 hr and is then stored at 4° overnight to allow the clot to retract. The clot is removed and the tube is centrifuged to sediment blood cells. The straw-colored serum is stored at −20° or −70° in convenient aliquots to avoid repeated freezing and thawing. Serum is tested for peptide and KLH titer by ELISA and for G-protein reactivity (and possible reactivity with other proteins) by Western immunoblotting. Virtually all rabbits develop antibodies to the peptide and to KLH that are detectable in the first bleed. The frequency of reactivity of the peptide antibodies with G proteins has, in our experience, ranged from 1 of 4 to 8 of 8 rabbits immunized with a single peptide conjugate. In some cases, reactivity with G protein is not easily detected until the second and succeeding bleeds. The antibody titer and/or affinity vary significantly between rabbits.

Additional Injections. We have found that booster injections of rabbits beyond the initial schedule outlined above is useful only after the antibody titer of an individual rabbit has dropped. The chronology of fluctuation of antibody titer is highly variable among rabbits but is easily monitored by ELISA. Once the titer has dropped significantly, the rabbit may be boosted by subcutaneous injection with Freund's incomplete adjuvant or intraperitoneal injection of the antigen adsorbed to alum. The results of such booster injection have varied widely from no effect at all to an increase in titer equal to or sometimes greater than that obtained initially.

Enzyme-Linked Immunosorbent Assay

Reactivity of whole antisera and purified antibodies with peptide and KLH can be determined by ELISA.[40] Antibody is bound to antigen that has been immobilized in wells of a multiwell polystyrene plate, and the antigen–antibody complex is detected with an enzyme conjugated to protein A. The complex is measured by addition of a chromogenic substrate. Peptide or KLH (50 μl; 10 μg/ml in 100 mM $NaHCO_3$, pH 9.6) is adsorbed

[40] E. Engvall, this series. Vol. 70, p. 419.

to the polystyrene wells of a 96-well plate (Immulon 2 from Dynatec Alexandria, VA) by incubation overnight at 4°. Plates are rinsed 4 time with a solution of 0.05% Tween 20 (v/v) in 0.9% NaCl (w/v). All furth incubations and rinses include detergent in solution, which allows ant gen–antibody complexes to form and persist but inhibits further adsorptio of protein to the plastic. Antisera are diluted (generally at volume ratio of 1 : 30, 1 : 100, 1 : 300, and 1 : 1000) with a solution of 0.1% gelatin, 0.05 Tween 20, and 0.02% NaN_3 in phosphate-buffered saline. Serum dilutio (50 μl) are incubated in duplicate coated wells for 2 hr at room temperatur During the incubation, the plate is stored in a sealed plastic box with wet paper towel below the plate to provide humidity.

The plate is rinsed as before and is then incubated for 2 hr with μl of alkaline phosphatase conjugated to protein A (Zymed, South Sa Francisco, CA) or to goat anti-rabbit IgG (diluted 1 : 1000 from 1 mg/m stock) in the same buffer used for the primary antibody incubation. Th plates are rinsed again before addition of 50 μl of the substrate, *p*-nitr phenyl phosphate, at 1 mg/ml in 0.2 *M* 2-amino-2-methyl-1,3-propanedi and 1 m*M* $MgCl_2$, pH 10.3. The reaction is stopped by the addition of 12 μl of 1 *N* NaOH when the development of yellow color in the wells obvious (15–20 min). The absorbance at 405 nm is determined with a pla scanner. In general, the antibody titer of whole serum against KLH greater than that against the synthetic peptide hapten.

Western Immunoblotting

Owing to greater sensitivity, immunoblotting rather than ELISA generally utilized to test the reaction of antisera with purified G protein The method is also useful for detection of G-protein subunits in tissue cultured cell membrane preparations because the complex mixture components is first resolved in one dimension on an SDS–polyacrylamic gel.[41] In general, 100 ng of purified G protein subunit or 50 μg of membra protein is loaded per lane of the gel. The proteins are transferred nitrocellulose (BA85, 0.45 μm, Schleicher and Schull, Keene, NH) in Hoeffer Transphor or equivalent apparatus overnight at 25–30 V (200–2 mA) in a solution of 20% (v/v) methanol, 25 m*M* Tris base, and 0.19 glycine, pH 8.3. Efficiency of transfer can be monitored by staining th gel with Coomassie blue and staining a piece of the nitrocellulose wit amido black dye (0.25 g naphthol blue black, 45 ml methanol, 10 ml acet acid, and 45 ml water). The nitrocellulose is destained with a solution 50% (v/v) methanol and 10% (v/v) acetic acid. Some, but not all, antise

[41] H. Towbin, T. Staehelin, and J. Gordon, *Proc. Natl. Acad. Sci. U.S.A.* **76,** 4350 (197

are unaffected in their ability to detect the antigen on the nitrocellulose if it has been stained with amido black.

The nitrocellulose blot is trimmed to the minimal size, and each piece of nitrocellulose is placed in a plastic box or tray that most closely approximates its size to minimize the required volume of antibody solution. Nonspecific protein binding sites on the nitrocellulose are blocked by incubation of the blot for 1 hr on an orbital shaker with Blotto [50 m*M* Tris, pH 8; 2 m*M* $CaCl_2$; 80 m*M* NaCl; 5% nonfat dry milk; 0.2% Nonidet P-40 (NP-40); and 0.02% NaN_3]. The Blotto solution is filtered through Whatman 4 filter paper just prior to use. The solution can be kept at 4° for daily use or frozen at −20° for storage. The Blotto used to block the nitrocellulose is discarded and replaced by antiserum diluted in Blotto (1 : 200 or 1 : 500 for initial screening purposes). The primary antiserum is generally incubated with the blot for 1–2 hr at room temperature (with shaking) or, if more convenient, overnight at 4°. Overnight incubation can increase sensitivity and/or the background reactivity, depending on the nature of the protein mixture to be probed and the antiserum. Typically, diluted primary antibody in Blotto is saved at 4° following incubation with a blot, since the solution can be reused 3–6 times over a period of 1 month or more.

Following three 10–15 min washes of the nitrocellulose with Blotto, it is incubated with the secondary antibody, ^{125}I-labeled goat anti-rabbit IgG $F(ab')_2$ (New England Nuclear, Boston, MA) at 5×10^5 cpm/ml of high-detergent Blotto. High-detergent Blotto is prepared by diluting conventional Blotto 1 : 10 with buffer A (50 m*M* Tris, pH 8; 80 m*M* NaCl; 2 m*M* $CaCl_2$) and supplementing it with NP-40 and SDS to final concentrations of 2 and 0.2%, respectively. Following a 1-hr, room temperature incubation of the blot with shaking in secondary antibody solution, the blot is washed 3 times (10–15 min each) with Blotto, rinsed quickly 2 times with buffer A, and washed 2 times (5–10 min each) in buffer A. The blot is air-dried, taped to backing paper, covered with Glad plastic wrap, and exposed to Kodak XAR-5 film overnight at −70° with one intensifying screen. A higher resolution autoradiogram is produced by increasing the exposure time approximately 5-fold in the absence of the screen.

Antibody Purification

For some purposes, particularly immunocytochemistry and immunoprecipitation, antibody purification may be necessary. Antibodies are purified from whole antiserum by affinity chromatography on peptide covalently linked to Sepharose. The peptide used for immunization is coupled via primary amino groups to CNBr-activated Sepharose 6MB in accordance with the manufacturer's instructions, detailed below (Pharmacia LKB Biotechnology, Inc., Piscataway, NJ).

Coupling of Peptide to Sepharose. Approximately 0.35 g of CNBr-activated Sepharose soaked in 3–5 ml of 1 m*M* HCl for 15 min will yield 1 ml of swollen Sepharose. The swollen Sepharose is poured into a 15-ml funnel with a course fritted disk filter and washed with 75 ml of 1 m*M* HCl by use of a vacuum flask. The vacuum is broken before the gel becomes dry, such that the gel can be transferred to a centrifuge tube with a pipette. The filter is rinsed with 1 m*M* HCl (without vacuum) to facilitate transfer of any remaining gel. The gel is sedimented in a clinical centrifuge for 30 sec, and the supernatant fraction is discarded. Peptide is dissolved at 3 mg/ml in coupling buffer (0.1 *M* $NaHCO_3$, 0.5 *M* NaCl, pH 8.3). A volume of peptide solution equal to that of the swollen CNBr-activated Sepharose is mixed with the Sepharose by gentle trituration with a pipette. The pH of the mixture is estimated with pH paper and should be neutral or slightly basic. The reaction is allowed to proceed at room temperature with mixing (A magnetic stirrer should not be used.) The gel is sedimented as before but the supernatant fraction is saved to determine the efficiency of coupling (UV absorption or fluorometric assay); this value usually approximates 90%. Unreacted sites on the CNBr-activated Sepharose are blocked by incubation of the gel with 3 ml of 0.2 *M* glycine (pH 8) per milliliter of gel at room temperature. The gel is then transferred to the fritted disk funnel for alternate washing 5 times with 3 ml each of coupling buffer and wash buffer (0.1 *M* sodium acetate, 0.5 *M* NaCl, pH 4). The gel is finally washed once with 0.2 *M* glycine, pH 2.2 (the buffer to be used for antibody elution) and once more with coupling buffer. If the gel is not used immediately, it should be preserved by addition of sodium azide to a final concentration of 0.02% (w/v).

Affinity Chromatography. Antiserum (1–5 ml) is diluted 3-fold with Tris buffer (20 m*M* Tris-HCl, pH 7.5; 100 m*M* NaCl) and is passed twice over 1 ml of peptide-Sepharose in a 0.7 × 4 cm column. The solution that flows through the second pass over the column is saved to assay for antibodies by ELISA and can be used as a negative control in future experiments. The matrix is washed with 25 ml of Tris buffer, and 1.5-ml fractions are collected. The last fractions should be devoid of absorbance at 280 nm. The peptide antibodies are eluted with 10 ml of 0.2 *M* glycine, pH 2.2. Fractions (0.5 ml) are collected into tubes containing 0.1 ml of *M* K_2HPO_4 to neutralize the glycine. Neutralized fractions that contain protein, as determined by absorbance at 280 nm, are pooled and dialyzed against Tris buffer. The yield of antibody activity, as determined by ELISA, is generally between 30 and 80%. KLH-reactive antibodies should be found only in the flow-through solution from the Tris buffer wash the column. The purified antibodies can be preserved by the addition of sodium azide (0.02%, w/v) and/or by freezing in aliquots. Numerous cycles of freezing and thawing should be avoided.

Acknowledgments

Work from the authors' laboratory was supported by United States Public Health Service Grant GM34497, American Cancer Society Grant BC555I, and the Raymond and Ellen Willie Chair of Molecular Neuropharmacology. We also acknowledge support from the Perot Family Foundation and The Lucille P. Markey Charitable Trust. Linda Hannigan provided excellent technical assistance.

[20] Quantitation and Purification of ADP-Ribosylation Factor

By RICHARD A. KAHN

Introduction

The ADP-ribosylation factor (ARF) of adenylyl cyclase (EC 4.6.1.1, adenylate cyclase) was originally identified as a protein cofactor required for the efficient activation of the regulatory subunit, G_s of adenylyl cyclase by cholera toxin (for reviews, see Kahn *et al.*[1] and Kahn[2]). The covalent attachment of the ADP-ribose moiety from NAD to the α subunit of G_s results in the irreversible activation of the G_s protein and consequently of adenylyl cyclase (Fig. 1). The covalent nature of the toxin-catalyzed activation of G_s allowed the first demonstration of the subunit dissociation model[3] for the activation of G_s in membranes.[4] One reflection of this activated form of $G_{s\alpha}$ is that GTP becomes a very potent activator of adenylyl cyclase, whereas prior to intoxication it is extremely weak. Endogenous levels of GTP are sufficient to activate cellular adenylyl cyclase maximally after exposure of cells to cholera toxin.

Work on ARF began as a result of interest in studying this unique

[1] R. A. Kahn, T. Katada, G. M. Bokoch, J. K. Northup, and A. G. Gilman, *in* "Posttranslational Modification of Proteins" (B. C. Johnson, ed.), p. 373. Academic Press, New York, 1983.

[2] R. A. Kahn *in* "G Proteins" (R. Iyengar and L. Birnbaumer, eds.), p. 201 Academic Press, Orlando, Florida, 1990.

[3] J. K. Northup, M. D. Smigel, P. C. Sternweis, and A. G. Gilman, *J. Biol. Chem.* **258,** 11369 (1983).

[4] R. A. Kahn and A. G. Gilman, *J. Biol. Chem.* **259,** 6235 (1984).

$$NAD + G_s\alpha\beta\gamma \xrightarrow[\text{toxin}]{\text{cholera}} ADP\text{-}Ri\text{-}G_s\alpha + G\beta\gamma + \text{Nicotinamide} + H^+$$

$$\uparrow \oplus$$

$$ARF{*}GTP + GDP \rightleftharpoons ARF{*}GDP + GTP$$

FIG. 1. Reactions involving ADP-ribosylation factor.

mechanism of activation of adenylyl cyclase in molecular detail and in developing a simple, quantitative assay for G_s; by monitoring the incorporation of [^{32}P]ADP-ribose into $G_{s\alpha}$. This led to the purification and characterization of ARF[5] as well as to the description of the details of the role of ARF in the cholera toxin reaction and the consequences of ADP-ribosylation on the properties on G_s.[4] The use of cholera toxin labeling as a quantitative assay for G_s remains a risky venture due to a number of problems, including the lack of specificity of the toxin and the difficulty in getting stoichiometric labeling of toxin substrates, even under optimal conditions. The use of $G_{s\alpha}$-specific antibodies now allows a more reliable means of quantitating G_s from crude cell or tissue extracts.

Interest in ARF as a regulatory protein independent of cholera toxin or adenylyl cyclase is increasing due to a number of recent observations. The observation that ARF is itself a GTP-binding protein and is active as a cofactor in the cholera toxin reaction only when in the activated, GTP-liganded form[6] is the first instance of two regulatory GTP-binding proteins interacting directly. Cloning and sequencing of ARF have demonstrated that ARF is a highly conserved protein with structural characteristics intermediate between the trimeric G protein α subunit (which includes $G_{s\alpha}$) and *ras* p21 families.[7] ARF activity and/or immunoreactivity have been found in every eukaryotic tissue examined including man, mouse, yeast, slime mold, *Drosophila,* and the plant, *Arabidopsis*.[8] The level of ARF can be as high as approximately 1% of cell protein in some tissues, for example, brain. Disruption of the ARF genes in yeast is lethal,[9] but cells capable of making human ARF, carried on a plasmid, remain viable.[10] These results all point to a much more fundamental role for ARF in cellular physiology than was originally supposed from studies on adenylyl cyclase.

[5] R. A. Kahn and A. G. Gilman, *J. Biol. Chem.* **259,** 6228 (1984).
[6] R. A. Kahn and A. G. Gilman, *J. Biol. Chem.* **261,** 7906 (1986).
[7] J. L. Sewell and R. A. Kahn, *Proc. Natl. Acad. Sci. U.S.A.* **85,** 4620 (1988).
[8] R. A. Kahn, C. Goddard, and M. Newkirk, *J. Biol. Chem.* **263,** 8282 (1988).
[9] T. Stearns, A. Hoyt, D. Botstein, and R. A. Kahn, *Mol. Cell. Biol.,* in press.
[10] R. A. Kahn, F. G. Kern, J. Clark, E. P. Gelmann, and C. Rulka, submitted.

ARF has a number of attractive features for studies concerning the role of amino-terminal myristoylation on protein subcellular localization and function. As with some of the G-protein α subunits (G_o and G_i proteins) ARF is myristoylated at the amino-terminal glycine.[8] It is much more abundant than several of the other known substrates of myristoyltransferase (e.g., $p60^{src}$) and has a defined biochemical activity. Finally, the genetic analyses of yeast ARF deletions have defined a number of ARF-related phenotypes[9] which can also be examined in determining the effect of myristoylation on ARF function *in vivo*. Though the factor was originally purified from crude membrane sources, it is now clear that ARF is found both tightly associated with membranes as well as in the cytosol. It is not yet clear what role myristoylation has in the determination of the subcellular distribution of ARF.

ARF purified from either rabbit liver membranes or bovine brain membranes runs as a closely spaced doublet on sodium dodecyl sulfate (SDS)–polyacrylamide gels. The doublet can be partially resolved by either preparative SDS–polyacrylamide gel electrophoresis or molecular sieve chromatography. The resolved bands appear to have very similar specific activities in the ARF assay. Preliminary data have revealed that only one of these bands in the ARF doublet appears to be myristoylated.[10] We are currently testing the hypothesis that there are two distinct ARF genes expressed in mammals, only one of which is myristoylated. This hypothesis was supported by the recent report of a second ARF gene cloned from a bovine retinal cDNA library.[11] Although 96% identical to the previously published bovine ARF1 cDNA-derived sequence,[7] the bARF2 sequence lacks the consensus sequence for myristoylation,[12] present in bovine ARF1. The possible presence of two ARF proteins in mammals with distinctive modifications is of undetermined significance. However, we have noted that the budding yeast, *Saccharomyces cerevisiae*, expresses two distinct ARF genes, which are 97% identical, and each is myristoylated.

Quantitation of ADP-Ribosylation Factor

There are currently a number of assays for ARF protein and activity, but each suffers from limitations of either availability of reagents or ease of processing a large number of samples, or both. Probably the best and

[10] C. Goddard and R. A. Kahn, unpublished observation.

[11] S. R. Price, M. Nightingale, S.-C. Tsai, K. C. Williamson, R. Adamik, H.-C. Chen, J. Moss, and M. Vaughan, *Proc. Natl. Acad. Sci. U.S.A.* **85,** 5488 (1988).

[12] D. A. Towler, S. R. Eubanks, D. S. Towery, S. P. Adams, and L. Glaser, *J. Biol. Chem.* **262,** 1030 (1987).

certainly the most versatile assay is the original one described by Schleifer *et al.*[13] in which purified G_s is used as a substrate in the cholera toxin reaction. For studies of cholera toxin itself this is the preferred assay as the end point is adenylyl cyclase activity, the target of the toxin *in vivo* rather than any of several artificial substrates. The advantages of this assay include both sensitivity and specificity. As little as 1 ng of ARF can be detected in crude cell or tissue lysates. Although there have recently been over 20, structurally related, small (20–25 kDa) GTP-binding proteins identified, only ARF has activity in this assay. The assay is performed under conditions in which the rate of ADP-ribosylation of $G_{s\alpha}$ is a linear function of the amount of ARF added. The extent of ADP-ribosylation is then monitored by determination of the GTP-dependent adenylyl cyclase activity after reconstitution of the modified G_s into *cyc*$^-$ membranes. Limitations include the need for purified G_s and *cyc*$^-$ membranes, each of which are both expensive and relatively difficult to obtain. To ensure meaningful quantitative results it is necessary to control for the addition of certain reagents, primarily detergents and lipids, which often are included in the samples being assayed. Both the activity and stability of ARF have been shown to be altered in the presence of different detergents.[5] What is really being measured is the reconstituted adenylyl cyclase activity in membranes devoid of endogenous G_s. Thus, the activity will depend on the specific activities of the G_s and the adenylyl cyclase. The unit defined for ARF in this assay is picomoles cAMP produced per minute per microgram G_s. Nevertheless, this assay has given reliable information on the ARF content of a wide variety of crude tissue extracts and cell lysates.

G_s *Reconstitution Assay*

Stage I: Cholera Toxin Reaction. A final reaction volume of 36 μl is used, containing 100–300 ng purified G_s, 1–50 ng ARF, 250 m*M* potassium phosphate, pH 7.5, 10 m*M* thymidine, 100 μ*M* GTP, 1 m*M* magnesium chloride, 0.1 m*M* EDTA, 3 m*M* L-α-dimyristoylphosphatidylcholine (DMPC), 0.1% sodium cholate, 0.01% Lubrol PX, and 100 μ*M* NAD. The reaction is begun with the addition of 1.5 μg activated cholera toxin. [Cholera toxin is activated by incubating a 1 : 1 mixture of 1 mg/ml cholera toxin with 50 m*M* potassium phosphate, pH 7.5, 40 m*M* dithiothreitol (DTT) at 37° for at least 15 min.] After 20 min at 30° the reaction is terminated by diluting 20-fold into ice-cold 20 m*M* sodium HEPES, pH 8.0, 2 m*M* $MgCl_2$, 1 m*M* EDTA, 0.1% Lubrol PX, and 100 μ*M* GTP.

The thymidine is included to inhibit poly(ADP)-ribosyltransferases and may be omitted when purified preparations of ARF are assayed. GTP is

[13] L. S. Schleifer, R. A. Kahn, E. Hanski, J. K. Northup, P. C. Sternweis, and A. G. Gilman, *J. Biol. Chem.* **257,** 20 (1982).

required in Stage I both to activate the ARF and to stabilize the ADP-ribosylated $G_{s\alpha}$. DMPC is prepared in a more concentrated form (30 mM) by sonication in a bath sonicator in 20 mM sodium HEPES, pH 8.0, 2 mM $MgCl_2$, 1 mM EDTA. After the solution becomes clarified (usually 5–15 min) it is stored at 30° until used in the assay.

Stage II: G_s Reconstitution into cyc$^-$ Membranes. Ten microliters of the diluted (stopped) Stage I reaction is added to 30 μl of purified *cyc*$^-$ membranes (2 mg/ml protein; prepared as described by Ross *et al.*[14]). The GTP-dependent adenylyl cyclase activity is then determined as described by Sternweis and Gilman[15] with 100 μM GTP as agonist.

Notes and Calculations

1. For the most accurate initial rate studies it is necessary to preincubate the Stage I reagents (all except cholera toxin) for 5 min at 30° prior to addition of toxin. This will prevent the slight (0.5–5 min) lag before a constant rate of ADP-ribosylation is attained. This preincubation allows the ARF to bind GTP and, once activated, to bind $G_{s\alpha}$.
2. Each assay must include a zero ARF control as the toxin-dependent activation of G_s is not absolutely dependent on ARF under these conditions. This value is then subtracted from the experimental values to determine the ARF-dependent activation of G_s by cholera toxin.
3. For this assay to be quantitative it is necessary that both Stage I and Stage II reactions be linear with respect to time. As a consequence of ADP-ribosylation the α subunit of G_s is released from $G_{\beta\gamma}$.[4] ADP-ribosylated $G_{s\alpha}$ is less stable than basal, trimeric G_s, and prolonged incubations can result in a loss of activity as a result. The G_s is stabilized by inclusion of GTP.
4. It is critical to control for the amount of detergent and lipid in the assay. Purified G_s is stored in 0.1% Lubrol PX and is the source of the 0.01% Lubrol in the Stage I assay. Stopping the Stage I reaction by diluting into Lubrol allows a well-defined detergent environment for reconstitution into *cyc*$^-$ membranes. ARF is generally purified and stored in 1% sodium cholate and is usually the source of the 0.1% cholate in the Stage I assay.

ADP-Ribose Incorporation into $G_{s\alpha}$

The above assay can be simplified and made independent of adenylyl cyclase by monitoring the extent of ADP-ribosylation of G_s directly. If [α-^{32}P]NAD is included in the Stage I reaction, it is possible to quantitate ARF by determining the ARF-dependent covalent incorporation of

[14] E. M. Ross, M. E. Maguire, T. W. Sturgill, R. L. Biltonen, and A. G. Gilman, *J. Biol. Chem.* **252,** 5761 (1977).

[15] P. C. Sternweis and A. G. Gilman, *J. Biol. Chem.* **254,** 3333 (1979).

[^{32}P]ADP-ribose into $G_{s\alpha}$. If the G_s and ARF are pure or there is no reason to question the specificity of the acceptor as $G_{s\alpha}$, then the incorporation of radioactivity into $G_{s\alpha}$ can be determined by the filter trapping method, as described by Bokoch *et al.*[16] When assaying crude samples it is advisable to demonstrate specificity (e.g., by autoradiography of dried SDS–polyacrylamide gels) to identify the specific acceptor as $G_{s\alpha}$. (Note: The cholera toxin A1 subunit can auto-ADP-ribosylate but in general does not contribute significantly to total acid-precipitable counts under these conditions.)

The sensitivity of this assay is set by the specific activity of NAD. We routinely use 10 μM NAD at a specific activity of 10,000 counts per minute (cpm)/pmol. This concentration of NAD is approximately 10-fold below the apparent K_m of the toxin; thus, if the goal is quantitative incorporation of ADP-ribose into $G_{s\alpha}$, these conditions may require prolonged incubations (several hours). It is important to keep in mind that Michaelis–Menten kinetics cannot be used to analyze this reaction as the enzyme is often the most abundant protein in the tube and one is forced to work at concentrations that are well below the apparent K_m values for each reactant.

The cholera toxin reaction is terminated with the addition of 0.5 ml of 1% SDS with 50 μg/ml bovine serum albumin, which acts as carrier. Proteins are precipitated by the addition of an equal volume of 30% trichloroacetic acid. After 30 min at 4° the samples are collected by filtration on 25-mm nitrocellulose filters under vacuum with 7 2-ml washes with ice-cold 10% tricholoracetic acid. Radioactivity incorporated into $G_{s\alpha}$ is then determined in a liquid scintillation counter. As the same reaction is being monitored, all of the above notes and cautions also apply in using this assay.

It has recently been possible to use bacterially expressed $G_{s\alpha}$[17] as the substrate in this assay. The expressed $G_{s\alpha}$ is an excellent substrate for cholera toxin, and thus this [^{32}P]NAD-labeling assay works quite well provided purified $G_{\beta\gamma}$ are included in the assay. Further, recombinant ARF proteins, purified from bacteria and thus lacking amino-terminal myristate, have recently been found to have specific activities identical to that of purified bovine brain ARF.[18]

Guanine Nucleotide Binding Assay

The simplest and most rapid assay for purified ARF is a radioligand binding assay, by the use of a modification of the rapid filtration technique

[16] G. M. Bokoch, T. Katada, J. K. Northup, M. Ui, and A. G. Gilman, *J. Biol. Chem.* **259,** 3560 (1984).

[17] M. Graziano, P. J. Casey, and A. G. Gilman, *J. Biol. Chem.* **262,** 11375 (1987).

[18] O. Weiss, J. Holden, C. Rulka, and R. A. Kahn, *J. Biol. Chem.* **264,** 21066 (1989).

described by Northup *et al.*[19] There is a strict correlation between the binding of guanine nucleotides and specific activity of ARF in the reconstitution assay, described above. Thus, binding data can be used to give an accurate estimate of the percentage of active protein in a preparation of ARF.

The conditions in which ARF binds stoichiometric amounts of guanine nucleotides are unique and include an absolute requirement for magnesium, phospholipid, and sodium cholate (ARF conditions).[6] Other GTP-binding proteins (e.g., G-protein α subunits, *ras* p21) also bind guanine nucleotides under these conditions. However, these other proteins will also bind GTP in the absence of lipid and cholate (non-ARF conditions). Thus, an estimate of the amount of ARF in a sample can be obtained by subtracting the amount of guanine nucleotide binding assayed under non-ARF conditions from that observed with ARF conditions. This method should only be used as an estimate as many of the other GTP-binding proteins show slightly elevated binding under ARF conditions than with non-ARF conditions, and thus using the difference between the two assays will overestimate the amount of ARF in the sample. Nevertheless, for estimates of ARF content and locating peaks of ARF on columns during purification, this assay has proved to be a useful way to speed the purification and save valuable reagents (G_s and *cyc*$^-$ membranes).

Nucleotide Binding Assay

The nucleotide binding assay for ARF has been described previously.[6] The nonhydrolyzable analog [^{35}S]GTPγS is routinely used as the radioligand, but GDP or GTP may be substituted. Binding is determined in a final volume of 50 μl containing 20 m*M* sodium HEPES, pH 8.0, 800 m*M* NaCl, 2 m*M* $MgCl_2$, 1 m*M* DTT, 3 m*M* DMPC, 0.1% sodium cholate, 1 μM [^{35}S]GTPγS (10,000 cpm/pmol). Equilibrium is usually reached within 60–90 min at 30°. The reaction is terminated by dilution with 2 ml ice-cold 20 m*M* Tris-Cl, pH 7.5, 100 m*M* NaCl, 5 m*M* $MgCl_2$, 1 m*M* DTT. The amount of bound nucleotide is then determined by filtration of samples through 25-mm BA85 nitrocellulose filters (Schleicher and Schuell, Keene, NH) with 7 2-ml washes with ice-cold stop buffer. Stoichiometries of greater than 0.7 mol nucleotide bound per mole ARF protein can be obtained with freshly prepared ARF. Nonspecific binding is 0.05% or less of the filtered radioactivity under these conditions.

Quantitative Immunoblot

A number of specific antisera directed against synthetic peptide sequences derived from bovine and yeast ARF proteins have proved useful

[19] J. K. Northup, M. D. Smigel, and A. G. Gilman, *J. Biol. Chem.* **257,** 11416 (1982).

TABLE I
PURIFICATION OF ARF FROM BOVINE BRAIN MEMBRANES[a]

Step	Volume (ml)	Activity (units/ml)	Protein (mg/ml)	Specific activity (units/mg)	Recovery (%)
Extract	1750	1360	3.65	372	100
DEAE-Sephacel	820	2170	1.93	1120	74
Ultrogel AcA 44	154	5240	0.93	5630	34
Heptylamine-Sepharose	153	2900	0.11	26,400	19
Preparative HPLC	1.8	200,000	1.17	171,000	15

[a] Extract was prepared from a crude membrane preparation obtained from two bovine brains. Recoveries shown are cumulative through the purification.

in immunodetection of ARF from a wide variety of tissues and species.[8] Most useful in this capacity is serum R-5, directed against amino acid residues 23–36 of the bovine protein. This antibody has been used to detect ARF in a wide variety of eukaryotic species; including plants, fungi, and man. Minor differences in primary sequence may dramatically affect the antibody–antigen recognition. To obtain quantitative information, a standard of purified ARF is also required. Thus, quantitative immunoblotting requires some knowledge of the protein sequence from the species being analyzed. In the vast majority of cases it is suggested that these antibodies be used in a qualitative way and confirmed or put into more quantitative terms with one of the above two assays.

Purification of ADP-Ribosylation Factor

The abundance and stability of ARF in bovine brain make this the tissue of choice for purification of large amounts of protein. The major drawback to this as a source of ARF protein is the abundance in crude brain membrane preparations of NAD^+ glycohydrolase (NADase) activity which degrades NAD and interferes with the ARF assay. This activity is largely resolved after the first step in purification and does not present a problem in subsequent steps. This method yields approximately 1–3 mg purified ARF from 10–15 g crude bovine brain membranes[5] (see Table I).

Preparation of Crude Bovine Brain Membranes. Crude brain membranes are prepared as described by Sternweis and Robishaw.[20] Bovine brains are obtained from freshly slaughtered animals and stored in ice-cold 10 m*M* Tris-Cl, pH 7.5, until used. Gray matter is crudely dissected

[20] P. C. Sternweis and J. D. Robishaw, *J. Biol. Chem.* **259,** 13806 (1984).

to remove brain stem and the larger portions of white matter. Tissue (150–200 g/brain) is homogenized at medium speed in 4 volumes 10 m*M* Tris-Cl, pH 7.5, 10% sucrose, 0.5 m*M* phenylmethylsulfonyl fluoride (PMSF) (buffer A). After filtration through cheesecloth, the membranes are collected by centrifugation at 20,000 *g* for 30 min. The membrane pellet is washed twice by resuspending in 5 volumes buffer A with a Potter–Elvehjem homogenizer and collecting by centrifugation at 20,000 *g* for 60 min. Finally, the membranes are resuspended in buffer A at a concentration of 10–20 mg protein/ml. Membranes are either extracted immediately or stored at −80°.

Detergent Extraction of Membranes. Extract (2 liters) is prepared from 12–15 g membrane protein by stirring for 1 hr at 4° in 20 m*M* Tris-Cl, pH 8.0, 1 m*M* EDTA, 1 m*M* DTT, 0.1 m*M* PMSF, 1% sodium cholate (TED/1% cholate). Extracted membranes are removed by centrifugation at 95,000 *g* for 1 hr. Approximately 40% of membrane protein is generally extracted in this procedure. Supernatants are pooled and activators[21] are added.

DEAE-Sephacel Column. The extract is applied to a 5 × 60 cm column of DEAE-Sephacel, which had been equilibrated with 3.4 liters of TED/0.9% cholate/5 m*M* $MgCl_2$/25 m*M* NaCl. Proteins are eluted with a linear gradient (2 liters) of NaCl (0–250 m*M*) at a flow rate of 200 ml/hr. Fractions of about 22 ml are collected and assayed for ARF activity. The peak of activity elutes at a salt concentration of around 100 m*M*. In the absence of magnesium, activity often appears in two peaks and recoveries are reduced.

Gel Filtration Column. The pooled DEAE peak (300–800 ml) is concentrated by ultrafiltration to 30 ml with an Amicon (Danvers, MA) YM30 membrane and applied to a 5 × 60 cm column of Ultrogel AcA 44, previously equilibrated in 1 bed volume TED/0.9% cholate/5 m*M* $MgCl_2$/100 m*M* NaCl. The column is developed at a flow rate of 70–90 ml/hr with the same buffer, and fractions of about 14 ml are collected. ARF activity elutes in a single peak with a K_d of 0.5.

Heptylamine-Sepharose Column. The AcA 44 peak is pooled and diluted with TED/100 m*M* NaCl/5 m*M* $MgCl_2$ to yield a sodium cholate concentration of 0.6%. The diluted ARF is applied to a 50 ml (1.5 × 27 cm) column of heptylamine-Sepharose (prepared as described by S.

[21] This purification scheme was designed to allow purification of several G proteins (G_s, $G_{i\alpha}$, $G_{o\alpha}$, and $G_{\beta\gamma}$) as well as ARF from the same starting material. Addition of 5 m*M* $MgCl_2$ to the membrane extract enhances the stability of ARF, presumably by maintaining the ARF in the GDP-liganded state. Inclusion of 20 μM $AlCl_3$ and 10 m*M* sodium fluoride with the metal will stabilize and enhance recoveries of the other proteins described.

Shaltiel[22]), that had been equilibrated with 250 ml TED/100 m*M* NaCl/5 m*M* $MgCl_2$/0.6% sodium cholate. The column is washed with 50 ml of equilibration buffer followed by 50 ml of TED/100 m*M* NaCl/5 m*M* $MgCl_2$/0.8% sodium cholate. ARF is eluted with 100 ml of TED/100 m*M* NaCl/5 m*M* $MgCl_2$/1.8% sodium cholate. The flow rate is about 50 ml/hr at each step. A 3- to 5-fold increase in specific activity can usually be achieved in this step, due primarily to the 70–80% of total proteins that do not adsorb to the column under these conditions.

Bio-Sil TSK Column. The pooled ARF from the heptylamine-Sepharose column is concentrated to 1–2 ml by ultrafiltration with an Amicon YM30 membrane. Concentrated ARF is then injected onto a Bio-Sil TSK (Bio-Rad, Richmond, CA) 250 column (2.15 × 60 cm) previously equilibrated with 50 m*M* sodium HEPES, pH 8.0, 100 m*M* NaCl, 5 m*M* $MgCl_2$, 1 m*M* DTT. The column is developed with the same buffer at a flow rate of 4 ml/min. ARF elutes at the position predicted for a 21-kDa protein. Although the ARF remains soluble and monodisperse throughout the molecular sieve chromatography, it is advisable to add sodium cholate (0.6–1.0%) to the purified ARF. The detergent appears to preserve activity and prevent aggregation that has been observed on storage of the purified protein.

Conclusions

The above purification scheme yields 1–3 mg of purified bovine brain ARF. A doublet is observed in SDS–polyacrylamide gels, but both bands have ARF activity. A nearly identical scheme has been employed to purify ARF from rabbit liver membranes and turkey erythrocytes, with success similar to that reported above. The purified bovine brain ARF appears to be quite stable ($t_{1/2}$ 30 days) at 4° when the protein concentration is above 0.5 mg/ml.

[22] S. Shaltiel, this series, Vol. 34, p. 126.

[21] Soluble Guanine Nucleotide-Dependent ADP-Ribosylation Factors in Activation of Adenylyl Cyclase by Cholera Toxin

By JOEL MOSS, SU-CHEN TSAI, S. RUSS PRICE, DAVID A. BOBAK, and MARTHA VAUGHAN

Introduction

The devastating diarrheal disease characteristic of cholera results in large part from the production by *Vibrio cholerae* of a protein toxin, known as choleragen or cholera toxin.[1-3] This enterotoxin exerts its effects on cells through the activation of the hormone-sensitive adenylyl cyclase (EC 4.6.1.1, adenylate cyclase), thereby increasing intracellular cAMP which causes the characteristic abnormalities in fluid and electrolyte fluxes. Activation of adenylyl cyclase results from the toxin-catalyzed ADP-ribosylation of the stimulatory guanine nucleotide-binding protein of the cyclase system, known as G_s. The α subunit of G_s, which is responsible for guanine nucleotide binding, is the site of modification (for review, see Refs. 1 and 2).

Under physiological conditions, G_s is regulated by hormone and neurotransmitter receptors, which, when activated by agonists, promote the release of bound GDP and binding of GTP. G_s with bound GTP dissociates to yield $G_{s\alpha} \cdot GTP$ and $G_{\beta\gamma}$. $G_{s\alpha} \cdot GTP$ is the species that is directly responsible for activation of the cyclase catalytic unit. Hydrolysis of bound GTP to GDP by a GTPase activity intrinsic to $G_{s\alpha}$ terminates the signal. Activation of G_s thus involves binding of GTP and dissociation of the α from $\beta\gamma$ subunits; inactivation is associated with the hydrolysis of GTP to GDP and formation of the $\alpha\beta\gamma$ heterotrimer (for review, see Refs. 1 and 2).

ADP-ribosylation of $G_{s\alpha}$ by cholera toxin promotes activation by a mechanism(s) different from those used by hormones and neurotransmitters. Cholera toxin-catalyzed ADP-ribosylation of $G_{s\alpha}$ alters the effects of guanine nucleotide on $G_{s\alpha}$ function. First, ADP-ribosylation increases sensitivity to the stimulatory effects of GTP,[4] presumably by inhibiting

[1] A. G. Gilman, *Annu. Rev. Biochem.* **56,** 615 (1987).

[2] J. Moss and M. Vaughan, "Advances in Enzymology," Vol. 61, p. 303. Wiley, New York, 1988.

[3] M. T. Kelly, *Pediatr. Infect. Dis.* **5,** 5101 (1986).

[4] S. Nakaya, J. Moss, and M. Vaughan, *Biochemistry* **19,** 4871 (1980).

$G_{s\alpha}$ GTPase activity and thus prolonging the half-life of the active species[5] second, ADP-ribosylation promotes release of bound GDP, thus facilitat ing binding of GTP[6]; and, third, ADP-ribosyl-G_s more readily dissociate to yield $G_{\beta\gamma}$ and the active ADP-ribosyl-$G_{s\alpha} \cdot$ GTP species.[7] Some of thes effects have been observed in membrane preparations, others with purifie $G_{s\alpha}$. It is unclear to what extent each contributes in intact cells to th activation of adenylyl cyclase by ADP-ribosyl-$G_{s\alpha}$.

ADP-ribosylation of G_s by cholera toxin is enhanced by membrane an soluble factors.[8–15] Kahn and Gilman purified 21-kDa proteins from live and brain membranes that were required for the ADP-ribosylation o purified G_s by cholera toxin[12,13]; the brain proteins were shown to bin guanine nucleotides.[13] In view of their ability to enhance toxin-catalyze ADP-ribosylation, these proteins were termed ADP-ribosylation factor (ARF). Three 19-kDa ARF-like proteins were purified to homogeneit from the membrane and soluble fractions of bovine brain.[14,15] Partial amin acid sequences and antibodies were obtained, and a putative cDNA clon was isolated from a retinal cDNA library.[16,17]

Purification of Soluble ADP-Ribosylation Factor from Bovine Brain

Based on mRNA hybridization and immunoblot analyses, it appear that neural tissues contain high levels of ARF-like proteins. In bovin brain, a significant proportion of the activity is found in the soluble fractio (sARF). To isolate the soluble ARF, fresh bovine cerebral cortex (400 g is minced and homogenized (Polytron, 30 sec) in 1.6 liters of buffer A [2 m*M* Tris-Cl, pH 8.0/1 m*M* EDTA/1 m*M* dithiothreitol/1 m*M* NaN_3/0. m*M* phenylmethylsulfonyl fluoride (PMSF)/10% sucrose]. This and a

[5] D. Cassel and Z. Selinger, *Proc. Natl. Acad. Sci. U.S.A.* **74,** 3307 (1977).
[6] D. L. Burns, J. Moss, and M. Vaughan, *J. Biol. Chem.* **257,** 32 (1982).
[7] R. A. Kahn and A. G. Gilman, *J. Biol. Chem.* **259,** 6235 (1984).
[8] K. Enomoto and D. M. Gill, *J. Biol. Chem.* **255,** 1252 (1980).
[9] H. Levine III and P. Cuatrecasas, *Biochim. Biophys. Acta* **672,** 248 (1981).
[10] M. O. Pinkett and W. B. Anderson, *Biochim. Biophys. Acta* **714,** 337 (1982).
[11] L. S. Schleifer, R. A. Kahn, E. Hanski, J. K. Northup, P. C. Sternweis, and A. G. Gilma *J. Biol. Chem.* **257,** 20 (1982).
[12] R. A. Kahn and A. G. Gilman, *J. Biol. Chem.* **259,** 6228 (1984).
[13] R. A. Kahn and A. G. Gilman, *J. Biol. Chem.* **261,** 7906 (1986).
[14] S.-C. Tsai, M. Noda, R. Adamik, J. Moss, and M. Vaughan, *Proc. Natl. Acad. Sci. U.S.A* **84,** 5139 (1987).
[15] S.-C. Tsai, M. Noda, R. Adamik, P. P. Chang, H.-C. Chen, J. Moss, and M. Vaughan, *Biol. Chem.* **263,** 1768 (1988).
[16] S. R. Price, M. Nightingale, S.-C. Tsai, K. C. Williamson, R. Adamik, H.-C. Chen, Moss, and M. Vaughan, *Proc. Natl. Acad. Sci. U.S.A.* **85,** 5488 (1988).
[17] S.-C. Tsai, R. Adamik, K. Williamson, J. Moss, and M. Vaughan, *J. Cell Biol.* **107,** 40 (1988).

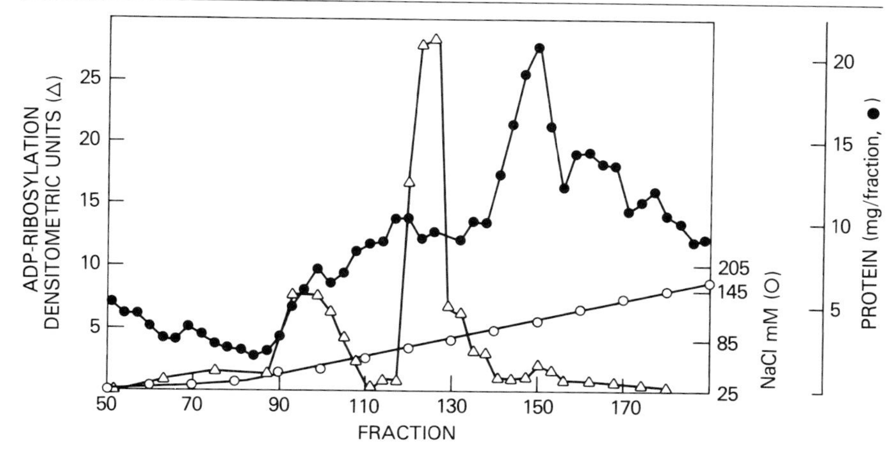

FIG. 1. Separation of sARF-I and sARF-II by chromatography on CM-Sepharose. Proteins that did not bind to CM-Sepharose at pH 7.0 were adjusted to pH 5.35 and subjected to chromatography on CM-Sepharose at pH 5.35 as described in the text. Fractions (10 ml) were adjusted to pH 7.0 with 1 *M* K_3PO_4. Samples (20 μl) of the indicated fractions were incubated with choleragen (25 μg), 500 μM ATP, 250 μM GTP, and 10 μM [^{32}P]NAD (2–3 μCi) for assay of CTA1 auto-ADP-ribosylation as described in the text.

subsequent steps in the purification are carried out and run at 4°. After centrifugation (15,000 *g*, 1 hr) of the homogenate, the supernatant (7.5 g of protein) is brought to 25% saturation by addition of solid $(NH_4)_2SO_4$, titrated to pH 7.5 with cold 0.75 *M* NH_4OH, and kept at 4° for 2 hr before centrifugation (15,000 *g*, 1 hr). The supernatant is adjusted to 70% saturation by further addition of solid $(NH_4)_2SO_4$ and, after 1 hr at 4°, is centrifuged (15,000 *g*, 1 hr). The precipitate is dissolved in 300 ml of buffer B (20 m*M* potassium phosphate, pH 7.0/1 m*M* EDTA/1 m*M* dithiothreitol/1 m*M* NaN_3/1 m*M* benzamidine) containing 0.25 *M* sucrose and dialyzed overnight against 8 liters of buffer B containing 0.5 *M* sucrose before centrifugation (105,000 *g*, 2 hr).

The precipitate is discarded, and the supernatant is applied to a column (5 × 31 cm, 608 ml) of CM-Sepharose Fast Flow (Pharmacia LKB Biotechnology, Piscataway, NJ) equilibrated in and eluted with buffer B containing 0.25 *M* sucrose. The first 220 ml of effluent is discarded; the next 540 ml (3.3 g of protein) is collected, adjusted to pH 5.35 with cold 1 *M* acetic acid, and centrifuged (15,000 *g*, 1 hr). The supernatant (525 ml, 2.8 g of protein) is applied to a column (4 × 33 cm, 414 ml) of CM-Sepharose Fast Flow equilibrated with buffer C (20 m*M* potassium phosphate, pH 5.35/1 m*M* EDTA/1 m*M* dithiothreitol/1 m*M* NaN_3/1 m*M* benzamidine/0.25 *M* sucrose) (Fig. 1). The column is washed with 500 ml of buffer C containing 25 m*M* NaCl and then eluted with a linear gradient of 25 to 200 m*M* NaCl

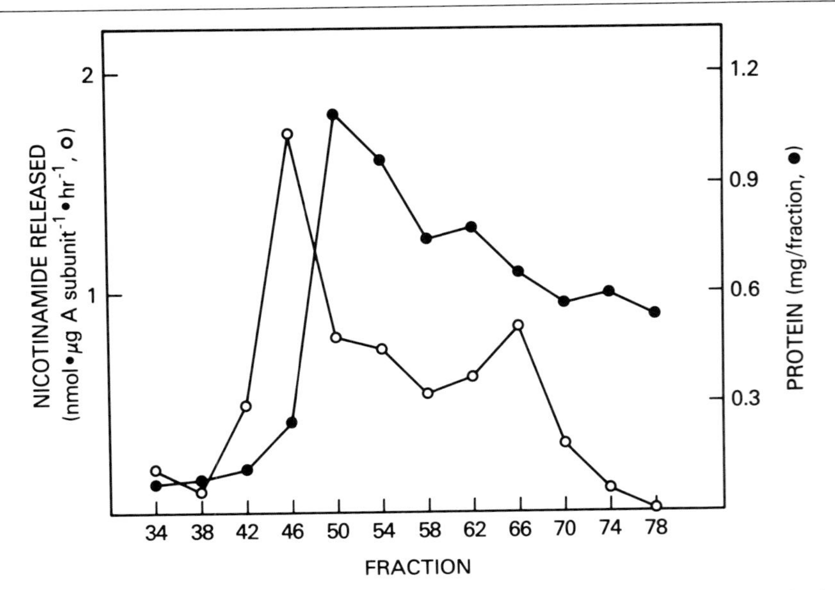

FIG. 2. Chromatography of sARF-II on hydroxylapatite. Fractions from the CM-Sepharose (pH 5.35) column that contained sARF-II were pooled, desalted on a column of Sephadex G-25, and applied to a column of hydroxylapatite, which was eluted as described in the text. Samples (40 μl) of the indicated fractions (3 ml) were assayed for NAD : agmatine ADP-ribosyltransferase activity and protein.

in buffer C (1.8 × 1.8 liters). Fractions are immediately adjusted to pH 7.0 with 1 *M* K_3PO_4 and assayed for sARF activity. Fractions corresponding to peaks of activity termed sARF-I and sARF-II are pooled separately. sARF-I (39 mg of protein) is eluted with approximately 33 m*M* NaCl and sARF-II (23 mg of protein) with approximately 77 m*M* NaCl (Fig. 1).

sARF-I and sARF-II are desalted on Sephadex G-25 (Pharmacia LKB Biotechnology) equilibrated and eluted with buffer D (20 m*M* Tris-Cl, pH 8.0/1 m*M* EDTA/1 m*M* dithiothreitol/1 m*M* NaN_3/0.25 *M* sucrose) with 1 m*M* benzamidine. Active fractions are pooled, concentrated to 60 ml with a YM10 membrane (Amicon, Danvers, MA), and sARF-I and sARF-II are separately purified on hydroxylapatite (Fig. 2) and Ultrogel AcA 54 (IBF, Savage, MD). The YM10 concentrate is applied to a column (20 ml) of hydroxylapatite (Bio-Rad, Richmond, CA) equilibrated with buffer D containing 5 m*M* $MgCl_2$ (included to stabilize the protein). After washing with 32 ml of the same buffer, elution is carried out with a linear gradient of potassium phosphate, pH 8.0 (0 to 50 m*M*), in buffer D (75 × 75 ml). Both sARF-I and sARF-II activity emerge at 20–30 m*M* potassium

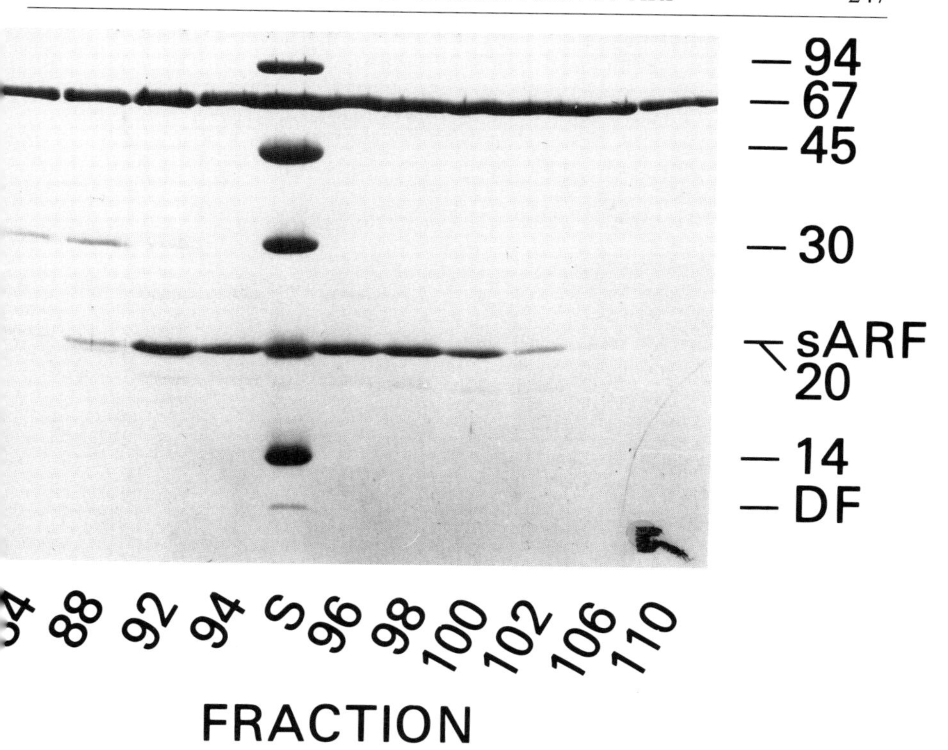

FIG. 3. Chromatography of sARF-II on Ultrogel AcA 54. Chromatography was carried out as described in the text. To samples (100 μl of fractions 84–92 and 106–110, 50 μl of fractions 94–104) were added 2.5 μg of bovine serum albumin and 2 ml of 7.5% trichloroacetic acid. Precipitated proteins were separated by denaturing electrophoresis in a 16% polyacrylamide minigel, which was stained with Coomassie blue. Standard proteins: phosphorylase *b*, 94K; bovine serum albumin, 67K, ovalbumin, 45K; carbonate dehydratase, 30K; soybean trypsin inhibitor, 20K; α-lactalbumin, 14K.

phosphate. Active fractions containing sARF-I (7.1 mg protein) or sARF-II (8.8 mg protein) are pooled, concentrated to 1.5 ml (YM10 membrane), and applied to a column (1.2 × 119 cm) of Ultrogel AcA 54 equilibrated and eluted with buffer D containing 0.1 *M* NaCl and 5 m*M* $MgCl_2$ (Fig. 3). Active fractions (1 ml) containing sARF-I (K_{av} 0.6) or sARF-II (K_{av} 0.56) are pooled and stored at $-20°$ in 0.5-ml portions to minimize repetitive freezing and thawing, which accelerate inactivation of the proteins. The results of a purification summarized in Table I are representative of many

TABLE I
PURIFICATION OF sARF-I AND sARF-II[a]

Fraction	Volume (ml)	Protein (mg)
Supernatant	1500	7500
CM-Sepharose, pH 7.0	540	3300
Supernatant, pH 5.35	525	2800
sARF-I		
CM-Sepharose, pH 5.35	416	39
Hydroxylapatite	28	7.1
Ultrogel AcA 54	11	1.6
sARF-II		
CM-Sepharose, pH 5.35	304	23
Hydroxylapatite	28	8.8
Ultrogel AcA 54	13	2.4

[a] Purification procedures are described in detail in the text. Data are from Ref. 15.

purifications performed independently by several members of our laboratory.

Immunological Characterization of ADP-Ribosylation Factor

To determine whether ARF-like proteins are immunologically conserved in different tissues and organ systems as well as across species, polyclonal antibodies are prepared against bovine brain sARF-II.[17] On immunoblots, the antibodies detect two major forms of ARF in the soluble fractions of bovine brain and retina with mobilities consistent with those of 20-kDa (sARF-II) and 19.5-kDa (sARF-I) proteins (Fig. 4). Two species similar in apparent size to sARF-I and sARF-II are present in the membrane fraction. Immunoreactivity is observed in all other bovine tissues; the predominant form behaves like a protein of 19.5 kDa. The rabbit antibodies against bovine ARF also react with ARF-like proteins in rat frog, and chicken tissues; in all, the molecular weights are quite similar and the proteins are most abundant in the brain soluble fraction. Two bands are observed in brain, whereas single bands, for the most part, are present in spleen and liver; immunoreactivity is very low in heart. The presence of ARF-like proteins in these tissues is supported by assays of ARF activity, specifically, the ability of a tissue extract to stimulate the auto-ADP-ribosylation of cholera toxin A1 (CTA1). Activity, in general parallels immunoreactivity, consistent with the hypothesis that the immunoreactive proteins are ARFs. Specific epitopes appear to be conserved across tissues and species lines.

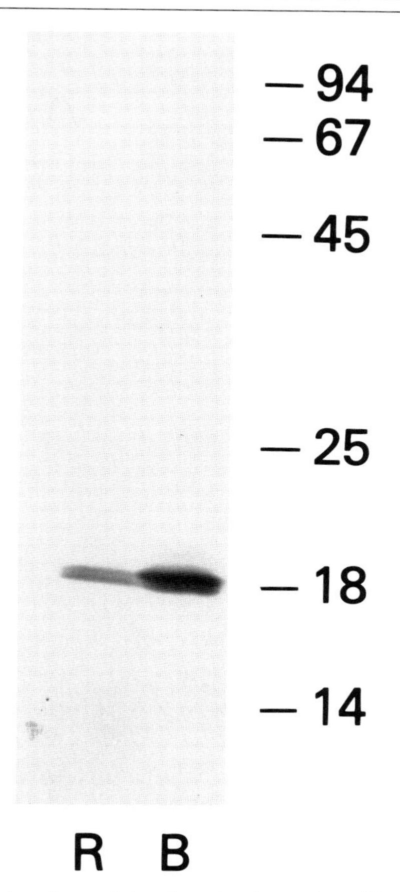

FIG. 4. Reaction of cytosolic proteins from bovine retina (R) and brain (B) with anti-sARF-II polyclonal antibodies. Cytosolic proteins (100 μg) from bovine retina and brain were separated by denaturing electrophoresis in a 16% polyacrylamide gel and transferred to nitrocellulose using a transblot apparatus as noted in H. Towbin, T. Staehelin, and J. Gordon, *Proc. Natl. Acad. Sci. U.S.A.* **76,** 4350 (1979). The blot was incubated for 16 hr at 23° with the IgG fraction from serum of a rabbit immunized with sARF-II from bovine brain (2 μg/ml). After washing, the blot was incubated with horseradish peroxidase-conjugated goat anti-rabbit IgG for 2 hr and developed with 4-chloro-1-naphthol and hydrogen peroxide.

Mechanism of Action of ADP-Ribosylation Factor on Cholera Toxin

To determine the mechanism of action of ARF, advantage was taken of the fact that cholera toxin, in addition to catalyzing the ADP-ribosylation of $G_{s\alpha}$,[18] also ADP-ribosylates (a) proteins unrelated to the cyclase

[18] J. K. Northup, P. C. Sternweis, M. D. Smigel, L. S. Schleifer, E. M. Ross, and A. G. Gilman, *Proc. Natl. Acad. Sci. U.S.A.* **77,** 6516 (1980).

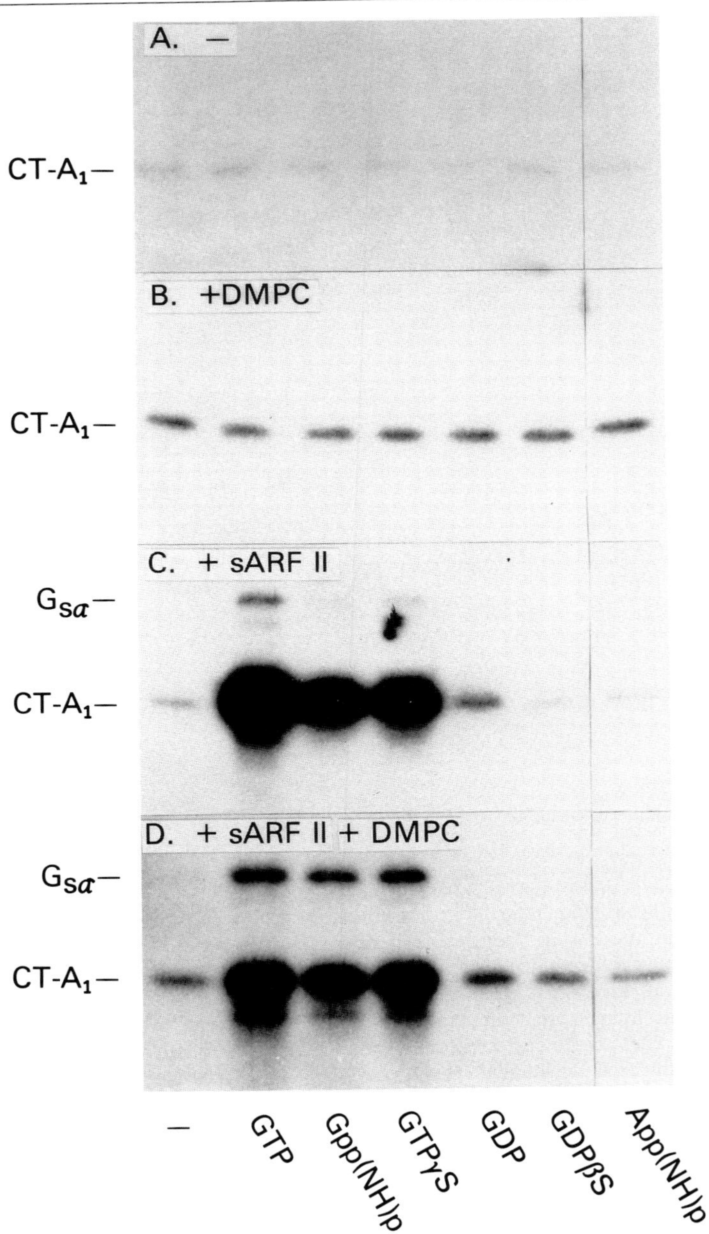
A. —
CT-A₁—
B. +DMPC
CT-A₁—
C. + sARF II
Gsα—
CT-A₁—
D. + sARF II + DMPC
Gsα—
CT-A₁—
—
GTP
Gpp(NH)p
GTPγS
GDP
GDPβS
App(NH)p

system,[19] (b) arginine and other simple guanidino compounds, for example, agmatine,[20] and (c) the toxin A1 catalytic unit (auto-ADP-ribosylation).[21,22] The toxin also possesses NAD^+ glycohydrolase (NADase) activity.[23] If the stimulatory effect of ARF on toxin-catalyzed ADP-ribosylation of $G_{s\alpha}$ were a result of ARF binding to $G_{s\alpha}$, thereby improving its ability to serve as a toxin substrate, ARF should have no effects on the ADP-ribosylation of simple guanidino compounds and proteins unrelated to the cyclase system or on auto-ADP-ribosylation. Alternatively, if ARF enhanced ADP-ribosylation by acting as an allosteric modifier of the toxin catalytic protein, then ARF might be expected to stimulate all toxin-catalyzed reactions. To address this question, two forms of ARF (designated sARF-I and sARF-II) were purified from the soluble fraction of bovine brain and reconstituted in an assay containing [*adenylate*-^{32}P]NAD, purified G_s, and cholera toxin in the presence or absence of dimyristoylphosphatidylcholine (DMPC) and various guanine and adenine nucleotides (Fig. 5). Auto-ADP-ribosylation of CTA1 was stimulated in a nucleotide-independent manner in the presence of DMPC. In the absence of DMPC, sARF-II enhanced the auto-ADP-ribosylation of CTA1 when GTP or a nonhydrolyzable analog such as GTPγS or Gpp(NH)p was present; GDP, GDPβS, and adenine nucleotides were ineffective.

Optimal ADP-ribosylation of $G_{s\alpha}$ required the presence of both DMPC and sARF-II in addition to GTP or an analog. The guanine nucleotide-dependent stimulation of CTA1 auto-ADP-ribosylation by ARF was not dependent on the presence of G_s. The fact that ARF stimulated the auto-ADP-ribosylation of CTA1 in the absence of G_s was consistent with the proposal that ARF directly activates the toxin. This hypothesis was further tested by assessing the effect of ARF on the ADP-ribosylation of agmatine, a simple guanidino compound, catalyzed by the CTA1 protein. In the presence of GTP, sARF-II stimulated the ADP-ribosylation of agmatine

[19] J. Moss and M. Vaughan, *Proc. Natl. Acad. Sci. U.S.A.* **75,** 3621 (1978).

[20] J. Moss and M. Vaughan, *J. Biol. Chem.* **252,** 2455 (1977).

[21] J. B. Trepel, D. M. Chuang, and N. H. Neff, *Proc. Natl. Acad. Sci. U.S.A.* **74,** 5440 (1977).

[22] J. Moss, S. J. Stanley, P. A. Watkins, and M. Vaughan, *J. Biol. Chem.* **255,** 7835 (1980).

[23] J. Moss, V. C. Manganiello, and M. Vaughan, *Proc. Natl. Acad. Sci. U.S.A.* **73,** 4424 (1976).

FIG. 5. Effect of nucleotides, DMPC, and sARF-II on ADP-ribosylation of $G_{s\alpha}$ and choleragen A1 subunit (CTA1). ADP-ribosylation of $G_{s\alpha}$ was carried out as described in the text except that the nucleotide (10 μM) was varied as indicated. Autoradiographs of gels (16%) are shown. Details are described in text. (A) Without additions, (B) with 1 mM DMPC, (C) with 1.5 μg sARF-II, and (D) with DMPC and sARF-II.

TABLE II
EFFECT OF NUCLEOTIDES ON NAD : AGMATINE ADP-RIBOSYLTRANSFERASE ACTIVITY OF CHOLERAGEN A SUBUNIT IN PRESENCE OF sARF-I OR sARF-II[a]

Nucleotide added (30 μM)	[*carbonyl*-^{14}C] Nicotinamide released (nmol/μg A subunit/hr)	
	sARF-I	sARF-II
None	2.10	1.22
GTP	5.48	2.12
GTPγS	4.70	1.91
GPP(NH)P	4.70	1.73
ATP	2.15	1.26
APP(NH)P	2.00	1.11
GDP	2.00	1.16
GDPβS	1.89	1.11

[a] Detailed description of the assay is given in the text. Assays contained choleragen A (1 μg), sARF-I (2.8 μg), or sARF-II (2.5 μg) with nucleotide as indicated. Data are from Ref. 15.

(Table II); stimulation by sARF-II required GTP or its analogs; GDPβS, GDP, and APP(NH)P were inactive (Table II). Based on Lineweaver–Burk analysis, the primary effect of ARF was to decrease the K_m, with little effect on V_{max}.[24] In the presence of the ionic detergent sodium dodecyl sulfate (SDS), ARF produced a somewhat larger decrease in K_m and increased V_{max}. Activation of ARF was also observed with certain other detergents and phospholipids such as cholate and DMPC/cholate.[25]

The K_a for GTP in ARF stimulation of the ADP-ribosylation of agmatine was quite sensitive to the detergent/phospholipid present in the assay. In SDS, the K_a for GTP was in the micromolar range, whereas in the presence of DMPC/cholate, it was approximately 2 orders of magnitude lower (Fig. 6).[25] In DMPC/cholate, high-affinity binding of GTPγS was observed with a K_d consistent with the K_a in the NAD : agmatine ADP-ribosyltransferase assay. In SDS, high affinity binding was not detected. These results are consistent with the conclusion that ARF is an allosteric activator of the

[24] M. Noda, S.-C. Tsai, R. Adamik, D. A. Bobak, M. M. Bliziotes, J. Moss, and M. Vaughan, *Biochim. Biophys. Acta* **1034,** 195 (1990).

[25] D. A. Bobak, M. M. Bliziotes, M. Noda, S. C. Tsai, R. Adamik, J. Moss, and M. Vaughan, *Biochemistry* **29,** 855 (1990).

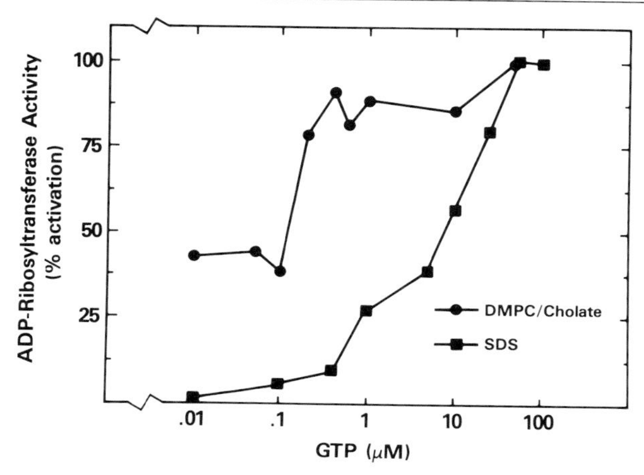

FIG. 6. Effect of DMPC/cholate or SDS on ARF-dependent activation of cholera toxin by GTP. Assays containing sARF-II (1 μg), cholera toxin A subunit (CTA) (1 μg), 100 μ*M* [*adenine*-^{14}C]NAD (60,000 cpm), 10 m*M* agmatine, 20 m*M* dithiothreitol, and the indicated concentration of GTP, with either 50 m*M* potassium phosphate, pH 7.5, 5 m*M* $MgCl_2$, 0.1 mg/ml ovalbumin, 0.003% SDS or 20 m*M* HEPES, pH 8, 1 m*M* EDTA, 6 m*M* $MgCl_2$, 3 m*M* DMPC, 0.2% sodium cholate, were incubated for 2 hr at 30°. Samples (50 μl) were withdrawn to determine ADP-ribosylagmatine formation.[15,24] [See also J. Moss and S. J. Stanley, *Proc. Nat. Acad. Sci. U.S.A.* **78,** 4809 (1981).] The data are means of triplicate assays [see also M. Noda, S.-C. Tsai, R. Adamik, D. A. Bobak, M. M. Bliziotes, J. Moss, and M. Vaughan, *24th U.S.–Japan Joint Cholera Conference*, Tokyo, Japan, in press.]

toxin, dependent on GTP for activity. Since both the activator of the toxin, ARF, as well as the toxin substrate, $G_{s\alpha}$, are guanine nucleotide-binding proteins, it would appear that a guanine nucleotide-binding protein cascade (Fig. 7) participates in the activation of adenylyl cyclase by cholera toxin.[15] The CTA1 protein is activated by ARF · GTP, enabling it to catalyze the ADP-ribosylation of $G_{s\alpha}$. ADP-ribosyl-$G_{s\alpha}$ in the presence of GTP enhances the activity of the catalytic unit of adenylyl cyclase, thereby stimulating the formation of cyclic AMP from ATP.

Assays for ADP-Ribosylation Factor-Like Activities

NAD : Cholera Toxin A1 Auto-ADP-ribosyltransferase

During purification, sARF-I and sARF-II activities are determined based on their ability to stimulate the auto-ADP-ribosylation of the choleragen A1 protein.[14,15] Assays at later stages of purification contain sARF or buffer, 10 μl of choleragen (25 μg in 30 m*M* dithiothreitol/75 m*M* glycine, pH 8.0, incubated at 30° for 10 min), 100 μ*M* GTP, 5 m*M* $MgCl_2$,

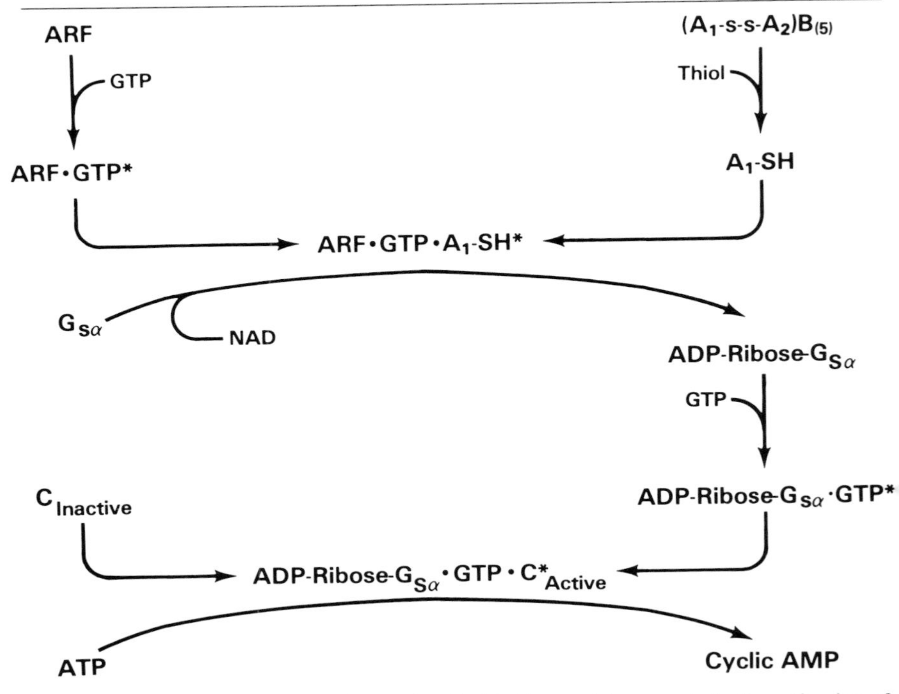

FIG. 7. Participation of a guanine nucleotide-binding protein cascade in the activation of adenylyl cyclase by choleragen. The choleragen subunit structure is represented as $(A_1$-s-s-$A_2)B_{(5)}$. The A subunit is composed of two proteins, A_1 and A_2, linked through a single disulfide bond. The B oligomer (B_5) is responsible for toxin binding to the cell surfaces. Addition of thiol results in release of A_1-SH which then associates with ARF · GTP. This active species catalyzes the ADP-ribosylation of $G_{s\alpha}$; ADP-ribosyl · $G_{s\alpha}$ is activated by GTP and stimulates the cyclase catalytic unit (C), thereby enhancing formation of cyclic AMP from ATP. [Reprinted by permission from S.-C. Tsai, M. Noda, R. Adamik, P. P. Chang, H.-C. Chen., J. Moss, and M. Vaughan, *Proc. Natl. Acad. Sci. U.S.A.* **85,** 5488 (1988).]

20 m*M* thymidine, 10 μ*M* [^{32}P]NAD (1–2 μCi), and 25 m*M* potassium phosphate, pH 7.5 (total volume 100 μl). ARF (or buffer), activated toxin, and the remainder of the components are added in that order followed by incubation for 60 min at 30°. After addition of 2 ml of cold 7.5% (v/v) trichloroacetic acid and 10 μg of bovine serum albumin, samples are kept at 4° for 30 min before centrifugation (2800 *g*, 30 min). The precipitated proteins are dissolved in 1% SDS (w/v)/5% mercaptoethanol (65°, 10 min) and subjected to electrophoresis in 15% SDS–polyacrylamide gels. Autoradiograms (Kodak X-Omat AR film) of gels are scanned by laser densitometry (LKB) to quantify auto-ADP-ribosylation of CTA1. Components of assay at early steps in purification (through CM-Sepharose) are slightly different as described in the legend for Fig. 1.

NAD : $G_{s\alpha}$ *ADP-Ribosyltransferase*

Assays are conducted as described for the NAD : CTA1 auto-ADP-ribosyltransferase assay except for the inclusion of 1 m*M* DMPC and G_s (0.2 μg in 10 μl of 20 m*M* Tris-Cl, pH 8.0/0.3% cholate/1 m*M* dithiothreitol/1 m*M* EDTA/5 m*M* $MgCl_2$/1 m*M* NaN_3) purified from rabbit liver membranes essentially as described by Sternweis *et al.*[26]

NAD : Agmatine ADP-Ribosyltransferase

Duplicate assays contain 50 m*M* potassium phosphate, pH 7.5, 5 m*M* $MgCl_2$, 100 μM GTP, 10 m*M* agmatine, 20 m*M* dithiothreitol, 0.1 mg/ml ovalbumin, 100 μM [*adenine*-^{14}C]NAD or [*carbonyl*-^{14}C]NAD [60,000 counts per minute (cpm)], and other additions (e.g., ARF or buffer) as indicated in a total volume of 0.3 ml. Reactions are initiated by the addition of the choleragen A subunit (1 μg). After 60 min at 30°, two 50-μl samples are transferred to columns (0.5 × 2 cm) of AG 1-X2 (BioRad, Richmond, CA), which are then washed 4 times with 1.2 ml of water. Eluates containing released [*carbonyl*-^{14}C]nicotinamide or [*adenine*-^{14}C]ADP-ribosylagmatine are collected for radioassay. Reaction rates are constant for 2 hr.

Characterization of ADP-Ribosylation Factor cDNA

An ARF cDNA coding for a protein of 181 amino acids was isolated from a bovine retinal library in λgt10.[16] Comparison of the deduced amino acid sequence of the ARF cDNA with sequences of tryptic peptides from purified brain sARF-II revealed only 2 differences out of a total of 60 amino acids. The deduced amino acid sequence of another ARF cDNA clone from bovine adrenal reported by Sewell and Kahn[27] is identical with the sequences of the tryptic peptides. It is possible that these two ARF clones are derived from different mRNAs encoding different proteins. The nucleotide sequences (coding regions) of the two bovine ARF clones are only 79% identical, although the deduced amino acid sequences differ in only 8 positions, 4 of these within the amino-terminal 11 amino acids. As noted above, ARF requires GTP (or suitable analogs) for activity. Comparison of the deduced amino acid sequence of ARF, $G_{o\alpha}$ (the α subunit of a guanine nucleotide-binding protein that may be involved in the regulation of ion fluxes), and c-Ha *ras* gene product p21 reveals similarities in regions believed to be involved in guanine nucleotide-binding

[26] P. C. Sternweis, J. K. Northup, M. D. Smigel, and A. G. Gilman, *J. Biol. Chem.* **256,** 11517 (1981).

[27] J. L. Sewell and R. A. Kahn, *Proc. Natl. Acad. Sci. U.S.A.* **85,** 4620 (1988).

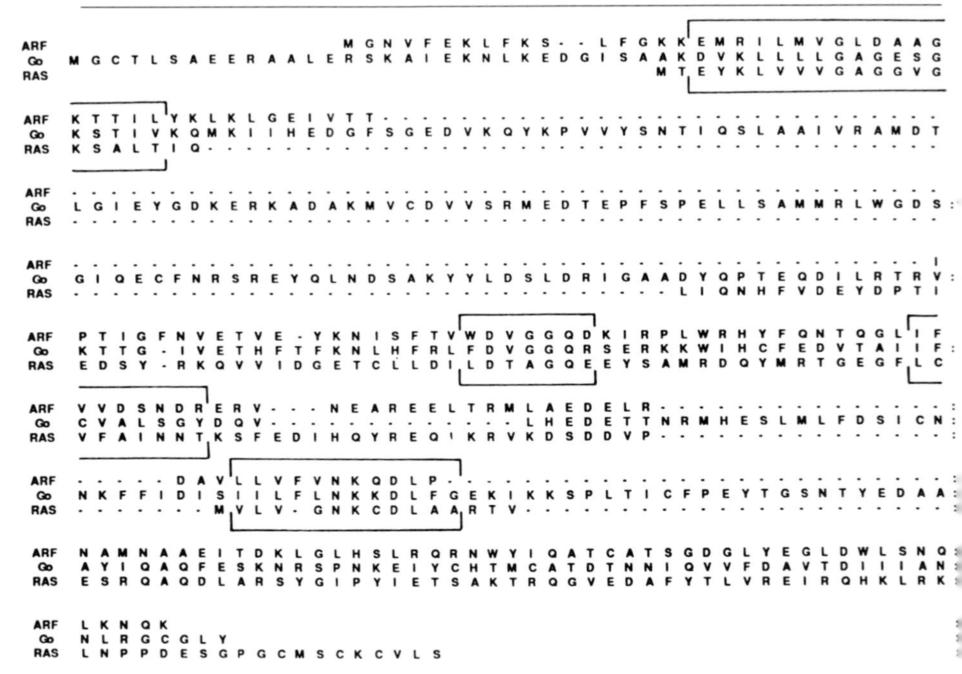

FIG. 8. Comparison of deduced amino acid sequences of bovine retinal ARF,[16] bovine retinal $G_{o\alpha}$,[28] and human c-Ha-*ras* (RAS) from Ref. 29. The alignment was performed using the method of S. B. Needleman and C. D. Wunsch [*J. Mol. Biol.* **48,** 443 (1970)] as modified by M. O. Dayoff ("Atlas of Protein Sequence and Structure," Vol. 5, Suppl. 3, p. 345, Natl. Biomed. Res. Found., Washington, D.C., 1978) and D. F. Feng, M. S. Johnson, and R. F. Doolittle [*J. Mol. Evol.* **21,** 112 (1985)] and was based in part on the alignment of G protein α subunits and *ras* protein by Masters *et al.*[30] Regions proposed to be involved in guanine nucleotide binding and hydrolysis are enclosed in brackets. Gaps indicated by hyphens were introduced to obtain maximal alignment. Single letter amino acid abbreviations are shown.

and GTP hydrolysis (Fig. 8).[28–30] ARF lacks sites analogous to those in the G protein α subunits that are ADP-ribosylated by cholera and pertussis toxins. The fact that ARF differs considerably from these other proteins in regions not obviously involved in guanine nucleotide-binding is consistent with a different role for this protein in cell function.

[28] K. P. Van Meurs, C. W. Angus, S. Lavu, H.-F. Kung, S. K. Czarnecki, J. Moss, and M. Vaughan, *Proc. Natl. Acad. Sci. U.S.A.* **84,** 3107 (1987).

[29] D. J. Capon, E. Y. Chen, A. D. Levinson, P. H. Seeburg, and D. V. Goeddel, *Nature (London)* **302,** 33 (1983).

[30] S. B. Masters, R. M. Stroud, and H. R. Bourne, *Protein Eng.* **1,** 47 (1986).

[22] ADP-Ribosylation of G Proteins with Pertussis Toxin

By GREGORY S. KOPF and MARILYN J. WOOLKALIS

Introduction

The guanine nucleotide-binding regulatory proteins (G proteins) occupy critical roles as signal transducing elements in coupling many ligand–receptor interactions with the generation of intracellular second messengers.[1] These heterotrimeric plasma membrane-associated proteins are composed of distinct α subunits, which contain a GTP-binding domain, and more highly conserved β and γ subunits. Activation of G proteins occurs upon GTP binding to the α subunit, causing dissociation of α–GTP from the $\beta\gamma$ dimer. These dissociated subunits then exert their regulatory actions on respective targets, ultimately resulting in ligand coupling to intracellular signalling systems. G proteins regulate the hormonally responsive adenylyl cyclase (EC 4.6.1.1, adenylate cyclase) of somatic cells,[2] the cyclic GMP phosphodiesterase of retinal rod outer segments,[3] and atrial K^+ channels,[4] and have been implicated in regulating phospholipase C,[5] phospholipase A_2,[6] and Ca^{2+} channels.[7] Since recent studies have also demonstrated the existence of intracellular G proteins involved in protein trafficking,[8,9] additional role(s) for G proteins in cellular function are only starting to be appreciated.

Bacterial toxins (cholera, pertussis) have been used as tools to identify and investigate the functional roles of G proteins in biological systems. These toxins covalently modify the α subunits of many G proteins by catalyzing the transfer of ADP-ribose from NAD^+ to specific amino acid residues[10] and, as a consequence, modify the biological activity of the target G protein. By using $[^{32}P]NAD^+$, it is possible to monitor the incorpo-

[1] P. J. Casey and A. G. Gilman, *J. Biol. Chem.* **263,** 2577 (1988).
[2] A. G. Gilman, *Annu. Rev. Biochem.* **56,** 615 (1987).
[3] L. Stryer, *Annu. Rev. Neurosci.* **9,** 87 (1986).
[4] R. Mattera, A. Yatani, G. E. Kirsch, R. Graf, K. Okabe, J. Olate, J. Codina, A. K. Brown, and L. Birnbaumer, *J. Biol. Chem.* **264,** 465 (1989).
[5] S. Cockcroft, *Trends Biochem. Sci.* **12,** 75 (1987).
[6] C. L. Jelsema and J. Axelrod, *Proc. Natl. Acad. Sci. U.S.A.* **84,** 3623 (1987).
[7] J. Hescheler, W. Rosenthal, W. Trautwein, and G. Schultz, *Nature (London)* **325,** 445 (1987).
[8] P. Melancon, B. S. Glick, V. Malhotra, P. J. Weidman, T. Serafini, M. L. Gleason, L. Orci, and J. E. Rothman, *Cell (Cambridge, Mass.)* **51,** 1053 (1987).
[9] N. Segev, J. Mulholland, and D. Botstein, *Cell (Cambridge, Mass.)* **52,** 915 (1988).
[10] J. Moss and M. Vaughan, *Adv. Enzymol.* **61,** 303 (1988).

ration of [^{32}P]ADP-ribose into these proteins and to identify them by autoradiography. Thus, only small amounts of crude protein extracts are required to identify and initially characterize general classes of G proteins. These toxins can also be used on intact cells to determine the effects of G-protein modification on cellular function.[11] It must be emphasized that the use of these toxins should not be considered the only tool with which to establish the identity and role(s) of G proteins in intracellular signal transduction.

Pertussis toxin (pertussigen; islet-activating protein), produced by strains of the bacterium *Bordetella pertussis,* exists in its native state as a hexamer of M_r 119,000 composed of five nonidentical subunits which comprise an enzymatically active A component (S1) and a binding B component (S2, S3, two S4, S5).[12] Treatment of the native complex with mild denaturing agents results in the dissociation of S1 (A protomer) from the rest of the complex. Dissociated S1, upon incubation with disulfide reducing agents, possesses both NAD^+ glycohydrolase (NAD^+ nucleosidase, EC 3.2.2.5) and ADP-ribosyltransferase activities. It is this dissociated, reduced form of the toxin that is used for *in vitro* ADP-ribosylation studies using NAD^+ as a cosubstrate. The function of the five-membered B-component is to attach the native toxin to the cell surface, which is required for the successful transfer of the S1 subunit into the cell. Once inside the cell, both enzymatic activities of S1 are activated by intracellular reduced glutathione, which permits the subsequent ADP-ribosylation of target proteins. Since the native holotoxin is required for binding to intact cells and internalization of the S1 subunit, only nonreduced, undissociated toxin can be used for studies on intact cells.

Pertussis toxin catalyzes the ADP-ribosylation of the α subunits of G_i,[13,14] transducin,[15,16] and G_o.[17,18] This covalent modification occurs at a cysteine residue located near the carboxy terminus of the protein,[19] and it

[11] R. D. Sekura, Y. L. Zhang, and M. J. Quentin-Millet, *in* "Pertussis Toxin" (R. D. Sekura, J. Moss, and M. Vaughan, eds.), p. 45. Academic Press, New York, 1985.

[12] M. Tamura, K. Nogimori, S. Murai, M. Yajima, K. Ito, T. Katada, M. Ui, and S. Ishii, *Biochemistry* **21,** 5516 (1982).

[13] T. Katada and M. Ui, *Proc. Natl. Acad. Sci. U.S.A.* **79,** 3129 (1982).

[14] T. Katada and M. Ui, *J. Biol. Chem.* **257,** 7210 (1982).

[15] C. Van Dop, G. Yamanaka, F. Steinberg, R. D. Sekura, C. R. Manclark, L. Stryer, and H. R. Bourne, *J. Biol. Chem.* **259,** 23 (1984).

[16] D. R. Manning, B. A. Fraser, R. A. Kahn, and A. G. Gilman, *J. Biol. Chem.* **259,** 749 (1984).

[17] P. C. Sternweis and J. D. Robishaw, *J. Biol. Chem.* **259,** 13806 (1984).

[18] E. J. Neer, J. M. Lok, and L. G. Wolf, *J. Biol. Chem.* **259,** 14222 (1984).

[19] R. E. West, J. Moss, M. Vaughan, T. Liu, and T. Y. Liu, *J. Biol. Chem.* **260,** 14428 (1985).

functionally uncouples the G protein from receptors.[2] There is no evidence to date which suggests that cofactors other than NAD^+ are required for the action of this toxin on its target G proteins. This is unlike the case with cholera toxin, which appears to require an ADP-ribosylation factor (ARF).[20] However, it appears that an intact G-protein heterotrimer is required for pertussis toxin-catalyzed ADP-ribosylation of the α subunit.[18] Thus, activation of the G protein, which results in subunit dissociation, prevents subsequent ADP-ribosylation and functional inactivation by the toxin. This particular property is useful in defining the specificity of the pertussis toxin-catalyzed ADP-ribosylation *in vitro* and for monitoring the degree of ligand-induced G-protein activation.[21]

Experimental Procedures

Preparation of Cell or Tissue Extracts for ADP-Ribosylation

Although a variety of methods have been used to prepare cell or tissue extracts for ADP-ribosylation, it is important to consider extraction conditions that minimize proteolysis. This is especially important when preparing extracts from secretory cells. The inclusion of protease inhibitors (e.g., phenylmethylsulfonyl fluoride, leupeptin, aprotinin, *p*-aminobenzamidine, soybean or lima bean trypsin inhibitors) in the extraction buffers does not interfere with pertussis toxin-catalyzed ADP-ribosylation. When performing ADP-ribosylations on extracts of spermatozoa, a battery of these protease inhibitors is included in all of the extraction steps,[22] since sperm extracts contain a trypsinlike protease which interferes with the inhibitory modulation (presumably through G_i) of hormonally sensitive adenylyl cyclases.[23] Since G proteins are, for the most part, associated with cell membranes, it is also advisable to develop extraction protocols that enrich for membrane fractions. This increases the relative concentrations of G proteins for subsequent labeling.

Additional fractionation might also be useful in preparing extracts for ADP-ribosylation. In initial experiments with crude spermatozoa membranes we were unsuccessful in observing any specific pertussis toxin-catalyzed ADP-ribosylation of proteins (Fig. 1, lane b). However, when membranes were first extracted with buffers containing 1% Lubrol PX and the resultant Lubrol PX supernatants then incubated with pertussis toxin

[20] R. A. Kahn and A. G. Gilman, *J. Biol. Chem.* **259,** 6228 (1984).
[21] L. F. Brass, M. J. Woolkalis, and D. R. Manning, *J. Biol. Chem.* **263,** 5348 (1988).
[22] G. S. Kopf, M. J. Woolkalis, and G. L. Gerton, *J. Biol. Chem.* **261,** 7327 (1986).
[23] R. A. Johnson, K. H. Jakobs, and G. Schultz, *J. Biol. Chem.* **260,** 114 (1985).

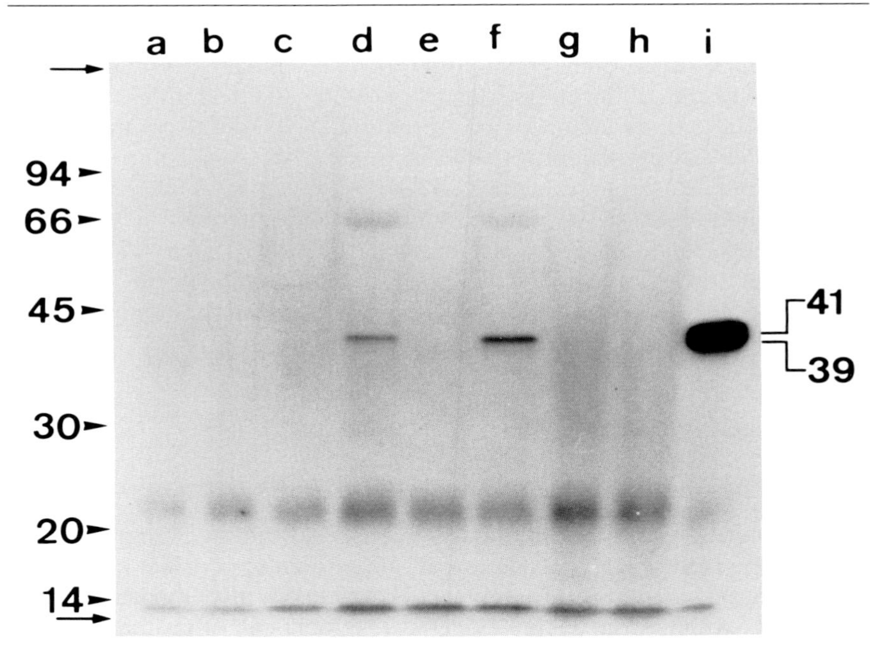

FIG. 1. Sodium dodecyl sulfate-polyacrylamide gel electrophoresis of abalone sperm extracts incubated in either the absence or presence of pertussis toxin. A 48,000 *g* pellet fraction was prepared and extracted with 1% Lubrol PX as previously described.[22] ADP-ribosylation was then carried out in either the absence (lanes a, c, e, and g) or presence (lanes b, d, f, h, and i) of pertussis toxin. Lanes a and b, 5 μg of 48,000 *g* pellet; lanes c and d, 5 μg Lubrol PX-extracted 48,000 *g* supernatant; lanes e and f, 9.5 μg Lubrol PX-extracted 48,000 *g* supernatant; lanes g and h, 5 μg Lubrol PX-extracted 48,000 *g* pellet; lane i, 25 μg of membranes from chicken embryo fibroblasts. Numbers at left represent the M_r ($\times 10^{-3}$) of the protein standards. Arrows indicate the origin and dye front of the gel. The M_r (10^{-3}) of the chicken embryo fibroblast pertussis toxin substrates are shown at right. (Reprinted with permission from the American Society of Biological Chemists, Inc.)[22]

and [^{32}P]NAD$^+$, a single M_r 41,000 pertussis toxin substrate was observed (Fig. 1, lanes d and f). The pellet fraction after the Lubrol PX extraction did not contain detectable pertussis toxin substrates (Fig. 1, lane h). Although it is not clear why extraction with this nonionic detergent enables one to detect a pertussis toxin substrate in spermatozoa, it is clear that G-protein heterotrimers are solubilized by the detergent, since the intact heterotrimer is required for successful ADP-ribosylation of the α subunit by the toxin. Detergent-induced alterations in the conformation of the G protein might account for these effects. Alternatively, Lubrol PX might inactivate and/or dissociate inhibitory factor(s).

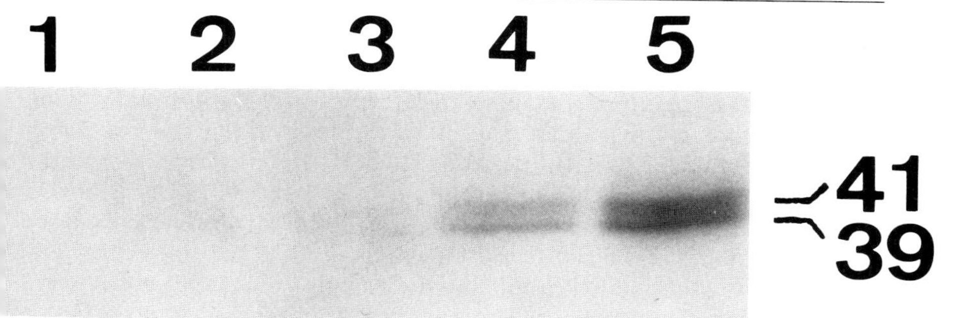

FIG. 2. The effects of pertussis toxin incubation with membranes on [^{32}P]ADP-ribosylation of proteins and the effects of different conditions of pertussis toxin activation on toxin-catalyzed [^{32}P]ADP-ribosylation of proteins. Membranes from chicken embryo fibroblasts were incubated in the absence (lane 1) or presence of pertussis toxin (lanes 2–5). Pertussis toxin, in a solution containing 50 m*M* HEPES, pH 8, and 1 mg/ml BSA, was activated at 30° for 30 min in either the absence (lane 2) or the presence of 20 m*M* DTT (lane 3), 250 m*M* DTT (lane 4), or 20 m*M* DTT plus 0.125% SDS (lane 5). Relevant portions of the autoradiogram are shown.

Activation of Pertussis Toxin for in Vitro ADP-Ribosylation

When studying broken cell preparations activation of pertussis toxin is not absolutely required but generally results in more extensive ADP-ribosylation of G proteins[24] (Fig. 2, compare lane 2 with lanes 3–5). Reduction and activation of the pertussis toxin S1 subunit for *in vitro* ADP-ribosylation studies is usually accomplished with dithiothreitol (DTT). This disulfide reducing agent activates pertussis toxin most effectively at high concentrations, 250 m*M* being significantly more efficacious than 20 m*M* (Fig. 2, compare lanes 3 and 4).[24] The presence of low concentrations of sodium dodecyl sulfate (SDS) (e.g., 0.125%) not only decreases the concentration of DTT required for toxin activation, but also amplifies the degree of toxin-specific ADP-ribosylation (Fig. 2, compare lanes 3 and 4 to lane 5). We standardly employ the following protocol for pertussis toxin activation: pertussis toxin (100 μg/ml; List Biological Laboratories, Inc., Campbell, CA) is activated by incubation in 50 m*M* HEPES, pH 8, 1 mg/ml bovine serum albumin (BSA), 20 m*M* DTT, and 0.125% SDS at 30° for 30 min.[25] This toxin mixture is then diluted 5-fold into the ADP-ribosylation assay so that the final concentrations of pertussis toxin and SDS are 20 μg/ml and 0.025%, respectively. Final SDS concentrations greater than 0.025% may decrease the extent of ADP-ribosylation.

[24] J. Moss, S. J. Stanley, D. L. Burns, J. A. Hsia, D. A. Yost, G. A. Meyers, and E. L. Hewlett, *J. Biol. Chem.* **258,** 11879 (1983).

[25] K. Enomoto and D. M. Gill, *J. Biol. Chem.* **255,** 1252 (1980).

ADP-Ribosylation Reaction

The ADP-ribosylation reaction is started by the addition of activated toxin to an assay mixture containing the crude extracts (5–100 μg protein). The final concentrations of all reagents in the assay are as follows: 1 m*M* EDTA, 5 m*M* DTT, 10 m*M* thymidine, 10 m*M* HEPES, pH 8, 0.2 mg/ml BSA, 0.025% SDS, 20 μg/ml pertussis toxin, and 5 μM [^{32}P]NAD$^+$ [~20,000 counts per minute (cpm)/pmol]. [^{32}P]NAD$^+$ is either purchased or synthesized.[22,26] The samples are incubated at 30° for 10–60 min, and the reaction is terminated by boiling the samples in sample buffer for 3 min.[27] The samples are then subjected to one-dimensional polyacrylamide gel electrophoresis (PAGE) as described below. Three control incubations help define the specificity of the pertussis toxin-catalyzed transfer of ADP-ribose to a G protein.

Toxin-Specific Reaction. A sample incubated without pertussis toxin, but with the toxin activation vehicle, allows discernment of proteins radio-labeled nonspecifically as a function of other glycohydrolases present in the membrane preparation (Fig. 2, lane 1). Poly(ADP-ribose) polymerase (NAD$^+$ ADP-ribosyltransferase, EC 2.4.2.30) is such an enzyme, which is often observed as a radioactive peptide of M_r 110,000; thymidine is included in the reaction mixture to diminish the catalytic activity of this enzyme.[28]

ADP-Ribosylated Protein. A sample incubated in the presence of excess nonradioactive NAD$^+$ (100 μM final concentration or greater) should greatly diminish or eliminate the detection of proteins susceptible to the enzymatic transfer of ADP-ribose from [^{32}P]NAD$^+$ (Fig. 3A, compare lane 1 to lanes 2–4).

GTP-Binding Protein. A sample incubated in the presence of guanosine 5′-*O*-3-thiotriphosphate (GTPγS; 100 μM final concentration) should result in the diminution of [^{32}P]ADP-ribosylated G-protein α subunit (Fig. 3B, compare lanes 1 and 2). GTP or its nonhydrolyzable analogs cause the dissociation of α subunits from the $\beta\gamma$ subunits, thereby preventing pertussis toxin-catalyzed ADP-ribosylation of the α subunit.

A number of experimental modifications of the above protocol may be useful for optimizing ADP-ribosylation of G proteins from different sources. The presence of the nonionic detergent Lubrol PX at final concentrations up to 0.5% in the reaction mixture can significantly increase the extent of ADP-ribosylation obtained with certain membrane preparations

[26] D. Cassel and T. Pfeuffer, *Proc. Natl. Acad. Sci. U.S.A.* **75,** 2669 (1978).

[27] U. K. Laemmli, *Nature (London)* **227,** 680 (1970).

[28] D. M. Gill and R. Meren, *Proc. Natl. Acad. Sci. U.S.A.* **75,** 3050 (1978).

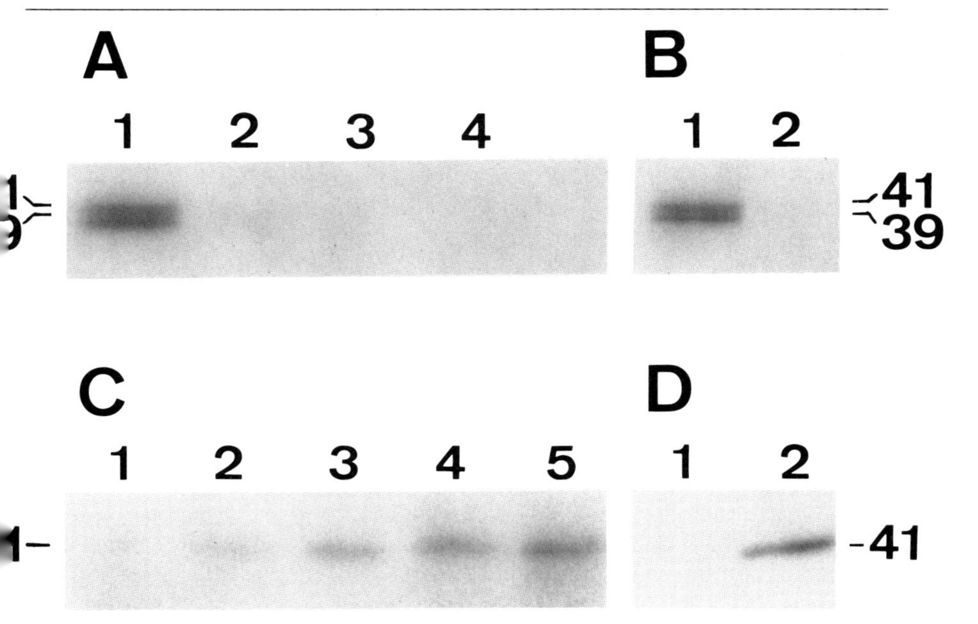

FIG. 3. (A) Inhibition of pertussis toxin-catalyzed [^{32}P]ADP-ribosylation of G proteins by nonradioactive NAD^+. Membranes from chicken embryo fibroblasts were incubated in the standard reaction mixture containing [^{32}P]NAD^+ in the absence (lane 1) or presence of 100 μM (lane 2), 500 μM (lane 3), and 1 mM nonradioactive NAD^+ (lane 4). (B) Inhibitory effect of GTPγS on pertussis toxin-catalyzed [^{32}P]ADP-ribosylation of G proteins. Membranes from chicken embryo fibroblasts were incubated in the standard reaction mixture containing [^{32}P]NAD^+ in the absence (lane 1) or presence of 100 μM GTPγS (lane 2). (C) The effect of Lubrol PX on pertussis toxin-catalyzed [^{32}P]ADP-ribosylation of G proteins. Mouse sperm extracts were incubated in the standard reaction mixture containing [^{32}P]NAD^+ in the absence (lane 1) or presence of 0.01% (lane 2), 0.05% (lane 3), 0.1% (lane 4), or 0.5% Lubrol PX (lane 5). (D) The effect of dimyristoylphosphatidylcholine on pertussis toxin-catalyzed [^{32}P]ADP-ribosylation of purified G_i. G_i purified from rabbit liver was incubated in the standard reaction mixture containing [^{32}P]NAD in either the absence (lane 1) or presence of 1 mg/ml dimyristoylphosphatidylcholine (lane 2). Relevant portions of the autoradiograms are shown.

that have not previously been extracted with detergent. An example of this effect is observed with sperm membranes (Fig. 3C). However, other membrane preparations (e.g., chicken embryo fibroblasts, 3T3 cells) display only minimal enhancement of ADP-ribosylation in the presence of this detergent (data not shown). The presence of phospholipids is probably important for effective ADP-ribosylation. When incubating purified preparations of G proteins, addition of a phospholipid such as dimyristoylphosphatidylcholine greatly increases the ability of the α subunit of the heterotrimer to be ADP-ribosylated by pertussis toxin. This is shown in Fig. 3D with G_i purified from rabbit liver (compare lanes 1 and 2). In crude membrane preparations endogenous phospholipid subserves this requirement.

Polyacrylamide Gel Electrophoresis

One-Dimensional SDS–PAGE. One-dimensional SDS–PAGE of the ADP-ribosylated proteins is carried out by the method of Laemmli.[27] Gels with a polyacrylamide concentration between 10 and 12% appear to be optimal for resolving ADP-ribosylated α subunits. We have found that the purity of SDS used in the electrophoresis buffers has pronounced effects on the resolution of pertussis toxin substrates in both somatic and germ cells.[29] For example, when using high-purity SDS throughout the electrophoresis system (99.9% lauryl sulfate; Bio-Rad, Richmond, CA, #161-0301; Polysciences, Warrington, PA, #3945), pertussis toxin substrates are resolved as a single radiolabeled band in membrane preparations from normal rat kidney cells and in extracts from mouse eggs and sperm. However, under identical ADP-ribosylation conditions, electrophoresis using SDS which contains 70% lauryl sulfate and the balance consisting of higher carbon chain homologs (Sigma, St. Louis, MO, #L5750) resolves two radiolabeled bands in each of the preparations. These pertussis toxin substrates are likely to reflect at least two distinct species, since (1) the two species resolved in the presence of Sigma SDS migrate with the same relative mobilities upon reelectrophoresis in the presence of Sigma SDS, (2) exchanging the Sigma SDS from the two resolved species for Bio-Rad SDS results in both species migrating with the same relative mobility, and (3) exchanging the Bio-Rad SDS from the single species for Sigma SDS now resolves two species. The altered mobilities of the pertussis toxin substrates that are observed with various SDS preparations may be due to differential interactions of the substrates with the multiple chain lengths present in the lower purity SDS. Such interactions do not occur when using more highly purified SDS.

Improved resolution of ADP-ribosylated α subunits by one-dimensional SDS–PAGE may also be effected by reduction of the sample with DTT in the presence of SDS followed by alkylation with reagents such as *N*-ethylmaleimide prior to the addition of sample buffer (although the results have been variable in our hands). It has been suggested that reductive alkylation prevents intrachain disulfide bond formation during electrophoresis.[17]

Additional information regarding the identity of ADP-ribosylated α

[29] J. Jones, G. S. Kopf, and R. M. Schultz, *FEBS Lett.* **243,** 409 (1989).

subunits can be obtained by peptide mapping with limited proteolytic digestion of the pertussis toxin substrate(s) after one-dimensional SDS–PAGE.[22] After electrophoresis the regions containing the [^{32}P]ADP-ribosylated bands are located by autoradiography of the wet gels. The bands are excised, applied to a second polyacrylamide gel, and proteolytic digestion and electrophoresis then carried out by the method of Cleveland *et al.*[30]

Two-Dimensional SDS–PAGE. The α subunits of G proteins in crude cell preparations are not well resolved by the O'Farrell procedure of two-dimensional SDS–PAGE.[31] The α subunits of G_i, in particular, tend to form a streak through the isoelectric focusing gel from the point of sample application to the region of their isoelectric points.[32] The following modifications of the O'Farrell procedure minimize the streaking of G protein α subunits.

The sample is solubilized in a solution containing 4% Lubrol PX, 0.2% sodium deoxycholate, 8 m*M* K_2CO_3, 70 m*M* DTT, and 3.3% ampholines for 1 hr on ice. Ampholines capable of generating linear gradients ranging from pH 4.5 to 6.5 are suitable for separation of the subunits from the better characterized G proteins which exhibit isoelectric points between pH 5 and 6. The sample is centrifuged at 50,000 *g* for 30 min (24°), and the supernatant is then adjusted with solid urea to a final concentration of 9.5 *M*. The samples are focused on gels that contain 2% Lubrol PX [in place of 2% Nonidet P-40] and 4% ampholines. The cathode buffer is 50 m*M* NaOH and the anode buffer is 30 m*M* H_3PO_4. Isoelectric focusing is carried out at 800 V for 16 hr. Electrophoresis in the second dimension is performed as described by O'Farrell.[31] Resolution between α subunits sharing identical isoelectric points but slightly different relative mobilities is frequently enhanced after two-dimensional SDS–PAGE, when compared with standard methods of one-dimensional SDS–PAGE; the reasons for this are not known. Comparison of the proteolytic fragments generated from differently charged ADP-ribosylated species of α subunits can be accomplished by first subjecting the sample to two-dimensional SDS–PAGE. After staining the SDS–PAGE gel with Coomassie blue, the region containing the [^{32}P]ADP-ribosylated proteins is excised (endogenous actin can be used as a marker) and then subjected to limited proteolytic digestion[21] by the method of Cleveland *et al.*[30]

[30] D. W. Cleveland, S. G. Fischer, M. W. Kirschner, and U. K. Laemmli, *J. Biol. Chem.* **252,** 1102 (1977).

[31] P. H. O'Farrell, *J. Biol. Chem.* **250,** 4007 (1975).

[32] M. J. Woolkalis and D. R. Manning, *Mol. Pharmacol.* **32,** 1 (1987).

Discussion

While pertussis toxin is a useful agent for the preliminary assessment of G-protein structure, it does not allow one to distinguish between different G protein subtypes (e.g., G_{i1}, G_{i2}, G_{i3}). It also does not allow for the detection of other G proteins, such as G_z,[33] that are not pertussis toxin substrates. At this time, a comprehensive analysis of G proteins requires the use of antibodies directed against either the purified α subunits or against peptide sequences that are unique or common among the different α subunits. To date, such immunological analyses in conjunction with pertussis toxin-catalyzed ADP-ribosylation offer the best approach to the study of G-protein structure and function.

The Experimental Procedures section describes the use of pertussis toxin for *in vitro* studies of G-protein identification and initial characterization. Variations of these methods can also be used to analyze G-protein function. Incubation of intact cells with pertussis toxin can alter the responsiveness of the cell to ligands, which is usually monitored by measuring levels of specific intracellular second messengers (i.e., cyclic AMP, inositol 1,4,5-trisphosphate).[34,35] Such approaches have been useful in establishing the potential role(s) of pertussis toxin-sensitive G proteins in signal transduction reactions leading to specific physiological responses. The ability of pertussis toxin to enter an intact cell and ADP-ribosylate G proteins using endogenous NAD^+ can be assessed as follows. Pertussis toxin-treated cells are permeabilized or extracted and then incubated in the standard ADP-ribosylation assay containing $[^{32}P]NAD^+$ and additional pertussis toxin. If the toxin has entered the intact cell and covalently modified the G protein, one should observe diminished *in vitro* $[^{32}P]$ADP-ribosylation.[36,37] Pertussis toxin-catalyzed $[^{32}P]$ADP-ribosylation has also been used in permeabilized cells to monitor receptor-coupled G-protein activation after the administration of specific ligands.[21] In these experiments, receptor-mediated G-protein activation and subunit dissociation results in a reduction of pertussis toxin-catalyzed $[^{32}P]$ADP-ribosylation.

[33] H. K. W. Fong, K. K. Yoshimoto, P. Eversole-Cire, and M. I. Simon, *Proc. Natl. Acad. Sci. U.S.A.* **85,** 3066 (1988).

[34] T. Katada and M. Ui, *J. Biol. Chem.* **257,** 7210 (1982).

[35] K. H. Krause, W. Schlegel, C. B. Wollheim, T. Andersson, F. F. Waldvogel, and P. D. Lew, *J. Clin. Invest.* **76,** 1348 (1985).

[36] J. C. Chambard, S. Paris, G. L'Allemain, and J. Pouyssegur, *Nature (London)* **326,** 800 (1987).

[37] Y. Endo, M. A. Lee, and G. S. Kopf, *Dev. Biol.* **119,** 210 (1987).

[23] Cholera Toxin-Catalyzed [^{32}P]ADP-Ribosylation of Proteins

By D. MICHAEL GILL and MARILYN J. WOOLKALIS

Introduction

Cholera toxin, the heat-labile enterotoxins (LTs) of *Escherichia coli*, and the enterotoxins of some other enteric bacteria are related multimeric proteins of subunit formula A5B. The A subunit of each, or, better, the sulfhydryl reduction product-reduced fragment A_1, catalyzes ADP-ribosylations of proteins: NAD^+ + protein → ADP-ribose–protein + nicotinamide + H^+.[1,2] The α subunit of G_s is generally the most readily modified substrate *in vitro* but the toxins also ADP-ribosylate a variety of particulate and soluble proteins more slowly. The ADP-ribose groups attach to arginine residues, and it may be that any arginine group is reactive to some extent. Indeed, unbound arginine or agmatine can be ADP-ribosylated.

For its maximal activity, cholera toxin needs to interact with a protein now known as the ADP-ribosylation factor (ARF),[3] formerly known as the "cytosolic factor,"[4] and in a membrane-bound active form as "S."[5] This is a "small G" protein that supports the activity of cholera toxin when bound to a guanine nucleoside triphosphate. In erythrocytes ARF is predominantly soluble, and the optimum activation procedure is to incubate membranes with a mixture of GPP(NH)P and a source of ARF. ARF[GPP(NH)P] forms more readily in the presence of membranes. Once formed, the ARF[GPP(NH)P] binds to the membranes which can be washed once without losing their ability to support ADP-ribosylation. Other membranes contain ample bound ARF after isolation, and for them it is necessary simply to add GPP(NH)P and incubate. Brain membranes are particularly rich in ARF.[3] GPP(NH)P, GPP(CH_2)P, and GTPγS are equally effective at activating ARF, but GTPγS is the least desirable of the three because at relatively high concentrations (>100 μM) it, alone, reduces the substrate activity of $G_{s\alpha}$ and will support less ADP-ribosylation than does GPP(NH)P.[5] Also, GTPγS is less stable than GPP(NH)P. GPP(CH_2)P, which is a poor activator of G_s, may be particularly useful

[1] D. M. Gill and R. Meren, *Proc. Natl. Acad. Sci. U.S.A.* **75,** 3050 (1978).
[2] D. M. Gill and S. Richardson, *J. Infect. Dis.* **141,** 64 (1980).
[3] O. Weiss, J. Holden, C. Rulka, and R. A. Kahn, *J. Biol. Chem.* **264,** 21066 (1989).
[4] D. M. Gill and R. Meren, *J. Biol. Chem.* **258,** 11908 (1983).
[5] J. Coburn and D. M. Gill, *Biochemistry* **26,** 6364 (1987).

when it is necessary to unscramble nucleotide effects on ARF and nucleotide effects on substrates.[5]

The substrates do *not* need to be bound to GTP. This should be stressed because the GTP requirement is commonly misunderstood. G_s is equally substrate competent when bound to GDP, and possibly when bound to no nucleotide, although without a nucleotide it may denature rather rapidly.

General Condition for ADP-Ribosylation *in Vitro*

Generation of Fragment A_1 of Cholera Toxin

The entire toxin multimer (A5B) allows ADP-ribosylation to take place inside whole cells, but for work with cell fractions its fragment A_1 is required. Fragment A_1 must be separated from the rest of the molecule, and it is separated from fragment A_2 by reduction of the linking disulfide. Its activity is increased further by the presence of sodium dodecyl sulfate (SDS), which separates A_1 from noncovalent association with other components of the toxin and probably also prevents aggregation.

Prepare a solution of 2 mg/ml toxin in 50% glycerol and store at $-20°$. The ADP-ribosyltransferase activity is stable. On the day of use, mix 10 μl of this stock solution with 90 μl of 0.5% SDS, 5 m*M* dithiothreitol, 130 m*M* NaCl, 10 m*M* HEPES, pH 7.3. Stopper and incubate at 37° for 10 min. Do not cool on ice or the SDS will precipitate. Use the activated toxin at a ratio of 1 to 20 so that the membranes are exposed to only 0.025% SDS, which is tolerated well. If a diluent is necessary, use the same activating buffer plus 0.1% ovalbumin. Discard any excess diluted toxin at the end of the day because it loses potency by oxidation.

Alternatively, prepare carboxymethyl-toxin which is stably activated. Dissolve 1 mg of toxin in 4 ml of 100 m*M* Tris-HCl, pH 8.8, 5 m*M* dithiothreitol, 0.5% SDS, 1 m*M* EDTA. Incubate 10 min at 37°. Add 7.4 mg iodoacetamide (giving 10 m*M*) and incubate for 30 min more at 37°. Dialyze against two changes of 0.1% SDS, 130 m*M* NaCl, 10 m*M* HEPES, pH 7.3, 25% glycerol. Bring up to 5 ml (i.e., 200 μg/ml toxin), aliquot, and store at $-20°$.

One-Step Method for ADP-Ribosylating Membrane Proteins

Suspend membranes in 10 to 20 volumes (avoid greater dilutions) of buffered saline (e.g., 130 m*M* NaCl, 10 m*M* HEPES, pH 7.3, 0.01% NaN_3, 0.01 trypsin inhibitor units/ml, aprotinin, and additional protease inhibitors if needed). Recover by centrifuging. Resuspend the pellet in 2 volumes of NAD-free cell cytosol, or a more purified source of ARF. The purification of ARF is described in Refs. 6 and 7 and also in [20] and [21], this volume. It is sufficient to use partially purified material such as that taken to the DEAE-Sephadex step in Ref. 8. If no ARF is to be supplied, compensate by using more guanine nucleotide, toxin, and NAD, for example, 3 times more of each. Alternatively use the endogenous ARF present in some membrane preparations (such as brain); in this case do not prewash the membranes.

For each gel slot, incubate 15 μl of membrane suspension for 30 min at 25° to 30° (avoid 37°) with 5 μl of the following mixture: 1 part of 200 μg/ml activated toxin, 2 parts of 50 μM [^{32}P]NAD [e.g., 10,000–50,000 counts per minute (cpm)/pmol], 1 part of guanine nucleotide, and 1 part of 200 mM thymidine (warmed to 37° to dissolve). The final concentrations are about 10 to 15 mg/ml of membrane protein, 10 μg/ml toxin, 5 μM NAD, 0.1 mM GPP(NH)P, 10 mM thymidine, and 100–200 mM NaCl.

Two-Step Method for ADP-Ribosylating Membrane Proteins

It is usually preferable to activate ARF before adding the toxin and NAD, since ARF is best activated at 37° while the toxin reaction is impaired at temperatures over 30°.[4] Incubate 15 μl of the membrane–ARF suspension with 100 μM GPP(NH)P for 10 min at 37°. Then add 4 μl of a mixture of toxin, [^{32}P]NAD, and thymidine in the proportions listed above and incubate for 30 min at 25° to 30°.

Gel Analysis and Interpretation

End the incorporation by adding 1 ml of ice-cold buffered saline, vortex, and recover the membranes by centrifugation. Before discarding the washings, 1-μl portions can be spotted onto thin-layer chromatography (TLC) plates for analysis of the soluble products (see below for details). If it is necessary to digest DNA before loading the gel, suspend the pellet in one drop of buffered saline containing 2 mM $CaCl_2$ and 3 units/ml of micrococcal nuclease. Incubate for 15 min at 37° and wash again.

Make the membrane pellet into a slurry by vortexing vigorously before adding SDS gel sample buffer. Without this precaution it is frequently difficult to resuspend the entire pellet. Fractionate on a 7.5–15% gradient SDS slab gel. Start at 100 V: the voltage can be increased when the dye front is about 1 cm past the stack. Stain with Coomassie blue. It helps to

[6] R. A. Kahn and A. G. Gilman, *J. Biol. Chem.* **259,** 6228 (1984).

[7] S.-C. Tsai, M. Noda, R. Adamik, P. P. Chang, H.-C. Chen, J. Moss, and M. Vaughan, *J. Biol. Chem.* **263,** 1768 (1988).

[8] M. Woolkalis, D. M. Gill, and J. Coburn, this series, Vol. 165, p. 246.

remove unincorporated [^{32}P]NAD and other nucleotides by including in the destain bath a nylon net bag containing about 5 g of a mixed-bed resin such as Dowex MR3. Dry the gel and prepare an autoradiogram.

The migration rate of ADP-ribosyl $G_{s\alpha}$ varies with the gel system. On a 7–15% gradient gel the fastest band occupies a position equivalent to 42,000 M_r. In erythrocytes this is the only band (Fig. 1a) but many tissues have an additional form of G_s which migrates at 46,000–48,000 M_r. This is often resolved into a doublet, and its migration is sometimes anomalous (Fig. 1c and d). The 42K and 46K/48K bands straddle the prominent stained band of actin (M_r 45,000) which is invariably present. On nongradient gels the G_s bands migrate relatively more slowly and may have apparent sizes of 44,000–45,000 and 49,000–52,000, respectively.

For quantitation, cut out gel bands after autoradiography and count them in Omnifluor. Depending on the situation, estimate the background from an adjacent area of the same gel track or the equivalent area of a companion track generated without toxin. When it is present at all, the 46K/48K species is usually more abundant than the 42K species, but it is ADP-ribosylated an order of magnitude more slowly. The ratio of the radioactivities in the two bands is therefore a measure of the overall efficiency of ADP-ribosylation.

The ADP-ribosylation of G_s purified using detergent requires phospholipid in addition.[9]

ADP-Ribosylation of Other G Proteins

G_s is ordinarily the only G protein that is ADP-ribosylated efficiently, but under certain circumstances some or all of the pertussis toxin substrates can be labeled, too. They all contain arginine at the site homologous to the target arginine in G_s. The relatively high abundance of G_i and G_o can offset their less efficient labeling.

Two conditions seem to be necessary. First, it is helpful, if not essential, for the G protein to be coupled to an occupied hormone receptor. Thus, opioids assist the labeling of a $G_{i\alpha}$ protein in NG-108-15 cells,[10,11] and FMLP assists the labeling of a $G_{i\alpha}$ in HL60 cells.[12,13] The coupling must not have been disrupted by pertussis toxin-catalyzed ADP-ribosylation,

[9] L. S. Schleifer, R. A. Kahn, E. Hanski, J. K. Northup, P. C. Sternweiss, and A. G. Gilman, *J. Biol. Chem.* **257**, 20 (1982).

[10] F.-J. Klinz and T. Costa, *Biochem. Biophys. Res. Commun.* **165**, 554 (1989).

[11] G. Milligan and F. R. McKenzie, *Biochem. J.* **252**, 369 (1988).

[12] T. Iiri, M. Tohkin, N. Morishima, Y. Ohoka, M. Ui, and T. Katada, *J. Biol. Chem.* **264**, 21394 (1989).

[13] P. Gierschik and K. H. Jakobs, *FEBS Lett.* **224**, 219 (1987).

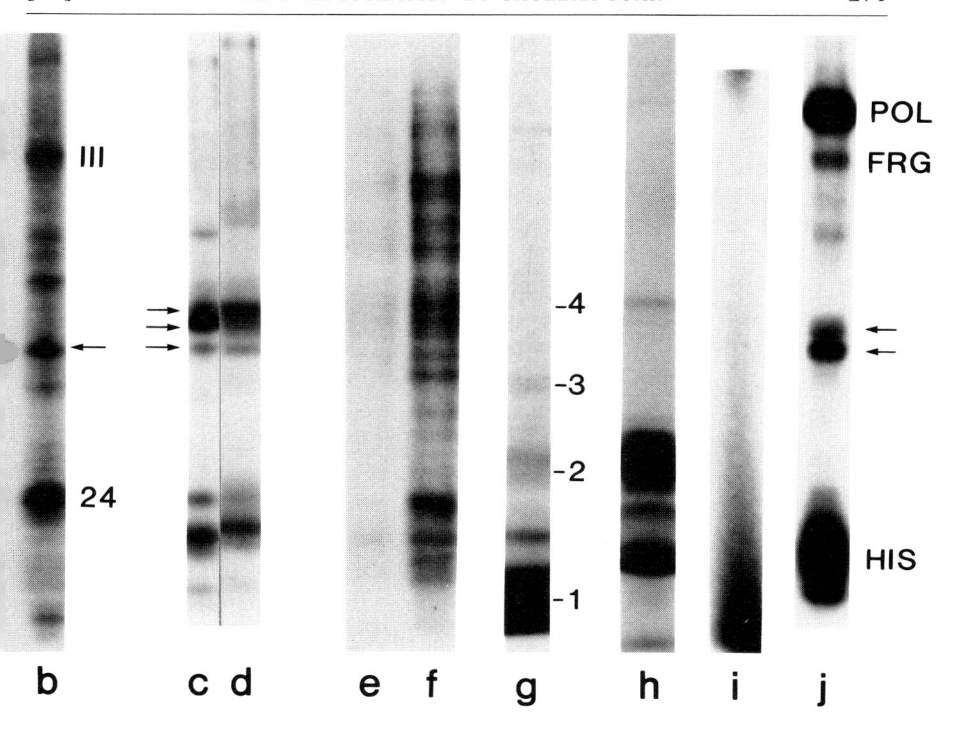

FIG. 1. Autoradiograms of [^{32}P]ADP-ribosylated products of the cholera toxin reaction and other processes. In lanes a–f the forms of G_s ADP-ribosylated by cholera toxin are indicated by arrows. (a) Pigeon erythrocyte membranes with only G_s (42,000 M_r) labeled. (b) The same, with secondary target proteins labeled as well. III, band III; 24, ~24,000 M_r. (c and d) Calf brain membranes in which the 46,000/48,000 M_r form of G_s is labeled as well as the 42,000 M_r form. The two lanes show different arrangements of the larger bands. (e and f) *cyc*$^-$ S49 lymphoma cell membranes without (e) or with (f) GTP. The membranes lack G_s, but the secondary target proteins are labeled when GTP is available. Lanes g–j illustrate various interfering reactions described in the text. (g) Nonenzymatic labeling of globin monomer, dimer, trimer, and tetramer in erythrocyte cytosol. (h) Toxin-independent labeling of tRNA molecules and half-molecules in bovine testis cytosol supplemented with GTP. (i) Radioactive smear obtained by running [^{32}P]ADP-ribose through an acrylamide gel; the gel itself is now radioactive. (j) Particulate fraction of wild-type S49 lymphoma cells in which G_s has been labeled by cholera toxin and all of the other bands have been labeled as a result of the activity of poly(ADP-ribose) polymerase. POL, Poly(ADP-ribose) polymerase itself; FRG, a proteolytic fragment of this polymerase; HIS, core histones. The incorporation represents only the residual formation of poly(ADP-ribose) in the presence of the inhibitor thymidine (10 m*M*).

although pertussis toxin may safely be used after the cholera toxin reaction. Second, the guanine nucleotide concentration must be selected with care. Occupation of the target G_i by GTP readily blocks the ADP-ribosylation,[11,13–18] but some guanine nucleotide must be added to satisfy the ARF requirement. A compromise must be struck, and in most reported cases the optimum overall guanine nucleotide concentration is very low, for example, 1 μM GPP(NH)P,[12] or, in many cases "no added nucleotide." Perhaps the best maneuver is to employ GPP(CH_2)P, which activates ARF well[5] but does not inactivate the G_i substrate.[12]

ADP-Ribosylation of Non-G Proteins

Other target proteins are ADP-ribosylated considerably more slowly than G_s, but when they are much more abundant they can dominate the labeling pattern (Fig. 1b). ARF and a guanine nucleotide appear necessary for the ADP-ribosylation of all substrates (Fig. 1e, f).

A constant finding is an ADP-ribosylated protein of 22,000–25,000 M_r which is at least 10 times more abundant than G_s. In erythrocytes this appears to be labeled in the soluble phase and then sticks to the membranes. Insulin has been reported to inhibit the labeling of a 25-kDa protein.[19] Most other minor bands, particularly those that are smaller than 40,000 M_r, are less predictable and vary in labeling intensity according to conditions. A high-mobility group nuclear protein, approximately 28,000 M_r, is enhanced in its labeling by high salt or SDS. In erythrocyte membranes band III is generally ADP-ribosylated to a small extent (Fig. 1b). The A_1 fragment of cholera toxin itself sometimes becomes ADP-ribosylated, but it is not totally clear that this arises from a direct transfer of ADP-ribose rather than by a nonenzymatic addition reaction with free ADP-ribose (see below).

Distinguishing G Proteins from Non-G Proteins as Substrates

Criteria by which the various forms of [^{32}P]ADP-ribosyl G_s can be recognized against a background of other toxin-dependent [^{32}P]ADP-ribosylated proteins include the following: (a) Molecular size. (b) G_s is the

[14] J. R. Owens, L. T. Frame, M. Ui, and D. M. F. Cooper, *J. Biol. Chem.* **260,** 15946 (1985).

[15] C. B. Graves, N. B. Klaven, and J. M. McDonald, *Biochemistry* **22,** 6291 (1983).

[16] R. R. Aksamit, P. S. Backlund, and G. L. Cantoni, *Proc. Natl. Acad. Sci. U.S.A.* **82,** 7475 (1985).

[17] G. Milligan, *Biochim. Biophys. Acta* **929,** 197 (1987).

[18] M. Verghese, R. J. Uhing, and R. Snyderman, *Biochem. Biophys. Res. Commun.* **138,** 887 (1986).

[19] C. M. Heyworth, A. D. Whetton, S. Wong, R. B. Martin, and M. Houslay, *Biochem. J.* **228,** 593 (1985).

most readily labeled and so is the first protein saturated as the toxin or NAD concentration is increased. (c) The labeling of G_s is blocked more readily by pretreatment of the membranes with the protein modifying reagents diethyl pyrocarbonate[20] or 3-chloromercuri-2-methoxypropylurea.[5,21] (d) Just as the labeling of G_i is blocked by nucleotides (see above) so, in certain cases, the labeling of G_s is reduced by GTPγS and a G_s-linked hormone. It is known that dissociated α subunits are poor substrates and that GTPγS causes the G_s subunits to separate in detergent solution, but it is not totally clear that subunit dissociation accounts for the GTPγS inhibition of membranous G_s labeling. The alternative has not been excluded that cholera toxin may simply need to void the guanine nucleotide site and that GTPγS is too difficult to dislodge.[22] In the case of pigeon erythrocyte membranes, 1 m*M* GTPγS alone somewhat reduces the number of G_s molecules that can be ADP-ribosylated [i.e., the extent of labeling with GPP(NH)P plus GTPγS may be lower than with GPP(NH)P alone],[22] and the application of GTPγS with the β-adrenergic agonist isoproterenol can reduce the labeling of G_s by 90%.[20]

Ethylenediaminetetraacetic Acid

The binding of guanine nucleotides to G_s, which can be detrimental to the labeling, has a much more stringent magnesium requirement than does the binding to ARF.[6] Probably for this reason, labeling is sometimes increased by EDTA. Even very high concentrations of EDTA (e.g., 300 m*M*) can be tolerated. EDTA also affects poly(ADP-ribose) formation (see below). Certain background reactions which depend on free ADP-ribose may be increased by EDTA, which inhibits the tissue ADP-ribose phosphodiesterase (see below).

Salts

Any concentration of salt above isotonic inhibits the activation of ARF. Moreover, the ADP-ribosylation reaction itself is blocked if there is no salt. Except for chlorides, the reaction is remarkably stimulated by high ionic strengths: the rate is still rising at 1 *M* potassium phosphate or potassium EDTA: K > Na > NH_4 > Tris, multivalent anions > monovalent. The ionic strength affects the toxin itself, not the substrate.[23,24] Not-

[20] J. Coburn and D. M. Gill, unpublished (1990).

[21] D. M. Gill and M. Woolkalis, *Ciba Found. Symp.* **112,** 57 (1985).

[22] P. Boquet and D. M. Gill, *in* "Structure Regulation and Activity of Bacterial Toxins" (J. E. Alouf and J. H. Freer, eds.), in press. Academic Press, San Diego, 1990.

[23] J. Moss and M. Vaughan, *J. Biol. Chem.* **252,** 2455 (1977).

[24] G. Soman, K. B. Tomer, and D. J. Graves, *Anal. Biochem.* **134,** 101 (1983).

withstanding the previous point, chloride ions above about 100–200 mE inhibit the ADP-ribosylation: 1 *M* NaCl inhibits by 80–90%. In general, chloride ions inhibit the ADP-ribosylation of G_s more than the ADP-ribosylation of other substrates.

Problem of NAD Degradation

It is commonly difficult to ADP-ribosylate G_s in a membrane preparation that exhibits a high endogenous NADase (NAD^+ glycohydrolase) activity. NADase (NAD → nicotinamide + ADP-ribose + H^+) is an ectoenzyme of the plasma membrane, particularly of macrophages,[25] and is therefore very active in membranes made from lymphoid tissues, brain, and gut. In certain species (e.g., rabbits) it is also very active on erythrocytes.[26] There is not usually a significant NAD^+ pyrophosphatase activity (NAD → NMN + AMP) which can anyway be controlled by EDTA. NAD^+ pyrophosphatase is found on the surfaces of many bacteria.

To assess the degradation of NAD during the toxin incubation, analyze 1-μl portions of the final washings by TLC. A suitable system is cellulose TLC plates (Eastman Kodak, Rochester, NY) developed with 60% ethanol/0.3 *M* ammonium acetate buffer, pH 5.0. Typical R_f values are ATP, 0.08; NAD, 0.15; ADPR, 0.40; AMP, 0.49; orthophosphate, 0.60; and nicotinamide, 0.94. Avoid the use of polyethyleneimine (PEI) cellulose plates because traces of [^{32}P]ADP-ribose react with the PEI cellulose itself to generate a radioactive smear that cannot be quantitated and is easily overlooked. Expose to X-ray film overnight with an intensifying screen.

ADP-Ribosylation in the Face of High NADase Activity

A combination of the following adjustments may be necessary.

1. Reduce the membrane concentration.
2. Increase the NAD concentration.
3. Increase the toxin concentration.
4. Supply 20 m*M* isonicotinic acid hydrazide (INH, isoniazid) and 1 m*M* 3-acetylpyridine adenine dinucleotide (3-APAD) to inhibit the NADase. This combination is most successful against the NADases of birds and ruminants, which are particularly sensitive to INH and in which 3-APAD acts synergistically with INH to preserve NAD.[27] The combination is less effective against the INH-insensitive NADases, for which

[25] M. Artman and R. J. Seeley, *Arch. Biochem. Biophys.* **195,** 121 (1979).
[26] S. G. A. Alivasatos and O. F. Denstedt, *Can. J. Biochem. Physiol.* **34,** 46 (1956).
[27] D. M. Gill and J. Coburn, *Biochim. Biophys. Acta* **954,** 65 (1987).

higher concentrations of the inhibitors may be tested. 3-APAD is a partial substrate for cholera toxin (~1.5% as efficient as NAD), so for quantitative work it is necessary to allow for the accumulation of nonradioactive ADP-ribosyl G_s from the 3-APAD.[27] The 3-APAD should be added at the same time as the NAD. INH-adenine dinucleotide (INHAD), which is a side product of vertebrate NADases but is not a good substrate for the toxins, runs with NAD on TLC unless benzaldehyde is added to form the hydrazone.[27] Use 50% ethanol, 0.3 *M* ammonium acetate buffer, pH 5.0, 10% benzaldehyde. Typical R_f values on cellulose plates are NAD, 0.41; ADPR, 0.49; orthophosphate, 0.63; INHAD, 0.72.

5. Supply 10 m*M* dithiothreitol as a further inhibitor of INH-sensitive and INH-insensitive NADases.
6. ADP-ribose, 20 m*M* is occasionally beneficial. It inhibits NADase but is also an inhibitor of the toxin reaction.
7. Use a two-step incubation so that ARF is already activated by the time that the NAD is supplied.

Despite all of these precautions we are still unable to control the NADase and efficiently ADP-ribosylate the membranes of rabbit gut.

Example: ADP-Ribosylation of G_s in Calf Brain Membranes

The following allows complete ADP-ribosylation with calf brain membranes preparations which have an active INH-sensitive NADase. As with all brain membrane preparations, these contain many closed vesicles, and a nonionic detergent must be supplied to allow access. Since the detergent tends to inactivate soluble ARF, the latter is generally omitted. There is much ARF activity in the brain membranes.

Suspend the membranes in buffered saline. Preincubate at 25° for 10 min with 0.2% Triton X-100 and 1 m*M* GPP(NH)P. To 10 μl of this membrane suspension (100 μg protein), immediately add 10 μl of a second mixture and incubate for another 60 min at 25°. The final concentrations are 20 μg/ml activated toxin, 20 μM [^{32}P]NAD (5000–15,000 cpm/pmol), 20 m*M* INH, 1 m*M* 3-APAD, 10 m*M* thymidine, 10 m*M* dithiothreitol, 0.1% Triton X-100, and 0.5 m*M* GPP(NH)P (Fig. 1c, d).

Comparison with Pertussis Toxin-Catalyzed ADP-Ribosylation of G_i in Calf Brain Membranes

The reaction conditions used for pertussis toxin are much simpler than those needed for cholera toxin (see [22],this volume). No ARF is involved so only one step has to be considered. ATP or an ATP analog is required, but not GTP.[28] Guanine nucleotides are best omitted for they inactivate the substrate.[29] Conversely, EDTA is beneficial, possibly because it maintains G_i in the toxin-reactive, but functionally inactive, configuration. The devices for limiting NADase action are the same as those developed for cholera toxin. The following concentrations seem to allow the total ADP-ribosylation of G_i in calf brain.

Incubate washed membranes (5 mg/ml of protein) with 10 μg/ml of activated pertussis toxin (preincubated at 37°, 30 min with 100 mM dithiothreitol; 0.1% SDS may also help), additional dithiothreitol for a final concentration of 10 mM, 100 μM [^{32}P]NAD (1000–5000 cpm/pmol), 50 mM EDTA, 20 mM INH, 1 mM 3-APAD, 1 mM ATPγS, 0.1% Triton X-100. After 30 min at 37°, process as for G_s. Correct the observed [^{32}P]ADP-ribose incorporation for additional nonradioactive ADP-ribosylation from the 3-APAD which is about 6% as efficient a substrate as is NAD.[27]

Other Competing Reactions and Artifactual Results

Nonenzymatic Addition of Free ADP-Ribose to Proteins and Other Macromolecules[30]

[^{32}P]ADP-ribose may be present in the starting [^{32}P]NAD sample (check by cellulose TLC), and more may be generated from the NAD by tissue NADase (plasma membrane) or by the combined action of poly(ADP-ribose) polymerase (nuclear) and poly(ADP-ribose) glycohydrolase (soluble) or by the phosphate-enhanced NAD^+ glycohydrolase activity of the cholera toxin itself. ADP-ribose reacts with primary amines; in principle it can react with any protein, and, although different proteins vary in their reaction rate, the reaction is most obvious for abundant proteins. In several tissues the main target is glyceraldehyde-3-phosphate dehydrogenase (subunit M_r 36,000, apparent product M_r ~37,000).[20] The labeling of globin and its multimers in lysed erythrocytes is shown in Fig. 1g. The nonenzymatic reaction is largely reversible and in our hands comes to equilibrium after about 24 hr at 37° with 1–2% of the ADP-ribose attached to protein. It cannot be blocked by adding an excess of unlabeled ADP-ribose. It is recognizable by the following criteria: (a) it continues for many hours, (b) it

[28] L-K. Lim, R. D. Sekura, and H. R. Kaslow, *J. Biol. Chem.* **260,** 2585 (1985).
[29] S. K. F. Wong, B. R. Martin, and A. M. Tolkovsky, *Biochem. J.* **232,** 191 (1985).
[30] E. Kun, A. C. Y. Chang, M. L. Sharma, A. M. Ferro, and D. Nitecki, *Proc. Natl. Acad. Sci. U.S.A.* **73,** 3131 (1976).

increases with increasing temperature whereas the reaction of cholera toxin is faster at 25° than at 37°, (c) it is not prevented by digesting the [^{32}P]NAD to [^{32}P]ADP-ribose, and (d) it may be reduced by tissue samples that contain the enzyme ADP-ribose phosphodiesterase (ADPR → AMP + ribose phosphate) providing magnesium is present in excess over EDTA and nucleotides.

To prepare [^{32}P]ADP-ribose, weigh 2 mg of pig brain acetone powder (sold by Sigma as NADase) into an Eppendorf tube. Soak it in water and wash several times by centrifugation. Discard any granules that float. Add 100 μl of 50 μM [^{32}P]NAD in 2 mM HEPES buffer, pH 7.3. Incubate at 37° with constant agitation for 10 min. Boil 2 min. Centrifuge. A portion should be analyzed by TLC (cellulose, not PEI) to ensure that no NAD remains and that the breakdown to [^{32}P]AMP and ortho[^{32}P]phosphate is modest. Other available mammalian NADases such as the one from bovine spleen may also be used, but the material sold as NADase from *Neurospora crassa* is very crude and should be avoided.

Generation of [^{32}P]tRNA from [^{32}P]ADP-Ribose

Most tissues contain ADP-ribose phosphodiesterase which rapidly converts [^{32}P]ADP-ribose to [^{32}P]AMP when magnesium is present.[31] The AMP may be dephosphorylated to yield ortho[^{32}P]phosphate or converted to [α-^{32}P]ATP at the expense of another nucleoside triphosphate. In the latter case, the radioactivity is not conjugated to proteins but is readily incorporated into the 3′ CCA terminal positions of tRNA molecules that have lost their final adenosine residue. During gel electrophoresis, the mixture of end-labeled [^{32}P]tRNA species gives a cluster of radioactive bands in the 21,000 M_r region and another cluster in the 12,000 M_r region which represents half-molecules of tRNA split in their anticodon loops (Fig. 1h).

Such incorporation has the following properties: (a) It depends on a nucleoside triphosphate such as GTP or UTP or, best of all, CTP. ATP will appear to be less effective since it reduces the specific activity of the product. Nonhydrolyzable nucleoside triphosphates do not work. (b) It is blocked by EDTA which inhibits ADP-ribose phosphodiesterase. (c) Depending on the mechanism of producing the [^{32}P]ADP-ribose, the incorporation is blocked by inhibitors of poly(ADP-ribose) synthesis or of NADases. (d) The product is alkali labile.

[31] J. M. Wu, M. B. Lennon, and R. J. Suhadolnik, *Biochim. Biophys. Acta* **520,** 588 (1978).

ADP-Ribosylation of Acrylamide Gels

If the incubated membranes are not washed well, any residual [^{32}P]ADP-ribose reacts with the polyacrylamide itself during gel electrophoresis and cannot be removed.[20] Over 0.1% of the input counts may attach to the gel. This effect generates a radioactive smear which is most intense toward the bottom of the gel where the migration of the [^{32}P]ADP-ribose is slowest (Fig. 1i). The smear is absent when the tissue samples contain ADP-ribose phosphodiesterase and magnesium ions.

ADP-Ribosylation Catalyzed by Endogenous Mono(ADP-ribosyl) Transferases

There are many descriptions of tissue enzymes that catalyze nonspecific ADP-ribosylations of proteins in broken cells.[32–34] These enzymes are particularly active in frog and turkey erythrocyte lysates, where they can mask toxin-dependent labeling. While it is evident that these enzymes must not be as active in the intact cell as they are *in vitro*, neither the natural inhibitory mechanism nor a sufficiently selective artificial inhibitor has been identified. The best cure is to purify the target membranes further.

Poly(ADP-ribosylation)

Membrane preparations usually contain nuclear fragments that contain both poly(ADP-ribose) polymerase and the broken DNA that this enzyme requires. It is advisable to add thymidine to inhibit this polymerase. Even so, there remains some ADP-ribose incorporation at 115,000 M_r (this is the polymerase itself), at various places between 70,000 and 90,000 M_r, which represent proteolytic fragments of the polymerase, and often also in the region below 20,000 apparent M_r, which represents core histones ADP-ribosylated as a result of poly(ADP-ribose) formation (Fig. 1j). If further inhibition is needed, supply 10 m*M* 3-aminobenzamide and 10 m*M* EDTA or 100 m*M* phosphate. EDTA both inhibits polymer formation and stimulates the activity of poly(ADP-ribose) glycohydrolase. In a two-step incubation, include EDTA only in the second step. Alternatively, extract the ADP-ribosylated products with a detergent that dissolves G_s but not the products of poly(ADP-ribose) polymerase. For example, use 0.5–0.7% Lubrol PX, 15 min at room temperature or 60 min on ice. For more complete extraction also add 0.15 *M* NaCl and 0.15% cetyltrimethylammonium bromide.

[32] D. A. Yost and J. Moss, *J. Biol. Chem.* **258,** 4926 (1983).

[33] Y. Tanigawa, M. Tsuchiya, Y. Imai, and M. Shimoyama, *J. Biol. Chem.* **259,** 2022 (1984).

[34] G. Soman and D. J. Graves, *Arch. Biochem. Biophys.* **260,** 56 (1988).

It has been common to use 3-aminobenzamide inhibition as an indicator of poly(ADP-ribosylation), but this reagent also inhibits the cellular mono(ADP-ribosyl) transferases. It is by far preferable to examine the product on a two-dimensional gel. A poly(ADP-ribosylated) product gives a chain of spots, rising slightly in size as they get more acidic. A substrate bound to a single ADP-ribose unit gives a single spot one charge unit more acidic than the original protein.

Cholera Toxin Activity Assays

The most sensitive assay of cholera toxin activity involves measuring the rise in adenylyl cyclase activity as a consequence of ADP-ribosylation. Freshly lysed pigeon erythrocytes are very sensitive and have a low basal cyclase activity. Wash the erythrocytes to remove serum and leukocytes, then suspend packed cells in 1 volume of HEPES-buffered saline using a glass tube. Freeze the sample in an ethanol–dry ice bath, then thaw in room temperature water. Supplement the lysate with 5 m*M* NAD and 5 m*M* GTP. Incubate 40-μl portions with activated toxin for 30 min at 25°. Wash the membranes in 1 ml of saline and measure the adenylyl cyclase activity, with GTP, by any standard method (see [1], this volume). The adenylyl cyclase activity is approximately proportional to the logarithm of the toxin concentration between 1 and 100 ng/ml.

A less sensitive but more rapid method involves trichloroacetic acid (TCA) precipitation of all of the ADP-ribosylated proteins that are formed in brain particles. We use the total microsomal fraction from bovine cortex which is preincubated with Triton X-100, to make the vesicles permeable, and GPP(NH)P. To packed brain membranes add 2 volumes of saline, giving a 1 in 3 slurry, Triton X-100 to 0.1%, and GPP(NH)P to 0.1 m*M*. Incubate 10 min at 25°. Return to ice and add additional ingredients to give a final 1 in 8 suspension and the following final concentrations: 1 m*M* 3-aminopyridine adenine dinucleotide (neutralized), 20 m*M* INH, 10 m*M* thymidine, 10 m*M* dithiothreitol, 10 m*M* EDTA, and lastly 1 *M* potassium phosphate, pH 7.6. The final pH is about 7.1. Mix well. Then mix in 1 μM [^{32}P]NAD (20,000–50,000 cpm/pmol) and immediately distribute 40-μl portions into chilled glass tubes.

Meanwhile, activate the toxin samples. For example, incubate 10 μl of toxin (~1 mg/ml) with 20 μl of activating mixture (0.5% SDS, 5 m*M* dithiothreitol, 1 mg/ml ovalbumin, in buffered saline) for 10 min at 37°, then dilute with 70 μl of buffered saline containing 0.1% ovalbumin. Prepare dilutions with 0.1% SDS/1 m*M* dithiothreitol/0.1% ovalbumin in the diluent.

To each membrane sample add 10 μl of activated toxin, or 10 μl of the toxin dilution buffer for the blank, and vortex immediately. Incubate at 25° for 60 min. Dilute with about 200 μl of ice-cold saline, add ice-cold 5% TCA, and vortex quickly to avoid the formation of protein lumps. Let this sit 10 min on ice and collect the protein by filtration on glass fiber circles. Wash the filter with 2.5% TCA. At 10 μg/ml, cholera toxin gives an incorporation of 0.2 pmol ADP-ribose, which is about 10 times the background level.

Acknowledgments

This work was supported by National Institutes of Health Grant AI 16928 and by National Institute of Diabetes, Digestive, and Kidney Diseases Grant AM 39428 to the Center for Gastrointestinal Research on Absorptive and Secretory Processes.

[24] Photoaffinity Labeling of GTP-Binding Proteins

By ROLF THOMAS and THOMAS PFEUFFER

Introduction

Purification of proteins, especially those which occur in low abundance, is greatly facilitated by identification methods which do not exclusively rely on enzymatic activity or ligand binding. Since it is desirable to monitor a protein under strongly denaturing conditions as well, for example, by SDS-gel electrophoresis,[1] only covalent modifications of the protein will be acceptable. The relevant method, affinity labeling, makes use of the fact that total binding free energy is usually sufficiently high to enable limited changes in the structure of a ligand to allow insertions of a protein-reactive chemical group without sacrificing the specificity and affinity of the native ligand. Among affinity labels, photogenerated reagents offer the advantage that an appropriate amino acid side chain is not

[1] MOPS, 3-(*N*-Morpholino)propane sulfonate; 4-azidoanilido-GTP, P^3-(4-azidoanilido)-P^1-guanosine triphosphate; GPP(NH)P, guanosine 5′-(β,γ-imido)triphosphate; GTPγS, guanosine 5′-(γ-thio)triphosphate; DEAE, diethylaminoethyl; TLC, thin-layer chromatography; G proteins, GTP-binding proteins; G_s, guanine nucleotide-binding regulatory component responsible for stimulation of adenylyl cyclase; G_i, guanine nucleotide-binding regulatory component responsible for inhibition of adenylyl cyclase; SDS, sodium dodecyl sulfate.

a stringent requirement, since the reactive species interacts with essentially any chemical group present at the binding site. Furthermore the reactive species, for instance, a nitrene or a carbene, can be generated at a time determined by the experimental design. Therefore, excessive reagent may be removed before photolysis (provided the affinity is high enough), thus minimizing nonspecific labeling.

Photoreactive GTP derivatives are of particular interest in light of the still growing family of GTP-binding proteins (G proteins) which serve as transducers of cellular signals. Most of these G proteins are membrane bound and coupled to receptors for hormones or growth factors. The most thoroughly investigated members among these are the heterotrimeric G proteins (α, β, γ), G_s and G_i, the stimulatory and inhibitory G proteins of adenylyl cyclase (EC 4.6.1.1, adenylate cyclase), respectively, and G_o (of hitherto unknown function). But many G proteins, for example, the *ras* proteins of yeast, the mammalian *ras* (N-*ras*, Ki-*ras*, and H-*ras*) proteins, other members of the p21 family, like ADP-ribosylating factors, and other "small" G proteins are seemingly not associated with β,γ subunits. Many if not all of these G proteins display GTPase activity which is exclusively located on the α moiety.[2]

In the course of a systematic screening of the GTP molecule we found[3] that removal of one negative charge of the terminal phosphate does not abolish binding or activation of the G_s protein. Although replacement of OH by F results in a less potent analog, more hydrophobic groups such as phenyl or naphthyl groups cause increased affinity and potency to activate adenylyl cyclase. Indeed, γ-phenyl esters and γ-anilidates of GTP stimulated adenylyl cyclase up to 60% of the activity displayed by the commonly used analog GPP(NH)P, with an affinity only 2–3 times lower. It is interesting to note that the order of potency for activating adenylyl cyclase, namely, GTPγS > GPP(NH)P > GTPγ esters > GTPγF, is considerably different for the bacterial elongation factor G (EF G). For this GTP-binding protein, GTPγF is much more potent than GTPγ esters or GTPγS.[4] However the same order exists for adenylyl cyclase and retinal phosphodiesterase.[5] This underscores the fact that the GTP-binding proteins G_s and transducin belong to a common family of coupling proteins. In analogy with GPP(NH)P, or GTPγS, the activation by the γ-phosphate esters or amidates could be shown to be quasi-irreversible, that is, resistant to repeated wash steps to remove nonbound analog. γ-Amidates are espe-

[2] A. G. Gilman, *Annu. Rev. Biochem.* **56,** 615 (1987).
[3] T. Pfeuffer and F. Eckstein, *FEBS Lett.* **67,** 354 (1976).
[4] F. Eckstein, W. Bruns, and A. Parmeggiani, *Biochemistry* **14,** 5225 (1975).
[5] G. Yamanaka, F. Eckstein, and L. Stryer, *Biochemistry* **25,** 6149 (1986).

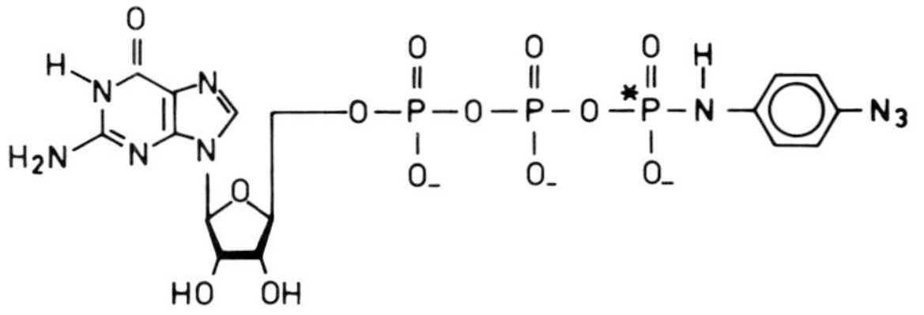

FIG. 1. 4-Azidoanilido-GTP.

cially easy to prepare in one step from GTP in aqueous solution through the use of condensing agents like water-soluble carbodiimides.

Preparation of P^3-(4-Azidoanilido)-P^1-5′-guanosine Triphosphate[6,7]

Guanosine triphosphate tetrasodium salt (0.1 mmol) and *N*-ethyl-*N*′-(3-dimethylaminopropyl)carbodiimide hydrochloride (0.4 mmol) are dissolved in 5 ml of 0.1 *M* triethanolamine buffer, pH 7.2, and mixed with 4-azidoaniline (0.8 mmol) in 2 ml of dioxane. 4-Azidoaniline is prepared according to Silberrad and Smart,[8] but it is now commercially available from Fluka (Buchs, Switzerland). Unreacted 4-azidoaniline is extracted with 3 portions of 10 ml each of peroxide-free diethyl ether. The water phase is applied to a DEAE-cellulose (DE-52, Whatman, Maidstone, UK) column (2.5 × 30 cm) in the carbonate form. Azidoanilido-GTP is eluted with a linear gradient (2 liters) from 0 to 0.5 *M* triethylammonium bicarbonate, pH 7.5, at 4°. The fractions, monitored by TLC under UV light (see below), containing 4-azidoanilido-GTP (see Fig. 1), are concentrated at reduced pressure, taken up in distilled water and stored as a concentrated solution (20 m*M*) at −20°. The yield is 60% based on GTP ($\varepsilon_{260\ nm}$ 30,600 M^{-1} cm^{-1} in phosphate buffer, pH 7.0). All operations are carried out in the dark.

[γ-^{32}P]GTP (28.1 TBq/mmol), 1 mCi, is dried down under reduced pressure and dissolved in 20 μl of 1 m*M* GTP in 0.2 *M* triethanolamine buffer, pH 7.2. This is followed by addition of 20 μl of 50 m*M* 4-azidoaniline hydrochloride (Fluka) and 40 μl of 25 m*M* NaOH. The reaction is started by addition of 2.5 μl of 400 m*M* *N*-ethyl-*N*′-(3-dimethylaminopropyl)carbodiimide hydrochloride in water. After 6 hr at 22° another 2.5 μl of the same solution is added and the reaction mixture left for a further 12 hr at 22°. The reaction is stopped by extraction of unreacted azidoaniline (3

[6] R. Thomas, Doctoral Thesis, University of Würzburg, Würzburg, West Germany, 1976.

[7] T. Pfeuffer, *J. Biol. Chem.* **252,** 7224 (1977).

[8] O. Silberrad and F. J. Smart, *J. Chem. Soc.* **89,** 170 (1906).

times with 200 μl each of water-saturated ethyl acetate). The remaining water phase is spotted onto two 20 × 20 cm polyethyleneimine (PEI)-cellulose thin-layer sheets, which are developed in 1 *M* LiCl. Following localization under UV light, the 4-azidoanilido-GTP-containing band is scraped off, eluted with 2 ml of 0.5 *M* NaCl, and the cellulose filtered off. Alternatively, 0.5 *M* triethylammonium bicarbonate, pH 7.5, is used for extraction. The buffer is removed by evaporation under reduced pressure, and the labeled nucleotide is dissolved in 1 ml of distilled water. The labeled nucleotide is stored in portions at −80°. (See also [25], this volume.)

Chemical and Physical Properties of the Affinity Label

The synthesized compound was identified as P^3-(4-azidoanilido)-P^1-guanosine triphosphate (4-azidoanilido-GTP) by the following criteria: it was photoreactive. On chromatography on PEI-cellulose and exposure to UV light (252 nm) it formed a dark spot which did not move on rechromatography. Furthermore, its IR spectrum revealed a strong, sharp band at 2150 cm^{-1} due to the —N_3 moiety.[9] The occurrence of the γ-^{32}P in the final product and its mobility similar to GDP during TLC (R_f for 4-azidoanilido-GTP, 0.5; R_f for GDP, 0.46) suggest that the product contains three phosphates and three negative charges. It was not degraded by alkaline phosphatase but yielded 5′-GMP on treatment with snake venom phosphodiesterase. When kept frozen (−20°) and protected from light the photoreactive GTP derivative was stable for many years.

Qualitatively, the 4-azidoanilido-GTP had properties comparable to other poorly hydrolyzed or nonhydrolyzable GTP analogs. Together with hormones it synergistically activated adenylyl cyclase as GPP(NH)P or GTPγS in a quasi-irreversible fashion. Binding to bulk GTP-binding proteins in pigeon erythrocyte membranes or rabbit myocardial membranes was likewise resistant to wash procedures. An apparent K_a for activation of pigeon erythrocyte adenylyl cyclase of 0.3–0.6 μ*M* and a K_D of 0.3 μ*M* for binding of 4-azidoanilido[^{32}P]GTP to pigeon erythrocyte membranes have been determined.[3,6,7] Similar affinities have been obtained for [^3H]GPP(NH)P.

Labeling of GTP-Binding Proteins in Avian Erythrocyte Membranes

Membranes (2–10 mg/ml) in buffer A (20 m*M* sodium phosphate, pH 7.4, 0.1 *M* NaCl, 1 m*M* EDTA, 10 m*M* $MgCl_2$, 10 μ*M* phenylmethylsulfonyl fluoride) are incubated with 0.1–0.5 μ*M* 4-azidoanilido[^{32}P]GTP (0.2

[9] L. J. Bellamy, "The Infrared Spectra of Complex Molecules," p. 273. Wiley, New York, 1960.

to 1.8 TBq/mmol) and 50 μM DL-isoproterenol in the presence and absence of 0.1 mM GPP(NH)P for 20 min at 37°. It is important to note that sulfhydryl compounds must be avoided because of interference with the azido moiety of the affinity label. Membranes are washed 3 times with 5 volumes of buffer A, adjusted to the original volume, and irradiated at 4° in quartz cuvettes (0.5 to 2 cm light path) under gentle stirring with a 200 W mercury lamp equipped with a cut-off filter at 345 nm for 40 min. Alternatively, samples are irradiated on ice at a distance of 5 cm for 20 min using a UV hand lamp (254 nm) (Hanau, type 406 AC, Hanau, Germany). This version allows the irradiation of several samples at the same time.

For monitoring covalent incorporation, 5- to 20-μl samples of the irradiated mixture are spotted onto Whatman GF/A glass-fiber disks, immersed in and washed with ice-cold 10% trichloroacetic acid. Following washing with ethanol and drying, filters are counted in a liquid scintillation counter. For electrophoretic evaluation of labeled GTP-binding proteins, membranes (250 μg) are solubilized in sample buffer by boiling for 5 min and subjected to SDS–polyacrylamide gel electrophoresis according to Laemmli.[10] Gels are dried and autoradiographed for varying times with Kodak XAR 5 film and a Siemens Titan 2HS intensifying screen. Protein standards are ^{125}I-labeled according to Ref. 11.

Solubilized membranes or soluble purified preparations are incubated with azidoanilido[^{32}P]GTP as described before, but in the absence of DL-isoproterenol. Excessive label is removed before irradiation by gel filtration on Sephadex G-25 (fine) in 10 mM MOPS buffer, pH 7.4, 100 mM NaCl, 1 mM Lubrol PX or Tween 60, with a centrifuge column procedure.[12]

Discussion

Results of an affinity labeling experiment with 4-azidoanilido[^{32}P]GTP and turkey erythrocyte membranes are shown in Fig. 2. The autoradiograph of a 12% SDS–polyacrylamide gel reveals three groups of specifically labeled proteins (lane A): a broad band at 86K, a doublet at 42K/40K, and a broad band at 25K which is sometimes resolved into a triplet. Note that virtually no labeling is observed when the binding of the 4-azidoanilido-GTP is performed in the presence of excessive GPP(NH)P (lane B). The 42K/40K doublet is most likely due to the labeling of the G_s and G_i proteins, since proteins of the same mobility can be immunostained by specific antibodies as demonstrated elsewhere in this volume (see [14]).

[10] U. K. Laemmli, *Nature (London)* **227,** 680 (1970).

[11] F. C. Greenwood, W. M. Hunter, and J. S. Glover, *Biochem. J.* **89,** 114 (1963).

[12] H. S. Penefsky, *J. Biol. Chem.* **252,** 2891 (1977).

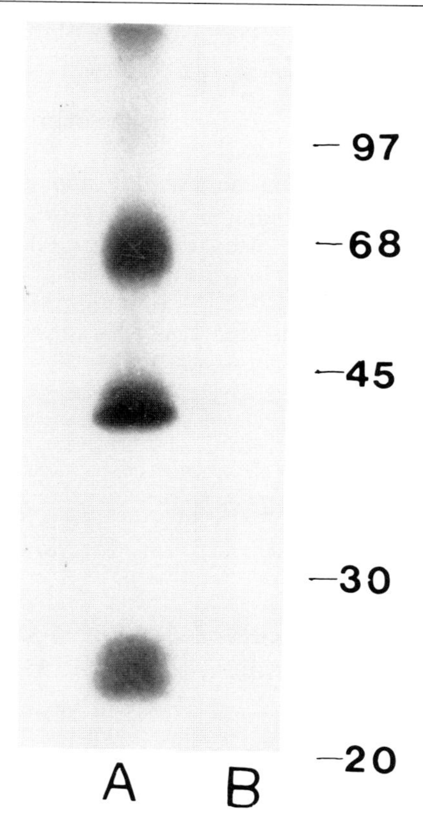

FIG. 2. Photoaffinity labeling of turkey erythrocyte membranes with 4-azidoanilido[^{32}P]-GTP. Turkey erythrocyte membranes (2 mg/ml in buffer A) were incubated with 0.5 μM 4-azidoanilido[^{32}P]GTP and 50 μM DL-isoproterenol in the absence (lane A) or presence (lane B) of 10^{-4} M GPP(NH)P, freed from nonbound affinity label, and photolyzed as described before. One hundred micrograms of membrane protein was subjected to electrophoresis on a 12% SDS-polyacrylamide gel. Bands were visualized by autoradiography (4 hr at $-80°$). Numbers refer to M_r values ($\times 10^{-3}$) of standard proteins: phosphorylase, 97K; bovine serum albumin, 68K; ovalbumin, 45K; carbonate dehydratase, 30K; soybean trypsin inhibitor, 21K.

The more strongly labeled 40K protein is in agreement with the excess of G_i over G_s. The two groups of GTP-binding proteins with an approximate M_r values of 40K and 25K can be observed in a variety of tissues, for example, brain, liver, and platelets (data not shown).

When compared with other protein-reactive GTP derivatives, for instance, periodate-oxidized GPP(NH)P[13] or 2-azidobenzoyl-GPP(NH)P,[7]

[13] T. Pfeuffer, unpublished observations (1976).

azidoanilido-GTP proved to be superior because of its higher affinity (1–2 orders of magnitude). This also seems to apply to the commercially available derivative 8-azido[^{32}P]GTP, which has been used to label pure $G_{s\alpha}$ and $G_{i\alpha}$,[14] but should be of no utility for labeling of crude systems because of interfering hydrolyzing activities.

4-Azidoanilido-GTP has been successfully applied to the labeling of $G_{s\alpha}$ in liver membranes[15] and labeling of $G_{s\alpha}$ and $G_{i\alpha}$ in rat synaptosomes.[16] Schäfer *et al.*[17] demonstrated cholecystokinin (CCK)-dependent incorporation of 4-azidoanilido[^{32}P]GTP into a 40K protein of rat pancreatic acinar membranes. Since CCK receptors (or a population of them) seem to be coupled to phospholipase C, the 40K protein could be a candidate for an α subunit of a G_p-type coupling protein.

[14] J. K. Northup, D. M. Smigel, and A. G. Gilman, *J. Biol. Chem.* **257,** 11416 (1982).
[15] S. K.-F. Wong and B. R. Martin, *Biochem. J.* **231,** 39 (1985).
[16] S. Hatta, M. M. Marcus, and M. Rasenick, *Proc. Natl. Acad. Sci. U.S.A.* **83,** 5439 (1986).
[17] R. Schäfer, A.-L. Christian, and I. Schulz, *Biochem. Biophys. Res. Commun.* **155,** 1051 (1988).

[25] Identification of Receptor-Activated G Proteins with Photoreactive GTP Analog, [α-^{32}P]GTP Azidoanilide

By STEFAN OFFERMANNS, GÜNTER SCHULTZ, and WALTER ROSENTHAL

Introduction

Many extracellular signals activate receptors coupled to heterotrimeric guanine nucleotide-binding regulatory proteins (G proteins)[1,2] attached to the inner face of the plasma membrane. Whereas 100 to 200 different receptors acting via G proteins appear to exist, the number of known G proteins is limited (<20). Interaction of receptors with G proteins is frequently demonstrated in membrane preparations by studying the effects of receptor agonists on the activity of high-affinity GTPase, the enzymatic activity ascribed to G-protein α subunits.[3] The interaction of receptors with G proteins in membrane preparations or reconstituted systems is also evident from the stimulatory effects of agonists on binding of GTP or its

[1] M. Freissmuth, P. J. Casey, and A. G. Gilman, *FASEB J.* **3,** 2125 (1989).
[2] L. Birnbaumer, A. Yatani, J. Codina, A. VanDongen, R. Graf, R. Mattera, J. Sanford, and A. M. Brown, *in* "Molecular Mechanisms of Hormone Action" (U. Gehring, E. Helmreich, and G. Schultz, eds.), p. 147. Springer-Verlag, Berlin, Heidelberg, 1989.
[3] D. Cassel and Z. Selinger, *Biochim. Biophys. Acta* **452,** 538 (1976).

analogs[4] or on the release of GDP.[5,6] In addition, binding of agonists to G-protein-coupled receptors is sensitive to guanine nucleotides.[7] These methods, however, usually do not allow differentiation between individual G proteins.

Photoreactive nucleotides have been widely applied for characterization of nucleotide-binding proteins.[8] Pfeuffer[9] introduced the GTP analog,

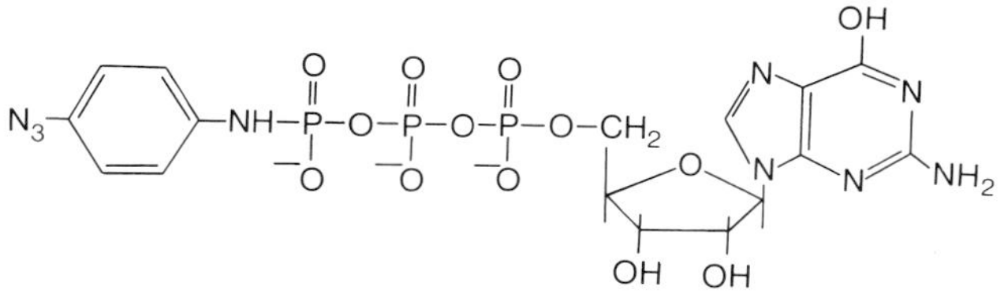

GTP azidoanilide

P[3]-(4-azidoanilido)-*P*[1]-5′-guanosine triphosphate (GTP azidoanilide) and used [γ-^{32}P]GTP azidoanilide for identification of the stimulatory G protein of adenylyl cyclase, G_s. Subsequently, GTP azidoanilide, ^{32}P-labeled in the α or γ position, has been applied to identify and to characterize various G proteins including G_s, G_i-type G proteins, G_o, and retinal G proteins (transducins).[10–13] GTP azidoanilide appears to be resistant to hydrolysis and causes a slow, persistent activation of G proteins, similar to that induced by other poorly hydrolyzable GTP analogs, namely, guanosine 5′-*O*-(3-thiotriphosphate) and 5′-guanylyl imidodiphosphate. Three groups reported that binding of [^{32}P]GTP azidoanilide to G-protein α subunits is promoted by activated receptors.[12,14–16]

[4] V. A. Florio and P. C. Sternweis, *J. Biol. Chem.* **264,** 3909 (1989).
[5] D. Cassel and Z. Selinger, *Proc. Natl. Acad. Sci. U.S.A.* **75,** 4155 (1978).
[6] T. Murayama and M. Ui, *J. Biol. Chem.* **259,** 761 (1984).
[7] M. Rodbell, *Nature (London)* **284,** 17 (1980).
[8] For general aspects of synthesis and application of photoreactive compounds, see Vol. 46, this series.
[9] T. Pfeuffer, *J. Biol. Chem.* **252,** 7224 (1977).
[10] S. K.-F. Wong and B. R. Martin, *Biochem. J.* **231,** 39 (1985).
[11] S. Hatta, M. M. Marcus, and M. M. Rasenick, *Proc. Natl. Acad. Sci. U.S.A.* **83,** 5439 (1986).
[12] O. Devary, O. Heichal, A. Blumenfeld, D. Cassel, E. Suss, S. Barash, C. T. Rubinstein, B. Minke, and Z. Selinger, *Proc. Natl. Acad. Sci. U.S.A.* **84,** 6939 (1987).
[13] J. H. Gordon and M. M. Rasenick, *FEBS Lett.* **235,** 201 (1988).
[14] R. Schäfer, A.-L. Christian, and I. Schulz, *Biochem. Biophys. Res. Commun.* **155,** 1051 (1988).
[15] S. Offermanns, R. Schäfer, B. Hoffmann, E. Bombien, K. Spicher, K.-D. Hinsch, G. Schultz, and W. Rosenthal, *FEBS Lett.* **260,** 14 (1990).

Here we describe the synthesis and purification of [α-^{32}P]GTP azidoanilide as well as a procedure suitable to study the stimulatory effects of receptor agonists on photolabeling of G-protein α subunits in membrane preparations. Since the mobilities of G-protein α-subunits differ in sodium dodecyl sulfate (SDS)–polyacrylamide gels, this method allows the identification of G proteins activated by a given receptor. (See also [24], this volume.)

Materials

Synthesis and Purification of [α-^{32}P]GTP Azidoanilide

Ortho[^{32}P]phosphoric acid (carrier-free) was from Du Pont–New England Nuclear (Dreieich, FRG). [α-^{32}P]GTP is synthesized according to Johnson and Walseth[17] (see [3] this volume) or purchased from Du Pont–New England Nuclear (specific activity 800–3000 Ci/mol; see Comments). 4-Azidoanilide was from Serva (Heidelberg, FRG), and *N*-(3-dimethylaminopropyl)-*N'*-ethylcarbodiimide hydrochloride was from Merck (Darmstadt, FRG). Triethylamine, purchased from Merck, is distilled prior to use. 1,4-Dioxane was purchased from Riedel de Haen (Seelze, FRG); peroxide-free dioxane is prepared by chromatography on dry neutral alumina (Merck). ODS Hypersil (particle size 5 μm; Shandon, Runcorn, UK) is filled into high-performance liquid chromatography (HPLC) columns (0.45 $\times$ 25 cm). Commercially available packed C_{18} columns are also suitable.[18] GTP was from Boehringer Mannheim (Mannheim, FRG). 2-(*N*-Morpholino)ethanesulfonic acid was from Serva.

Photolabeling of G-Protein α Subunits

Membranes from various cell lines are prepared by nitrogen cavitation and stored at $-70°$ in 10 m*M* triethanolamine (pH 7.0). G proteins are purified as previously described.[19,20] Nucleotides (GDP, ATP, GTP) were from Boehringer Mannheim. Urea was obtained from Bio-Rad (Munich, FRG).

[16] S. Offermanns, A. Schmidt, G. Schultz, and W. Rosenthal, *Naunyn-Schmiedebergs Arch. Pharmacol.* **340,** R 82 (1989).

[17] R. A. Johnson and T. F. Walseth, *Adv. Cyclic Nucleotide Res.* **10,** 135 (1979).

[18] C. W. Mahoney and R. G. Yount, *Anal. Biochem.* **138,** 246 (1984).

[19] J. Codina, W. Rosenthal, J. D. Hildebrandt, L. Birnbaumer, and R. D. Sekura, this series, Vol. 109, p. 446.

[20] W. Rosenthal, D. Koesling, U. Rudolph, C. Kleuss, M. Pallast, M. Yajima, and G. Schultz, *Eur. J. Biochem.* **158,** 255 (1986).

Required Equipment

Synthesis and Purification of [α-^{32}P]GTP Azidoanilide

1. Facilities to handle millicurie amounts of ^{32}P: Plexiglass (Lucite) shields, lead bricks, radiation hand monitor.

2. Facilities for protecting photoreactive compounds from direct light. For longer periods of time, these compounds have to be kept in the dark (e.g., in vessels wrapped with aluminum foil).

3. Small (~5 ml) silanized flask with a cone-shaped bottom (reaction vessel) which fits to a rotary system (e.g., a rotary evaporator).

4. HPLC system and fraction collector, preferably mounted in a cold room or refrigerator with a glass door. Since the solvents used are permanently gassed with CO_2, the HPLC system has to be equipped with two pumps (mixing of solvents on the high-pressure side of the pumps). For the same reason, traps for air bubbles have to be installed at the inlet of the pumps. The HPLC column (0.45 × 25 cm) has to be shielded by a plexiglass cylinder (wall diameter of at least 2 cm) surrounded by sheet lead (~2 mm thick). Other parts of the HPLC system, such as the sample loop and the injection valve, should be shielded in a similar way. An aperture is required to apply the sample. A disposable syringe (500 μl) should be used for application of the reaction mixture to the HPLC system. A hand monitor should be available for measuring the radioactivity in the collected fractions. Alternatively, a liquid scintillation counter is required to determine the radioactivity in aliquots of fractions.

5. System to gas solvents with CO_2 (for technical details see Ref. 18).

6. Freeze dryer (for lyophilization of [α-^{32}P]GTP and [α-^{32}P]GTP azidoanilide).

7. Freezer (preferably −70° or lower).

Photolabeling of G-Protein α Subunits

1. Incubator (30°).

2. Refrigerator (preferably with a glass door) or cold room (samples are irradiated with ultraviolet light at 4°).

3. Heavy metal plate covered with Parafilm and equipped with holders 3 cm distant (the size of the plate should be similar to that of the ultraviolet light source, see below).

4. Cooled (4°) centrifuge (10,000 to 12,000 *g*) for 1.5-ml Eppendorf-type test tubes (for experiments with membranes). Cooled (4°) centrifuge (3000 *g*) for glass tubes (10 × 75 mm; for experiments with purified G proteins).

5. Ultraviolet hand lamp. We have used a high-intensity ultraviolet

hand lamp (254 nm, 150 W; Vilber Lourmat, Torcy, France) with a light source area of 7 × 30 cm.

6. Facilities for SDS–polyacrylamide gel electrophoresis and autoradiography.

7. Liquid scintillation counter (for counting radioactivity in polyacrylamide gel slices).

8. Densitometer (for densitometric scanning of autoradiograms).

Synthesis of [α-^{32}P]GTP Azidoanilide

Synthesis of [α-^{32}P]GTP azidoanilide is carried out according to Pfeuffer[9] and Schäfer *et al.*[14] with modifications (see also [24] this volume).

1. [α-^{32}P]GTP (5–10 mCi) is freeze-dried in a silanized flask (see Equipment).

2. The freeze-dried sample is dissolved in 50 μl of 0.1 *M* 2-(*N*-morpholino)ethanesulfonic acid, pH 5.6 (adjusted with NaOH),containing 1.5 g (7.8 μmol) of *N*-(3-dimethylaminopropyl)-*N'*-ethylcarbodiimide hydrochloride.

3. Depending on the amount and on the specific activity of the radioactive [α-^{32}P]GTP and the desired specific activity of [α-^{32}P]GTP azidoanilide, GTP can be added in a volume of 10 μl of water (see Comments).

4. After 10 min, 2.4 mg of 4-azidoaniline (14.1 μmol), suspended in 30 μl of peroxide-free 1,4-dioxane, is added. At this stage, the pH value of the reaction mixture (90 μl) should be checked with litmus paper and, if necessary, corrected to pH 5–6 with 2-(*N*-morpholino)ethanesulfonic acid.

5. The reaction is allowed to procede for 4 hr at room temperature (20°–25°) in the dark. During the reaction, the incubation vessel is kept rotating. Occasionally, the reaction mixture may be cloudy. From our experience, this does not influence the conversion of [α-^{32}P]GTP to [α-^{32}P]GTP azidoanilide.

The reaction can be monitored by withdrawing small aliquots (<1 μl) with a micropipette from the reaction mixture and analyzing by thin-layer chromatography or by an HPLC system connected to an HPLC radioactivity monitor (see below).

Chromatographic Purification of [α-^{32}P]GTP Azidoanilide

Since only 50–70% of [α-^{32}P]GTP is converted to [α-^{32}P]GTP azidoanilide (Fig. 1), the photoreactive nucleotide has to be purified in order to yield a highly pure preparation for photolabeling studies. [α-^{32}P]GTP azidoanilide is isolated by ion-pairing chromatography on a C_{18} column

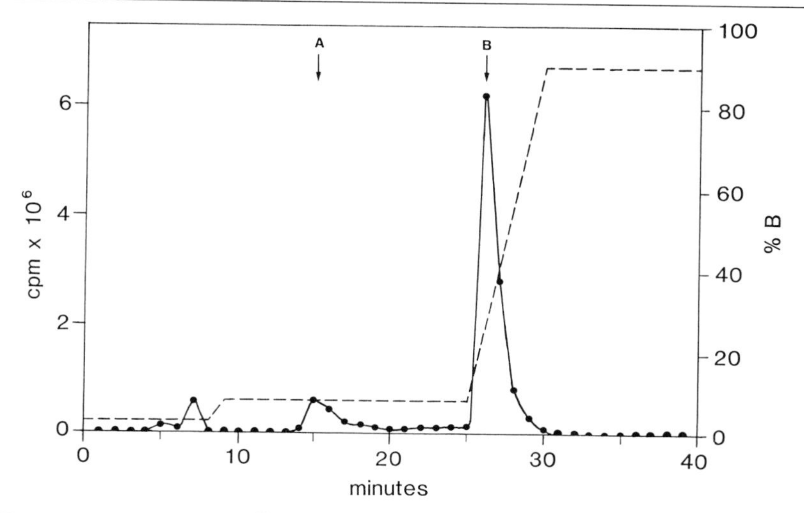

FIG. 1. Purification of [α-^{32}P]GTP azidoanilide by ion-pairing chromatography. Five millicuries of commercial [α-^{32}P]GTP (specific activity ~1000 Ci/mmol) was employed for synthesis of [α-^{32}P]GTP azidoanilide. The flow rate was 1 ml/min. One-milliliter fractions were collected. Solvents A and B were 100 m*M* triethylamine in water and 100 m*M* triethylamine in ethanol, respectively, each gassed with CO_2 to yield pH 7. The discontinuous gradient in solvent B is indicated by the dashed line. Retention times are indicated on the abscissa. Radioactivity in 1-ml fractions (2-μl aliquots counted in water) is shown on the ordinate. For further details, see text. Arrows A and B indicate the retention times of [α-^{32}P]GTP and [α-^{32}P]GTP azidoanilide, respectively. For further explanations, see text.

(0.45 × 25 cm) with volatile solvents.[18] We routinely perform HPLC at 4°; we have not tested whether [α-^{32}P]GTP azidoanilide can be isolated at higher temperatures.

1. Solvent A (100 m*M* triethylamine in water) and solvent B (100 m*M* triethylamine in ethanol) are gassed with CO_2 for about 30 min to obtain a pH of 7 at room temperature.
2. The C_{18} column is equilibrated with 97.2% solvent A and 2.8% solvent B at a flow rate of 1 ml/min.
3. The reaction mixture (90 μl) is diluted with solvent A to a final volume of 450 μl. It does not disturb the subsequent purification by HPLC if the reaction mixture remains slightly cloudy after the dilution (see above).
4. The injection is performed with a 500-μl disposable syringe.
5. For isolation of [α-^{32}P]GTP azidoanilide, a discontinuous gradient in solvent B from 2.8 to 90% (see Fig. 1) is applied. About 40 fractions of 1 ml each are collected.

6. The radioactivity of fractions can be determined with a hand monitor or by counting 2-μl aliquots in a liquid scintillation counter. Radioactivity elutes in three peaks within the gradient. [α-^{32}P]GTP azidoanilide elutes with a retention time of 22 to 29 min (depending on the HPLC system and the C_{18} column); the retention time of [α-^{32}P]GTP is shorter (14–18 min). If [α-^{32}P]GTP is used that is not purified after synthesis, an early radioactive peak (retention time below 10 min) is observed, of which a major portion represents [^{32}P]phosphoric acid and [^{32}P]GMP. Occasionally we observe a radioactive peak after the gradient has been completed; the identity of the late eluting radioactive material is not known. At least three-quarters of the radioactivity applied to the HPLC system is found in the fractions.

7. Fractions containing [α-^{32}P]GTP azidoanilide (~3 to 4) are pooled, lyophilized, and resuspended in water to yield a concentration of 1 μCi/μl.

8. [α-^{32}P]GTP azidoanilide is stored at $-70°$ or below. At this low temperature, the purity of the compound remains high ($\geq$95%) for several weeks. We have no experience with storage at higher temperatures.

9. The HPLC column is regenerated with and stored in 100% methanol.

Quality Control of [α-^{32}P]GTP Azidoanilide

Purified G proteins can be used to test the properties of the synthesized compound (Fig. 2). The assay is carried out in 1.5-ml Eppendorf-type test tubes. The assay volume is 60 μl. The final concentrations of buffer constituents are EDTA, 0.1 mM; Mg^{2+}, 10 mM; dithiothreitol, 2 mM; HEPES, 30 mM (pH 7.4).

1. Heterotrimeric G proteins are diluted with ice-cold 2-fold-concentrated incubation buffer. Thirty microliters of the diluted sample (which should contain about 1 μg of G protein) is employed per assay tube. Concentrations of detergents, which are constituents of G protein samples, should be kept as low as possible [e.g., Lubrol PX concentration $\leq$ 0.2% (w/v)].

2. Nucleotides, dissolved in water, are added in a volume of 10 μl.

3. Samples are incubated for 3 min at 30°.

4. [α-^{32}P]GTP azidoanilide (0.5–1 μCi), diluted in water, is added in a volume of 20 μl.

5. The incubation is continued for another 3 to 10 min.

6. The incubation is stopped by putting the sample tubes on ice or in a refrigerator.

7. Samples (except the controls) are irradiated with ultraviolet light at 4° as described below for membrane preparations.

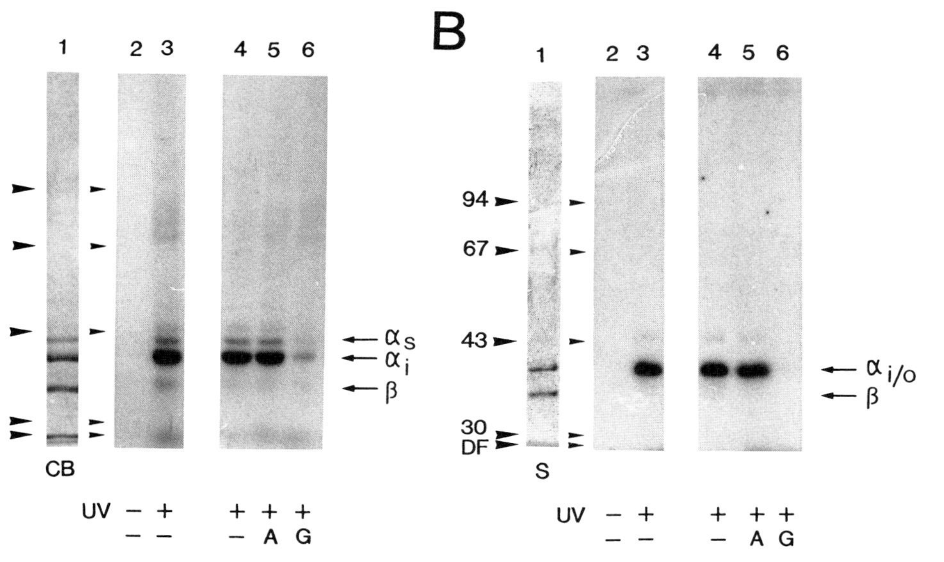

FIG. 2. Photolabeling of heterotrimeric purified G proteins with [α-^{32}P]GTP azidoanilide. G proteins were incubated with [α-^{32}P]GTP azidoanilide (1 μCi/tube) in a buffer consisting of 0.1 mM EDTA, 10 mM Mg^{2+}, 1 mM benzamidine, 2 mM dithiothreitol, and 30 mM HEPES (pH 7.4); the concentration of Lubrol PX was 0.2% (w/v). For further experimental details, see text. In (A), a mixture of G_s and G_i, purified from membranes of human erythrocytes, was employed (~12 μg of protein per sample). In (B), a mixture of G_i and G_o, purified from membranes of porcine brain, was employed (~1 μg of protein per sample). Lanes 1 show Coomassie blue-stained (CB; A) or silver-stained (S; B) SDS–polyacrylamide gels (10%, no urea). Lanes 2–6 (A and B) show autoradiograms of SDS–polyacrylamide gels. The samples were (+) or were not (−) irradiated with ultraviolet light; A, the assay mixture contained 10 μM ATP; G, the assay mixture contained 10 μM GTP. The numbers at left margins indicate the positions of molecular mass markers (kDa); DF, dye front. The positions of G-protein subunits are indicated at the right margins. α_s, α subunit of G_s; α_i, α subunits of G_i-type G proteins; α_o, α subunit of G_o; β, β subunits. Since urea was not present in these SDS–polyacrylamide gels, the α subunits of G_i-type G proteins and G_o are not resolved.

8. Samples are transferred to glass tubes, and proteins are precipitated by adding 540 μl (9 sample volumes) of ice-cold acetone.

9. After centrifugation at 4° (3000 g, 20 min) and removal of acetone, the precipitate is dissolved in Laemmli buffer and subjected to SDS–polyacrylamide gel electrophoresis in 10–12% SDS–polyacrylamide gels.[21] Heating of samples prior to electrophoresis should be avoided since it causes aggregation of G proteins.[19]

[21] U. K. Laemmli, *Nature (London)* **227,** 680 (1970).

10. The SDS–polyacrylamide gels are dried and autoradiographed for 12 hr to a few days with or without an enhancing screen.

Under these conditions, [α-^{32}P]GTP azidoanilide is specifically incorporated into G-protein α subunits. The reaction is inhibited by GTP but not by ATP. Stable incorporation of radioactivity into α subunits is only observed after irradiation of samples with ultraviolet light. In contrast to G-protein α subunits in membranes (see below), α subunits of purified G proteins incorporate detectable amounts of [α-^{32}P]GTP azidoanilide in the absence of Mg^{2+}. This may be explained by the lower background of autoradiograms compared with the background found for membrane preparations and by the relatively large amounts of G-protein subunits. Provided the Mg^{2+} concentration is high ($\sim$10 mM), incorporation of [α-^{32}P]GTP azidoanilide into G-protein β subunits (35/36 kDa) is very minor and not affected by unlabeled GTP. At low Mg^{2+} concentrations ($\leq$1 mM), considerable amounts of [α-^{32}P]GTP azidoanilide label are found in the G-protein β subunits and in other proteins (e.g., bovine serum albumin).

Agonist-Sensitive Photolabeling of G-Protein α Subunits in Membranes

The assay is performed in 1.5-ml Eppendorf-type test tubes. The assay volume is 60 μl. The final concentrations of buffer constituents are EDTA, 0.1 mM; $MgCl_2$, 0.5–5 mM; NaCl, 10–100 mM; GDP, 3–50 μM; HEPES, 30 mM (pH 7.4). The concentrations of $MgCl_2$, NaCl, and GDP have to be adjusted to the individual cell type (see Comments).

1. Sedimented membranes are suspended in 30 μl of a 2-fold concentrated ice-cold incubation buffer to yield a protein concentration of 50 μg/30 μl.
2. Sample tubes receive 30 μl of the membrane suspension.
3. Preincubation of the sample at 30° is started by the addition of receptor agonist (or carrier) added in a volume of 10 μl.
4. After 3 min, [α-^{32}P]GTP azidoanilide, diluted in water, is added in a volume of 20 μl (1 μCi/tube).
5. Samples are incubated for another 3 min.
6. The reaction is stopped by putting samples on ice or in a refrigerator. All subsequent procedures are performed at 4°.
7. Samples are centrifuged at 10,000 to 12,000 g for 5 min (for separation of bound and unbound [α-^{32}P]GTP azidoanilide).
8. The supernatant fraction is carefully removed, and the membrane pellet is resuspended in 60 μl of a modified GDP-free incubation buffer

supplemented with 2 m*M* dithiothreitol. The latter compound reduces unspecific photolabeling of membrane proteins.

9. Samples are transferred with a micropipette from the Eppendorf-type test tubes to Parafilm placed on a metal plate.

10. The samples (drops) are irradiated for 10 sec at 4° with an ultraviolet lamp (254 nm, 150 W) from a distance of 3 cm.

11. Samples are recollected into the Eppendorf-type test tubes and again centrifuged. After this, the supernatant fraction is carefully removed, and the pellets are dissolved in 20 μl of sample buffer.[21]

SDS–Polyacrylamide Gel Electrophoresis and Autoradiography

Samples are subjected to SDS–polyacrylamide gel electrophoresis without heating to avoid possible aggregation of G proteins.[19] SDS–polyacrylamide gel electrophoresis is performed according to Laemmli.[21] The size of polyacrylamide gels is 13.8 $\times$ 17.7 $\times$ 0.75 mm. The separating polyacrylamide gel contains 8% acrylamide, 0.21% bisacrylamide, and 4 *M* urea; the latter is crucial for the separation of the various G-protein α subunits. Polyacrylamide gels are run at constant current (15–20 mA, corresponding to voltages of 100 V at the beginning and 500 V at the end of the run). Standard procedures are used for staining of polyacrylamide gels with Coomassie blue, drying, and autoradiography (12 hr to a few days, with or without an enhancing screen).

Evaluation

Densitometric scans (performed preferentially with a laser densitometer to achieve resolution of subtypes of G-protein α subunits) may be valuable for semiquantitative evaluation of autoradiograms. A more precise quantitative evaluation is achieved if the 39–41 kDa regions of lanes of dried polyacrylamide gels are cut out, shaken in 1 ml of 30% (v/v) H_2O_2 for at least 1 hr, and counted for radioactivity after addition of 5 ml of scintillant.

If the identity of photolabeled G-protein α subunit cannot be derived from the mobility in SDS–polyacrylamide gels, the photolabeled sample is subjected to SDS–polyacrylamide gel electrophoresis in the usual manner, and the separated proteins are transferred to nitrocellulose filters by standard procedures.[22] After autoradiograms of the filters have been obtained, filters are incubated with appropriate antibodies; filter-bound antibodies are detected by a color reaction.[23] In various membrane preparations,

[22] J. Renart and I. V. Sandoval, this series, Vol. 104, p. 455.

[23] J. J. Leary, D. J. Brigati, and D. C. Ward, *Proc. Natl. Acad. Sci. U.S.A.* **80,** 40 (1983).

we found a close comigration of immunologically identified G-protein α subunits and photolabeled proteins. This method allows the identification of photolabeled α subunits of G_o, G_{i2}, and a G_i-subtype not further identified within one sample (see below). Provided precipitating antisera are available, the photolabeled membrane proteins may be solubilized and immunoprecipitated for identification.

Application to Various Cell Types

In membranes from various cell types, including myeloid differentiated HL-60 cells,[15] neuroblastoma × glioma hybrid cells (108CC15; N×G cells),[16] and pituitary GH_3 cells, photolabeling of G-protein α subunits is not observed in the absence of Mg^{2+} (1 m*M* EDTA-containing, Mg^{2+}-free buffer). Mg^{2+} ions (employed up to 30 m*M*) stimulate photolabeling. Stimulatory effects of receptor agonists on G_i-type G proteins and G_o are observed at all Mg^{2+} concentrations tested. Photolabeling of purified G proteins is also stimulated by Mg^{2+}.

GDP inhibits photolabeling of unstimulated G proteins to a greater extent than photolabeling of receptor-activated G proteins. A suitable concentration (between 3 and 50 μM) has to be determined for each cell type. In neuroblastoma × glioma hybrid cells (108CC15), a stimulatory effect of the synthetic opioid, [D-Ala^2,D-Leu^5]enkephalin, is only observed in the presence of GDP (Fig. 3). Depending on the cell type, NaCl or KCl (10–300 m*M*) may promote the increase in photolabeling induced by receptor agonists, mainly by inhibiting basal photolabeling.

Figure 4 shows differential effects of receptor agonists on photolabeling of G-protein α subunits in membranes from N×G cells. Autoradiograms were analyzed with a laser densitometer. The synthetic opioid, [D-Ala^2, D-Leu^5]enkephalin, stimulates to equal extents photolabeling of the 39- and 40-kDa proteins comigrating with the α subunits of G_o and G_{i2}, respectively. In contrast, the receptor agonist bradykinin stimulates photolabeling of the 40- but not of the 39-kDa protein.

Comments[24]

Specific Activity and Purification of [α-^{32}P]GTP Used for Synthesis of [α-^{32}P]GTP Azidoanilide

The specific activity of [α-^{32}P]GTP synthesized in our laboratory was 200 to 1000 Ci/mmol.[17] We routinely employed 5–10 mCi [α-^{32}P]GTP for

[24] "Rather than indulging in time-consuming determinations of quantum yields and radiation doses, the more empirical approach is almost always sufficient." Quoted from "*Photoaffinity Labeling*," by H. Bayley and J. R. Knowles, this series, Vol. 46, p. 108.

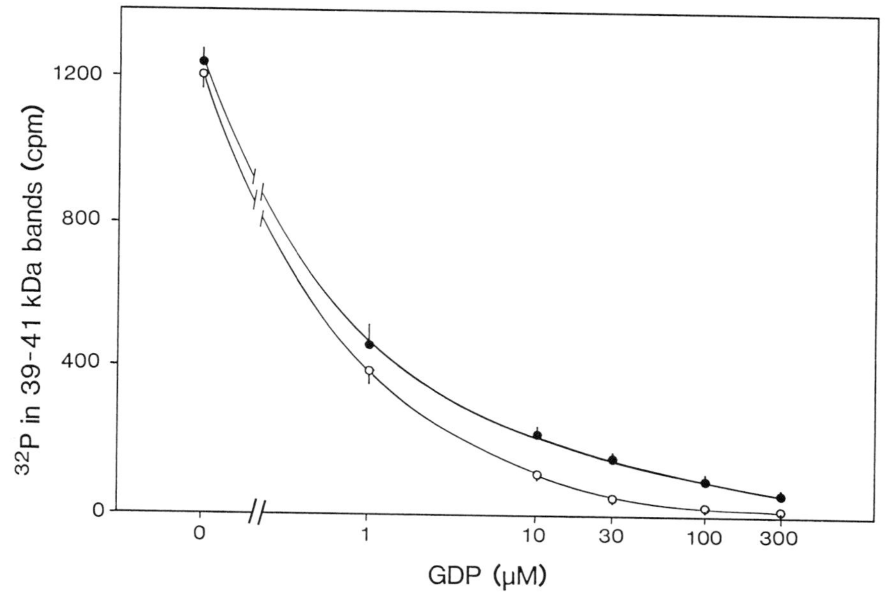

FIG. 3. Influence of GDP on basal and opioid-stimulated incorporation of [α-^{32}P]GTP azidoanilide into 39- to 41-kDa proteins of membranes from N × G cells. Membranes were incubated with [α-^{32}P]GTP azidoanilide in the absence (○) or presence of 1 μM [D-Ala^2,D-Leu^5]enkephalin (●). The concentrations of Mg^{2+}, NaCl, and benzamidine were 5, 100, and 1 mM, respectively. 5′-Adenylyl imidodiphosphate (0.1 mM) was employed solely to adjust the experimental conditions to those suitable for determining GTPase activity.[5] For further experimental details, see text. The GDP concentration is indicated on the abscissa, the radioactivity incorporated into the 39–41 kDa region of SDS–polyacrylamide gels (8% acrylamide, 4 M urea) on the ordinate. Values represent means of triplicate determinations ± standard deviations.

synthesis of [α-^{32}P]GTP azidoanilide. Thus, the amount of [α-^{32}P]GTP in the reaction mixture was 1 to 10 nmols, corresponding to a concentration of 11 to 110 μM. If commercial [α-^{32}P]GTP is employed, a product of high specific activity (800–3000 Ci/mmol) should be ordered. While commercial [α-^{32}P]GTP can be employed directly, we recommend purification of [α-^{32}P]GTP synthesized by the method of Johnson and Walseth[17] in order to remove buffer (25 mM Tris-HCl, pH 9.0) and to avoid formation of by-products. [α-^{32}P]GTP can be purified by chromatography on Dowex 1-X4 (200–400 mesh, Cl^- form)[17] or by ion-pairing chromatography as described above.[18]

Specific Activity of [α-^{32}P]GTP Azidoanilide

The specific activity of [α-^{32}P]GTP azidoanilide should be similar to that of [α-^{32}P]GTP employed for synthesis.[9] If the specific activity of

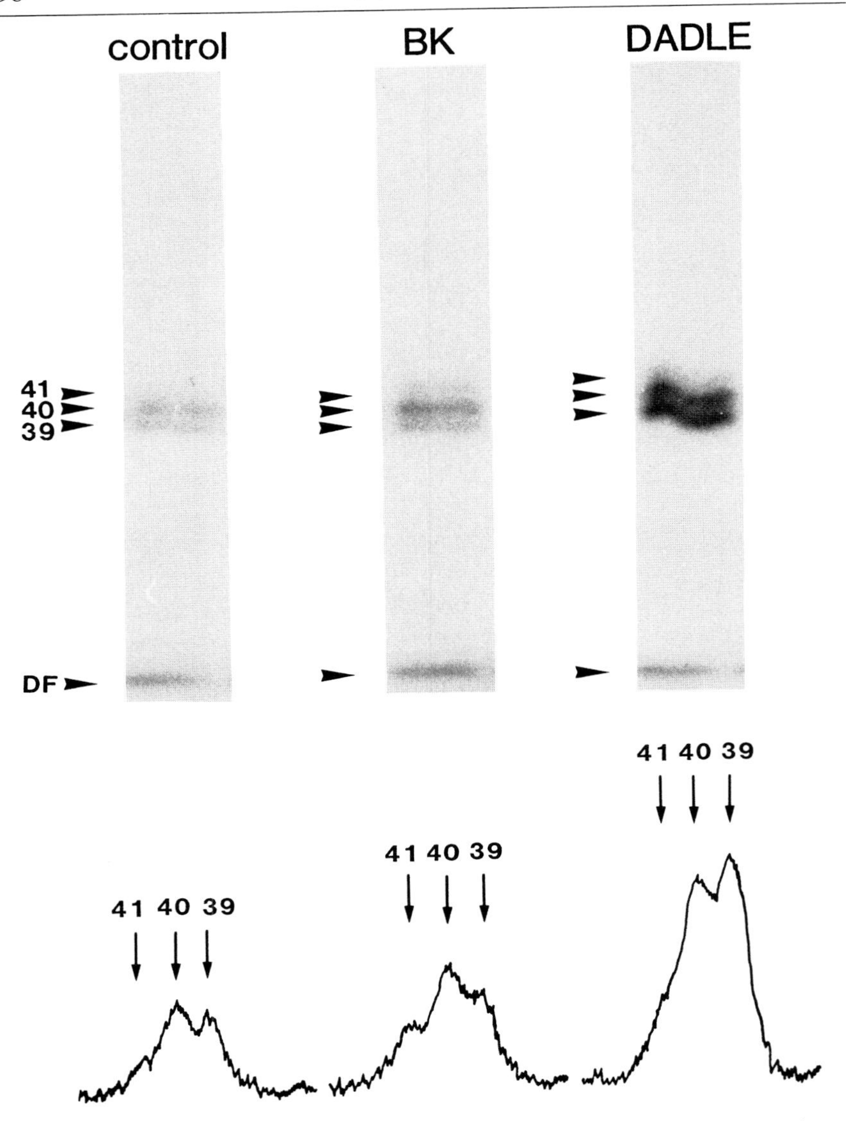

FIG. 4. Differential effects of receptor agonists on photolabeling of proteins in membranes from N × G cells. Membranes from N × G cells were incubated with [α-^{32}P]GTP azidoanilide in the absence of receptor agonists (control), with 1 μM bradykinin (BK), or with 1 μM [D-Ala^2,D-Leu^5]enkephalin (DADLE). The concentrations of constituents of the incubation mixture were Mg^{2+}, 5 mM; GDP, 30 μM; NaCl, 100 mM; benzamidine, 1 mM; 5′-adenylyl imidodiphosphate, 0.1 mM (see legend to Fig. 3). For further experimental details, see text. At top are shown autoradiograms of 8% polyacrylamide gels which contained 4 M urea; at bottom are densitometric scans of the autoradiograms. Numbers indicate molecular masses (kDa). The 39-, 40-, and 41-kDa proteins comigrated with the immunologically identified α subunits of G_o, G_{i2}, and a G_i subtype not further identified, respectively.

[α-^{32}P]GTP employed for synthesis is 200 to 1000 Ci/mmol (see above),[17] the concentration of [α-^{32}P]GTP azidoanilide is of the order of 3–15 nM under the assay conditions described above. [α-^{32}P]GTP azidoanilide of higher specific activity may be obtained by using commercially available [α-^{32}P]GTP of 3000 Ci/mmol (see above).

Alternative Purification Procedures

Schäfer *et al.*[14] used thin-layer chromatography of [α-^{32}P]GTP azidoanilide on polyethyleneimine-cellulose with 0.8 M triethylammonium bicarbonate (pH 7.5) as mobile phase. A detailed description of the procedure applied for analytical purposes is given by Pfeuffer.[9] We found R_f values of 0.60 for [α-^{32}P]GTP azidoanilide, 0.29 for GTP, 0.50 for GDP, and 0.62 for GMP. Thus, whereas the R_f values for GMP and GDP are similar to that of [α-^{32}P]GTP azidoanilide, thin-layer chromatography allows a complete separation of [α-^{32}P]GTP and [α-^{32}P]GTP azidoanilide. Purification of GTP azidoanilide by conventional ion-exchange chromatography on DEAE-cellulose is described by Pfeuffer.[9]

Incubation Time

Photolabeling of G-protein α subunits in membrane preparations increases for about 20 min of incubation. The ratio of agonist-stimulated to basal photolabeling of G-protein α subunits is greatest at short incubation times (Fig. 5). An incubation time of about 3 min is appropriate since (1) it allows sufficient incorporation of radioactivity into proteins, which can be accurately quantified by counting polyacrylamide gel slices into a liquid scintillation counter, and (2) the ratio of agonist-stimulated to basal photolabeling is satisfactory.

Photolysis

The appropriate time for irradiation of samples depends, among other factors, on the intensity of the light source. Since we used a high-energy ultraviolet light (150 W, 254 nm, 3 cm distance from sample), the time of irradiation was very short (10 sec).[25] After this time of irradiation, photolabeling of G-protein α subunits in membrane preparations was 60% of maximum, which was reached after about 4 min. Photolabeling for longer periods of time resulted in an overproportional increase in radioactivity incorporated into proteins other than those in the 40-kDa region (i.e., G-protein α subunits). In addition, the bands on the SDS–poly-

[25] H. Bayley and J. R. Knowles, this series, Vol. 46, p. 69.

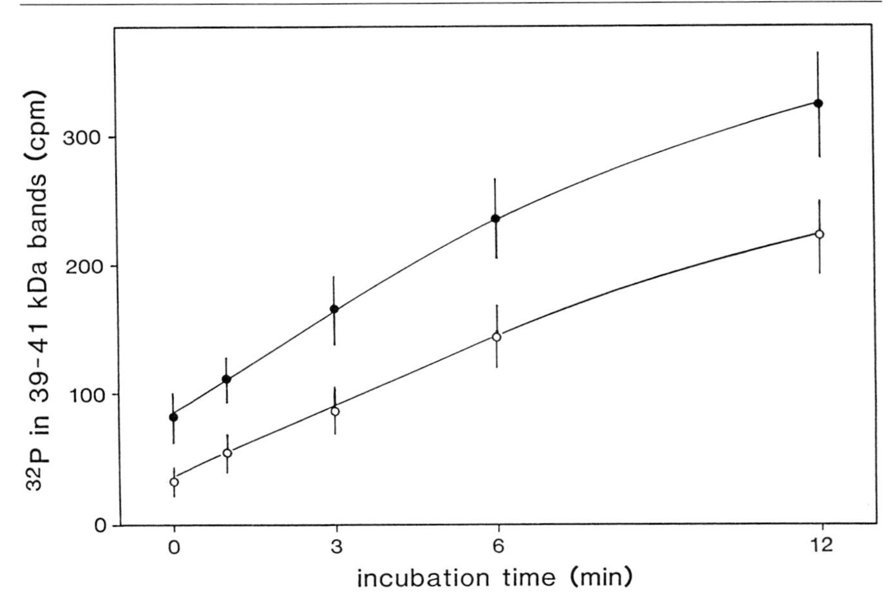

FIG. 5. Time course of photolabeling of proteins in membranes from N × G cells. Membranes from N × G cells were incubated with [α-^{32}P]GTP azidoanilide in the absence (○) or presence of 1 μM [D-Ala^2,D-Leu^5]enkephalin (●). Experimental conditions were as described in Fig. 4. For further experimental details, see text. The incubation time is indicated on the abscissa, the radioactivity incorporated into the 39–41 kDa region of SDS–polyacrylamide gels (8% acrylamide, 4 M urea) on the ordinate. Values represent means of triplicate determinations ± standard deviations.

acrylamide gels and on the autoradiograms became fuzzier and more Coomassie blue-stainable, and radioactive material was found in the dye front.

Alternative Procedure for Removal of Unbound [α-^{32}P]GTP Azidoanilide

Prior to photolysis, samples may be dotted onto small pieces of nitrocellulose filters, which are washed and subsequently irradiated; the filters with the irradiated sample are then inserted into wells of an SDS–polyacrylamide gel.[12] In a preliminary series of experiments we found that the electrophoretic transfer of proteins of nonirradiated membranes from filters into the gel is incomplete. An essentially complete recovery of the sample was achieved if nitrocellulose filters were boiled in sample buffer which was subsequently applied to the well.

SDS–Polyacrylamide Gel Electrophoresis

G-protein α subunits modified by pertussis toxin exhibit a decreased mobility in SDS–polyacrylamide gels,[26,27] particularly if urea ($\geq 4\ M$) is present.[28]

Protease Inhibitors

Protease inhibitors may be required in some membrane preparations. We have frequently used benzamidine (1 m*M*), which does not interfere with photolabeling of G proteins.

Photolabeling of G_s

We applied the method mainly in membrane preparations in which the pertussis toxin-sensitive G proteins of the G_i type and G_o occur in great excess over G_s (membranes from HL-60 cells, N × G cells, GH_3 cells). Photolabeling of proteins comigrating with the G_s α subunit was observed only after prolonged exposure of dried polyacrylamide gels to X-ray films. Clearly, [^{32}P]GTP azidoanilide is suitable for photolabeling of G_s (see Introduction). Since the abundance of G_s is low and the affinity of [^{32}P]GTP azidoanilide to G_s is lower than that to G_i-type G proteins and G_o,[13] a high specific activity of [^{32}P]GTP azidoanilide may be required in order to demonstrate effects of receptor agonists. In a preliminary series of experiments performed with membranes of GH_3 cells, we found that vasoactive intestinal peptide stimulated photolabeling of a 45-kDa protein comigrating with the immunologically identified G_s α subunit.

Acknowledgments

We thank Dr. Rainer Schäfer for initial help in establishing the procedures described above, Evelyn Bombien and Anke Schmidt for the performance of some experiments, Rosemarie Krüger for help in preparing the manuscript, and Monika Bigalke for drawing the figures and photography. We also thank Anke Schmidt for helpful discussions and critical reading of the manuscript. This work was supported by the Deutsche Forschungsgemeinschaft.

[26] B. Eide, P. Gierschik, and A. Spiegel, *Biochemistry* **25,** 6711 (1986).
[27] S. Mumby, I.-H. Pang, A. G. Gilman, and P. C. Sternweis, *J. Biol. Chem.* **263,** 2020 (1988).
[28] F. A. P. Ribeiro-Neto and M. Rodbell, *Proc. Natl. Acad. Sci. U.S.A.* **86,** 2577 (1989).

[26] Quantitative Immunoblotting of G-Protein Subunits

By DONNA J. CARTY, RICHARD T. PREMONT, and RAVI IYENGAR

Introduction

In recent years, the application of molecular biology techniques to signal transduction research has led to the isolation of the cDNAs for many G-protein subunits that had previously not been biochemically identified. The use of antipeptide antisera based on deduced sequences has allowed us to distinguish among closely related G proteins and has facilitated their purification. The laboratories of Gilman and Spiegel have developed a battery of antipeptide antisera[1,2] that have proved very useful for signal transduction research. We have used these antisera to isolate three forms of G_i[3] and two forms of G_o. We have also been able to quantify G-protein subunits before and after desensitization.[4] In addition, we have used the antipeptide antisera raised against the deduced amino acid sequence of the carboxy terminus of $G_{\alpha z}$ to identify the $G_{\alpha z}$ in human erythrocytes.[5] Several regions of G-protein α subunits have been used for making specific antisera. These regions are summarized in a schematic fashion in Fig. 1. In this chapter, we describe the use of these antisera in quantitative immunoblotting to determine the relative proportions of G α subunits in human erythrocytes, as well as to compose the levels of $\beta\gamma$ dimers in liver membranes.

Materials

Nitrocellulose sheets (BA85) were from Schleicher and Schuell (Keene, NH). Affinity-purified goat antibodies to rabbit IgG were from Calbiochem (San Diego, CA). Ovalbumin, Lubrol, or Triton X-100 were from Sigma (St. Louis, MO). Materials for sodium dodecyl sulfate-poly-

[1] S. Mumby, I. M. Pang, A. G. Gilman, and P. C. Sternweis, *J. Biol. Chem.* **263,** 2020 (1988).
[2] P. Goldsmith, K. Rossiter, A. Carter, W. Simonds, C. G. Unson, R. Vinitsky, and A. Spiegel, *J. Biol. Chem.* **263,** 6476 (1988).
[3] D. J. Carty, E. Padrell, J. Codina, L. Birnbaumer, J. D. Hildebrandt, and R. Iyengar, *J. Biol. Chem.* **265,** 6268 (1990).
[4] R. T. Premont and R. Iyengar, *Endocrinology* **125,** 1151 (1989).
[5] R. T. Premont, A. Buku, and R. Iyengar, *J. Biol. Chem.* **264,** 14960 (1989).

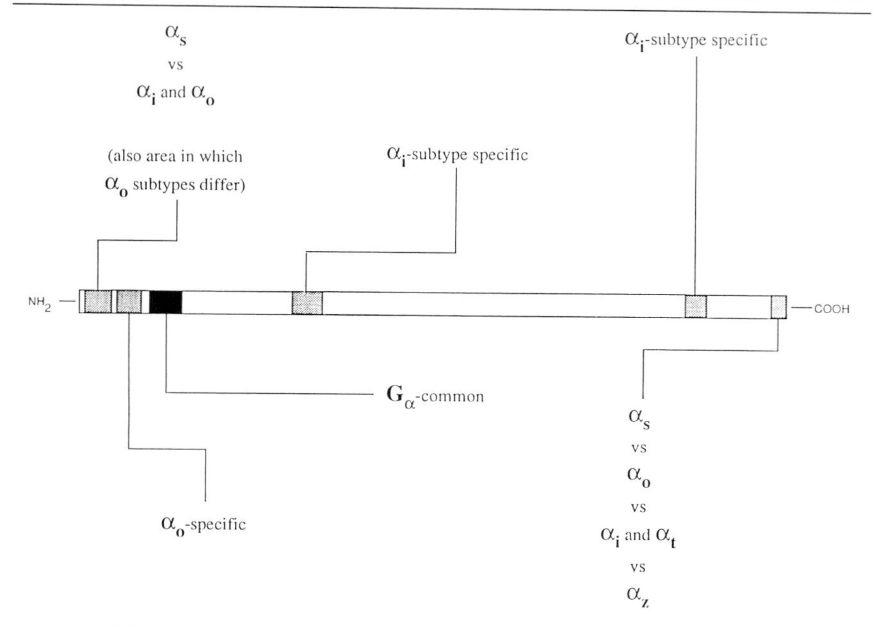

FIG. 1. Schematic summary of the various regions of the α subunits of G proteins utilized for the generation of sequence-specific antisera.

acrylamide gel electrophoresis (SDS–PAGE) were purchased from Bio-Rad (Richmond, CA). A Bio-Rad protein transfer apparatus was also used.

Experimental Procedures

^{125}I-Labeling of Goat Anti-Rabbit IgG

Goat anti-rabbit IgG is iodinated with Na^{125}I and chloramine-T to 10^8 cpm/μg. Five micrograms of IgG, 2 mCi of ^{125}I, and 10 μl of 1 mg/ml chloramine-T are mixed in a final volume of 25 μl in a solution containing 200 m*M* potassium phosphate buffer, pH 7.5. The reaction is allowed to proceed for 1 min at room temperature and then quenched by the addition of 25 μl of 2.5 ng/ml sodium metabisulfite. The mixture is then diluted to 500 μl and chromatographed over a prepacked Sephadex G-25 [e.g., Pharmacia (Piscataway, NJ) PD-10] column preequilibrated with 50 m*M* Tris-HCl, 150 m*M* NaCl, and 1% (w/v) ovalbumin, pH 8.5. The column is eluted with the same buffer, and 500-μl fractions are collected in the void volume.

Ovalbumin is stored as a 15% solution in 50 m*M* Tris-HCl, 150 m*M* NaCl, and 0.02% sodium azide, pH 8.5. Ovalbumin is stirred overnight at room temperature to achieve the 15% solution. After stirring, the suspension is centrifuged for 30 min at 10,000 *g*, and the clear supernatant is stored in 20-ml aliquots at −20°.

SDS Gel Electrophoresis and Transfer to Nitrocellulose

Gel electrophoresis is carried out according to the procedure of Laemmli.[6] When purified G proteins are used, we typically electrophorese 50–500 ng. When mixtures of pure G proteins are used, 4–5 μg of total G protein is electrophoresed. This corresponds to 2–2.5 μg of each subunit. When membranes are electrophoresed, we load no more than 100–150 μg of total membrane protein. Higher amounts (200–300 μg) results in uneven transfer to the nitrocellulose paper. Electrophoretic transfer of proteins from the polyacrylamide gel to nitrocellulose is carried out at 40 V according to the procedure of Towbin *et al.*[7]

Immunoblotting

After the transfer, the nitrocellulose sheets are treated with the blocking buffer (50 m*M* Tris-HCl, 150 m*M* NaCl, 3% (w/v) ovalbumin, and 0.1% sodium azide, pH 8.5) for 1–3 hr at room temperature to block all the additional protein-binding sites. The strips are then incubated with the appropriate dilution of the antisera (1 : 100–1 : 2000) in blocking buffer. This incubation is allowed to go overnight (~18 hr) at room temperature with gentle rocking on a platform shaker. Since the antisera are generally difficult to obtain, this step is performed in a minimal volume (~10 ml), and the antisera-containing solution is collected after the incubation and stored at −70° for reuse. For most antisera, we are able to use the diluted antisera 3–4 times before there is a noticeable (~20%) decrease in band intensity for a calibrated standard.

After incubation with the antisera, the nitrocellulose strips are washed individually, once in washing buffer (50 m*M* Tris-OH, 150 m*M* NaCl, pH 8.5) and then again in washing buffer containing 0.1% (v/v) Lubrol. This is done by vigorous shaking with 300–500 ml at room temperature on a platform shaker for 10 min. It is essential that nitrocellulose strips exposed to different antisera be kept separate from one another and washed in separate containers during the 2 washes. The detergent wash is followed by 2 more washes with washing buffer during which all the strips can be washed together. Further incubations and washes can also be carried out with all the strips in a single container.

The nitrocellulose strips are then incubated again for 1–3 hr with

[6] V. K. Laemmli, *Nature (London)* **227,** 680 (1970).
[7] H. Towbin, T. Staehelin, and J. Gordon, *Proc. Natl. Acad. Sci. U.S.A.* **76,** 4350 (1979).

blocking buffer to reblock the protein-binding sites, after which they are incubated with ^{125}I-labeled goat anti-rabbit IgG. Typically we use 20–50 ml at 1×10^6 counts per minute (cpm)/ml. After a 3-hr incubation at room temperature, the nitrocellulose strips are washed as before, once in washing buffer, once in washing buffer containing 0.1% Lubrol, and then 2 more times with the washing buffer. During the washes, the strips are shaken vigorously. Typical wash volumes are 50–100 ml per two-lane strip for each wash. The vigorous shaking during the wash is essential to dislodge the minute particles of protein in the blocking buffer that settle on the paper and lead to the appearance of black spots on the autoradiogram. It is advisable to use freshly ^{125}I labeled IgG since the major reason for weak bands appears to be the quality of the labeled IgG.

After the final washes, the nitrocellulose strips are air-dried and exposed to X-ray film for the appropriate length of time. Figures 2–10 are autoradiograms.

Standards for Quantification of G Proteins

Currently the most unequivocal standards that can be used for immunoblotting are the G-protein subunits expressed in *Escherichia coli*. Several G proteins have been expressed in *E. coli*. As shown in Fig. 2, the particulate fraction of the bacteria contains sufficient amounts of the expressed protein to be visualized by SDS–PAGE electrophoresis and Coomassie blue staining.

Use of Antipeptide Antisera for G-Protein Quantification

Recombinant α subunits are used to characterize the antisera for specificity as shown in Fig. 3. When three antisera raised to recognize unique sequences in $G_{\alpha i1}$, $G_{\alpha i2}$, and $G_{\alpha i3}$ were immunoblotted against the recombinant α_{i1}, α_{i2}, and α_{i3}, all three antisera displayed the expected specificity. Therefore, the three antisera were used to immunoblot three preparations of G proteins made over several years. As shown in Fig. 4, all three preparations contained all three forms of G_i. A cursory examination of the autoradiographs shown in Fig. 4 might give the impression that the three G_i proteins are present in comparable amounts. However, this will be an erroneous impression unless the three α_i-subtype-specific antisera have the same sensitivity. To answer this question, we determined the autoradiographic intensity of known amounts of the G_i proteins expressed in *E. coli*. This was done in two sets of measurements. First the amount of G-protein α subunits in the *E. coli* particulate fraction was determined. This was done by electrophoresing varying amounts of the G-protein-containing lysates and known amounts of bovine serum albumin (BSA)

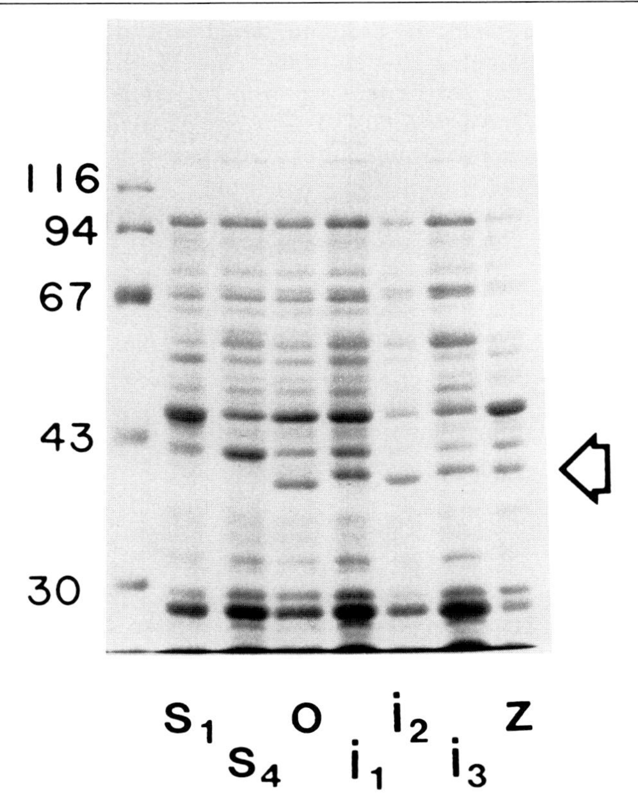

FIG. 2. Coomassie blue staining profile of the 100,000 g pellet proteins from *E. coli* expressing the indicated recombinant α subunits, indicated at the region of the open arrow. The recombinant α subunits are s_1, α_s short form; s_4, α_s long form; o, i_1, i_2, i_3, and z, α_o, α_{i1}, α_{i2}, α_{i3}, and α_z, respectively. Visual inspection of the 40K region indicates the presence of the appropriate α subunit.

protein on a single gel and then staining the gel with Coomassie blue (Fig. 5). The BSA and the desired α subunit were scanned with a densitometer. Using the BSA standard curve, we could estimate the amount of G-protein subunit in the mixture of *E. coli* proteins. This is shown in Table I.

Having determined the amount of α_i protein in the *E. coli* particulate fractions, we electrophoresed varying volumes of a detergent (SDS) solution of the *E. coli* particulate proteins, transferred them to nitrocellulose, and blotted with the three α_i-subtype-specific antisera. At constant dilution (1 : 200), the antisera showed different sensitivities. All three antisera recognized their appropriate protein α subunits in a proportional manner (Fig.

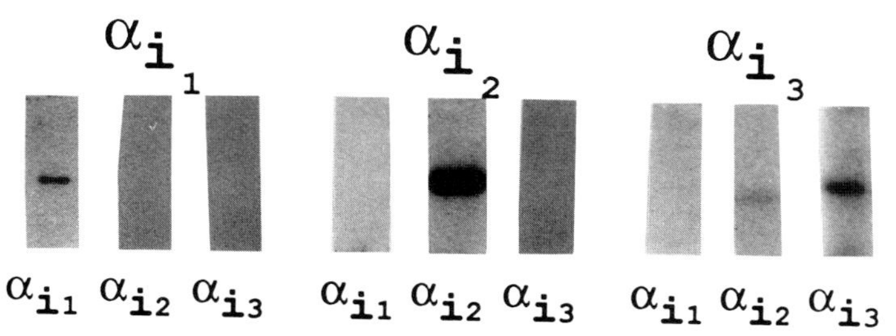

FIG. 3. Specificity of the interaction of the α_i-subunit-specific antisera with the recombinant α_{i1}, α_{i2}, and α_{i3}. The *E. coli* lysates were resolved by SDS-gel electrophoresis, transferred to nitrocellulose paper, and then blotted with α_{i1}-, α_{i2}-, and α_{i3}-specific antisera. The symbols above the autoradiograms show the presumed specificities of the antisera. The symbols below the autoradiograms indicate the identities of the recombinant G-protein α subunits used for the immunoblots. These blots show that the α_i-subtype-specific antisera have the expected specificities.

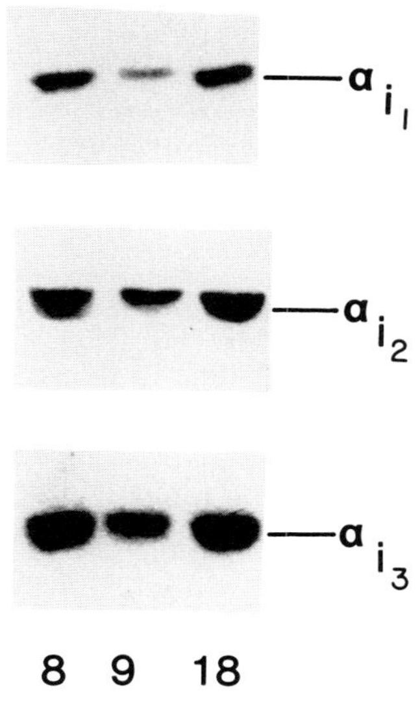

FIG. 4. Immunoblotting of human erythrocyte G-protein preparations 8, 9, and 18 with α_{i1}-, α_{i2}-, and α_{i3}-specific antisera, respectively. The G proteins are obtained after the second DEAE-Sephacel chromatography, when they are free of other contaminants but not resolved from each other. It can be seen that all three preparations contain all three $G_{i\alpha}$ proteins.

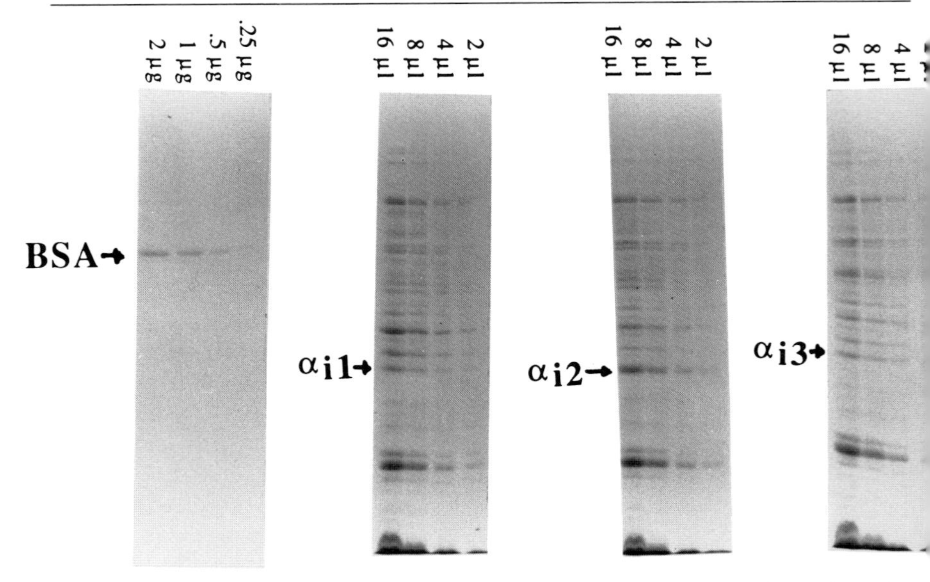

FIG. 5. Coomassie blue staining profile of varying concentrations of BSA and varying volumes of the *E. coli* particulate fraction in Laemmli sample buffer. Indicated amounts of BSA and varying volumes of *E. coli* particulate fractions in Laemmli buffer were analyzed by SDS gel electrophoresis, stained with Coomassie blue, and then destained before densitometry. The positions of the recombinant α_{i1}, α_{i2}, and α_{i3} proteins are indicated. Densitometric scans and values of protein derived from these experiments are shown in Table I.

6). The α_{i2}-specific antiserum was able to detect recombinant mouse α_{i2} protein in the range of 5–100 ng. The α_{i1}-specific antiserum was also sensitive, though not as sensitive as the α_{i2}-specific antiserum. The α_{i1}-specific antiserum was linear in the range of 10–100 ng. In contrast, the α_{i3}-specific antiserum had low sensitivity. It recognized recombinant human α_{i3} in the 100–400 ng range.

Using such calibration curves, we set out to determine the relative proportions of the G_i proteins in two human erythrocyte preparations (Table II). Initial comparison of values obtained with recombinant mouse α_{i2} protein as a standard with the Coomassie blue and ADP-ribosylation staining profile of human erythrocyte G proteins (Fig. 7) showed a very large discrepancy between the two, since the use of the recombinant protein standard indicated that a very low amount of 40K α_{i2} was present (Table II, middle column, values in parentheses), while the Coomassie blue staining and ADP-ribosylation indicated that the amount of 40K G_α was approximately half that of 41K G_α.

This discrepancy was not totally unexpected since in the 15 amino acid

TABLE I
AMOUNT OF RECOMBINANT α_i-SUBUNIT FUSION PROTEIN IN INSOLUBLE CELLULAR FRACTIONS OF *Escherichia coli* K-38 CELL LYSATES[a]

Sample	Experimental		Calculated	
	Volume (μl)	Densitometric scan (Arbitrary units)	Protein (μg/lane)	Concentration (μg/μl)
BSA standard				
0.25 μg	20	1066		
0.50 μg	20	2014		
1.00 μg	20	2988		
2.00 μg	20	4561		
r-α_{i1}	2	635	0.15	0.75
	4	1109	0.26	0.65
	8	2098	0.56	0.70
	16	2978	1.00	0.62
r-α_{i2}	2	1559	0.38	0.190
	4	2637	0.82	0.205
	8	4201	1.77	0.221
	16	Off scale	—	
r-α_{i3}	2	450	0.11	0.055
	4	736	0.18	0.045
	8	1680	0.41	0.051
	16	2833	0.92	0.058

[a] The insoluble cellular fractions of infected K-38 cell lysates were dissolved in Laemmli sample buffer containing 1% SDS and 20% 2-mercaptoethanol. Indicated volumes of the three α_i-containing solutions were loaded onto SDS-acrylamide gels and analyzed electrophoretically. Varying known concentrations of BSA were also electrophoresed as standards. The gels were stained with Coomassie blue and then destained; BSA standards and resolved proteins are shown in Fig. 5. The wet gels containing the BSA standards and the α_i subunits were scanned by a densitometer. By determining the staining density of the known amounts of BSA, the concentrations of α subunits were determined.

stretch (110–125) used to make the synthetic peptide antigen there are 3 amino acids that are different in the human protein as compared to the mouse protein. Since the antigen used was a peptide encoding the mouse sequence and since recombinant α_{i2} fusion protein is also derived from a mouse cDNA, the α_{i2}-specific antiserum recognized the recombinant protein with high efficiency. The efficiency with which the anti-mouse α_{i2} antiserum would recognize human α_{i2} was not yet known. To resolve the apparent discrepancy, we used known amounts of purified and apparently homogeneous human erythrocyte G_{i2} to construct a second standard curve.

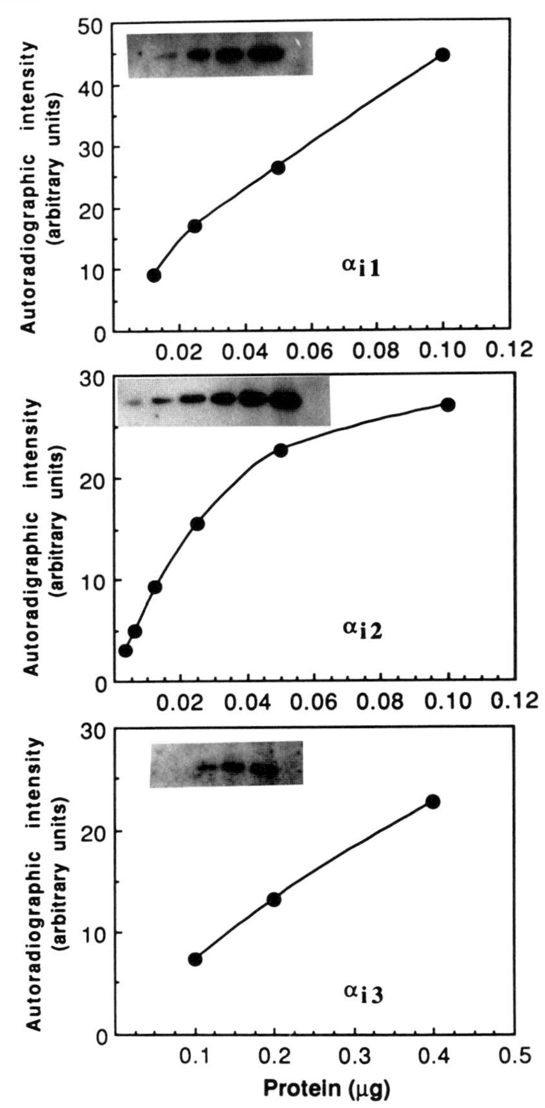

FIG. 6. Correlation of the autoradiographic intensity of the immunoblots with the varying amounts of recombinant α_i proteins for the three α_i-subtype-specific antisera.

TABLE II
RELATIVE AMOUNTS OF THREE G-PROTEIN α_i SUBUNITS IN HUMAN ERYTHROCYTE PREPARATIONS[a]

Preparation number	α subunit present (μg)				Ratio		
	α_{i1}	α_{i2}		α_{i3}	α_{i3}/α_{i1}	α_{i2}/α_{i1}	α_{i3}/α_{i2}
20	0.014	(0.011)	0.36	0.51	36	27	1.41
21	0.012	(0.012)	0.46	0.66	55	38	1.43

[a] Two partially purified preparations of human erythrocyte G proteins were immunoblotted against 1 : 200, 1 : 500, and 1 : 200 dilutions of α_{i1}-, α_{i2}-, and α_{i3}-specific antisera, respectively, in duplicate. The autoradiograms from the immunoblot were scanned with the densitometer, and the protein value corresponding to the autoradiographic intensity was calculated from the curves shown in Fig. 6. For α_{i2} the use of the middle curve in Fig. 6 gave grossly erroneous values. Hence, a standard curve for α_{i2} using highly purified human α_{i2} was constructed similar to the one shown in Fig. 8. For each of the preparations of G proteins, 2 μg total protein was loaded on the gel, resulting in 1 μg of α subunits.

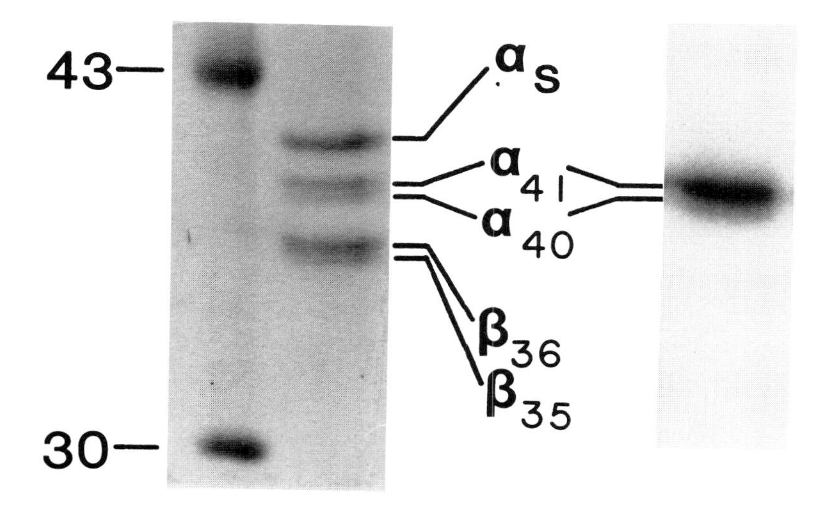

FIG. 7. Coomassie blue staining profile of the 41K and 40K (α_{i3} and α_{i2}, respectively) proteins in a purified human erythrocyte G-protein preparation. The right panel shows that the 41K and 40K proteins are ADP-ribosylated by pertussis toxin.

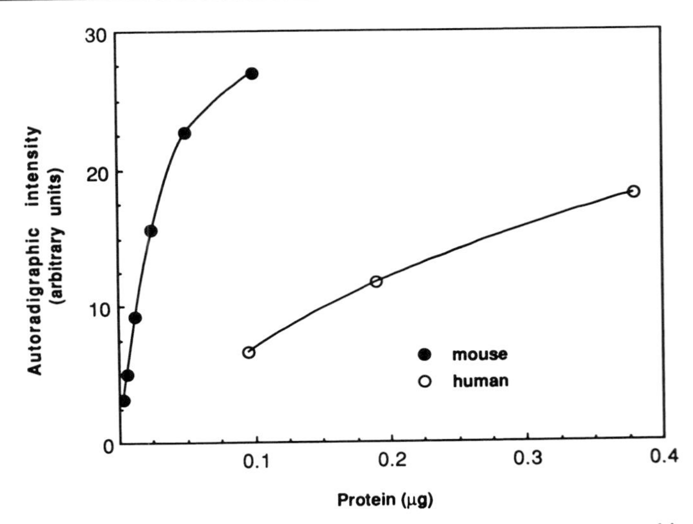

FIG. 8. Differential sensitivities of the α_{i2}-specific antiserum to the recombinant mouse α_{i2} protein and human erythrocyte α_{i2} protein.

Comparison of the two curves shows that the α_{i2} antiserum recognizes mouse α_{i2} much better than the human α_{i2} (Fig. 8). If the human α_{i2} standard curve is used for estimation of the protein concentration of α_{i2} in mixtures of G_i, the result is similar to that predicted by Coomassie blue staining of the 41K and 40K G-protein α subunits.

It is noteworthy that the α_{i2}-specific antiserum retained its subunit specificity in spite of the amino acid changes and the loss of sensitivity. Hence, care should be taken to ascertain that spuriously high or low values of G-protein subunits are not obtained due to the species differences resulting in amino acid substitutions in the region of interest.

In spite of these concerns, the quantification of the G_i proteins has been useful since it has shown that human erythrocyte G_i proteins are primarily G_{i2} and G_{i3} species. Very little G_{i1} was present.

Uses of Quantitative Immunoblotting

We have used quantitative immunoblotting to show that the G_z α subunit is a trace protein in human erythrocytes. A calibration curve using recombinant α_z protein along with a determination of the amount of G_z in 5 μg of various human erythrocyte preparations is shown in Fig. 9. They indicate that there is only 60–100 ng of G_z per microgram of G_i protein in human erythrocyte preparations.[5]

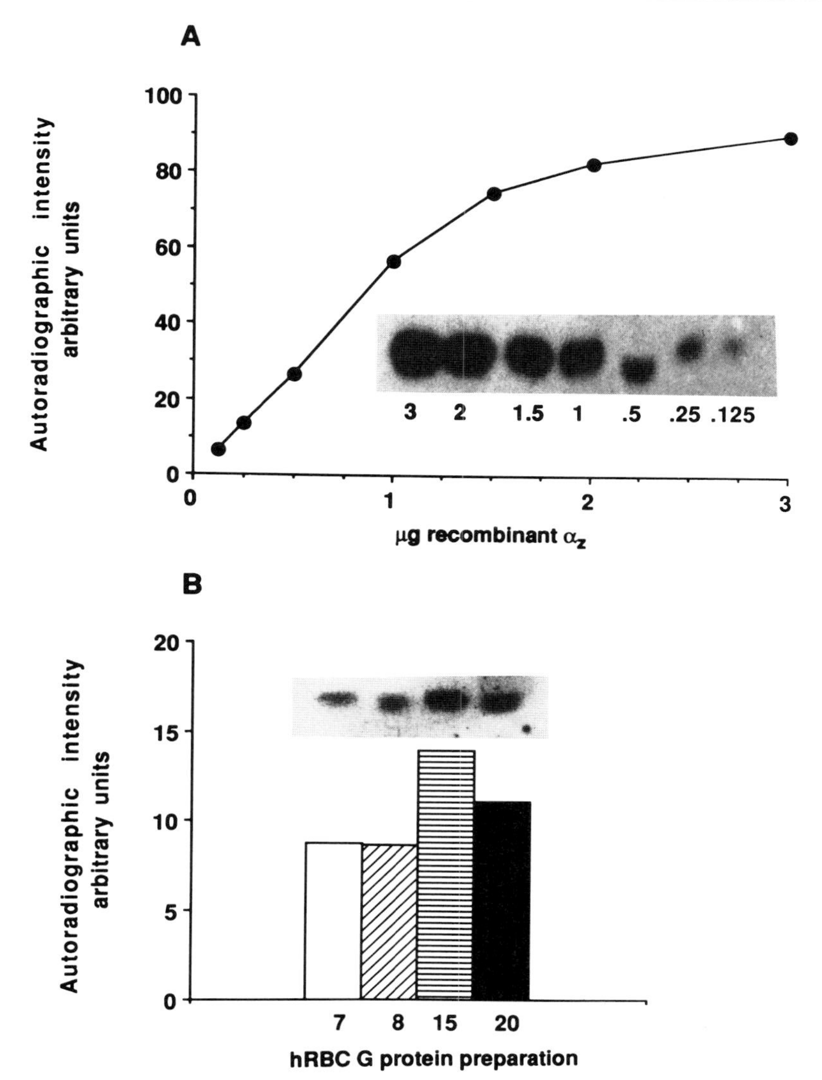

FIG. 9. Quantification of α_z in human erythrocyte G-protein preparations by immunoblotting. (A) Autoradiographic intensity corresponding to various amounts of recombinant α_z subunit. (B) Autoradiographic intensity when 5 μg (total protein) of the indicated human erythrocyte (hRBC) G-protein preparations are used. Using the calibration in (A), we conclude that there is 60–100 ng of α_z per microgram of 40/41K α subunits.

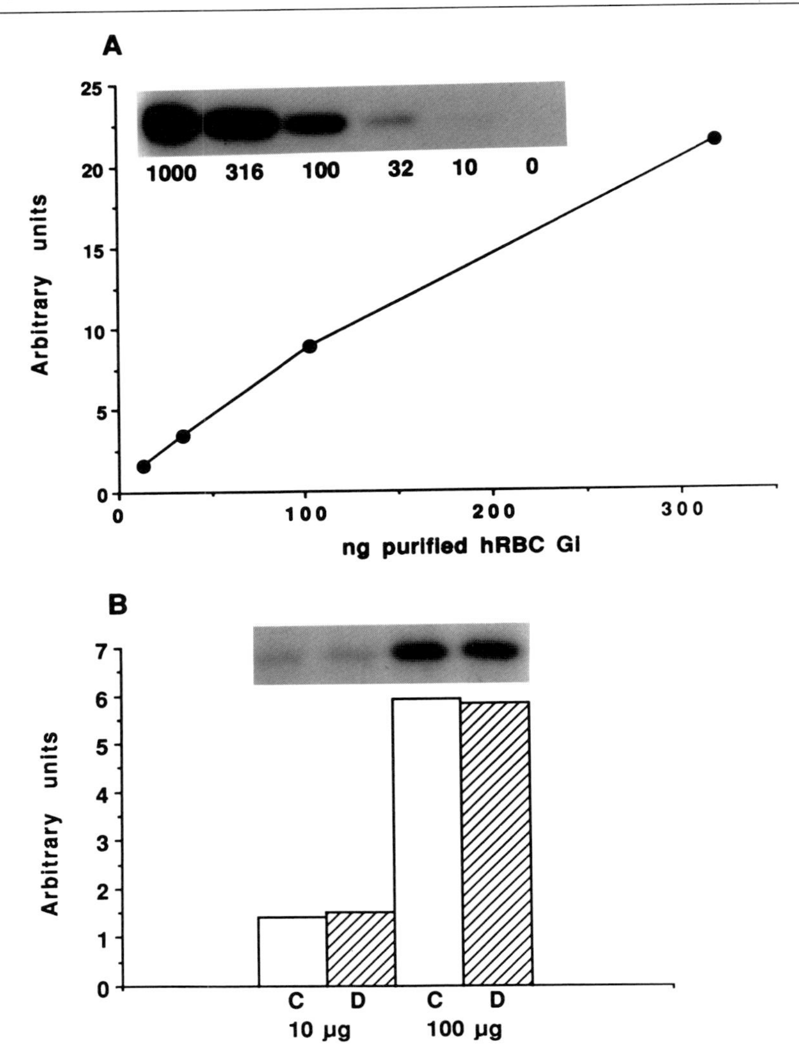

FIG. 10. Quantitative immunoblotting of $\beta\gamma$ dimers in membranes from untreated and glucagon-treated hepatocytes. The β-subunit-specific antiserum was used. (A) Standard curve using varying amounts of purified human erythrocyte G_i. The concentration of $\beta\gamma$ dimers is half that of the total G_i protein. (B) Immunoblotting intensity when 10 and 100 μg of control (C) and treated (D) membranes are used. The experiment shows that there is no change in the levels of $\beta\gamma$ subunits after desensitization. The amount of $\beta\gamma$ dimers in chick hepatocyte membranes can also be determined. It is about 35 ng of $\beta\gamma$ protein per 100 μg of total membrane protein.

We have also used quantitative immunoblotting to determine the amount of β subunits in chick liver membranes after postreceptor heterologous desensitization induced by treatment with glucagon.[3] One possible mechanism of postreceptor desensitization is an increase in $\beta\gamma$ dimers at the cell surface. This would attenuate G_s stimulation of adenylyl cyclase. Hence, we compared the levels of $\beta\gamma$ dimers before and after the induction of desensitization using a β-specific antiserum. Since we do not have recombinant β subunits expressed in *E. coli*, we used purified G_i protein from human erythrocytes as standards. By Coomassie blue staining, the β subunits are about 50% of the G_i proteins.[8] When varying amounts of G_i were electrophoresed, transferred to nitrocellulose papers, and blotted with a β-subunit-specific antiserum, we found that the signal was proportional to the amount of β subunit present (Fig. 10A). Under these linear conditions, we compared equivalent amounts of membrane protein from control and desensitized chick hepatocytes and found that the levels of $\beta\gamma$ dimers were essentially unaltered (Figure 10B). It is noteworthy, however, that even when crude membranes are used, the signals are in the proportional range (compare the 10 and 100 μg membrane proteins). Thus, it has been possible to quantify G-protein subunits even in crude membrane if antisera of appropriate sensitivity are available.

Acknowledgments

Research in our laboratory is supported by National Institutes of Health Grants DK 38761 and CA 44998. We thank Drs. A. G. Gilman and A. S. Spiegel for many of the antisera used in these studies.

[8] J. Codina, G. D. Hildebrandt, R. D. Sekura, M. Birnbaumer, J. Bryan, C. R. Manclark, R. Iyengar, and L. Birnbaumer, *J. Biol. Chem.* **259,** 5871 (1984).

[27] Assay of G-Protein $\beta\gamma$-Subunit Complex by Catalytic Support of ADP-Ribosylation of $G_{o\alpha}$

By PATRICK J. CASEY, IOK-HOU PANG, and ALFRED G. GILMAN

Interactions between the α and $\beta\gamma$ subunits of G proteins are believed to lie at the mechanistic nucleus of their role in transmembrane signaling.[1] The majority of the evidence that supports this assumption has come from

[1] P. J. Casey and A. G. Gilman, *J. Biol. Chem.* **263,** 2577 (1988).

analysis of the ability of a purified G-protein subunit to modulate not only the activities of its partner but also those of other proteins in the signaling pathway. Thus, assays for the subunits based on protein–protein interactions provide an accurate estimate of the concentration of "active" protein in the preparation. The chief problem with an assay of this type is the complicated nature of the techniques involved, rendering it unsuitable for routine use. However, the observations that the G-protein $\beta\gamma$-subunit complex is required to support ADP-ribosylation of the α subunit of G_o ($G_{o\alpha}$) by pertussis toxin[2,3] and the relative ease of purification of $G_{o\alpha}$[4] suggested a simple assay for the activity of the $\beta\gamma$ complex.[5] This chapter describes this procedure; its use enables one to measure subpicomole amounts of functional $\beta\gamma$ accurately and relatively rapidly.

Preparation of Proteins

Both $G_{o\alpha}$[4] and $\beta\gamma$[6] are readily purified from bovine brain membranes. For complete resolution of either protein from its respective partner, it is often necessary to perform an additional step after the final heptylamine-Sepharose columns. This step involves chromatography on a Pharmacia (Piscataway, NJ) fast protein liquid chromatography (FPLC) Mono Q column in the presence of $AlCl_3$, $MgCl_2$, and NaF (AMF); the procedure has been described for the separation of $G_{i\alpha}$ from $\beta\gamma$.[5] Briefly, the preparation is injected onto a 0.5 × 5.0 cm Mono Q column equilibrated in 50 m*M* Tris-HCl (pH 8.0), 1 m*M* EDTA, 1 m*M* dithiothreitol (DTT), 1.0% sodium cholate, 20 μM $AlCl_3$, 6 m*M* $MgCl_2$, and 10 m*M* NaF. The column is eluted with a 30-ml gradient of 0–250 m*M* NaCl in the same buffer. $G_{o\alpha}$ elutes from this column at about 50 m*M* NaCl, while $\beta\gamma$ elutes at approximately 170 m*M*. If desired, 1.0% octylglucoside can be substituted for sodium cholate in the chromatography buffers. If no FPLC is available, a 2-ml column of DEAE-Sephacel or equivalent can be substituted. The purified subunits are concentrated to greater than 1 mg/ml by pressure filtration through an Amicon PM30 membrane or by centrifugation through a Centricon 30 concentrator (Amicon, Danver, MA). Removal of AMF and exchange into Lubrol 12A9 are accomplished by gel filtration through Sephadex G-50 equilibrated in 50 m*M* Tris-HCl (pH 8.0), 1 m*M* EDTA, 1 m*M* DTT, and 0.05% Lubrol 12A9. The purified proteins are aliquoted,

[2] R. M. Huff and E. J. Neer, *J. Biol. Chem.* **261,** 1105 (1986).
[3] T. Katada, M. Oinuma, and M. Ui, *J. Biol. Chem.* **261,** 8182 (1986).
[4] P. C. Sternweis and J. D. Robishaw, *J. Biol. Chem.* **259,** 13806 (1984).
[5] P. J. Casey, M. P. Graziano, and A. G. Gilman, *Biochemistry* **28,** 611 (1989).
[6] D. J. Roof, M. L. Applebury, and P. C. Sternweis, *J. Biol. Chem.* **260,** 16242 (1985).

flash-frozen, and stored at $-70°$. However, both proteins are quite stable to freezing and thawing if done quickly, and the proteins are held on ice during use.

ADP-Ribosylation Assay

Materials

Stock Solutions

1 *M* Tris-HCl, pH 8.0: store at 2°
0.1 *M* Sodium EDTA, pH 8.0: store at 2°
1 *M* DTT: store at $-20°$ in aliquots
10% Lubrol 12A9 (or Lubrol PX): prepared and deionized with mixed-bed resin AG 501 (Bio-Rad, Richmond, CA); store at 2°
1 *M* $MgCl_2$: store at 2°
100 μM NAD (Sigma, St. Louis, MO): store at $-20°$ in aliquots
10 m*M* GDP (Sigma): store at $-20°$ in aliquots
30 m*M* Dimyristoylphosphatidylcholine (DMPC): suspend 10 mg of DMPC (Sigma) in 0.5 ml of 20 m*M* sodium HEPES (pH 8.0), 2 m*M* $MgCl_2$, 1 m*M* EDTA; store at $-20°$ and sonicate 5 min at room temperature before each use
100 μg/ml Pertussis toxin (PT, List Biologicals, Campbell, CA): add 0.5 ml of 2 *M* urea, 100 m*M* potassium phosphate, pH 7.0, to one vial (50 μg) of PT; the solution is stable for at least 2 months when stored at 2°
[^{32}P]NAD (New England Nuclear, Boston, MA), #NEG 023X
Stop solution: 2% SDS, 50 μM NAD; store at $-20°$
30% and 6% solutions of trichloroacetic acid (TCA): store at room temperature

Working Solutions

Buffer A: 50 m*M* Tris-HCl (pH 8.0), 1 m*M* sodium EDTA, 1 m*M* DTT, 0.025% Lubrol 12A9
$G_{o\alpha}$: dilute stock $G_{o\alpha}$ to 1.5 μM in buffer A; the protocol requires 10 μl per assay tube; hold on ice during use
$\beta\gamma$: prepare solutions of 0.02 and 0.2 μM $\beta\gamma$ by dilution of stock protein in buffer A; hold on ice during use
PT mix, 400 μl (for 25-tube assay, prepare fresh):
20 μl of 1 *M* Tris-HCl, pH 8.0
2.2 μl of 1 *M* DTT
26.7 μl of 100 μM NAD
2 μl of 1 *M* $MgCl_2$

10 μl of 10 mM GDP
53.3 μl of 100 μg/ml PT
19 μl of 30 mM DMPC
[^{32}P]NAD, approximately 2×10^7 counts per minute (cpm)
Water to make 400 μl

Procedure

The assay is conducted by mixing $G_{o\alpha}$ (10 μl, 15 pmol) with the solution containing $\beta\gamma$ (standards or samples) in 12 × 75 mm polypropylene tubes on ice. Buffer A is added to a final volume of 25 μl. The range of $\beta\gamma$ used for the standard curve is 0.05 to 2.0 pmol. Appropriate dilutions (in buffer A) of the samples to be analyzed are processed with the standards. At timed intervals, 15 μl of PT mix is added to each tube (final volume 40 μl), and tubes are then transferred to a 30° water bath. The reaction is stopped after 20 min by the addition of 0.5 ml of the stop solution at room temperature. The proteins are then precipitated by the addition of 0.5 ml of 30% TCA to each tube. Ice-cold 6% TCA (2 ml) is added to the sample, which is immediately filtered through a BA85 nitrocellulose filter (Schleicher & Schuell, Keene, NH). The filters are washed with an additional 14 ml of 6% TCA, dried, suspended in scintillation cocktail, and subjected to liquid scintillation spectrometry. The specific activity of the [^{32}P]NAD is quantitated by spotting 5 μl of the PT mix on a dry filter and counting this with the samples (5 μl = 33.3 pmol NAD).

Results

Figure 1 shows a typical standard curve for the assay. Since $\beta\gamma$ acts catalytically to support the ADP-ribosylation of $G_{o\alpha}$, it is important that the incubations are timed correctly. Although the stoichiometry of ADP-ribosylation of $G_{o\alpha}$ varies somewhat between preparations of this protein, the ability to quantitate $\beta\gamma$ by this procedure is essentially unaffected. The most accurate results are obtained when the samples contain 0.05–0.4 pmol of $\beta\gamma$, since the ADP-ribosylation shows essentially a linear dependence on $\beta\gamma$ in this range. This assay does not appear to discriminate between different molecular species of $\beta\gamma$, since the protein purified from bovine brain, which contains two different β subunits and multiple forms of γ, gives the same result as that purified from retina, which contains a single species of β and γ (Fig. 1, see also Ref. 5).

The assay described above has proved extremely useful in determining the concentration of active $\beta\gamma$, not only in samples that contain little or no PT-substrate G protein α subunits but also in samples that contain

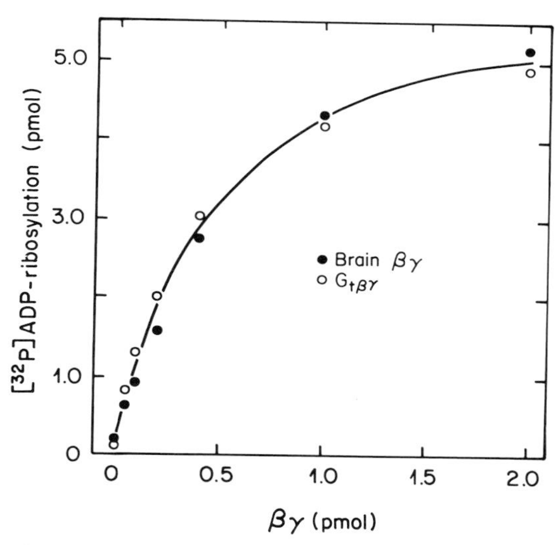

FIG. 1. Standard curve of $\beta\gamma$-supported ADP-ribosylation of $G_{o\alpha}$ by pertussis toxin. The $\beta\gamma$ used was purified from either bovine brain G proteins or from the retinal G protein, G_t. See text for procedure. (Reproduced with permission from Ref. 5.)

significant amounts of these polypeptides. As shown in Fig. 2, the ADP-ribosylation of $G_{o\alpha}$ at a fixed concentration of $\beta\gamma$ plateaus at relatively low concentrations of $G_{o\alpha}$ when the 20-min incubation period is used. Thus, by conducting the assay at concentrations of $G_{o\alpha}$ above 10 pmol/25 μl, interference caused by ADP-ribosylation of α subunits in the unknown sample is minimized. This is more clearly demonstrated in Fig. 3, which shows $\beta\gamma$ standard curves conducted in the presence of either 10 or 20 pmol of $G_{o\alpha}$. Though not identical, these standard curves are very similar, especially at levels of $\beta\gamma$ less than or equal to 0.3 pmol. The difference between the two curves can be viewed as an overestimation introduced by the additional 10 pmol of $G_{o\alpha}$. This error, at $\beta\gamma$ levels no higher than 0.3 pmol, is always less than 30%/10 pmol additional $G_{o\alpha}$ (i.e., <3%/pmol additional $G_{o\alpha}$). Since this overestimation is increased at higher concentrations of $\beta\gamma$ (Fig. 3), this region of the standard curve should not be used when samples that contain significant concentrations of PT-substrate α subunits are to be assayed. With this precaution in mind, one can accurately assay $\beta\gamma$ in samples that contain 3-fold excesses of PT-substrate α subunits; furthermore, if a 10% error is tolerable, the assay is useful for samples that contain 10 times more α subunit than $\beta\gamma$. If the supply of $G_{o\alpha}$ is not a limiting factor, performance of the assay in the

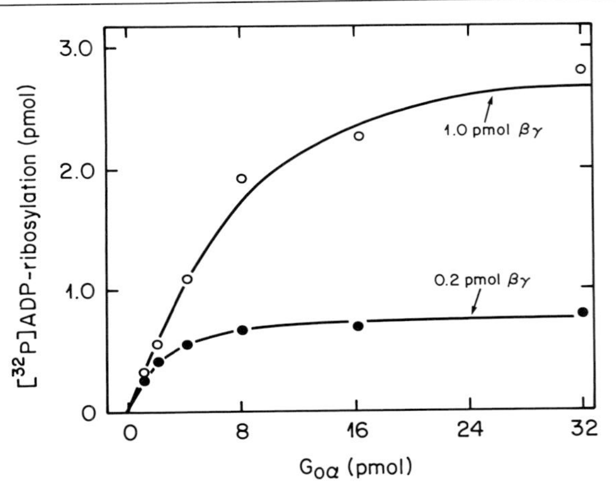

FIG. 2. Effect of the concentration of $G_{o\alpha}$ on $\beta\gamma$-supported ADP-ribosylation. (Reproduced with permission from Ref. 5.)

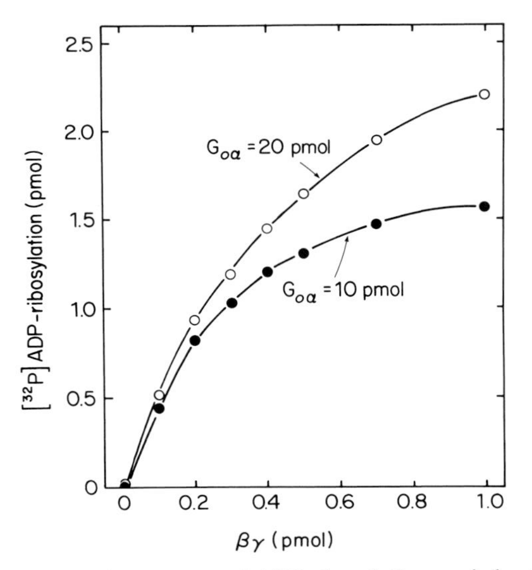

FIG. 3. Standard curves of $\beta\gamma$-supported ADP-ribosylation carried out in the presence of either 10 or 20 pmol of $G_{o\alpha}$.

presence of 25 pmol or more of $G_{o\alpha}$ results in an essentially α-independent $\beta\gamma$ response up to 1.0 pmol of $\beta\gamma$ in the sample.

Since we have utilized these assays primarily to determine functional concentrations of $\beta\gamma$ in highly purified preparations of these proteins, we have not conducted a systematic analysis of the interfering effects of detergents or other agents. However, owing to the sensitivity of this assay (50 fmol of $\beta\gamma$ yields a readily detectable signal), dilution of samples into buffer A obviates the majority of these concerns. Assessment of the effect of components of the sample on the signal generated by the standards and the ability to detect $\beta\gamma$ spiked into the sample will quickly allow one to determine if this procedure will work with the sample in question.

Acknowledgments

Work from the authors' laboratory was supported by United States Public Health Service Grant GM34497, American Cancer Society Grant BC555I, and the Raymond and Ellen Willie Chair of Molecular Neuropharmacology. P. Casey is the recipient of National Institutes of Health Postdoctoral Fellowship GM11982. We also acknowledge support from the Perot Family Foundation and The Lucille P. Markey Charitable Trust.

[28] Tryptophan Fluorescence of G Proteins: Analysis of Guanine Nucleotide Binding and Hydrolysis

By TSUTOMU HIGASHIJIMA and KENNETH M. FERGUSON

Introduction

Cell proliferation, vision, protein synthesis, secretion, and the response of cells to certain hormones are among the processes influenced by the action of guanine nucleotide-binding regulatory proteins.[1-5] When bound to GTP, these modulators cause certain target proteins to perform their specific function. This action of the regulatory proteins is terminated when the bound GTP is hydrolyzed to GDP. Thus, analysis of guanine nucleotide-dependent processes frequently involves the study of the switching between the GTP- and GDP-bound forms of guanine nucleotide-

[1] M. Barbacid, *Annu. Rev. Biochem.* **56,** 779 (1987).
[2] L. Stryer, *Annu. Rev. Neurosci.* **9,** 87 (1986).
[3] K. Moldave, *Annu. Rev. Biochem.* **54,** 1109 (1985).
[4] A. Salminen and P. J. Novick, *Cell (Cambridge, Mass.)* **49,** 527 (1987).
[5] A. G. Gilman, *Ann. Rev. Biochem.* **56,** 615 (1987).

binding regulatory proteins. For some of these proteins, the change in state is accompanied by a change in fluorescence.[6-10] Proteins that contain one or more tryptophan, tyrosine, or phenylalanine residues are fluorescent.[11,12] The observed emission spectrum depends on the number and type of fluorescent amino acids that the protein contains. Changes in the relative position of neighboring amino acids can alter the emission spectrum of a particular fluorescent amino acid. Interactions with ligands or with other proteins may perturb the structure of the protein under study, thereby changing its fluorescence spectrum and providing a way to monitor the interactions.

Certain types of receptors convey their signals to the cell by means of a particular class of guanine nucleotide-binding proteins known as G proteins. These proteins are oligomers of a guanine nucleotide-binding subunit (G_α) and two other subunits (G_β and G_γ). Biochemical studies have identified several classes of G proteins, among them G_s, G_i, and G_o, whose properties have been reviewed.[5] The nucleotide-binding subunits of the G proteins contain relatively few tryptophan residues, and at least one of these residues is affected by nucleotide binding. $G_{o\alpha}$ contains two tryptophan residues, and its fluorescence intensity increases when GTP binds. Extrapolation from the X-ray diffraction-determined structures of human H-*ras* and of elongation factor Tu, which are guanine nucleotide-binding proteins homologous to the G proteins, suggests that the tryptophan at position 212 of $G_{o\alpha}$ lies near the nucleotide-binding pocket and thus may be responsible for this fluorescence signal.[13,14] The presence of fluorescent residues in addition to those that are affected by structural changes decreases the relative fluorescence change, an effect that is evident among different kinds of G proteins. The magnitude of the nucleotide-induced change in fluorescence is correlated inversely with the total number of tryptophans in the protein. The relative increase in fluorescence intensity is greatest for $G_{o\alpha}$, less for $G_{i\alpha}$, which has three tryptophan residues, and the least for $G_{s\alpha}$, which has four tryptophan residues. Similarly, the α subunits of the G proteins display a greater relative fluorescence increase than do the oligomeric forms because the $G_{\beta\gamma}$ complex contains tryptophan

[6] K. Arai, T. Arai, M. Kawakita, and Y. Kaziro, *J. Biochem.* (*Tokyo*) **81,** 1335 (1977).

[7] M. P. Graziano, M. Freissmuth, and A. G. Gilman, *J. Biol. Chem.* **264,** 409 (1989).

[8] T. Higashijima, K. M. Ferguson, P. C. Sternweis, E. M. Ross, M. D. Smigel, and A. G. Gilman, *J. Biol. Chem.* **262,** 752 (1987).

[9] W. J. Phillips and R. A. Cerione, *J. Biol. Chem.* **263,** 15498 (1988).

[10] R. A. Kahn and A. G. Gilman, *J. Biol. Chem.* **261,** 7906 (1986).

[11] F. W. J. Teale and G. Weber, *Biochem. J.* **65,** 480 (1957).

[12] R. F. Chen, *Anal. Lett.* **1,** 37 (1967).

[13] S. Masters, R. M. Stroud, and H. R. Bourne, *Protein Eng.* **1,** 47 (1986).

[14] S. R. Holbrook and S.-H. Kim, *Proc. Natl. Acad. Sci. U.S.A.* **86,** 1751 (1989).

residues whose fluorescence is not affected by nucleotide binding to G_{α}. The changes in the fluorescence of G proteins can be used to measure the kinetics of nucleotide binding and of GTP hydrolysis.

Fluorescence Measurements of G Proteins

The excitation spectrum of tryptophan has a maximum at 280 nm while the maximum for tyrosine occurs at 275 nm and for phenylalanine at 260 nm.[11] For experiments other than the measurement of the fluorescence spectrum, an excitation wavelength of 290 nm is used as this provides an adequate stimulation of fluorescence tryptophan while minimizing the excitation of tyrosine. The fluorescence emission of tryptophan attains a maximum between 340 and 350 nm, depending on the particular conditions.

The following procedures were developed using a Spex Fluorolog 211 spectrophotometer equipped with a 450-W xenon arc lamp, a double-excitation monochromator, and single-emission monochromator. The optical bandwidth of the excitation monochromator was set at 2.25 nm and that of the emission monochromator was set at 4.5 nm. The sample cuvette was maintained at a 20°. Any similarly equipped fluorescence spectrophotometer should be adequate.

The intensity of light emitted by the lamp of the spectrophotometer changes with time and depends on the wavelength selected. To compensate for this variation, all measurements are made in ratio mode, wherein the signal from the sample is compared with the signal from a reference beam. When spectra are recorded, a rhodamine solution (8 g/liter rhodamine dissolved in 1,2-propanediol) is used in the reference chamber. The intensity of the signal from the rhodamine solution decreases with time; therefore, during experiments where the rate of change of the fluorescence intensity is measured, the rhodamine solution is replaced with a neutral density filter that has an optical density of 3.

Various size quartz cuvettes are available. The typical 10 × 10 mm cuvettes are convenient because the solution can be mixed efficiently with a stir bar and a magnetic stirrer during the fluorescence measurements. Such cells require about 1 ml of G-protein solution, however, so substitution of a 5 × 5 mm cuvette, which requires 400 μl of solution, or a 3 × 3 mm cuvette, which requires 200 μl, may be preferred. When 3 × 3 mm cells are used, photobleaching of the sample during prolonged observations occurs, presumably because the sample is mixed less efficiently in these cuvettes. These cells can be used if the exposure to the excitation beam is minimized by making observations for short periods (2 sec) at 1-min intervals and blocking the beam between measurements. A series of short records rather than a continuous record is obtained. If continuous records are required, the 5 × 5 mm cell is used, as there is no noticeable bleaching of the sample.

All reagents should be free from fluorescent contaminants. Water comparable to that provided by a Milli-Q (Millipore, Bedford, MA) purification system and typical reagent-grade chemicals are usually adequate.

The optical density of the sample at wavelengths corresponding to the fluorescence emission maximum of the protein and to its absorbance maximum should be kept low to prevent attenuation of the fluorescence emission or attenuation of the excitation energy. A sufficient precaution is to limit the optical density of the protein in the buffer solution to less than 0.3. This corresponds to G-protein concentrations of less than 10 μM at 290 nm excitation or 3 μM at 280 nm in a 1 × 1 cm cuvette. Avoid high concentrations of nucleotides because they absorb a portion of the excitation energy, which reduces the fluorescence emission intensity. The fluorescence of liposome-containing samples can be reliably measured if the optical density is less than 0.3 and the light scattering is constant.[9]

Prepare G_α by any of the published procedures.[15–18] For the procedures below, the buffer solution (buffer A) is composed of 50 nM sodium HEPES, pH 8.0, 1 mM sodium EDTA, 1 mM dithiothreitol (DTT), and 0.1% Lubrol.

Measurement of Fluorescence Spectra of G_α

1. Prepare three 400-μl samples of 400 nM G_α, one in buffer A, one in buffer A containing 1 μM GTPγS [guanosine 5′-(3-*O*-thio)triphosphate, a hydrolysis-resistant guanine nucleotide analog], and one in buffer A containing 1 μM GTPγS and 10 mM $MgSO_4$. Incubate the samples at 20° for 30 min.

2a. For the emission spectrum, set the excitation monochromator at 290 nm. Place each sample in turn in a 5 × 5 mm quartz cuvette in the spectrophotometer and record its emission spectrum by changing the emission monochromator from 300 to 400 nm at a rate of 2 nm/sec. Repeat the procedure using 400 μl of each buffer to obtain a record of the emission spectrum of the buffer without G protein.

2b. For the excitation spectrum, set the emission monochromator at

[15] G. M. Bokoch, T. Katada, J. K. Northup, M. Ui, and A. G. Gilman, *J. Biol. Chem.* **259,** 3560 (1984).

[16] P. C. Sternweis, J. K. Northup, M. D. Smigel, and A. G. Gilman, *J. Biol. Chem.* **256,** 11517 (1981).

[17] P. C. Sternweis and J. D. Robishaw, *J. Biol. Chem.* **259,** 13806 (1984).

[18] J. Codina, W. Rosenthal, J. D. Hildebrandt, L. Birnbaumer, and R. D. Sekura, this series, Vol. 109, p. 446.

340 nm. Record the excitation spectrum of each sample by changing the excitation monochromator from 250 to 335 nm at a rate of 2 nm/sec. Repeat the procedure with each buffer solution.

3. Subtract the appropriate buffer spectrum from each protein sample spectrum to obtain the fluorescence spectrum of the G protein. The buffer spectrum will contain a peak caused by Raman scattering of the excitation beam by water molecules. This peak is independent of the protein concentration and has a maximum at a wavelength that differs from the excitation wavelength by a constant amount. Therefore, the Raman peak appears at 320 nm when the sample is irradiated at 290 nm ($1/\lambda_{\mathrm{Raman}} = 1/\lambda_{\mathrm{excitation}} - 0.00033\ \mathrm{nm}^{-1}$).[19]

Measurement of Rate of GTPγS Binding to G_α

1. Prepare 400 μl of a 180 n*M* G_α solution in buffer A containing 6 m*M* $MgSO_4$ and place in a 5 × 5 mm cuvette in the spectrophotometer. Stir the sample during the following steps.

2. Set the excitation monochromator at 290 nm and the emission monochromator at 340 nm. Record the fluorescence intensity for 1–2 min.

3. Add 4 μl of 100 μ*M* GTPγS to the cuvette and continue recording.

4. Determine the rate of GTPγS association using the information on data analysis presented below.

Measurement of GTP Hydrolysis by G_α

Method A

1. Set the excitation monochromator at 290 nm and the emission monochromator at 340 nm.

2. Place 400 μl of a 200 n*M* G_α solution in buffer A in a 5 × 5 mm cuvette in the spectrophotometer and stir the sample in the cuvette continuously. Record the fluorescence intensity for 4 min.

3. Add 4 μl of 1 m*M* GTP and continue recording for 9 min.

4. Add 4 μl of 1 *M* $MgSO_4$ and continue recording for an additional 10 min.

5. Analyze the changes in the fluorescence intensity using the procedures described below.

Method B

1. Set the excitation monochromator at 290 nm and the emission monochromator at 340 nm.

[19] D. M. Jameson, *in* "Fluorescence Hapten: An Immunological Probe" (E. Voss, ed.), p. 23. CRC Press, Boca Raton, Florida, 1984.

2. Place 400 μl of a 200 nM G_α solution in buffer A containing 10 mM $MgSO_4$ in a 5 × 5 mm cuvette in the spectrophotometer and stir the sample continuously. Record the fluorescence intensity for 4 min.

3. Add 4 μl of 100 μM GTP and continue recording for 5 min.

4. Add 4 μl of 1 mM GDP and continue recording for an additional 10 min.

5. Repeat Steps 2 and 3. Add 4 μl of 1 mM GTP and continue recording for an additional 10 min.

6. Analyze the changes in the fluorescence intensity using the procedures described below.

Analysis of Results

Fluorescence Spectra

The spectra of G_α indicate that tryptophan residues are the primary contributors to the fluorescence of the protein. The maximum fluorescence excitation occurs at 280 nm, and the maximum fluorescence emission occurs at 347 nm for the GDP-bound protein. The fluorescence emission maximum shifts to 343 nm when the protein binds GTPγS in the presence of Mg^{2+}, and the fluorescence intensity increases.[8]

Guanine Nucleotide Binding to G_α

The usual methods of purifying G proteins yield molecules whose nucleotide-binding site contains GDP.[20] The rate of dissociation of GDP is much less than the rate of nucleotide binding to nucleotide-free G protein and, in certain types of experiments, determines the rate of any observed fluorescence changes. For example, the rate of GTPγS binding is the same as the rate of GDP dissociation.[20] The analysis of GTP binding is more complex because the hydrolysis of the nucleotide must be considered. The model below embodies these features of the interaction of G_α with nucleotides and Mg^{2+}.[21,22]

$$\begin{array}{ccc} G_\alpha + Mg^{2+} + GTP + GDP & \underset{k_{-o}}{\overset{k_o}{\rightleftharpoons}} & G_\alpha \cdot GTP + Mg^{2+} \\ k_2 \downarrow\uparrow k_{-2} & & k_1 \downarrow\uparrow k_{-1} \\ G_\alpha \cdot GDP + Mg^{2+} + P_i & \underset{k_{cat}}{\longleftarrow} & G_\alpha \cdot GTP \cdot Mg^{2+} \end{array}$$

[20] K. M. Ferguson, T. Higashijima, M. D. Smigel, and A. G. Gilman, *J. Biol. Chem.* **261,** 7393 (1986).

[21] T. Higashijima, K. M. Ferguson, M. D. Smigel, and A. G. Gilman, *J. Biol. Chem.* **262,** 757 (1987).

[22] T. Higashijima, K. M. Ferguson, P. C. Sternweis, M. D. Smigel, and A. G. Gilman, *J. Biol. Chem.* **262,** 762 (1987).

The binding of Mg^{2+}, GTP, and GTPγS is relatively rapid at the concentrations employed in these experiments, so either GDP dissociation or GTP hydrolysis is rate limiting. In addition, these concentrations exceed the apparent affinity constants of the G proteins for each of these ligands.

GTPγS Binding to G_α

The rate of binding of GTPγS to the protein is limited by the rate of GDP dissociation. When the nucleotide is added, the fluorescence increases exponentially [Eq. (1)].[8] I_{obs} is the observed fluorescence intensity, I_{GDP} is the intensity of the GDP-bound protein, and $I_{GTP\gamma S}$ is the intensity of the GTPγS-bound protein. The rate constant can be obtained by fitting the data to the model [Eq. (1)] with a nonlinear least-squares procedure. Alternatively, the rate constant can be obtained by a linear least-squares analysis of transformed data [Eq. (2)]. The rate of GTPγS binding to, and therefore GDP dissociation from, $G_{o\alpha}$ determined by this method is 0.3 min^{-1}, which is consistent with direct binding measurements.[20]

$$I_{obs}(t) = I_{GDP} + (I_{GTP\gamma S} - I_{GDP})(1 - e^{-kt}), \qquad k = k_{-2} \tag{1}$$

$$\ln[(I_{GTP\gamma S} - I_{obs})/(I_{GTP\gamma S} - I_{GDP})] = -kt \tag{2}$$

GTPase Activity of G_α

The procedure described in Method A has two phases. During the first phase, the protein and nucleotides achieve a steady state of GTP-bound protein (Fig. 1). The second phase is initiated when Mg^{2+} is added, which causes a rapid increase in fluorescence intensity. This is thought to represent the formation of a $G_\alpha \cdot GTP \cdot Mg^{2+}$ complex from the preexisting $G_\alpha \cdot GTP$. Subsequently, the fluorescence intensity exponentially declines as the bound GTP is hydrolyzed. The rate constant characterizing this exponential decline is $(k_{cat} + k_{-2})$ [Eq. (3)] and is obtained by either nonlinear least-squares analysis or by linear least-squares analysis of transformed data. I_{GTP} is the intensity of the $G_\alpha \cdot GTP \cdot Mg^{2+}$. A new steady state is reached which is determined by the ratio of the catalytic rate constant and the dissociation rate of GDP. The rate constant describing the fluorescence decline of $G_{o\alpha}$ is 2.2 min^{-1}, and the calculated hydrolysis rate is 1.9 min^{-1}.[21]

$$I_{obs}(t) = I_{GTP} - (I_{GTP} - I_{GDP})[k_{cat}/(k_{-2} + k_{cat})](1 - e^{-kt}); \qquad k = (k_{cat} + k_{-2}) \tag{3}$$

When Mg^{2+} and GTP are added simultaneously, the system approaches steady state with a rate constant equal to $(k_{cat} + k_{-2})$ [Eq. (4)]. After this steady state is reached, the addition of GDP prevents the rebinding of GTP, which causes the fluorescence intensity to decrease exponentially

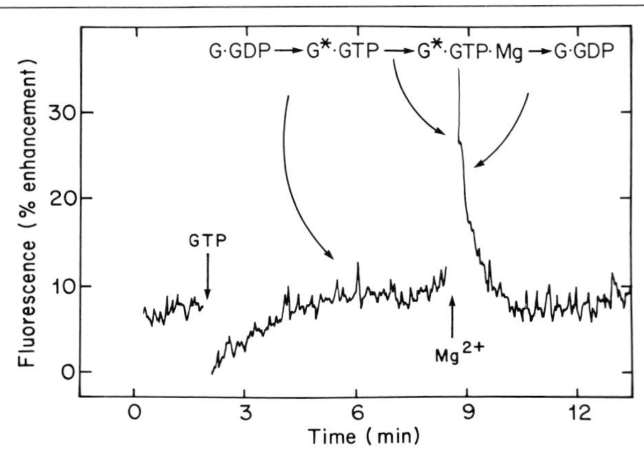

FIG. 1. GTP-induced change in the fluorescence of $G_{o\alpha}$. The fluorescence emission of $G_{o\alpha}$ is measured as described in Method A. The fluorescence of the protein (G · GDP) is recorded for 2.5 min and then 4 μl of 1 m*M* GTP is added, which causes an immediate decline in the fluorescence intensity due to its absorption of the excitation energy. As the protein binds GTP its fluorescence intensity increases and reaches a steady state (G* · GTP). Upon the addition of 4 μl of 1 *M* $MgCl_2$ the fluorescence intensity rapidly increases (forming G* · GTP · Mg^{2+}) and subsequently declines as the GTP is hydrolyzed. (Modified from Ref. 21 and used with permission.)

with a rate constant of $(k_{cat} + k_{-x})$ [Eq. (5)]. In this case, either hydrolysis or GTP dissociation reduces the fluorescence. The model shown above does not include the possibility of GTP dissociation from the Mg^{2+}-bound protein, and the constant k_{-x} is introduced to describe the rate of this process. Whether GTP dissociates appreciably from the Mg^{2+}-bound protein is unknown, because of the difficulty of measuring the dissociation of GTP during a Mg^{2+}-dependent hydrolysis reaction. The magnitude of k_{-x} is probably small because the rate of the fluorescence change measured by either of these procedures is the same. The addition of GTP rather than GDP should not affect the steady state. Its absorbance at 290 nm will alter the fluorescence intensity slightly by reducing the intensity of the excitation beam.

$$I_{obs}(t) = I_{GDP} + (I_{GTP} - I_{GDP})[k_{-2}/(k_{-2} + k_{cat})](1 - e^{-kt}); \quad k = (k_{-2} + k_{cat}) \tag{4}$$

$$I_{obs}(t) = I_{GDP} + (I_{GTP} - I_{GDP})[k_{-2}/(k_{-2} + k_{cat})](e^{-kt}); \quad k = (k_{cat} + k_{-x}) \tag{5}$$

[29] Reconstitution of Receptors and G Proteins in Phospholipid Vesicles

By RICHARD A. CERIONE and ELLIOTT M. ROSS

Introduction

Reconstitution of complex multiprotein systems is a standard method for both detecting and quantitating the activities of the components, as well as for studying the mechanism of their interaction. This approach has been central to the study of membrane-bound, G protein-mediated signaling systems such as adenylyl cyclase (EC 4.6.1.1, adenylate cyclase).

The development of a particular reconstituted assay system is largely dictated by the experimental goals. In general, these include the enumeration and identification of the components; the quantitative study of their interactions and of the contribution of partial reactions to overall activity; and the study of how the physicochemical properties of the membrane influence these reactions. An additional consideration is that, in the cases of most hormone receptors and G proteins, the ability to monitor functional interactions between these signaling proteins absolutely requires that they be incorporated into a lipid environment (phospholipid vesicles) that contains little or no detergent.

This chapter stresses the techniques of reconstitution and the data available from reconstitutive assays. A longer review by Ross[1] discusses the rationale and strategies of this approach. Other reviews on this topic have also appeared.[2–4]

General Considerations

Prior to developing a reconstituted signaling system, the investigator must consider such quantitative questions as the sensitivity of the assay that will be used to monitor reconstitution and the concentrations of the protein components to be reconstituted. The need for a sensitive and specific assay is crucial because initial attempts at reconstituting an activ-

[1] E. M. Ross, *in* "The β-Adrenergic Receptor" (J. P. Perkins, ed.), in press. Humana Press, Clifton Park, New Jersey, 1990.

[2] R. Cerione, *in* "Receptor-Effector Coupling: A Practical Approach" (E. C. Hulme and N. J. M. Birdsall, eds.), p. 59. IRL Press, Oxford, England, 1990.

[3] A. Levitzki, *Biochim. Biophys. Acta* **822,** 127 (1985).

[4] R. J. Lefkowitz, R. A. Cerione, J. Codina, L. Birnbaumer, and M. G. Caron, *J. Membr. Biol.* **87,** 1 (1985).

ity or regulatory effect rarely yield high activities or efficient modulation. Further development of the reconstitution protocol thus depends on having a sensitive assay for the relevant activity. The absolute specific activity of a reconstituted system is a less crucial issue, especially at first. If criteria of specificity are met and if the background is low, an initially inefficient reconstitution will usually provide a valid starting point for further study.

The protein concentration is important for several reasons. First, a protein molecule that is associated with one vesicle presumably cannot interact with a protein in a different vesicle. Thus, if receptor and G protein at low concentration are reconstituted into vesicles at a high concentration of total lipid, the chances that the two proteins will be present in the same vesicles and directly couple may be low.[5] The need to work at low concentrations of lipid and protein leads to problems of scale. One such problem is adsorption to dialysis bags, gel-filtration matrices, and detergent-adsorbing resins. For reconstitution by gel filtration (see below), it has been necessary to minimize the column volume, to coat surfaces with siliconizing compounds or lipids, and to maintain the highest feasible concentrations of phospholipid and protein. Even so, low yields of receptor or G_s activities during reconstitution have almost always reflected the physical loss of the protein itself during reconstitution rather than its denaturation.

Second, reconstituting detergent-solubilized proteins into vesicles concentrates them from the total aqueous volume into the far smaller annular volume of the lipid bilayers. In a typical reconstitution, functional concentrations will be increased about 5×10^4-fold (see Ref. 5). It is likely that many protein–protein interactions that apparently occur "only after reconstitution" are actually limited by the low concentration of each protein when solubilized.

Methods of Reconstitution

The basic technology of reconstituting protein–phospholipid vesicles was developed in the late 1960s and, conceptually, has changed very little.[6] In general, one mixes detergent-solubilized proteins with a dispersion of phospholipids and allows the lipids and proteins to associate slowly as detergent is removed. Experimental choices include the state of purity of the proteins, the identities of the detergent and lipids, and the means of removal of detergent. Protocols for reconstituting integral membrane

[5] S. E. Pedersen and E. M. Ross, *Proc. Natl. Acad. Sci. U.S.A.* **79,** 7228 (1982).

[6] E. Racker, "Reconstitution of Transporters, Receptors, and Pathological States." Academic Press, New York, 1986.

proteins are still essentially empirical, but experience has yielded some good starting points and strategies. We review the reconstitution of receptor–G protein systems as described by Cerione *et al.*,[7-9] Haga *et al.*,[10] Florio and Sternweis,[11] Levitzki and co-workers,[12,13] Citri and Schramm,[14] and Ross and co-workers.[4,15-18]

Purity of Protein

Most investigators have found it easiest to develop reconstitution protocols using relatively crude solubilized preparations because the stability of the proteins of interest is usually higher and adsorptive loss of proteins is lower. Another advantage of using relatively crude protein preparations is that any unidentified important components, including lipids, will not have been removed. As the important proteins are identified and as experience with the reconstitution accumulates, it becomes easier to purify the relevant components and retain confidence in the procedures. This stepwise approach allows the investigator to optimize and standardize assays for proteins that may be unstable during purification. In the case of the β-adrenergic receptor, the Ross group spent several years using crude or partially purified receptor preparations in reconstitution-based experiments,[4,15] and the Lefkowitz group developed their protocols for reconstitution by fusing pure receptors with intact cells.[19] Significant mechanistic information can be obtained from reconstitution studies using partially purified receptors or G proteins provided that specific assays are

[7] R. A. Cerione, J. Codina, J. L. Benovic, R. J. Lefkowitz, L. Birnbaumer, and M. G. Caron, *Biochemistry* **23,** 4519 (1984).
[8] R. A. Cerione, C. Staniszewski, J. L. Benovic, R. J. Lefkowitz, M. G. Caron, P. Gierschik, R. Somers, A. M. Spiegel, J. Codina, and L. Birnbaumer, *J. Biol. Chem.* **260,** 1493 (1985).
[9] R. A. Cerione, P. Gierschik, C. Staniszewski, J. L. Benovic, J. Codina, R. Somers, L. Birnbaumer, A. M. Spiegel, R. J. Lefkowitz, and M. G. Caron, *Biochemistry* **26,** 1485 (1987).
[10] K. Haga, T. Haga, and A. Ichiyama, *J. Biol. Chem.* **261,** 10133 (1986).
[11] V. A. Florio and P. C. Sternweis, *J. Biol. Chem.* **260,** 3477 (1985).
[12] D. Feder, M.-J. Im, H. W. Klein, M. Hekman, A. Holzhofer, C. Dees, A. Levitzki, E. J. M. Helmreich, and T. Pfeuffer, *EMBO J.* **5,** 1509 (1986).
[13] M. Hekman, D. Feder, A. K. Keenan, A. Gal, H. W. Klein, T. Pfeuffer, A. Levitzki, and J. M. Helmreich, *EMBO J.* **3,** 3339 (1984).
[14] Y. Citri and M. Schramm, *Nature (London)* **287,** 297 (1980).
[15] D. R. Brandt, T. Asano, S. E. Pedersen, and E. M. Ross, *Biochemistry* **22,** 4357 (1983).
[16] D. R. Brandt and E. M. Ross, *J. Biol. Chem.* **261,** 1656 (1986).
[17] T. Asano, S. E. Pedersen, C. W. Scott, and E. M. Ross, *Biochemistry* **23,** 5460 (1984).
[18] D. C. May, E. M. Ross, A. G. Gilman, and M. D. Smigel, *J. Biol. Chem.* **260,** 15829 (1985).
[19] R. A. Cerione, B. Strulovici, J. L. Benovic, R. J. Lefkowitz, and M. G. Caron, *Nature (London)* **306,** 562 (1983).

available. The formidable undertaking of purifying the individual signaling proteins of interest to homogeneity is therefore not always necessary.

Detergent

The choice of the detergent or detergents used in the reconstitution procedure is almost invariably dictated by the stability and solubility of the proteins. The only criterion for detergents that is unique to reconstitution is the ease with which they can be removed. Removal of detergents by gel filtration or dialysis is essentially dependent on using detergents that have a high critical micelle concentration (CMC) and/or a small micelle size. The classic example is cholate, but CHAPS and octylglucoside are also suitable choices. Small micelle size facilitates detergent removal because small micelles may pass through dialysis tubing of intermediate pore size or will be included during gel filtration on a resin that has an appropriate pore size. For a detergent that forms a relatively large micelle, the rate of removal will be proportional to the concentration of monomer, which is equal to the CMC.[20] Thus, Lubrol 12A9 will dialyze approximately 100 times more slowly than will cholate. If detergent is to be removed by adsorption to a hydrophobic matrix, there is little to recommend one detergent over another. Adsorption of detergents to hydrophobic beads seems equally applicable to ionic and nonionic detergents with high and low CMC values.

A frequently overlooked function of detergent in a reconstitution experiment is to maintain the dispersion of lipid as well as protein. This can allow lipid–protein association to occur prior to formation of a membrane or can add adequate disorder to an already formed acceptor membrane to allow an integral protein to insert. For example, Asano *et al.*[17] were unable to coreconstitute partially purified β-adrenergic receptor and G_s with phospholipids unless the lipid was dispersed with added deoxycholate. Similar experiences were reported by Cerione *et al.*,[8] who used octylglucoside to disperse preformed liposomes.

Detergent Removal

Techniques for removing detergent from a mixture of proteins and phospholipids include dilution, dialysis, gel filtration, or adsorption to a hydrophobic matrix. Each has proved useful in the reconstitution of receptor–effector systems. The removal of detergent is a first-order process. Thus, removing 99% of the detergent takes twice as much effort as removing 90%. If one begins with a 5-fold excess of detergent over lipid,

[20] C. Tanford, "The Hydrophobic Effect." Wiley, New York, 1980.

even removal of 99% of the detergent will yield a 5% contamination. This may be tolerable for some purposes, but it is inadequate for studies relating to the structure of the reconstituted bilayer.

Dilution. Because bilayer-forming phospholipids are essentially insoluble in water and most detergents display a CMC above 10^{-5} *M*, dilution of a mixture of detergent, lipid, and protein generally leads to the formation of vesicles into which integral proteins are incorporated. Vesicles form because the detergent is monomeric below the CMC while the concentration of phospholipids is still sufficient to maintain a structured bilayer. Although significant detergent will contaminate the vesicles that are formed, centrifugation of the vesicles allows the preparation of substantially detergent-free vesicles with relatively little manipulation.

Dilution has been useful for the reconstitution of transport proteins and for some of the enzymes of oxidative phosphorylation,[6] but it has not been used extensively with signal transducing proteins. The low concentrations of lipids used in these reconstitutions makes adequate dilution unwieldy, and adsorptive losses of lipid and protein decrease yield. However, dilution was successfully applied to the rhodopsin–transducin system by Tsai *et al.*[21] Witkin and Harden[22] showed that digitonin-solubilized β-adrenergic receptor could be assayed using iodopindolol if the preparation was diluted prior to incubation with the ligand. Separation of free and bound ligand was then performed by filtration on glass-fiber filters. Dilution followed by centrifugation also proved useful for promoting the reassociation of G_s with adenylyl cyclase in phosphatidylcholine vesicles.[23] Regardless, dilution is easy to test and may work well.

Gel Filtration. Removal of detergents by gel filtration is widely used for reconstituting receptors and associated proteins. Because this method involves the constant removal of a concentration of detergent about equal to the CMC, vesicles form slowly as detergent is depleted during chromatography. Gel filtration usually yields unilamellar vesicles of 1000–5000 Å diameter.[4,18] Recovery of activities and total protein can be quite good. Seventy-five percent of total G_s and over 50% of β-adrenergic receptor that is applied to such a column can be recovered in the vesicle fraction.

Reconstitution of phospholipid vesicles by gel filtration has usually been performed on a hydrophilic matrix such as Sephadex G-25 or G-50[4,10,17] or on a matrix with a larger pore size to remove soluble proteins.[11,24] Vesicles that form during gel filtration are eluted in the void

[21] S. Tsai, R. Adamik, Y. Kanaho, E. L. Hewlett, and J. Moss, *J. Biol. Chem.* **259,** 15320 (1984).

[22] K. M. Witkin and T. K. Harden, *J. Cyclic Nucleotide Res.* **7,** 235 (1981).

[23] E. M. Ross, *J. Biol. Chem.* **256,** 1949 (1981).

[24] R. C. Rubenstein, S. K.-F. Wong, and E. M. Ross, *J. Biol. Chem.* **262,** 16655 (1987).

volume, presumably with any contaminating water-soluble proteins that are large enough to be excluded from the matrix. Because many detergent micelles are also large enough to be partially excluded, a large column is used relative to the sample size, usually 10- to 20-fold by volume. It is likely that the efficiency of gel filtration is enhanced by adsorption of detergent to the nominally hydrophilic resin, and this effect is probably responsible for successful reconstitution from detergents with low CMC values and large micelles (Lubrol 12A9, digitonin, lauroyl sucrose). In these cases, the column volumes were not of theoretically adequate size to remove detergent if an amount equal only to the CMC were being depleted.

Dialysis. Detergent dialysis, the classic technique in the field, has been of limited utility in working with receptor–effector systems, probably because of adsorptive losses. The abundance of the rhodopsin and transducin has overcome this barrier, and these proteins have been efficiently reconstituted by dialysis.[25–27]

Adsorption. Preferential adsorption of detergents from the detergent–lipid–protein mixtures has now become the second standard technique for reconstituting receptor–G protein vesicles. Two resins are commonly used, Biobeads [from Bio-Rad, Richmond, CA); a washed Dowex-like, styrene–divinylbenzene copolymer] and Extractigel (from Pierce, Rockford, IL). The resin is added to the mixture of protein and lipid, or the mixture may be chromatographed through a small column of the resin. Removal of detergent can be efficient (95–98%) and is relatively rapid. Recovery of β-adrenergic receptor and G_s has been above 75%.[7,8,12,28,29]

Choice of Lipid

Crude lipid mixtures give the best results in developing reconstituted assay systems. It is assumed that a crude mixture of lipids is more likely to contain any essential lipid or more closely accommodate any particular lipid requirement than would a defined mixture of highly purified lipids. A crude phospholipid fraction prepared either from soybeans (asolectin) or egg yolk is often used.[5,6] Soy lipids have been widely used for β-adrenergic

[25] B. K.-K. Fung, *J. Biol. Chem.* **258,** 10495 (1983).

[26] K. Hong, P. J. Knudsen, and W. L. Hubbell, this series, Vol. 81, p. 144.

[27] K. Hong and W. L. Hubbell, *Biochemistry* **12,** 4517 (1973).

[28] R. A. Cerione, J. W. Regan, H. Nakata, J. Codina, J. L. Benovic, P. Gierschik, R. L. Somers, A. M. Speigel, L. Birnbaumer, R. J. Lefkowitz, and M. G. Caron, *J. Biol. Chem.* **261,** 3901 (1986).

[29] R. A. Cerione, J. Codina, B. F. Kilpatrick, C. Staniszewski, P. Gierschik, R. L. Somers, A. M. Spiegel, L. Birnbaumer, M. G. Caron, and R. J. Lefkowitz, *Biochemistry* **24,** 4499 (1985).

receptor systems.[7,12-14,30] An alternative is a total phospholipid extract of the membranes from which the proteins to be reconstituted were initially isolated. This was essential for the successful reconstitution of the IgE receptor[31] and for initial reconstitution of partially purified β-adrenergic receptor and purified G_s.[15,17]

If detailed biochemical or biophysical experiments are planned, the use of a single phospholipid considerably simplifies analysis of the physical structure of the reconstituted vesicles. However, such systems have not yet been developed for receptor–G protein–effector systems. For example, efficient coupling of the β-adrenergic receptor, G_s, and adenylyl cyclase has been reconstituted by the use of a mixture of either phosphatidylethanolamine (PE) plus phosphatidylserine (PS) or phosphatidylcholine (PC) plus phosphatidylglycerol (PG),[12,16,18,24] but we have been unable to eliminate the second component. We have also been unable to replace natural phospholipids totally with synthetic lipids that have a defined acyl chain composition without diminishing both coupling and recovery of activity. Probably, the simplest strategy for simplifying the phospholipid mixture is to begin with the mixture of natural lipids that was used in a successful reconstitution and then to remove those lipids that are found to be unnecessary.

Can an *in vitro* reconstitution protocol define structural or functional requirements of a membrane protein for specific lipids or merely define operationally which lipids yield efficient reconstitution? This question is not wholly semantic. One could imagine that a functionally suboptimal combination of lipids might be required for efficient reconstitution in a specific protocol. It seems reasonable, however, that those lipids which yield high activities in reconstituted systems should be similarly optimal *in vivo*. A more profound question is whether specific associations between the membrane lipids and signal transducing proteins influence these activities.

Two Reconstitution Protocols

This section contains detailed descriptions of two protocols for reconstituting receptors and G proteins developed by the authors for the β-adrenergic receptor and G_s. They are suitable for working with small amounts of protein and have been adapted for use with other receptors or G proteins.

[30] J. Kirilovsky, S. Steiner-Mordoch, Z. Selinger, and M. Schramm, *FEBS Lett.* **183,** 75 (1985).

[31] B. Rivnay and H. Metzger, *J. Biol. Chem.* **257,** 12800 (1982).

Detergent Adsorption[7]

A mixture is prepared that contains the following:

- Approximately 5 pmol β-adrenergic receptor purified from guinea pig lung[32] in about 10 μl of 50 m*M* Tris-Cl (pH 7.4), 50 m*M* NaCl, 0.2% digitonin
- Approximately 5 pmol G_s purified from human erythrocytes[33] in about 10 μl of 10 m*M* HEPES (pH 8.0), 1 m*M* EDTA, 20m*M* 2-mercaptoethanol, 30% ethylene glycol, 150 m*M* NaCl, 150 μg/ml bovine serum albumin (BSA), containing 0.5–12% Lubrol 12A9
- 1.7 mg soy phosphatidylcholine in 10 μl of 10 m*M* Tris-Cl (pH 7.4), 100 m*M* NaCl, sonicated to clarity under N_2
- 25 μl of 17% octylglucoside
- 10 m*M* Tris-Cl (pH 7.4), 0.1 *M* NaCl, to yield a final volume of 0.5 ml

Lipid, octylglucoside, and buffer are combined and incubated at 0° for 30 min. Proteins are added, and the mixture is applied to a 1-ml column of Extracti-gel (Pierce) that has been washed with 10 m*M* Tris-Cl (pH 7.4), 100 m*M* NaCl, 2 mg/ml BSA and then equilibrated in the same buffer without the albumin (elution buffer). The protein and vesicles that are eluted from the column with 1.5 ml of elution buffer are incubated for 5 min at room temperature with polyethylene glycol (MW 6000–8000; 12% final concentration), diluted about 10-fold with elution buffer and centrifuged for 1.5 hr at 250,000 g_{max}. The resuspended vesicles should contain 0.5–1 pmol (each) of β-adrenergic receptor and G_s protein. The size of the receptor–G_s vesicles is reasonably homogeneous, with an average diameter of approximately 1000 Å according to electron micrographs. Each vesicle is calculated to contain 1–5 molecules each of receptor and G protein based on the usual assumptions regarding the packing of phosphatidylcholine bilayers,[34] random insertion of proteins into the vesicles, and no loss of phospholipid during vesicle isolation.

This protocol calls for stoichiometric amounts of receptor and G protein and was designed for the study of the high-affinity agonist–receptor–G protein complex (see below). When isoproterenol binding to such vesicles was examined, the best results from a number of experiments indicated that 30% of the receptor bound the agonist with high affinity (K_d 2 n*M*) and 70% of the total receptor was in a low-affinity state (K_d 300 n*M*). In

[32] J. L. Benovic, R. G. L. Shorr, M. G. Caron, and R. Lefkowitz, *Biochemistry* **23,** 4510 (1984).

[33] C. Codina, J. D. Hildebrandt, R. D. Sekura, M. Birnbaumer, J. Bryan, R. Manclark, R. Iyengar, and L. Birnbaumer, *J. Biol. Chem.* **259,** 5871 (1984).

[34] C. Huang and J. T. Mason, *Proc. Natl. Acad. Sci. U.S.A.* **75,** 308 (1978).

some cases the addition of GPP(NH)P was able to convert essentially 100% of the receptors to a low-affinity state.

The amounts of lipid, octylglucoside, and polyethylene glycol are critical, and little change in their concentrations can be tolerated. The lipid can be varied by about 2-fold around the suggested level. However, decreasing the amount of lipid tends to cause a reduction in the amount of receptor that is precipitated, and increasing the amount of lipid decreases the number of receptor molecules per lipid vesicle. The amount of octylglucoside should not be reduced because it is required for efficient incorporation of receptors into the vesicles.[35] Increasing the amount of octylglucoside by 50% can be tolerated. Decreasing the amount of polyethylene glycol reduces the aggregation (fusion) of the vesicles and thus reduces the recovery of vesicles by centrifugation. Increasing the amount of polyethylene glycol causes interference in assays of receptor–G protein coupling.

The amounts of receptor and G protein can be varied from 2 to over 20 pmol with no significant changes in the efficiency of incorporation. However, if the volume of digitonin added to the initial reconstitution mixture is over 20 μl of a 0.2% solution, the detergent trapped in the vesicles will interfere with receptor–G protein coupling. Similarly, when stock G_s preparations are in 5–12% Lubrol, the volume of the G_s solution that can be added to the reconstitution incubation is limited to less than 20 μl. The latter consideration is rarely a problem, however, because G protein can be routinely stored in 0.1% Lubrol.

Gel Filtration

The following protocol was developed for use with β-adrenergic receptor purified from turkey erythrocytes and G_s purified from rabbit liver.[16,17] It has been used with few modifications for β-adrenergic receptor from S49 lymphoma cells[36] and for porcine brain muscarinic cholinergic receptor and G_i from rabbit liver.[10]

An initial mixture is prepared at 0° of the following:

- 0.1–0.2 volumes of 0.3–1.0 μM β-adrenergic receptor in 10 mM sodium HEPES (pH 8.0), 100 mM NaCl, 0.1 mM EDTA, approximately 0.1% digitonin
- 0.1–0.3 volumes of 0.5–3.0 μM G_s in 10 mM sodium HEPES (pH 8.0), 0.1 mM dithiothreitol, 1 mM EDTA, 0.1% Lubrol 12A9

[35] R. A. Cerione, B. Strulovici, J. L. Benovic, C. D. Strader, M. G. Caron, and R. J. Lefkowitz, *Proc. Natl. Acad. Sci. U.S.A.* **80,** 4899 (1983).

[36] C. P. Moxham, E. M. Ross, S. T. George, and C. C. Malbon, *Mol. Pharmacol.* **33,** 486 (1988).

0.5 volumes of 0.6 mg/ml bovine brain PE, 0.4 mg/ml bovine PS, 0.4% sodium deoxycholate, 0.04% sodium cholate in 20 m*M* sodium HEPES (pH 8.0), 3 m*M* $MgCl_2$, 1 m*M* EDTA, 100 m*M* NaCl; lipids are added as a $CHCl_3$ solution, dried under Ar, resuspended and sonicated to clarity under Ar at room temperature in a bath sonicator

Extra 20 m*M* sodium HEPES 3 m*M* $MgCl_2$, 1 m*M* EDTA, 100 m*M* NaCl (column buffer) is added to give 1.0 volume. The mixture is immediately applied to a column of Sephadex G-50 (20–50 μm bead diameter) in column buffer at 0°, and vesicles are recovered in a void volume fraction that is 4 times the volume of the initial mixture. Routinely, 50 μl of reconstitution mixture is applied to a 3 × 150 mm column; 600 μl can be applied to a 10 × 200 mm column with similar results. Yields are typically 30–50% of receptor and 50–70% of G_s. Yields are improved if the column is exposed to a lipid suspension corresponding to a mock reconstitution before it is first used. Columns are then regenerated repeatedly by washing with column buffer and stored in column buffer that contains 1 m*M* NaN_3. Column buffer must be degassed.

This procedure yields unilamellar vesicles of 100 nm average radius, varying over the range 20–200 nm. There are few if any multilamellar vesicles. It is convenient to prepare vesicles that contain a calculated average of 1–5 molecules of receptor and 10–100 molecules of G_s protein, calculated according to a 100-nm radius. Note, however, that the surface area of the vesicle varies as the square of its radius. Inhomogeneity in size is therefore magnified as inhomogeneity in the average protein content of a vesicle.

The procedure may be varied to suit individual experimental needs as follows:

1. The concentrations of the two proteins may be varied to yield different receptor : G_s ratios. Low ratios (~1–2) are useful for studying the high-affinity agonist–receptor–G_s complex,[18] but much higher ratios (~100) are needed in some studies of the regulation of GTPase intermediates.[16]

2. The lipids can be substituted within some limits. The PE : PS mixture gives optimal recovery and coupling in our hands, but a 3 : 2 mixture of PC : PG is nearly as good and eliminates lipids with free amino groups. Cholesterol can be added to stabilize thiol-reduced receptor.[37] Cholesterol at about 5 mol% of total lipid enhances coupling slightly, and 20–30 mol%

[37] S. E. Pedersen and E. M. Ross, *J. Biol. Chem.* **260,** 14150 (1985).

is markedly inhibitory. We have not noted the effects of neutral lipids that have been described by Kirilovsky, Schramm, and co-workers.[30,38,39]

3. CHAPS can be substituted for the deoxycholate–cholate mixture at the same total concentration with little loss of recovery or coupling.

4. Ultrogel AcA 34 can be substituted for the Sephadex G-50 with a slight loss in recovery. However, it is useful for separating nonreconstituted proteins, particularly free G-protein α subunits, from the vesicles.[40]

5. The G protein may be labeled with [α-^{32}P]GDP by incubation with [α-^{32}P]GTP before reconstitution. The concentration of $MgCl_2$ in the column buffer should be reduced to 2 mM to minimize dissociation.

6. Purified adenylyl cyclase can be included in the initial mixture to prepare vesicles that display GTP-dependent stimulation of adenylyl cyclase activity by agonists.[18]

Assays of Receptor–G Protein Coupling

A number of different assays can be used to assess reconstituted receptor–G protein coupling. The G protein-induced high-affinity binding of agonist to receptor indicates the stoichiometric formation of an agonist–receptor–G protein complex. The agonist-stimulated activation of adenylyl cyclase, the binding or release of guanine nucleotides, or the steady-state GTPase activity all measure the regulatory capacity of the receptor. Each assay gives different but complementary information.

High-Affinity Binding of Agonists. A receptor–G protein complex displays a higher affinity for agonist than does receptor alone. The increase in affinity can exceed 1000-fold, the effect is selective for agonists (i.e., not antagonists), and the complex is dissociated by the binding of guanine nucleotide to the G protein (see Ref. 41 for review). Although formation of this complex has been measured in reconstituted systems, high-affinity agonist binding has not been a common assay for receptor–G protein coupling except in the case of the muscarinic cholinergic receptor. First, careful determination of affinity is relatively tedious. It is often necessary to determine affinity in the presence and absence of guanine nucleotide to establish clearly whether the complex has been formed. Second, only a fraction of the receptor typically displays high affinity, perhaps reflecting

[38] J. Kirilovsky, S. Eimerl, S. Steiner-Mordoch, and M. Schramm, *Eur. J. Biochem.* **166,** 221 (1987).

[39] J. Kirilovsky and M. Schramm, *J. Biol. Chem.* **258,** 6841 (1983).

[40] P. C. Sternweis, *J. Biol. Chem.* **261,** 631 (1986).

[41] E. M. Ross, *Neuron* **3,** 141 (1989).

a modest affinity of receptor for G protein, random orientation of the proteins with respect to the membrane, or poor permeability of agonist or nucleotide.

Reconstitution of GTP-sensitive, high-affinity binding has been achieved for purified β-adrenergic receptor and G_s[35] and for the muscarinic cholinergic receptor and either G_o or G_i.[10,11] For the muscarinic receptor, the relative increase in affinity was similar to that observed in native membranes. Although a vast excess of G protein was needed to increase affinity maximally, the effective concentrations of G protein were similar to the estimated concentrations in cerebral cortical membranes. For the β-adrenergic receptor and G_s, the results were less striking but still formed an adequate demonstration of coupling. The ability of the muscarinic receptor to couple with either G_o or G_i suggested that it may regulate more than one signaling pathway *in vivo*.

Receptor-Catalyzed Nucleotide Exchange. Agonist-liganded receptors facilitate G-protein activation by accelerating GDP release and GTP binding,[41] thereby also accelerating steady-state GTP hydrolysis. The regulation of nucleotide exchange by the receptor thus offers several sensitive assays for coupling.

Steady-State GTPase

Measurement of receptor-stimulated steady-state GTPase activity in receptor–G protein vesicles is generally the easiest and most sensitive assay of coupling and is the most widely used.[15,16,42] With [γ-^{32}P]GTP as substrate and with attention to the linearity of the time course, sensitivity is practically limited only by basal activity and the contamination of substrate with [^{32}P]P_i. Although stimulation of GTPase activity is a convenient measure of the restoration of coupling, it provides little mechanistic information. Unless one can independently ascertain that all of the receptor and G protein in the vesicles is uniformly distributed and available for coupling, one cannot derive molecular parameters from the GTPase rates. In fact, because a fraction of reconstituted G protein is typically uncoupled, it is generally necessary to express specific GTPase activity according to the fraction of G protein that can couple to receptor (mol P_i released/min/mol of coupled G protein), which can be significantly less than half of the total. The relationship between steady-state GTPase and the rates of the partial reactions of the GTPase cycle has been discussed by Brandt, Ross and co-workers.[16,42]

[42] D. R. Brandt and E. M. Ross, *J. Biol. Chem.* **260,** 266 (1985).

GTP and GTPγS Binding

Receptors facilitate GTP binding to G proteins both directly and by promoting GDP release.[41] Measurement of agonist-stimulated GTP binding thus reflects both of these partial reactions of the GTPase cycle. Because GTP is rapidly hydrolyzed, a nonhydrolyzable analog such as GTPγS is more commonly used, and the [^{35}S]GTPγS–G protein complex can be measured directly.[17,43] In the case of G_s, the activity of the G_s–GTPγS complex can also be assayed according to its ability to activate adenylyl cyclase, which allows the study of impure reconstituted systems that may have spurious nucleotide-binding activity.[4] This assay is also about 5-fold more sensitive than the measurement of [^{35}S]GTPγS binding. A third assay is based on the observation that activation of a G-protein α subunit causes an increase in its intrinsic tryptophan fluorescence[44,45] (see also [28], this volume). The principal advantage of this assay is the ability to obtain continuous data over short time intervals. However, background fluorescence is frequently high, and this assay works best with reconstituted systems composed of highly purified receptors and G proteins.

For any of these assays, the mechanistically meaningful parameter is the receptor-stimulated rate of GTPγS binding, which is usually expressed as a first-order rate constant k_{obs}. The dependence of k_{obs} on the concentration of agonist-liganded receptor is predicted to be linear, and it frequently is. The slope of a plot of k_{obs} versus the concentration of receptor is a measure of the regulatory efficiency of the receptor that is usually referred to as k_{cat}. A plot of k_{obs} versus the concentration of agonist yields a typical hyperbolic saturation curve.

An additional advantage of GTP binding assays is that the data indicate directly the number of reconstituted G-protein molecules that can be regulated by the receptor. This provides practical information on the fraction of G protein that is accessible to receptor. The size of this fraction presumably reflects the efficiency of reconstitution, the distribution of receptor and G protein among vesicles, and their orientation with respect to the inside and outside of the vesicles. A possible disadvantage of the GTPγS binding assay is that the concentration of free $\beta\gamma$ subunits in the vesicles may increase during the course of the reaction, potentially altering the course of the reaction.

[43] J. K. Northup, M. D. Smigel, and A. G. Gilman, *J. Biol. Chem.* **257,** 11416 (1982).
[44] H. Higashijima, K. M. Ferguson, P. C. Sternweis, E. M. Ross, M. D. Smigel, and A. G. Gilman, *J. Biol. Chem.* **262,** 752 (1987).
[45] W. J. Phillips and R. A. Cerione, *J. Biol. Chem.* **263,** 15498 (1988).

GDP Release

If G proteins are incubated with radiolabeled GTP prior to reconstitution, then the receptor-catalyzed dissociation of labeled GDP can be used as another measure of coupling.[16] Release should again be a first-order process, with the rate constant reflecting the activity of agonist-liganded receptor. If the assay is performed in the presence of an excess of unlabeled GDP, then equilibrium is maintained during the release of label. This can be contrasted to the measurement of GTPγS binding, where activated α subunit is formed and free $\beta\gamma$ subunits may be released. The measurement of GDP release also allows the study of two α subunits simultaneously if one is labeled with ^{3}H and the other with ^{32}P.

Section III

Guanylyl Cyclase

A. Assay of Guanylyl Cyclase
Article 30

B. Purification and Characterization of Guanylyl Cyclase Isozymes
Articles 31 through 40

C. Regulation of Guanylyl Cyclases
Articles 41 through 44

[30] Assay of Guanylyl Cyclase Catalytic Activity

By STEVEN E. DOMINO, D. JANETTE TUBB, and DAVID L. GARBERS

The current assay methods used in our laboratory for guanylyl cyclase (EC 4.6.1.2, guanylate cyclase) are summarized based on modifications of previously published methods.[1] Two basic approaches for the assay of guanylyl cyclase activity are (1) the use of radiolabeled substrates, [α-^{32}P]GTP or [8-^{3}H]GTP, and the measurement of cyclic [^{32}P]GMP or cyclic [^{3}H]GMP formation, respectively, and (2) the use of GTP as a substrate and the estimation of cyclic GMP formation by radioimmunoassay.

[α-^{32}P]GTP Assay

The following method uses the radiolabeled substrate, [α-^{32}P]GTP, and detects the amount of [^{32}P]cGMP formed.[2] [^{32}P]cGMP is separated from unreacted substrate and other ^{32}P-containing compounds by coprecipitation of 5′-nucleotides with zinc carbonate and subsequent column chromatography on neutral alumina (aluminum oxide.)[3,4] The sensitivity of the method depends on the specific activity of GTP, the amount of radioactive substrate added per assay reaction mixture, the background radioactivity, and the assay time. Given a GTP specific activity of 10,000 counts per minute (cpm)/nmol and 10 pmol cGMP formed/min by guanylyl cyclase, one would detect about 100 cpm of [^{32}P]cGMP per 1-min assay. If the background is 100 cpm, then a 2-fold increase in [^{32}P]cGMP could theoretically be measured after 1 min. The guanylyl cyclase activities of tissues such as rat lung are approximately 120 pmol/min/mg protein for soluble extracts and 70 pmol/min/mg for detergent-solubilized particulate fractions when assays are performed at 37° in the presence of 600 μM MnGTP.[5]

Reagents

2.0 mM [α-^{32}P]GTP (substrate) at a specific activity of 2000 to 10,000 cpm/nmol. Add 10 to 200 nmol [α-^{32}P]GTP (~100,000–500,000 cpm)

[1] D. L. Garbers and F. Murad, *Adv. Cyclic Nucleotide Res.* **10,** 57 (1979).
[2] J. P. Durham, *Eur. J. Biochem.* **61,** 535 (1976).
[3] K. H. Jakobs, W. Saur, and G. Schultz, *J. Cyclic Nucleotide Res.* **2,** 381 (1976).
[4] R. Gerzer, F. Hofmann, and G. Schultz, *Eur. J. Biochem.* **116,** 479 (1981).
[5] T. D. Chrisman, D. L. Garbers, M. A. Parks, and J. G. Hardman, *J. Biol. Chem.* **250,** 374 (1975).

per assay tube. [α-^{32}P]GTP can be purchased at relatively high specific activity (~3000 Ci/mmol, 10 mCi/ml, New England Nuclear/Du Pont, Boston, MA) and unlabeled GTP added to lower the specific activity to the desired level. To prepare a 10 m*M* GTP stock solution, initially add 120–130 mg of GTP (sodium salt) to 18 ml of water. Neutralize the solution with a few drops of 0.1 *N* NaOH. Dilute a small aliquot 100-fold with water and measure the absorbance at 253 nm in a 1-cm path length quartz cuvette. Given an extinction coefficient of 13.7 mM^{-1}, a 100-fold diluted aliquot from a 10 m*M* stock solution should have an absorbance of 1.37. Dilute the stock solution with water to give a 10.0 m*M* GTP concentration.

200 m*M* buffer [such as 2-(*N*-morpholino)ethanesulfonic acid (MES), Tris, or triethanolamine] at the desired pH. Prepare 200 m*M* MES by adding 3.91 g of MES to 90 ml of water. Adjust the pH to 6.5 with NaOH and add water to 100 ml. For 200 m*M* Tris buffer, add 2.42 g of Tris base to 90 ml of water, adjust the pH to 7.6 with HCl, and add water to 100 ml. For 200 m*M* triethanolamine, add 3.71 g of triethanolamine hydrochloride to 90 ml of water, adjust pH to 8.0 with NaOH, and add water to 100 ml.

100 m*M* $MnCl_2$ or 100 m*M* $MgCl_2$ (required cofactors). Guanylyl cyclase activity is usually higher with the substrate MnGTP than with MgGTP. However, differences in activity above basal caused by various treatments may be more or less apparent, dependent on the metal cofactor used. For 100 m*M* stock solutions of Mn^{2+} or Mg^{2+}, add 1.98 g of $MnCl_2 \cdot 4H_2O$ or 2.03 g of $MgCl_2 \cdot 6H_2O$ per 100 ml of water. Check the actual metal ion concentration by atomic absorption spectroscopy.

10 m*M* 1-methyl-3-isobutylxanthine (MIX) or 100 m*M* theophylline (phosphodiesterase inhibitors). Add 222 mg of MIX or 1.80 g of theophylline per 100 ml of water. Stock solutions will need to be warmed to dissolve the MIX or theophylline.

10 m*M* nonradioactive cGMP ("cold trap" to minimize loss of [^{32}P]cGMP). Add approximately 100–110 mg of cGMP (sodium salt, 98% pure) to 18 ml of water. Dilute a small aliquot 100-fold with water and measure the absorbance at 254 nm in a 1-cm path length cuvette. Given an extinction coefficient of 13.0 mM^{-1}, a 100-fold diluted aliquot from a 10 m*M* stock solution should have an absorbance of 1.30. Dilute the stock solution with water as needed.

500 m*M* KF and/or 2.5 m*M* sodium orthovanadate (for phosphatase inhibition, if needed). Add 4.71 g of $KF \cdot 2H_2O$ per 100 ml of water or add 46.0 mg of Na_3VO_4 per 100 ml of water.

10 mg/ml bovine serum albumin (to stabilize the enzyme, if needed).

50 m*M* dithiothreitol (to maintain the reduced state of the enzyme, if needed). Add 1.5 mg to 200 μl of water.

120 m*M* zinc acetate (to stop guanylyl cyclase reaction). Add 26.3 g $Zn(C_2H_3O_2)_2 \cdot 2H_2O$ per 1000 ml of water.

144 m*M* sodium carbonate (to coprecipitate zinc carbonate and 5′-nucleotides). Add 15.3 g Na_2CO_3 per 1000 ml of water.

Neutral alumina (AG 7, 100–200 mesh, Bio-Rad Laboratories, Richmond, CA).

100 m*M* Tris, pH 7.5. Add 12.1 g of Tris base to 900 ml of water, adjust the pH with HCl, and add water to 1000 ml.

Assay Procedure

Reaction mixtures, containing a buffer, [α-^{32}P]GTP, a metal cofactor in concentrations greater than GTP, cyclic nucleotide phosphodiesterase inhibitor, and an enzyme source, are usually assayed in a volume of 100–150 μl, at 30° or 37°, for up to 15 min.

1. For a 50-tube assay, prepare 3 ml of an assay buffer stock solution containing (1) 600 μl of 200 m*M* MES, Tris, or triethanolamine buffer, (2) 180 μl of 100 m*M* $MnCl_2$ or $MgCl_2$, (3) 600 μl of 10 m*M* MIX, (4) 600 μl of 10 m*M* cGMP, (5) 300 μl of 10 mg/ml bovine serum albumin, and (6) 720 μl of water. The assay buffer may contain, if needed, KF and sodium orthovanadate (used with sea urchin sperm enzyme preparations to inhibit dephosphorylation of the enzyme[6]) and dithiothreitol. If desired, add 120 μl of 500 m*M* KF, 120 μl of 2.5 m*M* sodium orthovanadate, 60 μl of 50 m*M* dithiothreitol, and 420 μl of water in place of 720 μl of water to the assay buffer stock solution. Some assays may also require addition of creatine kinase/creatine phosphate to regenerate GTP. For a 50-tube assay, add 1.5 mg creatine kinase and 33.1 mg of creatine phosphate (disodium salt) to the assay buffer stock solution. Add 50 μl of the final stock solution to each 12 × 75 mm glass or plastic assay tube.

2. Add 25 μl of 2.0 m*M* GTP (50 nmol) containing 300,000–500,000 cpm [α-^{32}P]GTP per assay tube.

3. Warm tubes at the assay temperature for 3–5 min. Start the assay with the addition of 25 μl of the enzyme source. In time zero controls, stop the assay reactions with 500 μl of 120 m*M* zinc acetate before addition of enzyme. The time zero controls are used to subtract background radioactivity when calculating the amount of cGMP formed (see Calculations). Nonenzymatic formation of cGMP may be tested by adding 25 μl of the buffer from the enzyme source or 25 μl of the enzyme boiled in a 100° water bath for 5 min in separate control reaction mixtures.

[6] C. S. Ramarao and D. L. Garbers, *J. Biol. Chem.* **260,** 8390 (1985).

4. Stop assay reactions with the addition of 500 μl of 120 m*M* zinc acetate. Assay reactions must be terminated within the period of linear product formation with respect to time and amount of enzyme added. Place samples in an ice bath.

Separation of Radiolabeled Substrate and Product

1. Fill glass or plastic columns (0.7 cm internal diameter × 15 cm, Econo-Columns from Bio-Rad, or similar columns) to a height of 2.5 cm with dry alumina. Add 6–8 ml of 100 m*M* Tris, pH 7.5, and resuspend the powder within the column. After the buffer has dripped through the columns, wash with an additional 2 ml of 100 m*M* Tris, pH 7.5. Columns can be prepared before the start of the assay.

2. Once all of the assay reactions have been stopped, add 600 μl of 144 m*M* sodium carbonate to precipitate 5′-nucleotides, including unreacted [α-^{32}P]GTP. Centrifuge at 2000 *g* for 5–10 min. Samples may be frozen and thawed prior to centrifugation, if desired, to form a more tightly packing precipitate.

3. Pour samples over neutral alumina columns. Elute [^{32}P]cGMP with 5 ml[7] of 100 m*M* Tris, pH 7.5, into scintillation vials. Add 10 ml scintillation fluid or water and determine the radioactivity of the samples in a scintillation counter. Background values depend on the amount of radioactivity added to the columns (usually 500–3000 cpm[8]). Recoveries can be determined by measuring the absorbance at 252 nm of aliquots of each sample before and after column separation[1] or with a tracer amount of cyclic [^{3}H]GMP[9] (ICN Biomedicals, Costa Mesa, CA) in control samples. Recoveries are usually between 60 and 70%.

Calculations

1. Subtract average time zero background radioactivity (i.e., 1000 cpm) from average duplicate samples. Duplicate samples should be within 10% of each other.

2. Calculate the amount of cGMP obtained in nanomoles (i.e., if each

[7] Individual batches of alumina may need to be tested in pilot studies.

[8] Background values can be reduced to less than 100 cpm and, consequently, the sensitivity of the assay increased by using a two column procedure to isolate radiolabeled product described below (Purification of Cyclic GMP with Dual Columns).

[9] Commercially prepared [^{3}H]cGMP may be purified before use: (1) mix 200 μl of [^{3}H]cGMP (0.2 mCi) with 800 μl of 0.1 *N* HCl, (2) apply the sample to 24 × 0.7 cm (i.d.) column containing Dowex 50W-X8 previously washed with 0.1 *N* HCl, (3) elute with a total of 27 ml of 0.1 *N* HCl, discarding the first 12 ml and collecting the next 15 ml, (4) neutralize with 1 ml of 1 *M* Tris, pH 7.4, and by dropwise addition of approximately 1 ml of 1 *N* NaOH, and (5) dilute to 20,000 cpm per 25 μl and freeze.

reaction mixture contained 50 nmol and 500,000 cpm of GTP, then 10,000 cpm of product equals 1 nmol of cGMP).

3. Correct the activity for recovery. Dividing the amount of cGMP obtained by the recovery of individual samples or average recovery (i.e., divide by 0.65 for 65% recovery).

4. To determine the specific activity of guanylyl cyclase, divide the activity (pmol) by the time of incubation (min) and amount of protein (mg of either protein or wet weight).

[^{3}H]GTP Assay

If the activity of guanylyl cyclase in the tissue being measured is greater than 10 pmol/min, an assay using [^{3}H]GTP as a substrate may be used.[5,10,11] [^{3}H]cGMP is separated from other ^{3}H-labeled compounds by zinc carbonate coprecipitation of 5′-nucleotides and polyethyleneimine-cellulose column chromatography.

Reagents. In addition to various reagents from the [α-^{32}P]GTP assay, additional reagents are required:

2.0 m*M* [8-^{3}H]GTP at a specific activity of 5000–10,000 cpm/nmol [purchased at a specific activity of 10–50 Ci/mmol from Amersham (Arlington Heights, IL) or New England Nuclear/Du Pont]. Use 200,000–1,000,000 cpm of [^{3}H]GTP (10–200 nmol) per assay tube.

Zinc acetate–cGMP solution. To prepare 200 m*M* zince acetate, add 43.9 g $Zn(C_2H_3O_2)_2 \cdot 2H_2O$ per 1000 ml of water. For a 50-tube assay, prepare 15 ml of zinc acetate–cGMP solution by adding 12 mg of cGMP to 15 ml of 200 m*M* zinc acetate.

200 m*M* sodium carbonate. Add 21.2 g Na_2CO_3 per 1000 ml of water.

50 m*M* acetic acid. Add 2.88 ml of glacial acetic acid per 1000 ml of water.

20 m*M* LiCl. Add 848 mg of LiCl per 1000 ml of water.

Polyethyleneimine-cellulose resin [0.05–0.10 mm particle size, Accurate Chemical and Scientific (Westbury, NY)].

Assay Procedure

Assay reaction mixtures are prepared similarly to experiments using [α-^{32}P]GTP as a substrate. Assay reactions are stopped by the addition of 250 μl of zinc acetate–cGMP solution and must be terminated within the period of linear product formation with respect to time and amount of enzyme added.

[10] D. L. Garbers, J. G. Hardman, and F. B. Rudolph, *Biochemistry* **13,** 4166 (1974).

[11] K. Nakazawa and M. Sano, *J. Biol. Chem.* **249,** 4207 (1974).

Separation of Radiolabeled Substrate and Product

1. Wash the polyethyleneimine-cellulose resin by stirring the dry resin with water. Allow the resin to settle, and pour off the fines. Wash the resin 5 additional times. Fill glass or plastic columns (0.7 cm internal diameter) to a height of 5–10 cm with the washed polyethyleneimine-cellulose. Condition columns before use by washing with 15 ml of 50 m*M* acetic acid. Columns can be prepared before the start of the assay.

2. Once all of the assay reactions have been stopped and placed in an ice bath, add 250 μl of 200 m*M* sodium carbonate to precipitate 5′-nucleotides. Samples may be frozen and thawed, if desired. Centrifuge at 2000 *g* for 5–10 min.

3. Pour samples over the polyethyleneimine-cellulose columns. When the samples have entered the column resin, add 1 ml of 50 m*M* acetic acid. Allow the liquid to drip through the columns. Add an additional 15 ml of 50 m*M* acetic acid and then 15 ml of water. Next, depending on the bed volume used, cGMP can be eluted by washing with approximately 4 ml of 20 m*M* LiCl and collecting the next 5 ml of 20 m*M* LiCl in scintillation vials. Since the characteristics of each batch of resin may vary, determine the volume of 20 m*M* LiCl and bed volume needed for separation of GTP and cGMP in pilot experiments.

4. Recoveries can be determined by measuring the absorbance at 252 nm of aliquots of the samples before and after column separation. Scintillation fluid (Amersham ACS) is added before determination of the radioactivity of the samples. Background (time zero) controls are usually 500–3000 cpm.

Calculations

The calculations for determining guanylyl cyclase activity using [^{3}H]GTP as a substrate are basically the same as when using [α-^{32}P]GTP.

Assay with GTP as Substrate and Radioimmunoassay for Cyclic GMP

Guanylyl cyclase activity may be measured by using GTP as a substrate and estimating the amount of cGMP formed by radioimmunoassay (RIA).[12,13] When the samples are purified[14] and cGMP acetylated[15] to

[12] A. L. Steiner, D. M. Kipnis, R. Utiger, and C. Parker, *Proc. Natl. Acad. Sci. U.S.A.* **64,** 367 (1969).

[13] A. L. Steiner, C. W. Parker, and D. M. Kipnis, *J. Biol. Chem.* **247,** 1106 (1972).

[14] K. H. Jakobs, E. Böhme, and G. Schultz, *in* "Eukaryotic Cell Function and Growth: Regulation by Intracellular Cyclic Nucleotides" (J. E. Dumont, B. L. Brown, and N. J. Marshall, eds.), p. 295. Plenum, New York, 1976.

[15] J. F. Harper and G. Brooker, *J. Cyclic Nucleotide Res.* **1,** 207 (1975).

increase the sensitivity of the RIA, femtomole amounts of cGMP can be detected. This guanylyl cyclase assay method can reliably detect activity as low as 0.1 pmol/min. The same radioimmunoassay method can also be used to measure tissue concentrations of cGMP.

Reagents. In addition to various reagents from the [α-^{32}P]GTP assay, several other reagents are required for use in the radioimmunoassay:

0.5 *N* perchloric acid. Add 52.8 ml of 70% $HClO_4$ per 1000 ml of water.

Nonradioactive cGMP standards, at concentrations of 0.1, 0.2, 0.4, 0.8, 1.0, 2.0, and 4.0 n*M* cGMP.

Triethylamine (Aldrich, Milwaukee, WI, 99%) and acetic anhydride (Aldrich, 99+%).

250 and 50 m*M* sodium acetate–acetic acid buffers, pH 4.75. For a 1 *N* stock solution of sodium acetate–acetic acid buffer, pH 4.75, add 68.0 g of $NaC_2H_3O_2 \cdot 3H_2O$ to 500 ml of water. Since acetate ion concentration is critical, adjust the pH to 4.75 with 1 *N* acetic acid (57.5 ml of glacial acetic acid per 1000 ml of water).

^{125}I-Labeled tryrosine methyl ester 2′-*O*-succinyl-cGMP.[16]

Antibody raised in rabbits to thyroglobulin-linked 2′-*O*-succinyl-cGMP and stored in small aliquots at −70°. Ten microliters of the rabbit serum is diluted 400-fold with 0.1% γ-globulin before use and stored in 400-μl aliquots at −70° to minimize freeze/thaw cycles.

12% polyethylene glycol in 50 m*M* sodium acetate–acetic acid buffer, pH 6.2. Add 120 g of polyethylene glycol per 1000 ml of 50 m*M* sodium acetate–acetic acid buffer, pH 6.2. For a 1 *N* stock solution of sodium acetate–acetic acid buffer, pH 6.2, add 68.0 g of $NaC_2H_3O_2 \cdot 3H_2O$ to 500 ml of water and adjust the pH to 6.2 with 1 *N* acetic acid.

Human plasma (expired plasma from local blood bank) containing 0.15% EDTA. Prepare 80 m*M* EDTA by adding 2.98 g per 100 ml of water. Add 0.5 ml of 80 m*M* EDTA to 9.5 ml of plasma.

Assay Procedure

Assay conditions for the use of GTP as a substrate are similar to those for [α-^{32}P]GTP. However, the stopping procedure uses 1.0 ml of 0.5 *N* perchloric acid. For time zero controls, stop the assay reactions with 1.0

[16] Commercially prepared radioligand may be purchased (ICN Biomedicals) or tyrosyl methyl ester 2′-*O*-succinyl-cGMP (Sigma) can be iodinated with iodine-125 (Amersham) catalyzed by chloramine-T (Eastman Kodak, Rochester, NY) and purified by QAE-Sephadex chromatography with a method similar to that of G. Brooker, this series, Vol. 159, p. 45. Fractions collected from the QAE-Sephadex column are assayed in a pilot RIA for competitive binding, and the active fractions are combined.

ml of 0.5 *N* perchloric acid prior to addition of enzyme. Time zero controls represent background values subtracted from the amount of cGMP formed during the assay incubation.

Purification of Cyclic GMP with Dual Columns

1. Acidified alumina columns, prepared from dry neutral alumina (AG 7, 100–200 mesh, Bio-Rad) are used to remove potential interfering compounds. Then, Dowex 50W-X8 (hydrogen form, Bio-Rad) columns are used to separate cAMP and cGMP, as well as GTP and other nucleotides. Acidified alumina columns are prepared by adding dry alumina to glass or plastic columns (0.7 cm internal diameter) to a height of 2.5 cm and washing first with 2 ml of 200 m*M* ammonium formate, pH 6.6, and then 20 ml of water. Dowex 50W-X8 resin is prepared by washing with water to remove fines and storing in 0.1 *N* HCl. To prepare Dowex columns, add Dowex in 0.1 *N* HCl to glass or plastic columns (0.7 cm internal diameter) to a height of 4.5 cm and wash with 20 ml of water.

2. Samples that have been stopped with perchloric acid, including several control samples for estimating recovery with purified [^{3}H]cGMP are centrifuged at 2000 *g* for 20 min to pellet precipitated materials. Samples are then poured over individual alumina columns. After the samples have entered the resin, columns are sequentially washed with (a) 10 ml of water, (b) 5 ml of 0.6 *N* HCl in 95% ethanol, (c) 5 ml of 50% ethanol, and (d) 2 ml of water.

3. Alumina columns are then "piggy-backed" over Dowex 50W-X8 columns. Three milliliters of 200 m*M* ammonium formate is added to the alumina columns to elute nucleotides into the Dowex resin. The piggy-backed columns are placed over vials, and 1 ml of 200 mM ammonium formate is added. After the liquid has dripped through the alumina resin, the columns are unstacked and the alumina discarded. cGMP is further eluted from the Dowex resin with 3 ml of water.[17] Samples are then frozen overnight and lyophilized. Dowex resin can be regenerated for repeated use.[18] Recoveries of cGMP may be averaged between the control samples and are usually 35–45%. Alternatively, if the samples contain high enough concentrations of cGMP, a tracer amount of purified [^{3}H]cGMP (~1000 cpm) is added before purification to determine recovery of each sample individually.

[17] For experiments measuring cAMP as well as cGMP, the Dowex columns are placed over separate collection vials following collection of cGMP, and an additional 5 ml of water is added to elute cAMP.

[18] Wash 1 kg of Dowex resin sequentially with (1) 8 liters of 0.5 *N* NaOH, (2) 8 liters of water, (3) 12 liters of 2 *N* HCl, (4) 12 liters of water, and (5) 4 liters of 0.1 *N* HCl. Store in 0.1 *N* HCl.

TABLE I
TYPICAL DATA OBTAINED FOR STANDARD CURVE FROM CYCLIC GMP RADIOIMMUNOASSAY[a]

Standard concentration (fmol)	Radioactivity (cpm)	Net radioactivity (cpm)	Reference/bound
5	3436	3186	1.19
5	3731	3481	1.09
10	2790	2540	1.50
10	2955	2705	1.41
20	2246	1996	1.91
20	2610	2360	1.61
40	1814	1564	2.43
40	1487	1237	3.08
50	1419	1169	3.25
50	1354	1104	3.45
100	947	697	5.46
100	985	735	5.18
200	698	448	8.49
200	634	384	9.91

[a] Average nonspecific binding, 250 cpm; average total binding, 4055 cpm; "Reference" is net radioactivity of the total binding (3805 cpm); "bound" is net radioactivity of each standard (384–3481 cpm).

Radioimmunoassay Procedure

1. Resuspend experimental unknowns in a volume of water (0.5 ml minimum) to achieve an estimated concentration of cGMP of 10–100 fmol per 50 μl. Pilot studies may be needed to determine the approximate concentration range of the unknown samples. Acetylate 0.5-ml aliquots of the unknowns, standards of 0.1, 0.2, 0.4, 0.8, 1.0, 2.0, and 4.0 n*M* cGMP, and water (5–10 aliquots of 0.5 ml), by the addition of 10 μl triethylamine and 5 μl acetic anhydride in rapid succession at room temperature.

2. Pipette 50 μl into two 12 × 75 mm glass tubes for each acetylated cGMP standard, containing 5, 10, 20, 40, 50, 100, and 200 fmol of cGMP per 50 μl. The most accurate range of the standard curve will therefore be 10–100 fmol cGMP. Pipette 25 and 50 μl of each sample into 12 × 75 mm tubes. Add 25 μl "acetylated" water to the tubes containing only 25 μl of sample. Include quadruplicate samples containing 50 μl "acetylated" water for total binding (reference) and duplicate tubes containing "acetylated" water for nonspecific binding. The values for total and nonspecific binding are used in preparing a standard curve for the radioimmunoassay.

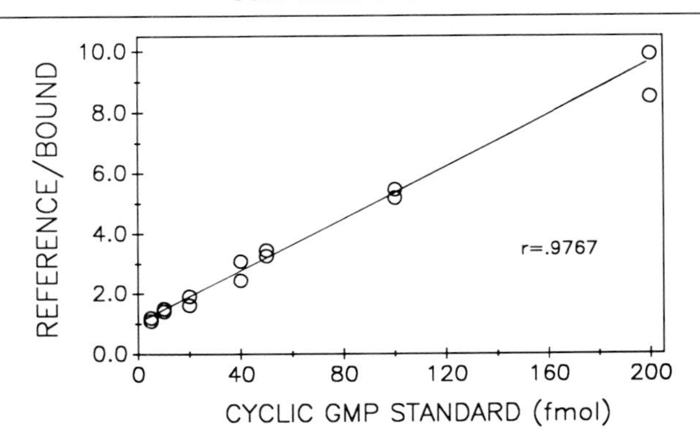

FIG. 1. Linear plot (weighted regression analysis) of reference/bound versus cyclic GMP concentration. Data plotted are from Table I.

Store the remainder of the acetylated samples at 4° until the assay has been analyzed.

3. Add 10 μl of 250 m*M* sodium acetate buffer, pH 4.75, to each tube, mix, and place tubes in an ice bath.

4. Dilute [^{125}I]cGMP with 50 m*M* sodium acetate buffer, pH 4.75, to approximately 16,000 cpm per 40 μl.[19] Add 40 μl of diluted [^{125}I]cGMP to each tube and mix.

5. Dilute antibody (stored at a 400-fold dilution) with 0.1% γ-globulin 8.3-fold in order to achieve a final dilution in the assay of 20,000-fold.[20] Add 20 μl of diluted antibody to all tubes, except two tubes to estimate nonspecific binding. To the nonspecific binding controls, add 50 μl of 0.1% γ-globulin. Gently mix each tube.

6. Allow samples to incubate overnight (14–18 hr) at 4°.

7. Dilute human plasma containing EDTA 2.5-fold with 50 m*M* sodium acetate buffer, pH 4.75. Add 50 μl of diluted human plasma to each tube and mix.

8. Precipitate the antibody complex by the addition of 1 ml of cold 12% polyethylene glycol in 50 m*M* sodium acetate buffer, pH 6.2.

9. Incubate 1 hr at 4°. Centrifuge samples for 30 min at approximately 2000 *g*.

10. Gently aspirate supernatant fractions. Layer 1 ml of 12% polyethylene glycol in each tube, but do not resuspend. Incubate on ice 20 min. Centrifuge 30 min at 2000 *g* and aspirate supernatant fractions.

11. Count radioactivity in a gamma counter.

[19] The amounts of radioligand and antibody added vary depending on the binding characteristics of the ligand and antibody titer of the undiluted serum.

[20] Dilution of the antibody required may vary from 12,000- to 20,000-fold.

Calculations

1. Subtract the averaged nonspecific binding (i.e., 250 cpm) from the averaged total binding (i.e., 4055 cpm) and from the standards and unknowns (see Table I).
2. Prepare a standard curve of the total binding (reference) divided by the radioactivity of each standard (bound) (see Fig. 1).
3. Determine the amount of cGMP (i.e., 5–200 fmol) in each sample by extrapolation from the standard curve. Samples with concentrations higher than 200 fmol will need to be diluted and assayed again. The acetylated samples and standards may be used repeatedly for several days if stored at 4°.
4. Correct for recovery the amount of cGMP detected in unknown samples and time zero control (i.e., divide by 0.40 for 40% recovery).
5. Subtract time zero background (i.e., 0.03 fmol) from unknowns (5–200 fmol).
6. Determine activity (pmol) in the original volume of the unknown samples (divide by 25 or 50 μl assayed/500-μl sample).
7. To determine specific activity of guanylyl cyclase, divide the activity (pmol) by the time of incubation (min) and amount of protein (mg).

Acknowledgments

Supported by National Institutes of Health Grants HD10254, HD05797, GM07347, and a grant from the United States Department of Agriculture.

[31] Preparative Polyacrylamide Gel Electrophoresis Apparatus for Purification of Guanylyl Cyclase

By ALEXANDER MÜLSCH and RUPERT GERZER

Introduction

Since the first use of preparative polyacrylamide gel electrophoresis in the purification of proteins in the early 1960s, the method is still outstanding for chromatographic resolution and purification of enzymes. Several modifications of the original method have been described concerning the composition of gels and buffers, the geometry of electrophoresis devices, and the harvesting of separated proteins.[1-3] Following electrophoretic separation, the part of the preparative gel containing the protein of interest (which has to be visualized by special techniques) is usually excised and then eluted by diffusion or by electrophoresis with specially designed devices. However, this procedure is affected by several problems: detection methods could destroy enzyme activity, the resolution is dependent on cutting performance, and recovery relies on elution procedures that have their own inherent limitations. Therefore, electrophoresis apparatuses providing the opportunity for continuous elution of electrophoretically separated proteins have been developed and are commercially available (e.g., by Colora/Isco, Lorch, FRG; Camag, West Berlin; Schütt, Göttingen, FRG).[4]

The preparative gel electrophoresis apparatus from Colora/Isco designed by Nees was first used in the purification of guanylyl cyclase by Garbers.[4,5] However, this and other commercially available apparatuses did not provide optimal conditions for the purification of the mammalian soluble enzyme. Therefore, we have developed modified apparatuses that were used successfully in the purification of soluble and of particulate guanylyl cyclases.[6-10] The latest version is presently used for the large-scale purification of heme-containing soluble guanylyl cyclase as well as for the purification of soluble guanylyl cyclase from various sources such as platelets.[11] Since the principle of these modifications of preparative gel electrophoresis techniques might also be useful for the purification of other proteins, we give a detailed description of the preparative gel electrophoresis apparatuses used for the purification of soluble guanylyl cyclase.

[1] A. T. Andrews, *in* "Electrophoresis" (A. R. Peacocke and W. F. Harrington, eds.), 2nd Ed., p. 178. Oxford Univ. Press, Oxford, 1986.

[2] H. R. Maurer, *in* "Disc Electrophoresis" (K. Fischbeck, ed.), 2nd Ed., p. 117. de Gruyter, Berlin, 1971.

[3] H. Stegmann, *in* "Electrophoresis '79" (B. J. Radola, ed.), p. 571. de Gruyter, Berlin, 1980.

[4] S. Nees, *in* "Electrophoresis and Isoelectric Focusing in Polyacrylamide Gel" (R. C. Allen and H. R. Maurer, eds.), p. 189. de Gruyter, Berlin, 1974.

[5] D. L. Garbers, *J. Biol. Chem.* **254,** 240 (1979).

[6] R. Gerzer, F. Hofmann, and G. Schultz, *Eur. J. Biochem.* **116,** 479 (1981).

[7] A. Mülsch, Ph.D. Thesis, University of Heidelberg, Heidelberg, West Germany (1986).

[8] A. Mülsch and R. Gerzer, this volume [34].

[9] A. Mülsch, E. Böhme, and R. Busse, *Eur. J. Pharmacol.* **135,** 247 (1987).

[10] E. W. Radany, R. Gerzer, and D. L. Garbers, *J. Biol. Chem.* **258,** 8346 (1983).

[11] K. U. Schmidt, *Naunyn-Schmiedebergs Arch. Pharmacol.* **337,** R3 (1988).

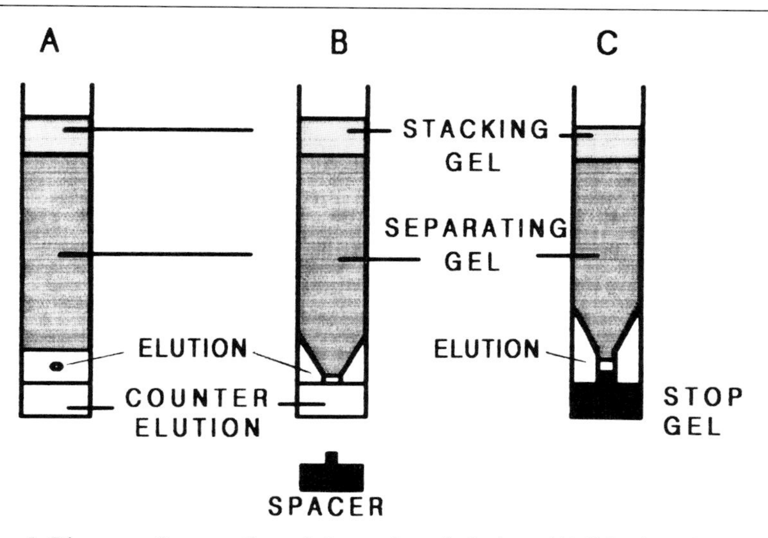

FIG. 1. Diagrams of preparative gel electrophoresis devices. (A) Side view of the original device.[4] The enzyme migrates through the separating gel into the elution chamber. Mobility toward the anionic side is delayed by a counterelution chamber. (B) Side view of the initial modification.[6] The lower end of the separating chamber is narrowed; thus, the elution chamber is only 2 mm wide. For pouring of the separating gel, a spacer is inserted to allow formation of the elution space. (C) Side view of the final version.[7] A silicon tube is inserted before gel polymerization to allow formation of the elution space and removed after polymerization.

Principle of Devices Used

Ultraphor System

The original "Ultraphor" electrophoresis device from Colora/Isco consists of a vertical glass cuvette (8 × 120 × 200 mm) bearing the polyacrylamide gel, an elution device fixed to the bottom of the cuvette, a counterelution chamber, and a supporting frame (Fig. 1A).[4] The cuvette contains a separating (height 30 mm) and a stacking polyacrylamide gel (height 10 mm). The elution device consists of two compartments formed by a pile of plates with dialysis membranes and ring seals in between. The upper compartment collects the proteins emerging from the lower end of the separating gel, whereas the lower compartment carries the counterelution buffer. To prevent clogging of the eluted proteins in the dialysis membrane, the ionic strength of the buffers in the individual compartments is adjusted according to the principle of "counterelution,"[4] that is, it is lowest in the elution buffer. By this procedure the electric field strength and consequently the electrophoretic mobility of the proteins is lower in this than in the other compartments.

As a major disadvantage of this apparatus, the volume of the elution chamber is rather high. Therefore, proteins are highly diluted during the elution process. This may not be a problem if large quantities of protein can be applied without the need for a high purification factor (the total purification factor for sea urchin sperm guanylyl cyclase is about 500[10]). However, it is unacceptable for labile enzymes that require a high purification factor (the total purification factor of heme-containing soluble guanylyl cyclase is about 15,000[6–9]). In addition, a dead space at the inlet and outlet, respectively, of the Colora/Isco apparatus alters the flow, thereby decreasing the separation quality. Furthermore, air bubbles may accumulate, which decrease the separation performance and render unattended electrophoresis risky.

Modified Systems

Based on the advantages and disadvantages of the Colora/Isco device, we have developed new apparatuses. A scheme of the apparatus that was used for the purification of heme-containing soluble guanylyl cyclase,[6] and later also of particulate guanylyl cyclase from sea urchin sperm,[10] is shown in Fig. 1B. The main advantage of this apparatus is the reduced size of the elution chamber (2×2 mm width and height). Thus, the flow of elution buffer in the chamber is high enough to reduce the danger of remixing of separated proteins. In addition, samples can be eluted in a small volume (usually <100 ml). Furthermore, because of the high flow rate and the reduced dead space, the danger of accumulation of air bubbles is virtually eliminated.

This device was either adapted to the system "Havana" from Desaga,[6] Heidelberg, FRG, or it was built completely new.[10] With the system Havana the yield of individual purifications was increased, as four separations could be done at the same time. In these versions, the counterelution systems were in principle the same as in the Colora/Isco apparatus.

In the most recent version of the apparatus,[8,9] the counterelution system is replaced by a stopping gel to make handling even simpler. A diagram of the apparatus is shown in Figs. 1C and 2. In this apparatus, a stop gel covered by a silicon tube is first polymerized at the lower part of the gel. Then the separating gel is polymerized on top of the silicon tube. After polymerization the silicon tube is withdrawn to form the space for the elution channel, and the gel is ready for use.

Operation of Apparatuses

Gels and buffers are identical in all systems with the exception that the stop gel is only used in the latest device.

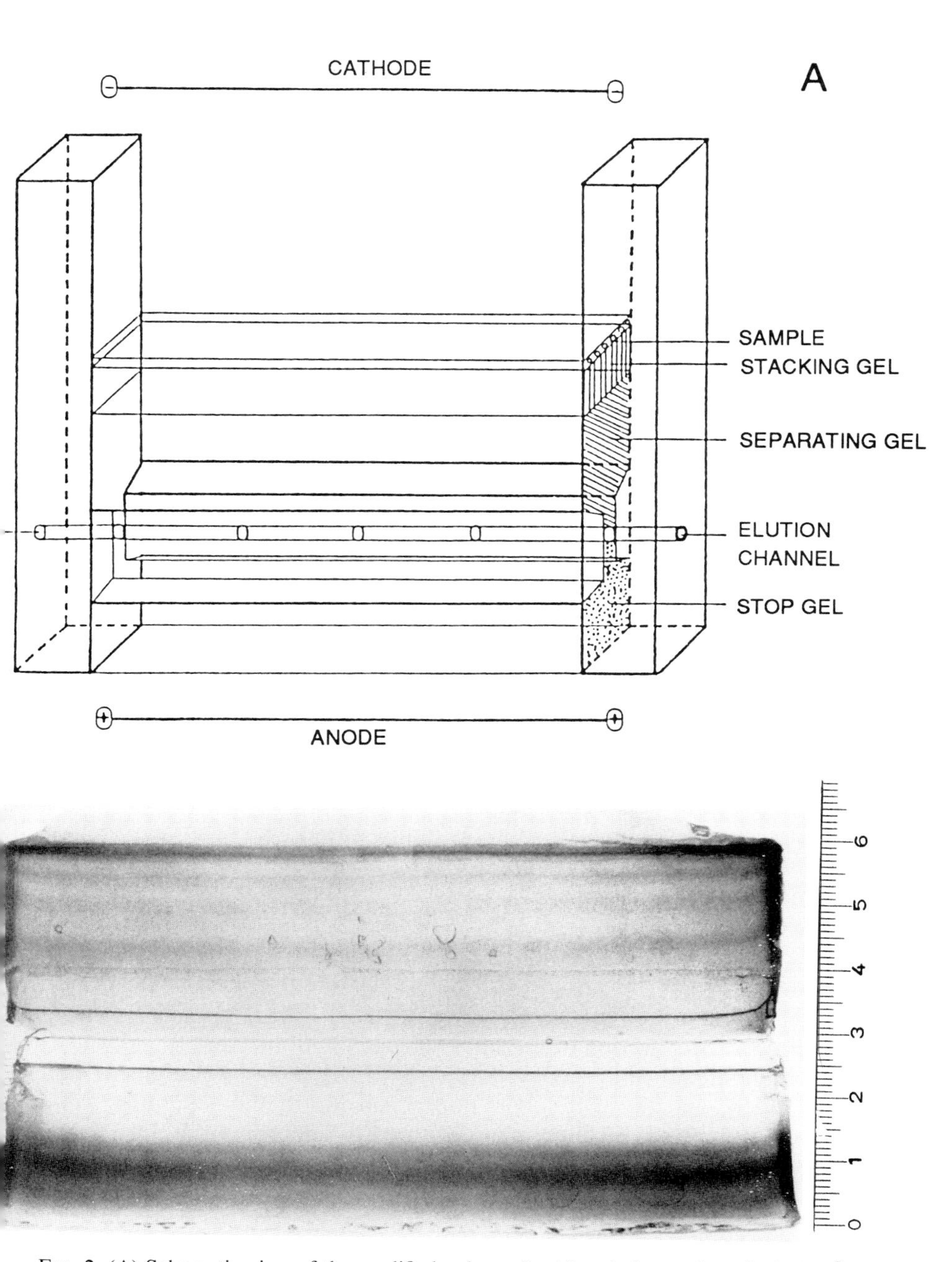

FIG. 2. (A) Schematic view of the modified polyacrylamide gel electrophoresis device.[7] For explanation, see text. Front and back plates as well as electrode compartments are omitted for clarity. (B) Preparative gel stained with Coomassie blue after elution of guanylyl cyclase. The gels are shown in the same orientation as in A.

Preparation of Gels

The composition of buffers is based on the discontinuous system of Davis.[12]

Stock Solutions (in Doubly Distilled Water)

Solution I: 1.000 *M* Tris(hydroxymethyl)aminomethane-HCl, pH 8.9
Solution II: 0.485 *M* Tris-HCl, pH 6.9
Solution A: 30.0% (w/v) acrylamide, 0.8% (w/v) *N,N'*-methylenebisacrylamide
Solution B: 10.0% (w/v) acrylamide, 2.5% (w/v) *N,N'*-methylenebisacrylamide

Fresh Solutions

Solution III: 3.0% (w/v) ammonium persulfate
Solution C: 4 mg/100 ml riboflavin (protect from light)

Gel Buffers

Stop gel: 9 ml of solution I, 15 ml of solution A, 5 ml water, 1 ml solution III, 0.1% (v/v) *N,N,N',N'*-tetramethylethylenediamine (TEMED)

Separating gel: 9 ml of solution I, 10 ml solution A, 10 ml water, 1 ml solution III, 0.1% (v/v) TEMED
Stacking gel: 1 ml of solution II, 2 ml solution B, 2 ml water, 1 ml solution C, 0.1% (v/v) TEMED

To prepare the stop gel or the separating gel, solution I is mixed with the respective volumes of solution A and water. The solution is degassed *in vacuo* for 15 min, after which TEMED and solution III are added. The solution is then poured quickly into the gel chamber and is overlayered carefully with water (1-ml syringe with a 1-mm stainless steel needle). After complete polymerization, as indicated by the appearance of a sharp border between the gel and the water, the water is decanted. Then, the next gel is poured on top of the first gel.

For the preparation of the stacking gel, solutions I and B and water are carefully mixed and degassed. Then, TEMED and solutions II and C are added, and the solution is poured on top of the separating gel and photopolymerized by illumination with a 50-W neon lamp. Before illumination, the solution may either be overlayered with water or a mold former may be introduced.

[12] B. J. Davis, *Ann. N.Y. Acad. Sci.* **121,** 404 (1964).

Operation with Counterelution

After polymerization (as indicated by the milky appearance of the stacking gel) the elution device is attached to the bottom of the cuvette, and both are placed into the electrophoresis device. Electrophoresis buffer (192 m*M* glycine, 25 m*M* Tris, 50 m*M* 2-mercaptoethanol, 0.2 m*M* EDTA, 0.2 m*M* benzamidine-hydrochloride, pH 8.6) is filled into the upper (cathodic) and lower (anodic) compartments. The concentration of Tris/glycine in the elution buffer and in the counterelution buffer is 0.8- and 3-fold, respectively, relative to the electrophoresis buffer. The buffers are filled into the respective compartments through silicon tubings, which connect one side of the compartment to a roller pump and the other side to the respective buffer reservoir. After preelectrophoresis for 4 hr at 2 W in a cold room, the concentrated guanylyl cyclase solution [<50 mg protein in 2 ml elution buffer containing 4% (w/v) sucrose] is layered below the electrophoresis buffer on top of the stacking gel by means of a long cannula. Preelectrophoresis is essential since without such a step guanylyl cyclase will lose its heme and enzyme activity during the electrophoretic separation.

Electrophoresis is performed at 2 W until the protein has entered the separating gel and at 4 W thereafter. The elution and counterelution compartments are continuously perfused (30–60 ml/hr), and the eluate is fractionated in aliquots of 2 ml. Fractions containing guanylyl cyclase activity are pooled, eventually concentrated by vacuum dialysis, and are stored with 50% (v/v) glycerol at $-70°$ under an atmosphere of nitrogen.

Electrophoresis with Stop Gel

The same solutions and buffers are used as described above. An elution buffer with lower pH (8.3) containing 17 m*M* Tris, 192 m*M* glycine, and 1 m*M* glutathione may also be used. After polymerization of the gels and removal of the silicon tubing that forms the elution channel, the elution channel is connected on one side to the elution buffer reservoir and on the other side to a roller pump with individual silicon tubings. Preelectrophoresis, enzyme application, and electrophoresis are done as described above. The elution buffer is passed through the elution chamber (60 ml/hr/gel) as soon as the yellow band of soluble guanylyl cyclase enters the lower third of the separating gel. If proteins other than soluble guanylyl cyclase are purified and therefore the protein of interest is not visible during migration through the gel, the elution system must be turned on during the whole procedure. The recovery of enzyme activity is enhanced, and the concentration of thiols in the elution buffer can be lowered, when the whole device is operated under an atmosphere of nitrogen. This latter procedure

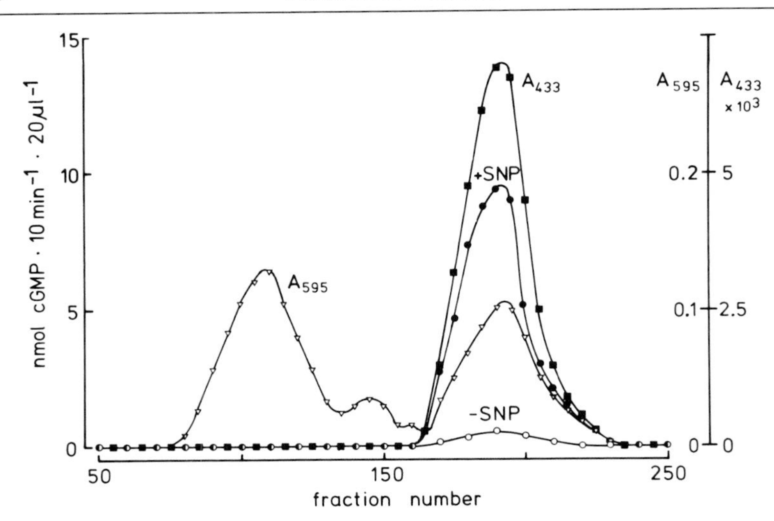

FIG. 3. Elution plot of a gel electrophoretic separation of soluble guanylyl cyclase [R. Gerzer, F. Hofmann, E. Böhme, K. Ivanova, C. Spiess, and G. Schultz, *Adv. Cyclic Nucleotide Protein Phosphorylation Res.* **14,** 255 (1981)]. The enzyme was prepurified from bovine lung by anion-exchange chromatography on DEAE-cellulose and affinity chromatography on Blue Sepharose CL-6B. Electrophoresis and elution were done as described in the text. Guanylyl cyclase activity (nmol cGMP/min/20 μl) was determined in individual fractions in the absence and presence of 50 μM sodium nitroprusside by formation of [^{32}P]GMP from [α-^{32}P]GTP. Protein was determined according to Bradford [M. M. Bradford, *Anal. Biochem.* **72,** 248 (1976)] and is expressed as absorbance at 595 nm. Expression of guanylyl cyclase activity in individual fractions coincided with a measurable absorbance at 430 nm which was due to the heme attached to the enzyme [R. Gerzer, F. Hofmann, and G. Schultz, *FEBS Lett.* **132,** 71 (1981)].

was required for preparation of thiol-free soluble guanylyl cyclase.[7,13] Effective cooling of the electrophoresis buffer is necessary. Figure 2B shows a preparative gel stained with Coomassie blue after guanylyl cyclase has been eluted.[7] Note the unstained section of the lower stop gel, which indicates the effectiveness of the elution procedure.

Conclusions

The procedures described here result in high purification factors of the respective enzymes. An example of purified soluble guanylyl cyclase is shown in Fig. 3. With the newly developed modified apparatuses, the

[13] F. Niroomand, R. Rössle, A. Mülsch, and E. Böhme, *Biochem. Biophys. Res. Commun.* **161,** 75 (1989).

enzyme elutes in a volume of 50 to 100 ml and does not need to be concentrated for further activity determinations. However, the enzyme needs to be concentrated for the determination of spectral characteristics. Concentrations by vacuum dialysis or centrifugation (polysulfone membranes with a 30-kDa cutoff, Millipore, Bedford, MA) have been found to produce the least loss of activity. In conclusion, we assume that our modifications of preparative gel electrophoresis may also be helpful for the purification of other macromolecules.

[32] Detergent Interactions and Solubilization Techniques for Membrane Guanylyl Cyclase

By ARNOLD A. WHITE and PUSHKARAJ J. LAD

Introduction

Nearly all membrane-bound guanylyl cyclases (EC 4.6.1.2, guanylate cyclase) demonstrate extremely low basal activities with MgGTP as substrate. Since for many years workers were unable to find agonists that would stimulate this activity, it became the practice to assay with MnGTP, which increased activity 10- to 20-fold.[1-3] However, even with MnGTP, the percentage of the total guanylyl cyclase activity that was particulate was underestimated, as Hardman *et al.*[4] and we[5] demonstrated, when whole homogenates or subcellular fractions were exposed to the nonionic detergent Triton X-100. By a mechanism which is still unclear, nonionic detergents revealed a large amount of cryptic activity in many tissues. While the simplest explanation is that this results from solubilization of enzyme which is not accessible to substrate, this is probably not a complete answer. The fact that some detergents also activate soluble guanylyl cyclase,[6] implies that at least that form of the enzyme must undergo some change in structure more subtle than solubilization.

With the discovery of receptor-mediated activation of membranous

[1] J. G. Hardman and E. W. Sutherland, *J. Biol. Chem.* **244,** 6363 (1969).
[2] G. Schultz, E. Böhme, and K. Munske, *Life Sci.* **8,** 1323 (1969).
[3] A. A. White and G. D. Aurbach, *Biochim. Biophys. Acta* **191,** 686 (1969).
[4] J. G. Hardman, J. A. Beavo, J.P. Gray, T. D. Chrisman, W. D. Patterson, and E. W. Sutherland, *Ann. N.Y. Acad. Sci.* **185,** 27 (1971).
[5] A. A. White, *Adv. Cyclic Nucleotide Res.* **5,** 353 (1975).
[6] P. J. Lad and A. A. White, *Arch. Biochem. Biophys.* **197,** 244 (1979).

guanylyl cyclase by the *Escherichia coli* heat-stable enterotoxin[7] and the atriopeptins,[8] workers are solubilizing those membranes in order to elucidate the activation mechanisms. However, the detergents used for solubilization will often themselves greatly stimulate activity, so that it is difficult to detect response to an agonist.[9] This situation is exacerbated when activity is determined with MnGTP rather than MgGTP, since the high basal activity seen with the former simply decreases the percent stimulation by the agonist[10] (and also by the detergent[11]). The choice of a suitable detergent to study receptor-mediated activation will have to be made empirically. However, there appear to be some generalizations one can make about both activation and solubilization of particulate guanylyl cyclase by detergents that would be useful for purification of the enzyme. This chapter summarizes results from a study on rat lung.[6]

Experimental Procedures

Enzyme Preparation

Sprague-Dawley male rats are obtained from Charles River Breeding Laboratories. They should be in the 100–125 g weight range at purchase, and should be used before they reach 250 g. The animals are sacrificed by decapitation, and the lungs are perfused *in situ* with 5 ml of cold homogenizing medium containing 0.25 *M* sucrose, 50 m*M* Tris-HCl (pH 7.6), 0.5 m*M* EDTA (Tris), 10 m*M* 2-mercaptoethanol and 1 m*M* $MgCl_2$, (STEMMg) by injection into the right ventricle of the heart. Usually the lungs from two animals are used, and these are removed, blotted dry, weighed, and placed in a volume of medium equal to 3 times their weight. This is contained in a 50-ml polycarbonate centrifuge tube, which is jacketed in ice. The tissue is homogenized for 10 sec at 10,000 rpm with a Willems Polytron equipped with a PT 20 ST generator, after which it is filtered through 40-mesh stainless steel wire screen. A 1-inch disk of the wire cloth is mounted in a polypropylene Swinny filter holder, and the holder fixed to the barrel of a polypropylene disposable syringe. The

[7] M. Field, L. H. Graff, Jr, W. J. Laird, and P. L. Smith, *Proc. Natl. Acad. Sci. U.S.A.* **75,** 2800 (1978).

[8] S. A. Waldman, R. M. Rapoport, and F. Murad, *J. Biol. Chem.* **259,** 14332 (1984).

[9] M. M. R. ElDeib, C. D. Parker, T. L. Veum, G. M. Zinn, and A. A. White, *Arch. Biochem. Biophys.* **245,** 51 (1986).

[10] M. M. R. ElDeib, C. D. Parker, and A. A. White, *Biochim. Biophys. Acta* **928,** 83 (1987).

[11] W. H. Frey, Jr., B. M. Boman, D. Newman, and N. D. Goldberg, *J. Biol. Chem.* **252,** 4298 (1977).

homogenate is poured into the syringe barrel and forced through the filter with gentle pressure from the syringe plunger.

Three types of particulate preparations are used. The first, termed "total particulates," is obtained by centrifuging the filtered homogenate at 102,900 g_{av} for 60 min (or 160,900 g_{av} for 39 min) and resuspending the pellet in the STEMMg medium. The second type is obtained by first centrifuging at 8000 g for 30 min, after which the supernatant is centrifuged as above. This pellet is resuspended in the homogenizing medium and termed "microsomes." In order to wash the microsomes free of soluble guanylyl cyclase without repetitive centrifugations, 3 ml of the microsomal suspension is applied to a 2.5 × 26 cm Sepharose 2B column, equilibrated by reverse flow with the STEMMg medium. The fractions in the void volume are pooled to form the third preparation, termed "washed microsomes." Centrifugations and chromatography are at 4°, and all preparations are kept in ice until assayed.

Detergent Treatment and Guanylyl Cyclase Assay

Twenty-five microliters of enzyme preparation are mixed in a 10 × 75 mm tube with 10 μl detergent and 30 μl containing all of the components of the guanylyl cyclase reaction mixture except for [α-^{32}P]GTP and cyclic GMP (cGMP). After 30 min in ice, the tube is warmed to bath temperature (30°) for 5 min, after which the reaction is initiated by the addition of [α-^{32}P]GTP and cGMP in 10 μl. The 75-μl reaction mixture contains 50 mM Tris-HCl, pH 7.6, 6 mM $MnCl_2$, 10 mM 2-mercaptoethanol, 5 mM cGMP, 1.2 mM [α-^{32}P]GTP, 15 mM creatine phosphate, and 10 units creatine phosphokinase. The reaction time is 6 min, and the reaction is terminated by the addition of 0.5 ml of 0.1 M HCl, containing about 30,000 counts per minute (cpm) [^{3}H]cGMP and by heating at 100° for 2 min in a heating block, after which the reaction tube is placed in an ice bath. When all reactions are terminated, 0.5 ml of 0.2 M imidazole is added to each tube, which serves to neutralize the HCl to pH 7.07. The contents of each tube are then applied to an aluminum oxide column, equilibrated with 0.1 M imidazole, 50 mM HCl, pH 7.07, for purification of the [^{32}P]cGMP formed.[12,13] The columns are eluted with the imidazole-HCl buffer, and the third and fourth milliliters are collected into scintillation vials to which 10 ml of a scintillation solution is added and counted. The recovery of [^{3}H]cGMP determined in each vial is used to correct the calculated [^{32}P]cGMP formation for losses during purification. Every determination is performed in triplicate or quadruplicate, and the mean result is reported.

[12] A. A. White and T. V. Zenser, *Anal. Biochem.* **41,** 372 (1971).
[13] A. A. White, this series, Vol. 38, p. 41.

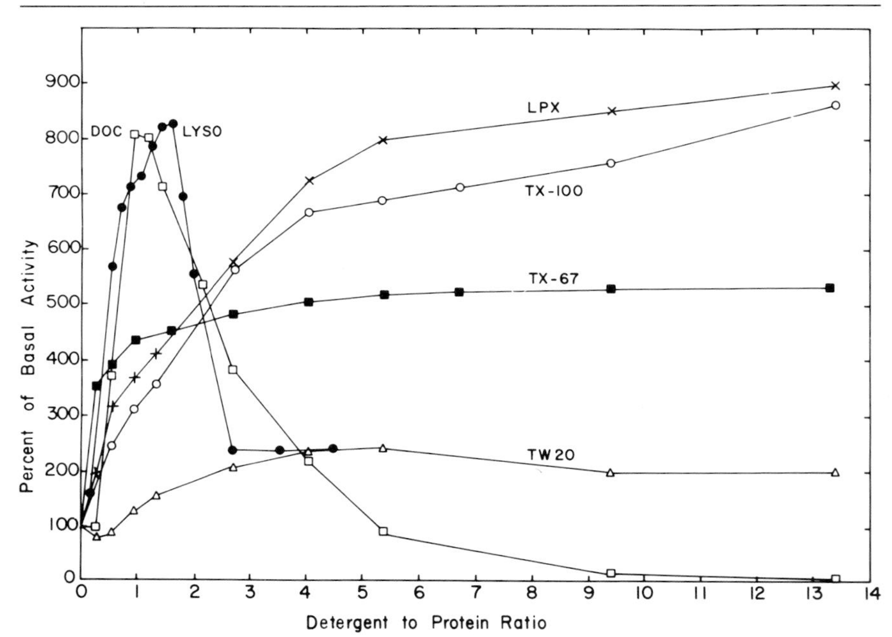

FIG. 1. Effect of certain detergents on particulate guanylyl cyclase. The detergents used are as follows: sodium deoxycholate (DOC), lysolecithin (LYSO), Lubrol PX (LPX), Triton X-100 (TX-100), Triton X-67 (TX-67), Tween 20 (TW20). Ratios of detergent to protein are weight-to-weight ratios. A washed microsomal preparation was used.

Specific activity is given as picomoles of cGMP formed per minute per milligram protein at 30°. (See also this volume, [30].)

Other Analytical Methods

Lubrol PX is quantitated by the method developed by Garewal for Triton X-100, and deoxycholate is determined by the method of Szalkowski and Mader.[14] Protein is determined by the method of Lowry et al.[15] after precipitation with 2.0 ml of silicotungstic acid.[16]

Figure 1 shows the differences in response of the particulate enzyme to a series of nonionic detergents (Tween 20, Triton X-67, Triton X-100, Lubrol PX) as compared with two anionic detergents (lysolecithin and deoxycholate). The nonionic detergents used were chosen from a large

[14] C. R. Szalkowski and W. J. Mader, *Anal. Chem.* **24,** 1602 (1952).

[15] O. H. Lowry, N. J. Rosebrough, A. L. Farr, and R. J. Randall, *J. Biol. Chem.* **193,** 265 (1951).

[16] J. B. Martin and D. M. Doty, *Anal. Chem.* **21,** 965 (1949).

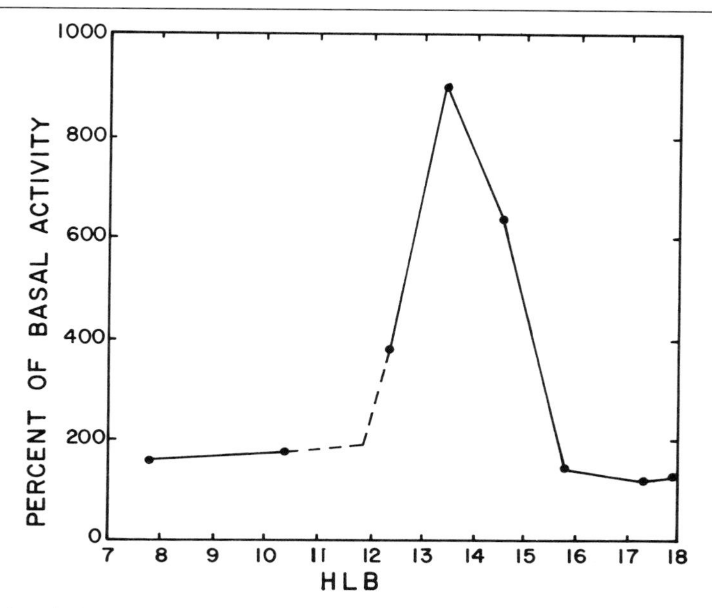

FIG. 2. Relationship of the HLB number of detergents on the Triton series to their effects on particulate guanylyl cyclase (microsomes). The Triton detergents used, and their HLB numbers, are as follows: X-15, 3.6; X-35, 7.8; X-45, 10.4; X-114, 12.4; X-100, 13.5; X-102, 14.6; X-165, 15.8; X-305, 17.3; X-405, 17.9 Activity is given as percent that of the untreated enzyme.

number tried in preliminary assays. Tween 20 (ICI) is described as polyoxyethylene 20 sorbitan monolaurate, while Lubrol PX (ICI) is a condensate of fatty alcohol with polyoxyethylene (9–10). Triton X-100 (Rohm and Haas) is polyoxyethylene (9–10) octylphenol while Triton X-67 (Rohm and Haas) is a polyoxyethylene ether of a fatty alcohol. The activation by these detergents varied widely, and, since the maximal activity achieved by each was relatively constant over a wide range of detergent-to-protein ratios, we conclude that it is the chemical structure of the detergent that determines percent activation. However, differences in activation will also be seen within a homologous series of compounds. Thus, with the Triton series of detergents we were able to show a correlation between HLB (hydrophilic/lipophilic balance) number and ability to activitate particulate activity (Fig. 2). The HLB number represents the relative ratio of hydrophilic to hydrophobic regions within a surfactant, a higher value corresponding to an increase in hydrophilic character. In this series the HLB number varies with the length of the ethylene oxide chain and can be calculated by dividing the weight percent of ethylene oxide by 5. Because of the regularity of the substitutions in this homologous series, we might expect to show structure–activity relationships not easily demonstrated with more structurally varied surfactants. This was indeed the case, with optimal activity appearing in a fairly narrow range (HLB 12.5–15). Similar optima were observed for the activation of D-alanine carboxypeptidase from *Bacillus subtilis* and succinate dehydrogenase from *Micrococcus luteus*.[17]

In the case of the anionic detergents deoxycholate and lysolecithin (Fig. 1), the same maximal activity was reached when deoxycholate-to-protein or lysolecithin-to-protein ratios were approximately 1.5. This maximum was similar to that achieved by Lubrol PX or Triton X-100 (i.e., 800–900%). However, unlike the case for nonionic detergents, higher concentrations of the anionic detergents sharply inhibited activity. The inhibition was reversible, since, when membranes treated with deoxycholate were in turn exposed to increasing concentrations of Lubrol PX, the same maximal activation achieved by Lubrol PX alone was seen, although at a higher Lubrol concentration (Fig. 3). We concluded that the shift to the right of the Lubrol response curve represented the amount of Lubrol required to sequester the deoxycholate previously inhibiting the enzyme. Such sequestration might be in the form of mixed micelles with Lubrol, and this was evident when the detergents were chromatographed separately and together on a Sephadex G-200 column (data not shown). We were able to achieve the same effect by removing the deoxycholate by gel filtration on a Sephadex G-50 column.

Table I (Exp. B) shows that, while deoxycholate at the concentration used stimulated particulate activity, that activity was doubled after gel filtration (deoxycholate-to-protein ratio in this experiment was 2.2, which while higher than the optimum was still in the stimulatory range). When activity was determined in the presence of Lubrol PX we found, as expected, equivalent activity before and after deoxycholate treatment. However, Lubrol-stimulated activity was lower after deoxycholate treatment and gel filtration. This may be due to the removal of lipids from the membranes under these conditions, and if so, implies that activation by detergents is not only a function of the chemical nature of the detergent, but may also be affected by the membrane lipid present. (It is interesting that Lubrol PX did not increase the activity of deoxycholate-treated membranes after gel filtration.) We concluded that Lubrol PX can form mixed micelles with excess deoxycholate, and it is the sequestration of this excess that allows measurement of enzyme activity. Table I (Exp. A) also shows the effect of deoxycholate on soluble guanylyl cyclase. While the

[17] J. N. Umbreit and J. L. Strominger, *Proc. Natl. Acad. Sci. U.S.A.* **70,** 2997 (1973).

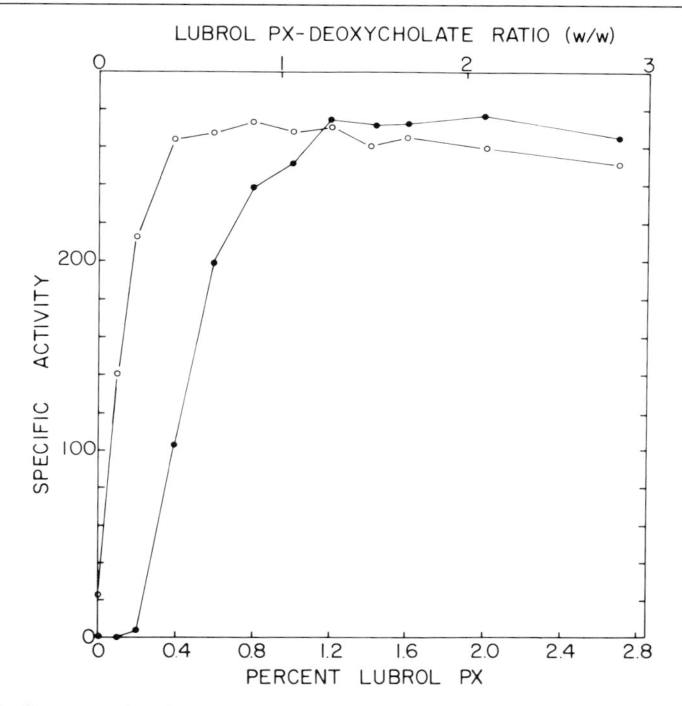

FIG. 3. Concentration-dependent reversal of deoxycholate inhibition by Lubrol PX. A microsomal pellet was resuspended in homogenizing medium and 1 ml of the suspension, containing 5.99 mg protein, was mixed with 0.4 ml of 10% sodium deoxycholate (●) while another 1 ml was mixed with 0.4 ml of water (○). After 30 min in ice, 25-μl aliquots were added to iced reaction tubes containing increasing concentrations of Lubrol PX in 10 μl. After another 30 min, each tube was warmed to 30° for 5 min and then assayed for guanylyl cyclase activity. The concentration of Lubrol PX shown was that present in the final reaction mixture. Also shown are the ratios of Lubrol PX to deoxycholate when both were present.

soluble enzyme is activated by Lubrol PX, we were never able to demonstrate anything other than inhibition by deoxycholate (and lysolecithin). In addition, neither nonionic detergents nor phospholipids (lysolecithin, lecithin, phosphatidylethanolamine) were able to restore the activity even after removal of deoxycholate by Sephadex G-50 chromatography. Therefore, the inhibition of soluble enzyme by anionic detergents appears irreversible.

The ability of Lubrol PX to reverse the inhibition of guanylyl cyclase activity by high concentrations of sodium deoxycholate allowed us to determine the ability of deoxycholate to solubilize particulate enzyme. The anionic detergents are usually more efficient solubilizers than the

TABLE I
EFFECT OF SODIUM DEOXYCHOLATE ON SOLUBLE AND PARTICULATE GUANYLYL CYCLASE[a]

Step	cGMP formation (pmol/min/mg protein)	
	−Lubrol PX	+Lubrol PX
Experiment A		
Supernatant fraction	282	604
Deoxycholate-treated supernatant fraction	2	4
Deoxycholate-treated supernatant fraction after Sephadex G-50	36	32
Experiment B		
Microsomal suspension	54	428
Deoxycholate-treated suspension	194	404
Deoxycholate-treated suspension after Sephadex G-50	382	347

[a] In two separate experiments, a microsomal pellet and a supernatant fraction were prepared from rat lung homogenates in STEMMg. In Experiment A, 1.4 ml supernatant fraction (containing 4.1 mg protein) was mixed with 0.4 ml of 2 *M* sucrose and 0.2 ml of 10% sodium deoxycholate. One milliliter of this solution was chromatographed on a 1 × 22 cm Sephadex G-50 column eluted with STEMMg in 1-ml fractions. The void volume fractions (7, 8, 9) were previously shown to contain all the guanylyl cyclase activity, and these were collected and assayed. Deoxycholate did not appear until fraction 16. In Experiment B, 1.4 ml of microsomal suspension (9.1 mg protein) was mixed with 0.4 ml of 2 *M* sucrose and 0.2 ml of 10% sodium deoxycholate. Again 1 ml of the latter was chromatographed on the Sephadex G-50 column. Guanylyl cyclase activity was determined with and without 0.8% Lubrol PX in the reaction mixture.

nonionic detergents, and the same was true of particulate guanylyl cyclase. As shown in Fig. 4, more than 95% of guanylyl cyclase activity was solubilized when deoxycholate-to-protein ratio was about 0.5, along with approximately 50% of the protein. A comparative experiment was performed with the more commonly used enzyme solubilizer, Triton X-100 (Fig. 5). With this detergent the optimum Triton X-100-to-protein ratio was 1, when 60–65% of the enzyme activity and 30–35% of the protein

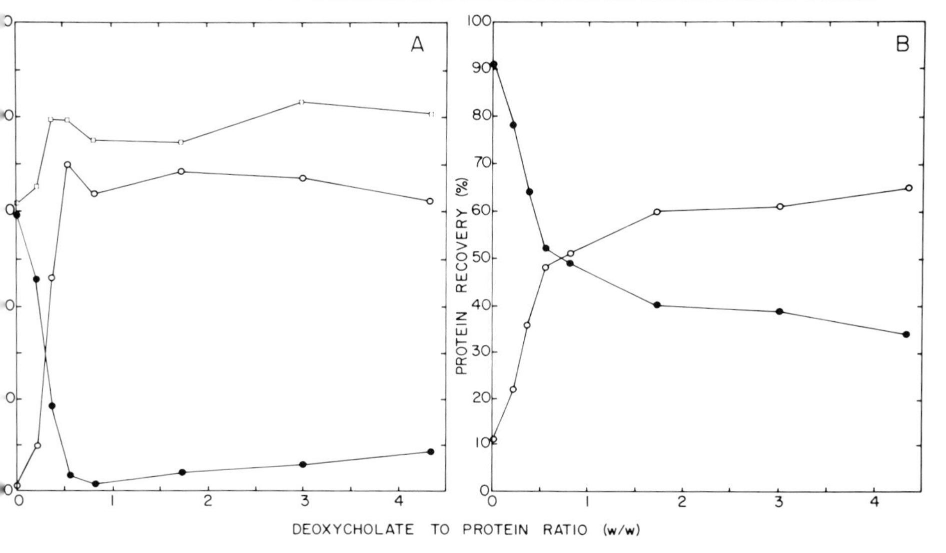

FIG. 4. Solubilization of guanylyl cyclase by deoxycholate. A total particulates fraction was resuspended in STEMMg medium and aliquots mixed, in centrifuge tubes, with increasing amounts of 10% deoxycholate in the same medium. After the volume in each tube was adjusted to 2.5 ml with medium, the tubes were kept in ice for 30 min, after which 0.4 ml was removed from each tube and saved, and the remainder of the solution was centrifuged at 100,000 *g* for 60 min. After each supernatant solution was decanted, the pellet was carefully rinsed with 0.2 ml medium and this added to the supernatant fluid. The pellets were each resuspended in 2.1 ml of medium. The original deoxycholate-treated suspension, the supernatant fractions, and the resuspended pellets were all assayed for protein and for guanylyl cyclase activity with 0.8% Lubrol PX in the reaction mixture. Guanylyl cyclase activity (A) is the total activity (pmol/min) calculated to be present in the 2.1 ml of each deoxycholate-treated suspension (□) and also in the supernatant fractions (○) and pellets (●) obtained by centrifugation. The protein recoveries (B) are expressed as percent total protein found in each pellet (●) and supernatant fraction (○).

was solubilized. Total solubilization of activity was not obtained even at Triton X-100-to-protein ratios of 3. We have observed, however, that repeated extraction with 1.2% Triton X-100 will solubilize 80–90% of the microsomal activity.

Helenius and Simons[18] showed that deoxycholate and Triton X-100 would bind to proteins with hydrophobic character (low density lipoproteins, Semiliki Forest virus envelope, human erythrocyte stroma) but would not bind to hydrophilic proteins. They suggested that the hydrophobic regions involved are normally occupied by lipid. Lipid removal by

[18] A. Helenius and K. Simons, *J. Biol. Chem.* **247,** 3656 (1972).

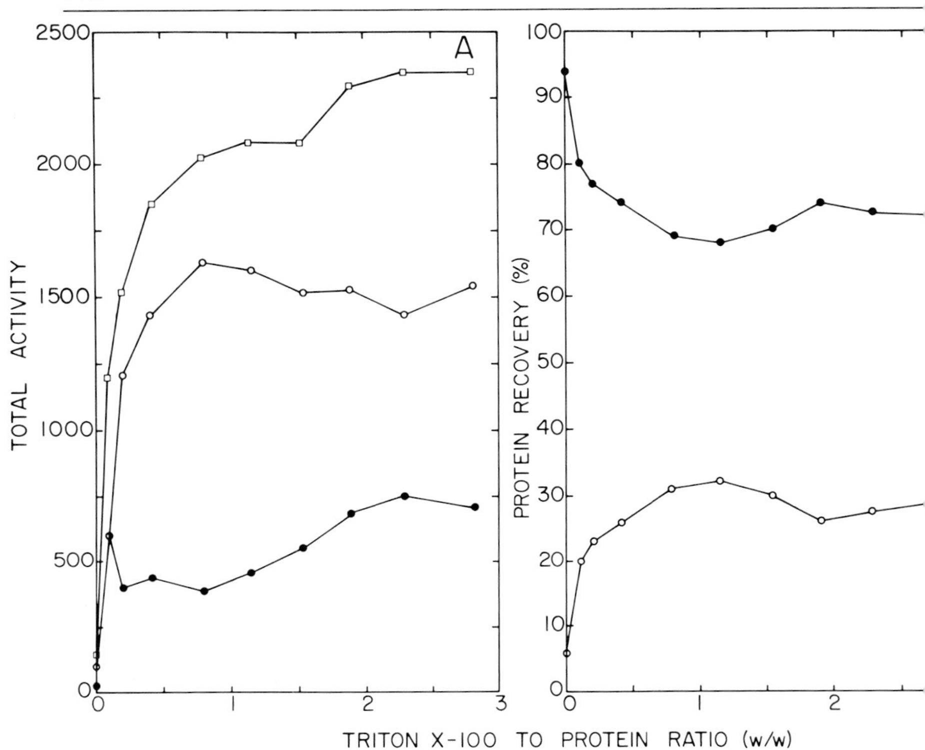

FIG. 5. Solubilization of particulate guanylyl cyclase by Triton X-100. The experiment was performed essentially the same as described in the legend to Fig. 4. However, no Lubrol PX was present in the reaction mixture, but rather Triton X-100 at the same concentration as used for solubilization. Guanylyl cyclase activity (A) and protein recovery (B) are expressed as in Fig. 4.

detergent was therefore an exchange of bound lipid for bound detergent, resulting in a complex with the hydrophilic part of the detergent molecule exposed to the aqueous medium, thus rendering the protein soluble. The enhanced ability of deoxycholate to solubilize guanylyl cyclase as compared with Triton X-100 may be related to the unique characteristics of bile salt micelles. Carey and Small[19] noted that they are smaller than those formed by other detergents, with aggregation numbers from 2 to 9. They form mixed micelles with a variety of soluble and insoluble lipids, and have a striking capacity to solubilize phospholipids. Furthermore, the resulting mixed micelles are able to solubilize long chain fatty acids and

[19] M. C. Carey and D. M. Small, *Arch. Intern. Med.* **130,** 506 (1972).

cholesterol, which would not ordinarily be solubilized by pure bile salt solutions. Some of the unique properties of bile salts have been preserved in the zwitterionic derivative 3-[(3-cholamidopropyl)dimethylammonio]-1-propane sulfonate,[20] known as CHAPS, and here again we have found that while CHAPS will inhibit guanylyl cyclase, this inhibition will be reversed by Lubrol PX.

[20] L. M. Hjelmeland, *Proc. Natl. Acad. Sci. U.S.A.* **77,** 6368 (1980).

[33] Purification of Membrane Form of Guanylyl Cyclase

By CHODAVARAPU S. RAMARAO and DAVID L. GARBERS

Introduction

Guanylyl cyclase (EC 4.6.1.2, guanylate cyclase) catalyzes the formation of 3′,5′-cyclic guanosine monophosphate (cGMP) from GTP in the presence of Mn^{2+} or Mg^{2+} ions. It occurs in both the soluble and particulate fractions of the cell with a wide range of distribution throughout the animal kingdom.[1] The particulate enzyme represents at least two major components: membrane and cytoskeletal in that detergents easily solubilize the enzyme associated with the plasma membrane but fail to solubilize the enzyme from gut microsomes and rod outer segments effectively.[2] The ratio of membrane to soluble forms of the enzyme varies widely depending on the tissue, and both forms differ from each other with respect to kinetics, metal ion sensitivity, apparent molecular weight, and their responses to specific activators.[3,4] It is now known that the two forms are homologous, based on cDNA cloning and the predicted amino acid sequences.[5,6]

Sea urchin spermatozoa are not only one of the richest sources of guanylyl cyclase, but the enzyme from this source is also almost entirely membranous in origin.[3,7] The sea urchin sperm enzyme exists in a highly

[1] C. K. Mittal and F. Murad, *in* "Handbook of Experimental Pharmacology" (J. A. Nathanson, and J. W. Kebabian, eds.), p. 225. Springer-Verlag, Berlin, 1982.
[2] D. Fleishman, *Curr. Top. Membr. Transp.* **15,** 109 (1981).
[3] D. L. Garbers, J. G. Hardman, and F. B. Rudolph, *Biochemistry* **13,** 4166 (1974).
[4] H. Kimura and F. Murad, *J. Biol. Chem.* **249,** 6910 (1974).
[5] S. Singh, D. G. Lowe, D. S. Thorpe, H. Rodriguez, W.-J. Kuang, L. J. Dangott, M. Chinkers, D. V. Goddel, and D. L. Garbers, *Nature (London)* **334,** 708 (1988).
[6] S. Schulz, M. Chinkers, and D. L. Garbers, *FASEB J.* **3,** 2026 (1989).
[7] J. P. Gray and G. I. Drummond, *Arch. Biochem. Biophys.* **172,** 31 (1976).

phosphorylated state under normal conditions,[8,9] but the enzyme is rapidly dephosphorylated if intact cells are treated with species-specific peptides[9] or if cells are homogenized at neutral pH in the absence of protein phosphatase inhibitors.[10] Dephosphorylation of the membrane form of the enzyme from sea urchin spermatozoa appears to represent a mechanism of desensitization.[11] Recent research showing that guanylyl cyclase is a cell surface receptor suggests that phosphorylation/dephosphorylation could play a major regulatory role as with other receptors. Both the phosphorylated[10] and dephosphorylated[12,13] forms of the enzyme from sea urchin spermatozoa have been purified to apparent homogeneity. A membrane form of guanylyl cyclase has been purified from rat lung,[14] rat adrenocortical carcinoma cells,[15] and other sources,[16,17] and methods have varied for each tissue.

Assay Methods

Detailed assay procedures for guanylyl cyclase activity are described elsewhere in this volume.[18] Typically guanylyl cyclase is assayed in the presence of GTP as the substrate, Mn^{2+} or Mg^{2+} as cofactors, and a phosphodiesterase inhibitor like caffeine or 3-isobutyl-1-methylxanthine.

Purification Procedure

The procedure described below (see also Table I) has been used with success to purify the phosphorylated, membrane form of the enzyme from sea urchin spermatozoa.[19]

Step 1: Collection and Homogenization of Spermatozoa. Spermatozoa

[8] G. E. Ward and V. D. Vacquier, *Proc. Natl. Acad. Sci. U.S.A.* **80,** 5578 (1983).

[9] N. Suzuki, H. Shimomura, E. W. Radany, C. S. Ramarao, G. E. Ward, J. K. Bentley, and D. L. Garbers, *J. Biol. Chem.* **259,** 14874 (1984).

[10] C. S. Ramarao and D. L. Garbers, *J. Biol. Chem.* **263,** 1524 (1988).

[11] J. K. Bentley, D. J. Tubb, and D. L. Garbers, *J. Biol. Chem.* **261,** 14859 (1986).

[12] D. L. Garbers, *J. Biol. Chem.* **251,** 4071 (1976).

[13] E. W. Radany, R. Gerzer, and D. L. Garbers, *J. Biol. Chem.* **258,** 8346 (1983).

[14] T. Kuno, J. W. Andresen, Y. Kamisaki, S. A. Waldman, L. Y. Chang, S. Saheki, D. C. Lietman, M. Nakane, and F. Murad, *J. Biol. Chem.* **261,** 5817 (1986).

[15] A. K. Paul, R. B. Marala, R. K. Jaiswal, and R. K. Sharma, *Science* **235,** 1224 (1987).

[16] R. Takayanagi, T. Inagami, R. M. Snajdar, T. Imada, M. Tamura, and K. S. Misono, *J. Biol. Chem.* **262,** 12104 (1987).

[17] S. A. Waldman, J. A. Lewicki, H. J. Brandwein, and F. Murad, *J. Cyclic Nucleotide Res.* **8,** 359 (1982).

[18] S. E. Domino, D. J. Tubb, and D. L. Garbers, this volume, [30].

[19] S. Singh and D. L. Garbers, this volume, [39].

TABLE I
PURIFICATION OF MEMBRANE FORM OF GUANYLYL CYCLASE[a]

Step	Protein (μg)	Total activity (μmol cGMP formed/min)	Specific activity (μmol cGMP formed/min/mg)	Purification (-fold)	Recovery (%)
1	41,250	4.5	0.11	—	100
2	5520	4.4	0.79	7	96
3	470	2.4	5.1	46	53
4	110	1.9	17.6	160	43
5	1.09	0.09	83.1	755	2

[a] From Ref. 10.

from the sea urchin *Arbacia punctulata* are collected by intracoelomic injection of 0.5 *M* KCl. Ten milliliters of fresh spermatozoa (approximate wet weight 4 g) is washed twice with artificial seawater, pH 6.5, buffered with [2-(2-amino-2-oxoethyl)amino]ethanesulfonic acid (10 m*M*), by centrifugation at 1000 *g* for 10 min at 0°–4° followed by resuspension. Washed sperm cells are diluted in 10 volumes of a buffer containing 25 m*M* 2-(*N*-morpholino)ethanesulfonic acid (MES), pH 6.2, 0.5% (v/v) Lubrol PX, 35% glycerol, 10 m*M* NaF, 10 m*M* benzamidine hydrochloride, 5 m*M* dithiothreitol, and 5 m*M* $MnCl_2$ (buffer A). Inclusion of 30–50% glycerol in all of the buffers stabilizes the phosphorylated form of the enzyme.[10]

Step 2: Recovery of Enzyme Activity from Homogenate. The sperm cell suspension is disrupted by three 10-sec sonications with a Branson sonifier (cell disruptor 185) at No. 7 setting and clarified by centrifugation at 30,000 *g* for 30 min at 0°. A majority of the membrane-bound enzyme activity is solubilized by this treatment.

Step 3: GTP-Agarose Affinity Chromatography. The supernatant fluid is next batch adsorbed to 15 ml of GTP-agarose (Sigma, St. Louis, MO) which was previously equilibrated with buffer A. After 2 hr, the mixture is centrifuged at 20,000 *g* for 15 min, and the GTP-agarose pellet is washed 2 times with buffer A (30 min each wash) by centrifugation. The enzyme is then eluted from GTP-agarose with 25 ml of a solution containing 25 m*M* MES, 0.1% Lubrol PX, 35% glycerol, 10 m*M* NaF, 10 m*M* benzamidine hydrochloride, 5 m*M* dithiothreitol, and 1m*M* EDTA (buffer B).

Step 4: DEAE Anion-Exchange Chromatography. After centrifugation at 20,000 *g* for 15 min, the solution containing the enzyme is added to DEAE-Sephacel which had been previously equilibrated with buffer B. After 1 hr, the DEAE-Sephacel is washed twice with 2 volumes of buffer B followed by 2 washes with buffer B containing 0.2 *M* NaCl. The enzyme is then eluted from DEAE with 2 volumes of buffer B containing 0.5 *M* NaCl.

Step 5: Concanavalin A-Sepharose Affinity Chromatography. The eluate is mixed with 5 ml of concanavalin A-Sepharose. The concanavalin A-Sepharose was previously equilibrated with a buffer containing 25 m*M* triethanolamine, pH 7.5, 35% glycerol, 0.1% Lubrol PX (v/v), 0.5 *M* NaCl, 10 m*M* NaF, 10 m*M* benzamidine hydrochloride, and 5 m*M* dithiothreitol. After 12 hr of mixing, the enzyme bound to the concanavalin A-Sepharose is washed with the same buffer 2 times, and guanylyl cyclase is finally eluted from the resin with a solution containing 25 m*M* MES, pH 6.2, 25% glycerol, 10% ethylene glycol, 1% Lubrol PX, 10 m*M* benzamidine, 10 m*M* NaF, 5 m*M* dithiothreitol, and 0.5 *M* α-methyl mannopyranoside.

Step 6: Concentration of Enzyme. The eluted enzyme is concentrated by several centrifugations at 5000 *g* for 30 min by the use of Centricon YM30 microconcentrators (Amicon, Danvers, MA) and stored at $-20°$ until use. Guanylyl cyclase is stable for at least 1 month under these storage conditions.

Properties

Molecular Weight and Homogeneity. Although recoveries from the lectin affinity chromatography are generally poor, guanylyl cyclase prepared by this method is retained in its more active, phosphorylated state. Specific activities in the range of 70–110 μmol/min/mg protein are severalfold higher than the specific activity of the dephosphorylated form of the enzyme. The purified enzyme has an apparent molecular weight of 160,000 as estimated on sodium dodecyl sulfate (SDS)-polyacrylamide gels. A faint 50,000 band also may be visible after silver staining. The major band comigrates with the phosphorylated form of guanylyl cyclase found in intact spermatozoa.

Effects of Dephosphorylation. Species-specific egg peptides cause a dephosphorylation of the enzyme in intact spermatozoa.[20] The purified, phosphorylated form of the enzyme can be dephosphorylated by either a crude preparation of a sperm phosphatase or calf intestinal alkaline phosphatase.[10] Dephosphorylation of guanylyl cyclase results in up to a 90% reduction in enzyme activity[20] and in a loss of positive cooperativity with respect to its substrate, MnGTP.[10] (See also [43], this volume.)

[20] C. S. Ramarao and D. L. Garbers, *J. Biol. Chem.* **260,** 8390 (1985).

Other Methods

After affinity chromatography and ion-exchange steps, Radany *et al.*[13] used a continuously eluting preparative polyacrylamide gel electrophoresis system as a final step in the purification of the enzyme (see [31], this volume). Recoveries obtained by this method are far superior to the concanavalin A-Sepharose elution described above, but the resulting enzyme is in the dephosphorylated form. Garbers[12] used BioGel A 0.5m column (Bio-Rad, Richmond, CA) chromatography as the final purification step, and Murad and co-workers[14] used phenyl-agarose and wheat germ agglutinin A affinity columns after GTP-agarose and DEAE-Sephacel steps to copurify the guanylyl cyclase and atrial natriuretic factor receptor activity from rat lung (see also [37] and [38], this volume). A single band of 120,000 was obtained as judged by SDS-polyacrylamide gels. Starting with membranes obtained from rat adrenocortical carcinoma cells, Sharma and co-workers[15] have purified guanylyl cyclase activity to apparent homogeneity by GTP-agarose and cGMP-agarose affinity columns. The resulting 180,000 protein possessed both guanylyl cyclase activity and atrial natriuretic factor receptor activity.[15,16]

[34] Purification of Heme-Containing Soluble Guanylyl Cyclase

By ALEXANDER MÜLSCH and RUPERT GERZER

Introduction

Guanylyl cyclases [EC 4.6.1.2, guanylate cyclase, GTP pyrophosphate-lyase (cyclizing)] catalyze the formation of 3′,5′-cyclic GMP (cGMP) from 5′-GTP. In vertebrates, at least two different forms, particulate and cytosolic guanylyl cyclase, exist (for reviews, see Refs. 1 and 2). The cytosolic enzyme from bovine lung,[3,4] rat lung,[5] and rat liver[6] and the

[1] F. Murad, S. A. Waldman, R. R. Fiscus, and R. M. Rapoport, *J. Cardiovasc. Pharmacol.* **8** (Suppl. 8), S57 (1986).
[2] J. Tremblay, R. Gerzer, and P. Hamet, *Adv. Second Messenger Phosphoprotein Res.* **22,** 319 (1988).
[3] R. Gerzer, F. Hofmann, and G. Schultz, *Eur. J. Biochem.* **116,** 479 (1981).
[4] L. J. Ignarro, K. S. Wood, and M. S. Wolin, *Proc. Natl. Acad. Sci. U.S.A.* **79,** 2870 (1982).
[5] J. A. Lewicki, H. J. Brandwein, C. K. Mittal, W. P. Arnold, and F. Murad, *J. Cyclic Nucleotide Res.* **8,** 17 (1982).
[6] P. A. Craven and F. R. DeRubertis, *Biochim. Biophys. Acta* **745,** 310 (1983).

particulate enzyme from sea urchin spermatozoa,[7] rat adrenocortical carcinoma,[8] and rat lung have been purified to apparent homogeneity.[9] The particulate form from sea urchin spermatozoa is a glycoprotein with partial homology in its primary sequence to protein kinases.[10] The particulate guanylyl cyclase from mammals is probably different from the sea urchin sperm enzyme and is activated by the atrial natriuretic factor and copurifies with the receptor of this factor.[1,2] The soluble form from bovine lung is a hemoprotein consisting of two subunits of about 75 and 70 kDa,[11,12] with the smaller subunit exhibiting partial homology with the sea urchin spermatozoan enzyme.[13] Soluble and particulate forms differ with respect to enzyme kinetics and susceptibility to activation and inhibition by biological and chemical compounds.[1,2] The soluble heme-containing guanylyl cyclase from bovine lung is directly activated by nitric oxide released from nitric oxide-containing vasodilators or by endothelium-derived nitric oxide (EDRF) released from endothelial cells.[11,14,15]

In this chapter we describe a procedure for obtaining highly purified, heme-containing soluble guanylyl cyclase from bovine lung. The original procedure for purification of soluble guanylyl cyclase from bovine lung[3] was modified to meet the requirements for larger amounts of this enzyme for EPR studies,[12,16] sequencing,[13] and the detection of EDRF.[14] The original procedure is outlined in principle and the modifications described in detail.

Determination of Guanylyl Cyclase Activity

Guanylyl cyclase is localized in individual fractions after each purification step by measuring the enzymatic conversion of [α-^{32}P]GTP to [^{32}P]cGMP. Briefly, aliquots of selected fractions are incubated for 10 min at 37° in the presence and absence of 100 μM sodium nitroprusside in a

[7] E. W. Radany, R. Gerzer, and D. L. Garbers, *J. Biol. Chem.* **258,** 8346 (1983).

[8] A. K. Paul, R. B. Marala, R. K. Jaiswal, and R. K. Sharma, *Science* **235,** 1224 (1987).

[9] L. J. Chang, B. J. Chang, S. A. Waldman, and F. Murad, *Fed. Proc., Fed. Am. Soc. Exp. Biol.* **44,** 1677 (1985).

[10] S. Singh, D. G. Lowe, D. S. Thorpe, H. Rodriguez, W. J. Kuang, L. J. Dangott, M. Chinkers, D. V. Goeddel, and D. L. Garbers, *Nature (London)* **334,** 708 (1988).

[11] R. Gerzer, E. Böhme, F. Hofmann, and G. Schultz, *FEBS Lett.* **132,** 71 (1981).

[12] A. Mülsch, Ph.D. Thesis, University of Heidelberg, Heidelberg, West Germany (1986).

[13] D. Koesling, J. Herz, H. Gausepohl, F. Niroomand, K. D. Hinsch, A. Mülsch, E. Böhme, G. Schultz, and R. Frank, *FEBS Lett.* **239,** 29 (1988).

[14] R. M. J. Palmer, A. G. Ferrige, and S. Moncada, Nature (*London*) **327,** 524 (1987).

[15] A. Mülsch, E. Böhme, and R. Busse, *Eur. J. Pharmacol.* **135,** 247 (1987).

[16] A. Mülsch and E. Böhme, *Naunyn-Schmiedebergs Arch. Pharmacol.* **325** (Suppl.), R31 (1984).

total volume of 100 μl containing 30 m*M* triethanolamine hydrochloride, pH 7.4, 3 m*M* magnesium chloride, 3 m*M* glutathione, 0.2 m*M* (0.2 μCi) [α-^{32}P]GTP, 0.1 m*M* cGMP, 5 m*M* creatine phosphate, 5 units creatine phosphokinase, 1 m*M* 3-isobutyl-1-methylxanthine, and 0.1 m*M* EGTA. The reaction is stopped by addition of 0.4 ml of 120 m*M* zinc acetate and 0.5 ml of 120 m*M* sodium carbonate. After centrifugation (10 min at 10,000 *g*), formed [^{32}P]cGMP is isolated by chromatography of the supernatant fraction on acid alumina.[15] Recovery of cGMP is about 50% as determined by the addition of [^{3}H]cGMP to representative samples.

Purification Procedure

General Outline. After preparation of a crude tissue extract, soluble guanylyl cyclase is purified by sequential application of anion-exchange chromatography on DEAE-Sepharose, ammonium sulfate precipitation, affinity chromatography on Blue Sepharose, and preparative polyacrylamide gel electrophoresis.[3] In the modified procedure, granular DEAE-Sepharose is replaced by DEAE-paper cartridges (Zetaprep-3200, Cuno, Mainz, FRG), the washing procedure for Blue Sepharose is reduced to one step, and a modified electrophoresis device is used.[17]

Step 1: Preparation of Tissue Extract. All steps are performed in a cold room. Four to five fresh bovine lungs (10–12 kg) are obtained from a local slaughterhouse and placed on ice for transportation. Connective tissue and fat are carefully removed. The lungs are then minced and homogenized in a Waring blendor (45 sec at 15,000 rpm) with homogenizing buffer (1 liter per 500 g mince) consisting of 10 m*M* potassium phosphate, pH 6.5, 0.2 m*M* benzamidine, and 6 m*M* EDTA. Following centrifugation (30 min at 14,000 *g*) the supernatants are filtered through glass wool and diluted 1 : 1 with homogenization buffer containing 100 m*M* 2-mercaptoethanol and 2 m*M* EDTA.

Step 2: Chromatography on DEAE-Cellulose. Five Zetaprep-3200 DEAE cartridges connected in series are preprocessed (flow rate 50 liters/hr) according to the following scheme: Regenerate with (1) 50 liters of 0.5 *M* acetic acid, pH 2–3; (2) 50 liters of 0.1 *M* sodium phosphate, pH 8; (3) repeat (1) and (2); (4) preequilibrate with 32 liters of 0.15 *M* potassium phosphate, pH 7; (5) equilibrate with 100 liters of 10 m*M* potassium phosphate, pH 6.5. The crude tissue extract (about 50 liters) is applied to the cartridges (20 liters/hr) by pressure with a roller pump. After 1 hr the first cartridge is disconnected from the perfusion because of high flow

[17] A. Mülsch and R. Gerzer, this volume [31].

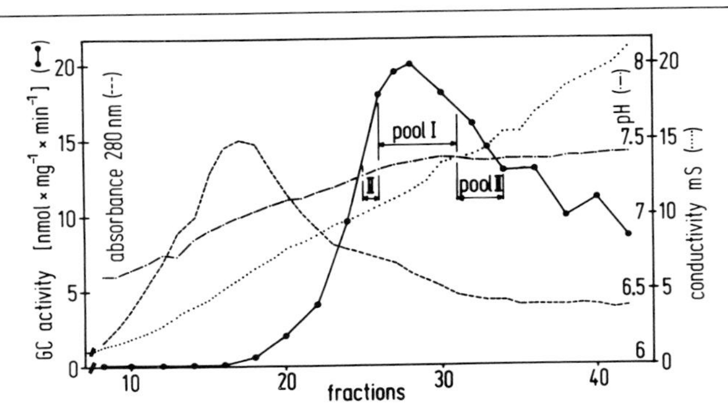

FIG. 1. DEAE-cellulose chromatography of bovine lung supernatant on Zetaprep-3200 fixed-bed cartridges. Five cartridges are perfused in series (6 liters/hr). Note the increase in pH during the development of the cartridges with the potassium chloride gradient. Pool I is used for the next purification step.

resistance. After loading the remainder of the extract onto cartridges 2 to 5, all cartridges are washed individually for at least 1 hr with modified homogenizing buffer containing 0.2 m*M* EDTA and 50 m*M* 2-mercaptoethanol with reversed flow (20 liters/hr). As soon as the wash is colorless, the cartridges are connected in series, and guanylyl cyclase is eluted (6 liters/hr) with a lienar gradient in potassium chloride (6 liters modified homogenizing buffer to 6 liters of the same buffer containing 0.5 *M* potassium chloride) and subsequently with 10 liters of 0.5 *M* potassium chloride. The eluate is fractionated (0.5 liter), and absorbance at 280 nm (protein content), pH, conductivity, and guanylyl cyclase activity are measured in individual fractions (Fig. 1). Fractions containing guanylyl cyclase activity are pooled as indicated in Fig. 1.

Step 3: Ammonium Sulfate Precipitation. Pool I from the DEAE eluate is saturated with solid ammonium sulfate to 50% (0.294 g/ml) by agitation with an overhead stirrer. The pH should not be allowed to decrease below 6.5 and may require adjustment with ammonium hydroxide. After stirring for 30 min the precipitate is collected by centrifugation (30 min at 14,000 *g*) and is redissolved in 0.5 liter dialysis buffer (10 m*M* triethanolamine hydrochloride, pH 6.5, 0.2 m*M* benzamidine, 0.2 m*M* EDTA, 50 m*M* 2-mercaptoethanol). The solution is dialyzed for 2 hr against 40 liters of 0.2 m*M* benzamidine, 0.2 m*M* EDTA, 50 m*M* 2-mercaptoethanol, and afterward against 40 liters of dialysis buffer until the conductivity of the dialyzate is below 1 mS (about 7 hr).

Step 4: Chromatography on Blue Sepharose. The dialyzate is checked for conductivity and pH (which must be 6.5 ± 0.1) and is stirred for 30 min with 2.5 liters Blue Sepharose (Sepharose CL-6B containing 1 μmol/

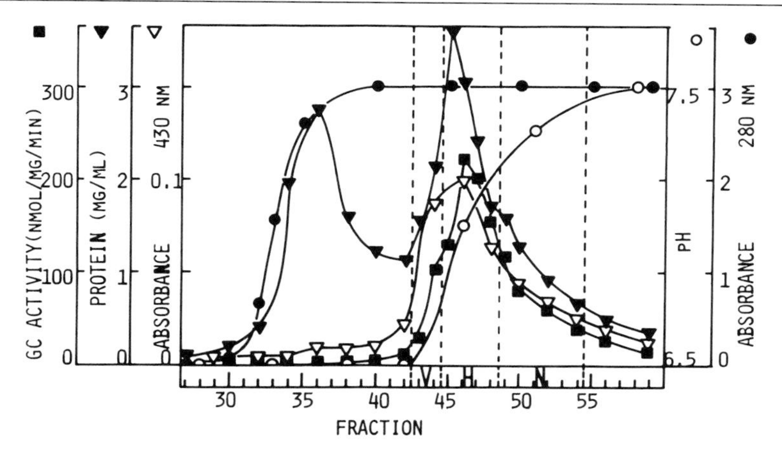

FIG. 2. Chromatography of ammonium sulfate dialyzate on Blue Sepharose. The column (10 × 30 cm) is developed (0.5 liter/hr) with 3 liters of buffer containing 0.5 m*M* GTP/$MnCl_2$, pH 7.5. Note the dissociation in the rise of the GTP concentration and the pH in the eluate, which results in elution of contaminating proteins before the guanylyl cyclase is desorbed. The peak of guanylyl cyclase activity coincides with the elution of a chromophore absorbing at 430 nm. Pool H is used for the next purification step.

ml covalently coupled Cibacron Blue F3G-A) preequilibrated with dialysis buffer. The gel is washed with 40 liters dialysis buffer on a Büchner funnel (keep the first 2 liters for a check of binding) and packed into a column (10 × 60 cm) by gravity. Guanylyl cyclase is eluted in one step (0.5 liter/hr) with 3 liters dialysis buffer with the pH adjusted to 7.5 and supplemented with 0.5 m*M* GTP and 0.5 m*M* manganese chloride. The eluate is continuously monitored for absorbance at 280 nm (for GTP) and 430 nm (for hemoprotein). It is fractionated (30-ml fractions), and guanylyl cyclase activity, protein (Coomassie blue method), and pH are measured in individual fractions (Fig. 2). Guanylyl cyclase activity elutes as soon as the pH in the eluate rises. Fractions containing guanylyl cyclase activity are pooled as indicated in Fig. 2 and concentrated by centrifugation in ultrafree concentration units (30-kDa cutoff polysulfone membranes, Millipore, Bedford, MA).

Step 5: Preparative Polyacrylamide Gel Electrophoresis. The concentrated pool (pool H) from the preceding step is diluted with electrophoresis elution buffer (154 m*M* glycine, 20 m*M* Tris, 50 m*M* 2-mercaptoethanol, 0.2 m*M* EDTA, 0.2 m*M* benzamidine hydrochloride, pH 8.6) containing 4% sucrose and is finally concentrated to 8 ml. Preparative electrophoresis is performed as described elsewhere in this volume.[17] The finally eluted

TABLE I
PURIFICATION OF SOLUBLE GUANYLYL CYCLASE FROM BOVINE LUNG[a]

Step	Protein (mg)	Specific activity (nmol/mg/min)		Total activity (nmol/min)		Recovery (%)	
		+SNP	−SNP	+SNP	−SNP	+SNP	−SNP
Supernatant	240,000	0.03	0.003	7200	720	—	—
DEAE-cellulose	15,000	16.00	0.400	240,000	6000	100	100
AS dialyzate	6530	44.00	1.000	287,300	6530	120	109
Blue Sepharose	330	310.00	6.000	102,300	1980	43	33
Electrophoresis	10	2140.00	38.000	21,400	380	9	6

[a] Data are from a representative purification. Guanylyl cyclase activity was determined as described in the text. +SNP, Enzyme activity with 100 μM sodium nitroprusside; −SNP, basal activity; AS, ammonium sulfate. The calculation of recovery starts with the DEAE-cellulose step, since guanylyl cyclase activity is inhibited in the supernatant by hemoproteins and other factors.

enzyme is stored in 50% glycerol at −70° under an atmosphere of nitrogen. No loss of activity has been found during storage up to 1 year.

Comments on Purification

The modified procedure shown in Table I yields about 10 mg soluble guanylyl cyclase from 5 bovine lungs within 3 days. The advantage of the modified method as compared to the original procedure is speed and capacity. This is achieved by the Zetaprep-3200 DEAE cartridges, which allow high flow rates, by Blue Sepharose containing less covalently coupled Cibacron Blue per volume of Sepharose, which permits a simplified washing and elution procedure, and the upscaling of the preparative electrophoresis device. The loading and washing of the Blue Sepharose critically depend on the pH, as a pH below 6.4 leads to diminished activation of guanylyl cyclase by nitric oxide due to loss of the heme,[11] whereas at a pH above 6.6 the binding capacity decreases.[12] Also, the preparative polyacrylamide gel should not be overloaded with protein, as this results in clogging of the protein onto the gel surface. Though the enzyme was purified to apparent homogeneity according to analytical polyacrylamide gel electrophoresis (Fig. 3), as well as isoelectric focusing (p*I* 5.9), the specific activity with sodium nitroprusside (about 2.1 units/mg) was lower than reported previously.[3] Since the heme binding to the enzyme is very labile,[11] and the stimulation of soluble guanylyl cyclase depends on the presence of heme,[3,4,6,7] the lower activation by sodium nitroprusside may be due to the lower heme content of the present preparation.[16] According

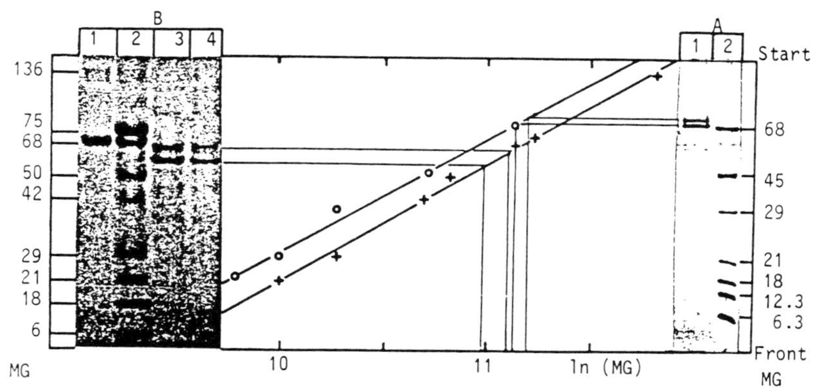

FIG. 3. Determination of the subunit composition of purified soluble guanylyl cyclase from bovine lung. Apparent relative electrophoretic mobilities are plotted versus ln[molecular weight ($\times 10^{-3}$)] of marker proteins. Note the higher relative mobility of soluble guanylyl cyclase subunits in the SDS–urea gel, which may be due to carbamylation of the protein by cyanate, which is in equilibrium with urea. (A) SDS–polyacrylamide gradient gel (12–18% acrylamide) prepared according to U. K. Laemmli, *Nature* (*London*) **227,** 680 (1970). Lane 1, 1 μg soluble guanylyl cyclase from the preparative electrophoresis step; lane 2, marker proteins. Silver staining was performed according to W. Wray, T. Boulikas, V. P. Wray, and R. Hancock, *Anal. Biochem.* **118,** 197 (1981). (B) SDS–urea–polyacrylamide gel (6% acrylamide) prepared according to W. W. Minuth and K. Tiedemann, *Histochemistry* **68,** 147 (1980). Lanes 1 and 2, marker proteins; lanes 3 and 4; 10 and 5 μg, respectively, of soluble guanylyl cyclase from the preparative electrophoresis step. The gel was stained with Coomassie blue. MG, Molecular weight.

to analytical polyacrylamide gel electrophoresis, sucrose density centrifugation, and size-exclusion chromatography (AcA 34, Pharmacia, Piscataway, NJ; TSK 3000), the native enzyme has an apparent molecular weight of 145,000. Under denaturing conditions it dissociates into two subunits of 75,000 and 70,000 (Fig. 3).

Addendum

In recent preparations we used 10 liters of DEAE-Sepharose Fast Flow (Pharmacia, Freiburg, FRG) instead of the Zetaprep cartridges, since the latter lose binding capacity after prolonged use. The lung extract was cleared by passage through a 10-μm filter cartridge (Cuno) and was loaded (10 liters/hr) on a column (15 $\times$ 30 cm) of DEAE-Sepharose Fast Flow, which was subsequently washed for 1 hr with 30 liters of 10 mM potassium phosphate, pH 7. Guanylyl cyclase was then eluted with a salt gradient as described above.

[35] Preparation of Soluble Guanylyl Cyclase from Bovine Lung by Immunoaffinity Chromatography

By PETER HUMBERT, FERAYDOON NIROOMAND, GABRIELA FISCHER, BERND MAYER, DORIS KOESLING, KLAUS-DIETER HINSCH, GÜNTER SCHULTZ, and EYCKE BÖHME

Introduction

Guanylyl cyclases [GTP pyrophosphate-lyase (cyclizing); EC 4.6.1.2, guanylate cyclase] catalyze the formation of cyclic GMP from GTP in the presence of a divalent cation such as Mg^{2+} or Mn^{2+}. At least two cGMP-forming isoenzymes exist, a soluble, cytosolic and a particulate, plasma membrane-bound enzyme form differing in their regulation, in biochemical and physicochemical properties,[1,2] as well as in amino acid sequence.[3–5] The soluble enzyme is activated by NO-containing compounds, for example, sodium nitroprusside, and nitric oxide itself. Nitric oxide or a labile NO-containing compound seems to be an ubiquitous intra- and intercellular signal molecule, with soluble guanylyl cyclase acting as an effector system.

Soluble guanylyl cyclase is purified from different tissues by various methods to apparent homogeneity.[2] However, the biochemical and physicochemical properties of the enzyme are still controversial. The holoenzyme has an apparent molecular mass of about 150 kDa and consists of two subunits, one with an M_r of 70,000 and a larger one with a reported M_r value between 72,000 and 82,000. Until now, only one immunoaffinity chromatographic method for rapid purification of soluble guanylyl cyclase has been developed, which uses monoclonal antibodies against a larger subunit of 82K of the enzyme.[6] The enzyme isolated by this procedure,

[1] C. K. Mittal and F. Murad, *in* "Handbook of Pharmacology" (J. A. Nathanson and J. W. Kebabian, eds.), Vol. 58/I, p. 225. Springer-Verlag, Berlin, 1982.

[2] S. A. Waldman and F. Murad, *Pharmacol. Rev.* **39,** 163 (1987).

[3] D. Koesling, J. Herz, H. Gausepohl, F. Niroomand, K.-D. Hinsch, A. Mülsch, E. Böhme, G. Schultz, and R. Frank, *FEBS Lett.* **239,** 29 (1988).

[4] M. Nakane, S. Saheki, T. Kuno, K. Ishii, and F. Murad, *Biochem. Biophys. Res. Commun.* **157,** 1139 (1988).

[5] M. Chinkers, D. L. Garbers, M.-S. Chang, D. G. Lowe, H. Chin, D. V. Goeddel, and S. Schulz, *Nature (London)* **338,** 78 (1989).

[6] Y. Kamisaki, S. Saheki, M. Nakane, J. A. Palmieri, T. Kuno, B. Y. Chang, S. A. Waldman, and F. Murad, *J. Biol. Chem.* **261,** 7236 (1986).

however, was stimulated by sodium nitroprusside to a considerably lower degree than a conventionally purified enzyme.[7]

The preparation of soluble guanylyl cyclase from bovine lung described here is based on immunoaffinity chromatography with peptide antibodies against a synthetic C-terminal peptide of the 70K subunit. The elution of the antibody-bound holoenzyme is performed with an excess of the antigenic peptide. This specific elution appears to be superior to the described immunoaffinity chromatography method which uses high concentrations of urea for protein elution. The method described here implicates an entirely new strategy for large-scale purifications of homogeneous proteins with a (partially) known amino acid sequence by immunoaffinity chromatography. With this procedure about 2 mg of apparently homogeneous soluble guanylyl cyclase, stimulated up to 170-fold by sodium nitroprusside in the presence of Mg^{2+}, was purified from about 40 g of cytosolic proteins from bovine lung.[8]

Guanylyl Cyclase Assay

Guanylyl cyclase activity is estimated as formation of [^{32}P]cGMP from [α-^{32}P]GTP as previously described,[9] in the presence of 0.5 m*M* [α-^{32}P]GTP (specific activity 3–5 TBq/mmol, from Du Pont de Nemours, Dreieich, FRG), 3 m*M* $MgCl_2$ or $MnCl_2$, 1 m*M* cGMP, 0.5 mg/ml bovine serum albumin, 3 m*M* dithiothreitol, and 50 m*M* triethanolamine-HCl, pH 7.4, with and without sodium nitroprusside. Reactions are carried out in a total volume of 0.1 ml at 37° for 10 min after addition of 10 μl of enzyme, which is diluted as required with a 50 m*M* triethanolamine-HCl buffer, pH 7.4, containing 3 m*M* dithiothreitol and 0.5 mg/ml bovine serum albumin.

Preparation of Antibody-Sepharose

A peptide (SRKNTGTEETEQDEN) corresponding to the C-terminal deduced amino acid sequence of the 70-kDa subunit of soluble guanylyl cyclase is synthesized as described by Frank and Gausepohl.[10] Rabbits are immunized with the peptide coupled through an N-terminally added

[7] R. Gerzer, F. Hofmann, and G. Schultz, *Eur. J. Biochem.* **116,** 479 (1981).

[8] P. Humbert, F. Niroomand, G. Fischer, B. Mayer, D. Koesling, K.-D. Hinsch, H. Gausepohl, R. Frank, G. Schultz, and E. Böhme, *Eur. J. Biochem.* **190,** 273 (1990).

[9] G. Schultz and E. Böhme, *in* "Methods of Enzymatic Analysis" (H. U. Bergmeyer, ed.), Vol. 4, p. 379. Verlag Chemie, Weinheim, FRG, 1984.

[10] R. Frank and H. Gausepohl, *in* "Modern Methods in Protein Chemistry" (H. Tscheche, ed.), Vol. 3, p. 41. de Gruyter, Berlin, 1988.

cysteine to keyhole limpet hemocyanin as previously described.[11] Antisera are screened for their ability to recognize native soluble guanylyl cyclase by immunoprecipitation of guanylyl cyclase activity from crude enzyme preparations with protein A-Sepharose. For preparation of the antibody-Sepharose, the antiserum is supplemented with sodium azide, phenylmethylsulfonyl fluoride (PMSF), and EDTA to give final concentrations of 0.02% (w/v), 0.5 m*M*, and 2 m*M*, respectively. In order to obtain a crude fraction of IgG, precipitation of the antiserum with ammonium sulfate (40% saturation) is performed. The pellet is washed once and dialyzed against an 0.1 *M* sodium carbonate buffer, pH 8.3, containing 0.5 *M* NaCl. Proteins of this fraction are coupled to CNBr-activated Sepharose 4B according to the recommendations of Pharmacia. About 80 ml of antiserum, yielding approximately 1 g of precipitated proteins, is used for the preparation of 100 ml of antibody-Sepharose. For storage, the antibody-Sepharose is equilibrated with a 50 m*M* triethanolamine-HCl buffer, pH 7.4, containing 0.1 *M* NaCl, 0.2 m*M* EDTA, and 0.02% (w/v) sodium azide, and pepstatin A and PMSF are added to final concentrations of 0.5 μM and 0.5 m*M*, respectively.

Preparation of Soluble Guanylyl Cyclase

A two-step procedure for the purification of soluble guanylyl cyclase from bovine lung supernatant (160,000 g_{av} for 75 min) is described. The supernatant is subjected to anion-exchange chromatography on DEAE-Sepharose Fast Flow in order to reduce the amount of contaminating proteins as well as the sample volume. The DEAE fraction containing guanylyl cyclase is used for subsequent immunoaffinity chromatography. The described method has been successfully repeated at various scales. The results of one representative large-scale purification are summarized in Table I.

Preparation of 160,000 g_{av} Supernatant from Bovine Lung. Minced bovine lungs (5 kg) are mixed with 1 liter of a 300 m*M* triethanolamine-HCl buffer, pH 7.0, containing 60 m*M* EDTA, 12 m*M* reduced glutathione, 6 m*M* benzamidine, 360 kU/liter penicillin G, and 360 mg/liter streptomycin, and the tissue is homogenized by means of a Microcut MC 10 with 0.9 mm cutting rings (Stephan, Hameln, FRG). All subsequent procedures are carried out at 4°. The homogenate is centrifuged for 60 min at 10,000 g_{av}. The supernatant is supplemented with PMSF and pepstatin A (final

[11] K.-D. Hinsch, I. Tychowiecka, H. Gausepohl, R. Frank, W. Rosenthal, and G. Schultz, *Biochim. Biophys. Acta* **1013**, 60 (1989).

TABLE I

PURIFICATION OF SOLUBLE GUANYLYL CYCLASE FROM BOVINE LUNG

Fraction	Protein (mg)	Specific activity[a] (nmol/mg/min)						Total activity (μmol/min)			
		Mg^{2+}			Mn^{2+}			Mg^{2+}		Mn^{2+}	
		−SNP	+SNP	Factor[b]	−SNP	+SNP	Factor[b]	−SNP	+SNP	−SNP	+SNP
Supernatant	40,100	0.0087	1.42	163	0.057	1.26	22	0.35	56.9	2.30	50.5
DEAE fraction	1,540	0.11	40.0	352	1.59	34.9	22	0.18	61.1	2.45	53.9
Immunoaffinity	2.05	12.1	2,100	174	96.0	1,630	17	0.025	4.3	0.20	3.3

[a] Guanylyl cyclase activities were determined in the presence of 0.5 m*M* GTP, 3 m*M* Mg^{2+} or Mn^{2+}, 3 m*M* dithiothreitol, and 0.5 mg/ml bovine serum albumin in 50 m*M* triethanolamine-HCl buffer, pH 7.4, at 37° for 10 min with and without 0.1 m*M* sodium nitroprusside (SNP).

[b] Stimulation of guanylyl cyclase by sodium nitroprusside.

concentrations 0.5 mM and 1 μM, respectively), filtered through glass wool, and centrifuged for 75 min at 160,000 g_{av}. About 40 g of cytosolic proteins is obtained from 5 kg of bovine lung. Guanylyl cyclase activity is about 10 and 60 pmol/min/mg in the presence of Mg^{2+} and Mn^{2+}, respectively.

Anion-Exchange Chromatography on DEAE-Sepharose Fast Flow. The 160,000 g_{av} supernatant fraction is adjusted to a conductivity of 7 mS/cm by addition of demineralized water or 2 M NaCl and applied to a column (5 × 11 cm) of DEAE-Sepharose Fast Flow (Pharmacia LKB, Freiburg, FRG) which had been equilibrated with a 50 mM triethanolamine-HCl buffer, pH 7.0, containing 75 mM NaCl, 1 mM EDTA, 2 mM reduced glutathione, and 0.2 mM benzamidine. The column was washed with 4 bed volumes of this buffer; then the NaCl concentration is increased to 175 mM, and 25-ml fractions are collected. Fractions containing guanylyl cyclase activity (175 ml) are pooled, and reduced glutathione, PMSF, and pepstatin A are added at final concentrations of 2.0 mM, 0.5 mM, and 1 μM, respectively. This DEAE fraction contains 1.5 g of protein, and specific guanylyl cyclase activities are 0.11 and 1.6 nmol/min/mg in the presence of Mg^{2+} and Mn^{2+}, respectively. This represents about 15- and 30-fold purifications over the 160,000 g_{av} supernatant, with a yield of about 50 and 100% in the presence of Mg^{2+} and Mn^{2+}, respectively.

Immunoaffinity Chromatography. An aliquot of 160 ml of the DEAE fraction is applied onto the antibody-Sepharose column (4.4 × 11 cm which had been preequilibrated with a 50 mM triethanolamine-HC buffer, pH 7.5, containing 175 mM NaCl, 2 mM reduced glutathione and 1 mM EDTA. The DEAE fraction is circulated through the column for 6 to 9 hr with a flow rate of 240 ml/hr. Subsequently, the column is washed with 5 to 10 bed volumes of a 50 mM triethanolamine-HC buffer, pH 7.5, containing 75 mM NaCl, 2 mM reduced glutathione and 1 mM EDTA. For elution of bound soluble guanylyl cyclase, 2/3 bed volume (110 ml) of this buffer, supplemented with 0.1 mg/ml of the antigenic peptide, is circulated through the column for 6 hr at a flow rate of 240 ml/hr.

As soluble guanylyl cyclase is eluted as a rather dilute protein solution, the volume of the eluate has to be decreased considerably All conventional methods for concentration of protein solutions result in a nearly complete loss of enzyme activity. The eluted enzyme therefore, is directly bound to a small column of DEAE-Sepharose from which it could be eluted by increasing the salt concentration. For this purpose, the circulated sample (which contains guanylyl cyclase eluted from the antibody-Sepharose column) and another bed volume

(170 ml) of the elution buffer are directly passed over a small column (1 × 1.2 cm) of DEAE-Sepharose Fast Flow at a flow rate of 120 ml/hr, and soluble guanylyl cyclase bound to the DEAE material is subsequently eluted with a 50 mM triethanolamine-HCl buffer, pH 7.5, containing 175 mM NaCl, 2 mM reduced glutathione, and 1 mM EDTA. Fractions (0.5 ml) containing the major part of guanylyl cyclase activity are pooled, mixed with 1 volume of glycerol, and stored at −80° for several months without loss of enzyme activity. In several large-scale purifications about 2 mg of soluble guanylyl cyclase from 40 g of cytosolic proteins is recovered. Specific guanylyl cyclase activities are between 10 and 20 nmol/min/mg (Mg^{2+}) and between 80 and 100 nmol/min/mg (Mn^{2+}). The enzyme is purified between 1300- and 1500-fold, with a recovery of 7 and 10%, as calculated from enzyme activities determined in the presence of Mg^{2+} and Mn^{2+}, respectively.

Characterization of Soluble Guanylyl Cyclase by Electrophoresis

For monitoring the purity of enzyme preparations and for the determination of M_r values of the subunits of the soluble guanylyl cyclase, sodium dodecyl sulfate-polyacrylamide gel electrophoresis (SDS-PAGE) on 10% gels is performed. The purified enzyme shows only two major protein bands, with molecular masses of 70 and 73 kDa. Additionally, one minor band corresponding to a molecular mass of 43 kDa is observed. Analysis of parts of the amino acid sequence of the latter protein revealed that it is identical with actin. This was further confirmed by immunoblotting with antiactin antibodies which recognized the 43-kDa band. Actin could be separated from the enzyme by gel-permeation fast protein liquid chromatography (FPLC) and by electrophoresis under nondenaturating conditions.

Purified preparations of soluble guanylyl cyclase are analyzed by electrophoresis under nondenaturating conditions by the use of slab gels with a 4.5% (w/v) acrylamide stacking gel [100 mM Tris-HCl, pH 6.8, 3 mM dithiothreitol, 0.5% (v/v) Triton X-100] and an 8% (w/v) acrylamide separating gel [200 mM Tris-HCl, pH 7.8, 3 mM dithiothreitol, 0.5% (v/v) Triton X-100]. One lane of the gel is stained with Coomassie Blue R-250, and a nonstained lane of the gel is sliced and, after extraction, assayed for guanylyl cyclase activity. The enzyme activity comigrates with a single protein band. A small contamination by actin, which is not observed after protein staining, is separated from soluble guanylyl cyclase as revealed by immunoblotting with antiactin antibodies.

Properties of Soluble Guanylyl Cyclase

Molecular Mass. For determination of the native molecular mass, purified soluble guanylyl cyclase is analyzed by gel-permeation chromatography. The enzyme is injected onto a column (0.75 × 60 cm) of TSK G3000SW (Pharmacia LKB) which had been preequilibrated with a 50 mM sodium phosphate buffer, pH 6.5, containing 3 mM dithiothreitol. The column is eluted with the same buffer at 0.33 ml/min and 0.8 MPa. The enzyme activity elutes coincidently with one single protein peak corresponding to a molecular mass of about 150 kDa.

Spectroscopy. Monitoring of the UV/VIS absorption spectra in the course of FPLC gel-permeation chromatography by means of a photodiode-array detector (Waters 990, Millipore GmbH, Eschborn, FRG) shows that the eluted protein exhibits a spectrum typical for hemoproteins. The spectrum shows a Soret band at 430 nm with an absorption coefficient between 26 and 35 mM^{-1} cm^{-1}. The absorption ratio A_{280}/A_{430} is between 1.3 and 1.5.

Stimulation of Soluble Guanylyl Cyclase by Sodium Nitroprusside. As shown in Table I, guanylyl cyclase activities in the 160,000 g_{av} supernatant, DEAE fraction, and purified enzyme preparations are stimulated by sodium nitroprusside. Maximal specific enzyme activities are between 1.2 and 2.2 (Mg^{2+}) and between 1.0 and 1.8 μmol/min/mg (Mn^{2+}). Thus, the enzyme is maximally activated about 170- and 20-fold by sodium nitroprusside in the presence of Mg^{2+} and Mn^{2+}, respectively. As calculated from concentration–response curves, the EC_{50} values for sodium nitroprusside are 2.0 and 0.4 μM in the presence of Mg^{2+} and Mn^{2+}, respectively, and maximal activation of soluble guanylyl cyclase is reached between 10 and 100 μM sodium nitroprusside.

Comments

Soluble guanylyl cyclase was purified about 1400-fold from bovine lung, with a yield of about 7%. It should be mentioned that these data are based solely on the determination of enzyme activities which may be severely influenced by many factors present in various amounts at different steps of the purification. Furthermore, the specific enzyme activities often varied between different preparations of apparently homogeneous enzyme. Therefore, the degree of the calculated purification factor and of the enzyme recovery depends on the conditions chosen for the determination of enzyme activity, namely, basal or stimulated activity in the presence of Mg^{2+} or Mn^{2+}. It will remain unclear to what extent these calculated values reflect real values until an immunological method for the quantitative determination of the guanylyl cyclase is available.

Acknowledgments

The authors are grateful to Jürgen Malkewitz for valuable technical assistance. Synthesis of the peptides was kindly performed by Drs. Heinrich Gausepohl and Rainer Frank, European Molecular Biology Laboratory, Heidelberg, FRG. Bernd Mayer is a recipient of a research fellowship of the Alexander von Humboldt-Stiftung. Financial support of the Deutsche Forschungsgemeinschaft and of the Fonds der Chemischen Industrie is acknowledged.

[36] Immunoaffinity Purification of Soluble Guanylyl Cyclase

By SCOTT A. WALDMAN, DALE C. LEITMAN, and FERID MURAD

Introduction

Most tissues examined possess at least two forms of guanylyl cyclase [GTP pyrophosphate-lyase (cyclizing); EC 4.6.1.2, guanylate cyclase] located in cellular membranes (particulate) or the cytoplasm (soluble).[1-5] The guanylyl cyclase isoenzymes differ in their physicochemical, kinetic, and antigenic properties.[5-12] In addition, different factors regulate the activity of the guanylyl cyclase isoenzymes. The particulate isoenzyme is activated by atrial natriuretic peptides and the heat-stable enterotoxin

[1] H. Kimura and F. Murad, *J. Biol. Chem.* **249,** 6910 (1974).
[2] H. Kimura and F. Murad, *J. Biol. Chem.* **250,** 4810 (1975).
[3] H. Kimura and F. Murad, *Life Sci.* **17,** 837 (1975).
[4] H. Kimura and F. Murad, *Proc. Natl. Acad. Sci. U.S.A.* **72,** 1965 (1975).
[5] S. A. Waldman and F. Murad, *Pharm. Rev.* **39,** 163 (1987).
[6] J. M. Braughler, C. K. Mittal, and F. Murad, *Proc. Natl. Acad. Sci. U.S.A.* **76,** 219 (1979).
[7] D. L. Garbers, *J. Biol. Chem.* **254,** 240 (1979).
[8] J. A. Lewicki, H. J. Brandwein, S.A. Waldman, and F. Murad, *J. Cyclic Nucleotide Res.* **6,** 283 (1980).
[9] Y. Kamisaki, S. Saheki, M. Nakane, J. A. Palmieri, T. Kuno, B. Y. Chang, S. A. Waldman, and F. Murad, *J. Biol. Chem.* **261,** 7236 (1986).
[10] S. A. Waldman, J. A. Lewicki, L. Y. Chang, and F. Murad, *Mol. Cell. Biochem.* **57,** 155 (1983).
[11] S. A. Waldman, L. Y. Chang, and F. Murad, *Prep. Biochem.* **15,** 103 (1985).
[12] T. Kuno, J. W. Andresen, Y. Kamisaki, S. A.Waldman, L. Y. Chang, S. Saheki, D. C. Leitman, M. Nakane, and F. Murad, *J. Biol. Chem.* **261,** 5817 (1986).

produced by *Escherichia coli*.[5,12–15] The activation of guanylyl cyclase leads to increased intracellular cyclic guanosine 3′,5′-monophosphate (cyclic GMP, cGMP), which mediates vascular smooth muscle relaxation and secretory diarrhea produced by these peptides, respectively.[5,13–15] In contrast, nitrovasodilators and endothelium-dependent vasodilators selectively activate soluble guanylyl cyclase, resulting in the generation of cGMP and vascular smooth muscle relaxation.[16,17]

Elucidating the structural basis underlying differences in the physical and biochemical properties of these isoenzyme forms requires large quantities of purified enzymes. Although quantitative purification of particulate guanylyl cyclase from mammalian systems continues to be difficult and time-consuming, the soluble isozyme has been obtained in milligram amounts in relatively homogeneous preparations by single-step immunoaffinity purification with monoclonal antibodies[9] (see also [35], this volume).

Preparation of Immunogen

Soluble guanylyl cyclase is purified by classic chromatographic and electrophoretic techniques as described previously.[6,8] The 105,000 *g* supernatant fraction from homogenates of rat lung is subjected to isoelectric (pH 5.0) and ammonium sulfate (20–50%) precipitations, chromatography on GTP-Sepharose, and preparative electrophoresis on a 7% polyacrylamide gel. By the use of this protocol preparations with specific activities of 250–400 nmol of cGMP produced/min/mg of protein with MnGTP and 30–60 nmol/min/mg of protein with MgGTP are obtained, representing about 2000- to 3000-fold purification of the enzyme.[6,8] These preparations yield a single protein band that comigrates with enzyme activity on analytical polyacrylamide gels.[6,8] With this protocol about 100 to 200 μg of purified enzyme can be prepared from 250 g of rat lung.

Production of Monoclonal Antibodies

Monoclonal antibodies to soluble guanylyl cyclase are produced as described previously.[18] BALB/c mice are injected several times monthly with 5 to 10 μg of soluble guanylyl cyclase purified as described above.

[13] S. A. Waldman, R. M. Rapoport, and F. Murad, *J. Biol. Chem.* **259,** 14332 (1984).

[14] S. A. Waldman, T. Kuno, Y. Kamisaki, L. Y. Chang, J. Gariepy, G. Schoolnik, and F. Murad, *Infect. Immun.* **51,** 320 (1986).

[15] T. Kuno, Y. Kamisaki, S. A. Waldman, J. Gariepy, G. Schoolnik, and F. Murad, *J. Biol. Chem.* **261,** 1470 (1986).

[16] R. M. Rapoport, M. B. Draznin, and F. Murad, *Trans. Assoc. Am. Physicians* **96,** 19 (1983).

[17] S. A. Waldman and F. Murad, *J. Cardiovasc. Physiol.* (*Suppl.*), S11 (1989).

[18] H. J. Brandwein, J. A. Lewicki, and F. Murad, *Proc. Natl. Acad. Sci. U.S.A.* **78,** 4241 (1981).

Four days subsequent to the final immunization mice are sacrificed, spleens harvested, and splenocytes fused with SP 2/0 myeloma cells with polyethylene glycol as the fusing agent. Hybrids are selected by the use of hypoxanthine/aminopterin/thymidine medium, and hybrids secreting antiguanylyl cyclase are identified by a solid-phase plate-coat assay. Positive clones are subcloned several times to ensure that they are monoclonal. The monoclonal cell lines are grown in mass culture, harvested, and injected intraperitoneally into pristane-pretreated BALB/c mice to produce large quantities of antibody-rich ascites. Ascitic fluid is pooled, and monoclonal antibody is purified on an Affi-Gel protein A-Sepharose column.

Constructing Immunoaffinity Column

An immunoaffinity column using monoclonal antibodies to soluble guanylyl cyclase is constructed as described previously.[9] Cyanogen bromide-activated Sepharose 4B (33g) is treated with 1 m*M* HCl, washed extensively with the same solution, and then equilibrated with 0.1 *M* $NaHCO_3$ and 0.5 *M* NaCl (pH 8.0). Monoclonal antibody purified as described above (100 mg) is dialyzed against a similar buffer and incubated with the washed gel at 4° for 16 hr. The resin is then incubated with 1 *M* Tris-HCl (pH 7.6) at 25° for 2 hr in order to block the remaining CNBr-activated groups. The resulting immunoaffinity column is washed with 50 m*M* Tris-HCl (pH 7.6) containing 1 m*M* EDTA, 1 m*M* dithiothreitol (TED buffer), and 3 *M* urea. The column is equilibrated with TED buffer before each use.

Immunoaffinity Purification of Soluble Guanylyl Cyclase

Soluble guanylyl cyclase from rat lungs is purified by the use of the monoclonal antibody column constructed above as described previously.[9] Frozen rat lungs (300 to 500 g) are thawed and homogenized in 4 volumes of ice-cold TED buffer containing 2 m*M* phenylmethylsulfonyl fluoride (PMSF). All subsequent procedures are conducted at 4°. The homogenate is centrifuged at 105,000 *g* for 60 min. The resulting supernatant is applied to the immunoaffinity column and washed with 1.4 liters of TED buffer containing 0.5 *M* NaCl and 0.1 m*M* PMSF. The column is then washed with 100 ml of TED buffer containing 0.5 *M* NaCl, 1 *M* urea, and 0.1 m*M* PMSF. Guanylyl cyclase is eluted with 400 ml of TED buffer containing 3 *M* urea and 0.1 m*M* PMSF. Fractions containing guanylyl cyclase activity are pooled and immediately concentrated to approximately 10 ml with an Amicon (Danvers, MA) ultrafiltration apparatus fitted with a PM30

TABLE I
IMMUNOAFFINITY PURIFICATION OF SOLUBLE GUANYLYL CYCLASE FROM RAT LUNG[a]

	Mn^{2+}			Mg^{2+}		
Fraction[b]	SA[b]	Yield[c]	Purification[d]	SA	Yield	Purification
Supernatant	0.18	100	1	0.03	100	1
Immunopurified	432	53	2400	49.1	40	1820

[a] After Ref. 9.
[b] Rat lungs were processed and fractions prepared as outlined in the text.
[c] SA, Specific activity (nmol of cGMP produced/min/mg of protein).
[d] Yield, Total activity recovered as purified enzyme/total activity in the initial supernatant.
[e] Purification, Specific activity of purified enzyme/specific activity of supernatant.

filter. Samples are then dialyzed against TED buffer containing 20% glycerol.

Guanylyl Cyclase Assay

Guanylyl cyclase activity is determined as described previously.[9] Samples are incubated for 10 min at 37° in a 100-μl reaction mixture that consists of 50 mM Tris-HCl (pH 7.6), 4 mM $MnCl_2$ or $MgCl_2$, 1 mM GTP, 1 mM dithiothreitol, 0.5 mM isobutylmethylxanthine (IBMX), 0.1% (w/v) bovine serum albumin, and a GTP-regenerating system containing 15 mM creatine phosphate and 20 μg (135 units/mg) of creatine phosphokinase. IBMX and the GTP-regenerating system are not utilized in incubations of purified preparations. cGMP formed is determined by radioimmunoassay[1–4] (see [30], this volume).

Purification

The immunoaffinity purification of soluble guanylyl cyclase has been summarized previously (see Table I).[9] Guanylyl cyclase purified by the use of monoclonal antibodies had a final specific activity of 432 nmol cGMP produced/min/mg of protein with Mn^{2+} as the cation cofactor, representing a 2400-fold purification compared to crude supernatants with a recovery of 53%. Similarly, the specific activity was 49.1 nmol/min/mg with Mg^{2+} as the cation cofactor, representing an 1800-fold purification with a 40% recovery. These values compare favorably with those obtained with soluble guanylyl cyclase purified by conventional chromatographic and electrophoretic techniques.[5–9]

TABLE II
CHARACTERISTICS OF SOLUBLE GUANYLYL CYCLASE IMMUNOPURIFIED FROM RAT LUNG[a]

Characteristic	Value
Subunits	Heterodimer of 82K and 70K proteins
pI	5.7–6.1
M_r	150,000
K_m	
Mn^{2+}	10.5 μM
Mg^{2+}	130 μM

[a] After Ref. 9.

Characteristics of Purified Soluble Guanylyl Cyclase

The physical and biochemical characteristics of soluble guanylyl cyclase purified by immunoaffinity chromatography are summarized in Table II.[9] Purified preparations of soluble guanylyl cyclase were denatured with sodium dodecyl sulfate (SDS) and dithiothreitol and analyzed by SDS–polyacrylamide gel electrophoresis. With this technique two protein bands were visualized with subunit molecular weights of 82,000 and 70,000. These results are considerably different from previous reports which showed two subunits with molecular weights of 70K and 68K.[5-8] Indeed, if the protease inhibitors PMSF and EDTA are not utilized during purification, the 80K subunit is lost and the 68K subunit appears.[9] Thus, the 68K subunit observed in previous reports is presumably derived from the 80K subunit by proteolysis.[9] The 70K and 82K subunits appear to be different proteins without significant structural homology since their proteolytic maps do not demonstrate shared peptide fragments.[9] In addition, a variety of polyclonal and monoclonal antibodies have been generated against soluble guanylyl cyclase which recognize one but not both of these subunits, suggesting the presence of different antigenic epitopes.[9]

Analysis of immunoaffinity-purified soluble guanylyl cyclase by nondenaturing polyacrylamide gel electrophoresis revealed a single protein-staining band of 150K which precisely comigrated with enzyme activity. Also, gel-permeation high-performance chromatography of purified cyclase yielded a single protein peak which comigrated with enzyme activity. In addition, analysis of the gel-permeation fractions on denaturing polyacrylamide gels demonstrated that both the 82K and 70K subunits comigrated with the peak of protein and enzyme activities. Similarly, purified preparations migrated as a single symmetrical peak of enzyme activity with a pI of 5.7–6.1, with concordance between the migration of enzyme activity and the two subunits upon analysis by isoelectric focusing in Sephadex gels.

Soluble guanylyl cyclase purified by monoclonal immunoaffinity chromatography demonstrated classic Michaelis–Menten kinetics when incubated with increasing concentrations of substrate.[9] The apparent Michaelis constants were 10.5 and 130 μM of metal–GTP when Mn^{2+} or Mg^{2+}, respectively, was utilized as the cation cofactor. Double-reciprocal plots demonstrated a single straight line, suggesting a single substrate binding site which does not intereact in a cooperative fashion. In addition, sodium nitroprusside was a potent activator of these purified preparations, increasing enzyme activity by 10- to 20-fold. These results are in close agreement with previously published kinetics of soluble guanylyl cyclase.[5–9]

In conclusion, soluble guanylyl cyclase from rat lung has been purified to apparent homogeneity with a single-step affinity chromatography procedure employing monoclonal antibodies generated against the soluble enzyme. By the use of this technique milligram quantities of the enzyme can be prepared which retain similar physical, chemical, and functional characteristics compared with enzyme purified by standard chromatographic and electrophoretic techniques. Furthermore, the purified enzyme is activated by agents, such as sodium nitroprusside. The availability of large amounts of soluble guanylyl cyclase make it possible not only to identify the catalytic and regulatory properties but also to study the mechanisms by which nitrate-containing compounds activate the enzyme. The availability of large amounts of soluble guanylyl cyclase purified by immunoaffinity chromatography has permitted us to obtain a partial amino acid sequence, which was used to clone the cDNA for the 70K subunit[19] and the 82K subunit (S. A. Waldman, unpublished).

[19] M. Nakane, T. Kuno, S. Saheki, K. Ishi, and F. Murad, *Biochem. Biophys. Res. Commun.* **157,** 1139 (1988).

[37] Copurification of Atrial Natriuretic Peptide Receptor and Particulate Guanylyl Cyclase

By SCOTT A. WALDMAN, DALE C. LEITMAN, and FERID MURAD

Introduction

Atrial natriuretic peptides (ANP) are a family of low molecular weight, heat-stable peptides[1,2] produced in cardiac atria[1-3] and elsewhere[4] which directly regulate volume and blood pressure.[5,6] Secretion of ANP from atria is stimulated by a variety of conditions including fluid volume overload.[1-6] The effects of ANP encompass an integrated, pleiomorphic physiological response involving a variety of organs, including vascular smooth muscle relaxation,[7] natriuresis and diuresis,[8] and inhibition of aldosterone synthesis and secretion.[9] These responses all contribute to a decreasing circulating volume and blood pressure in the face of fluid overload.[3,5,6]

ANP exerts its physiological effects by first binding to specific cell surface receptors. Two populations of receptors have been described and purified: a 130-kDa monomer (designated ANP-R1) and a 130-kDa dimer that is comprised of two identical 66-kDa subunits joined by disulfide bonds (designated ANP-R2).[10-13] Cross-linking studies suggest that the ANP-R2 receptor may also exist in a 66-kDa monomeric form in the membrane.[12] Binding of ANP to receptors activates particulate guanylyl

[1] M. G. Currie, D. M. Geller, B. R. Cole, J. G. Baylon, W. YuSheng, S. W. Holmberg, and P. Needleman, *Science* **221,** 71 (1983).

[2] T. G. Flynn, M. L. deBold, and A. J.deBold, *Biochem. Biophys. Res. Commun.* **117,** 859 (1983).

[3] A. deBold, *Science* **230,** 767 (1985).

[4] D. L. Song, K. P. Kohse, and F. Murad, *FEBS Lett.* **232,** 125 (1988).

[5] B. J. Ballerman and B. M. Brenner, *J. Clin. Invest.* **76,** 2041 (1985).

[6] J. Genest and M. Cantin, *Circulation* **75** (Suppl. 1), 118 (1985).

[7] R. J. Winquist, E. P. Faison, S. A. Waldman, K. Schwartz, F. Murad, and R. M. Rapoport, *Proc. Natl. Acad. Sci. U.S.A.* **81,** 7661 (1984).

[8] A. deBold, H. B. Bornstein, A. T. Veress, and H. Sonnenberg, *Life Sci.* **28,** 89 (1981).

[9] T. Kudo and A. Baird, *Nature (London)* **312,** 756 (1984).

[10] C. C. Yip, L. P. Laing, and T. G. Flynn, *J. Biol. Chem.* **260,** 8229 (1985).

[11] T. Kuno, J. W. Andresen, Y. Kamisaki, S. A. Waldman, L. Y. Chang, S. Saheki, D. C. Leitman, M. Nakane, and F. Murad, *J. Biol. Chem.* **261,** 5817 (1986).

[12] D. C. Leitman, J. W. Andresen, T. Kuno, Y. Kamisaki, J. K. Chang, and F. Murad, *J. Biol. Chem.* **261,** 11650 (1986).

[13] D. Shenk, M. N. Phelps, J. G. Porter, R. M. Scarborough, G. A. McEnroe, and J. A. Lewicki, *Proc. Natl. Acad. Sci. U.S.A.* **84,** 1521 (1987).

cyclase and elevates intracellular concentrations of cyclic GMP.[7,11,14,15] This second messenger presumably mediates the cellular response to these peptides in some tissues, most notably vascular smooth muscle relaxation[16,17] and natriuresis in the kidney.[18] Studies comparing the kinetics of ligand binding, affinity labeling of the various receptors, and activation of guanylyl cyclase have demonstrated that only the 130-kDa monomer ANP-R1 is functionally coupled to enzyme activation.[12] The second messenger mediating the effects of ANP binding to the ANP-R2 receptor subtype and its resultant physiological action remain uncertain.[19] However, this receptor is coupled to phospholipase C activation and inositol phosphate and dioctylglycerol formation in some systems.[20]

In this chapter the copurification from rat lung membranes of particulate guanylyl cyclase and ANP-binding activities in a single transmembrane glycoprotein is described.[11]

Methods

Membrane Preparation and Solubilization

Rat lung membranes are prepared as described previously.[11] Frozen rat lungs (80–100 g) are thawed and homogenized with a Polytron (Brinkmann Instruments, Westbury, NY) in 5–10 volumes of ice-cold 50 m*M* Tris-HCl buffer, pH 7.6, containing 1 m*M* EDTA, 1 m*M* dithiothreitol (DTT), and 0.1 m*M* phenylmethylsulfonyl fluoride (PMSF). All procedures are conducted at 4°. Homogenates are centrifuged at 100,000 *g* for 60 min, and pellets are washed consecutively in (1) 50 m*M* Tris-HCl, pH 7.6, containing 1 m*M* EDTA, 1 m*M* DTT, 0.1 m*M* PMSF; (2) 50 m*M* Tris-HCl, pH 7.6, containing 1 m*M* EDTA, 1 m*M* DTT, 0.1 m*M* PMSF, and 0.6 *M* KCl; and (3) 50 m*M* Tris-HCl, pH 7.6, containing 1 m*M* EDTA, 1 m*M* DTT, and 1 m*M* PMSF. The protein concentration is adjusted to 2–6 mg/ml, and proteins are solubilized in 0.2% (v/v) Lubrol PX for 60 min. The extracted supernatant fraction is separated from the insoluble pellet by centrifuga-

[14] S. A. Waldman, R. M. Rapoport, and F. Murad, *J. Biol. Chem.* **259,** 14332 (1984).

[15] S. A. Waldman, R. M. Rapoport, R. R. Fiscus, and F. Murad, *Biochim. Biophys. Acta* **845,** 298 (1985).

[16] R. M. Rapoport and F. Murad, *J. Cyclic Nucleotide Protein Phosphorylation Res.* **52,** 352 (1983).

[17] R. M. Rapoport, S. A. Waldman, K. Schwartz, R. J. Winquist, and F. Murad, *Eur. J. Pharmacol.* **115,** 219 (1985).

[18] D. B. Light, E. M. Schwiebert, K. H. Karlson, and B. A. Stanton, *Science* **243,** 383 (1989).

[19] T. Maack, M. Suzuki, F. A. Almeida, D. Nussenzveig, R. M. Scarborough, G. A. McEnroe, and J. A. Lewicki, *Science* **238,** 675 (1987).

[20] M. Hirata, C. H. Chang, and F. Murad, *Biochim. Biophys. Acta* **1010,** 346 (1989).

tion and used for further purification of ANP receptors and particulate guanylyl cyclase.

Purification by Sequential Chromatography

Particulate guanylyl cyclase and ANP receptor are purified by sequential chromatography as described previously.[11] Detergent extracts are adjusted to 20% (w/v) glycerol, 0.025% (w/v) phosphatidylcholine, 2 m*M* $MnCl_2$, and 1 m*M* NaN_3 and loaded onto GTP-agarose previously equilibrated with 20 m*M* Tris-HCl, pH 7.6, containing 1 m*M* DTT, 0.1 m*M* PMSF, 0.1% Lubrol PX, 0.025% phosphatidylcholine, 2 m*M* $MnCl_2$, 1 m*M* NaN_3, and 20% glycerol (buffer A). The loaded column is washed with 200 ml of buffer A and eluted with 100 ml of buffer A containing 5 m*M* GTP. Fractions possessing guanylyl cyclase and ANP-binding activities are pooled and loaded onto a DEAE-Sephacel column equilibrated in buffer containing Tris-HCl, pH 7.6, 1 m*M* DTT, 0.1 m*M* PMSF, 0.03% Lubrol PX, 0.015% phosphatidylcholine, and 20% glycerol (buffer B). After the column is washed with 150 ml of buffer B, the protein is eluted with 80 ml of a 0–300 m*M* NaCl gradient in buffer B. Fractions possessing enzyme and receptor-binding activities are pooled, adjusted to 1 *M* $(NH_4)_2SO_4$, and loaded onto phenyl-agarose that was equilibrated with buffer B containing 1 *M* $(NH_4)_2SO_4$. The column is washed with 150 ml of buffer B containing 1 *M* $(NH_4)_2SO_4$. The protein is eluted with a 150-ml decreasing linear gradient (1–0 *M*) of $(NH_4)_2SO_4$. Fractions possessing guanylyl cyclase and ANP-binding activity are pooled and loaded onto a wheat germ agglutinin (WGA)-Sepharose column equilibrated in buffer containing 20 m*M* HEPES, pH 7.4, 1 m*M* DTT, 0.1 m*M* PMSF, 0.1% Lubrol PX, 0.025% phosphatidylcholine, 0.5 *M* NaCl, and 20% glycerol (buffer C). After washing with 150 ml buffer C, the protein is eluted with 0.25 *M* *N*-acetylglucosamine in buffer C.

ANP Receptor Binding Assay

ANP receptor binding assays are performed as described previously.[11] Reaction mixtures containing 25–100 μl of sample, 50 m*M* Tris-HCl, pH 7.6, 1 m*M* EDTA, 1 m*M* cystamine, 150 m*M* NaCl, 5 m*M* $MnCl_2$, 0.1 m*M* PMSF, 0.1% bacitracin, iodinated ANP, and the competing ligand in a final volume of 200 μl are incubated for 1 hr at 25°. Nonspecific binding is determined with 0.1 μM of unlabeled ligand. Bound ANP is separated from free ligand by vacuum filtration through Whatman, Reeves Angel 934 AH glass fiber filters pretreated with 0.3% polyethyleneimine. Filters are washed 10 times with 2 to 3 ml of phosphate-buffered saline, pH 7.2. Radioactivity on filters is determined by a Beckman Gamma 4000 counter.

Guanylyl Cyclase Assay

Particulate guanylyl cyclase activity is determined as described previously.[11,12,14–17] Reaction mixtures containing 5 μl of sample, 50 m*M* Tris-HCl, pH 7.6, 0.5 m*M* isobutylmethylxanthine (IBMX), 0.1% bovine serum albumin, and a GTP-regenerating system consisting of 15 m*M* creatine phosphate and 20 μg (135 units/mg) of creatine phosphokinase in a final volume of 100 μl are incubated for 10 min at 37°. When purified fractions are assayed, IBMX and the GTP-regenerating system are omitted from the incubation. Reactions are terminated with 900 μl of ice-cold sodium acetate, pH 4.0, and samples are incubated in a boiling water bath for 3 min. The generated cGMP is measured by radioimmunoassay as described previously (see [30], this volume).

Affinity Cross-Linking Procedure

Iodinated ANP is specifically cross-linked to receptors by a procedure described previously.[11] Purified samples (50 μl) are incubated with 1 n*M* ^{125}I-labeled ANP in buffer containing 20 m*M* HEPES, pH 7.4, 1 m*M* EDTA, 150 m*M* $MnCl_2$, 0.1 m*M* PMSF, 0.1% (w/v) bacitracin, and the nonradioactive competing ligand, where indicated, in a final volume of 100 μl for 60 min at 25°. Upon completion of the incubation, disuccinimidyl suberate (in dimethyl sulfoxide) is added to a final concentration of 0.1 m*M*, and the cross-linking reaction is allowed to proceed for 15 min at 25°. Reactions are terminated by the addition of 50 μl of 0.2 *M* Tris-HCl, pH 6.8, containing 15% (w/v) sodium dodecyl sulfate (SDS), 20% glycerol, 200 m*M* DTT, and 0.15 mg/ml of bromphenol blue, and heating for 3 min at 100°.

Results

Purification of ANP Receptor–Guanylyl Cyclase

Particulate guanylyl cyclase and ANP receptor-binding activities coextracted from rat lung membranes and precisely coeluted in each chromatographic step of the purification procedure (Table I). The final purified preparations contained 8% of the total particulate guanylyl cyclase and 10% of the total ANP-binding activities present in the original starting material (rat lung membranes washed with Lubrol PX). Particulate guanylyl cyclase is purified 15,000-fold to a final specific activity of 19 μmol cGMP produced/min/mg protein, whereas the ANP receptor is purified 19,000-fold to a final specific activity of 630 pmol ANP bound/mg protein. The chromatographic step yielding the greatest degree of purification is

TABLE I
COPURIFICATION OF ANP RECEPTOR AND PARTICULATE GUANYLYL CYCLASE FROM RAT LUNG[a]

Step	ANP receptor			Guanylyl cyclase		
	SA[b]	Recovery (%)	Purification (-fold)	SA[c]	Recovery (%)	Purification (-fold)
Membrane + Lubrol PX	0.033	100	1.0	1.3	100	1.0
Lubrol extract	0.064	67	1.9	2.9	77	2.2
GTP-agarose	1.5	34	45	53	31	41
DEAE-Sephacel	2.5	24	76	88	21	68
Phenyl-agarose	7.7	17	230	370	20	280
WGA-agarose	630.0	10	19,000	19,000	8	15,000

[a] After Ref. 11.
[b] SA (specific activity) = pmol ^{125}I-labeled ANP bound/mg protein.
[c] SA (specific activity) = nmol cGMP produced/min/mg protein.

WGA-Sepharose, in which guanylyl cyclase is purified about 50-fold whereas the ANP receptor is purified about 80-fold. This procedure yielded about 5–10 μg of purified protein from about 100 g of tissue.

Physical Characterization

Analysis by SDS–polyacrylamide gel electrophoresis (SDS–PAGE) of protein preparations purified by the above procedure resulted in the appearance of a single protein band with a molecular weight of 130,000. There is no evidence of protein at lower molecular weights, particularly at 66,000. The 66K protein is known to be present in the original rat lung original membranes, but unlike the 130K ANP receptor it does not bind to the GTP affinity column. In addition, when these preparations are specifically cross-linked with radiolabeled ANP, analyzed by SDS–PAGE, and subjected to autoradiography, a single band of 130K is specifically labeled. No labeling of a lower molecular weight band, particularly at 66K, is observed. The protein cross-linked with ^{125}I-labeled ANP precisely comigrated with the protein band visualized by silver staining. In addition, chromatography of these preparations by gel-permeation HPLC resulted in the precise comigration of ANP-binding and guanylyl cyclase activities, with a Stokes radius of 5.7 nm. Similarly, when these preparations are subjected to centrifugation through glycerol gradients, enzyme and receptor-binding activities comigrated with a sedimentation coefficient of 9.7 S. From these values an estimated molecular weight (M_r) of about 230,000

TABLE II
PHYSICAL CHARACTERISTICS OF ANP RECEPTOR AND PARTICULATE GUANYLYL CYCLASE PURIFIED FROM RAT LUNG[a]

Characteristic	Value
Subunit	130,000 monomer
Glycoprotein	Yes
$s_{20,w}$	9.7 ± 0.3 S
Stokes radius	5.7 ± 0.2 nm
M_r	230,000

[a] After Ref. 11.

is calculated for the enzyme–receptor protein. The physical characteristics of these preparations are listed in Table II.

Kinetics of Purified Enzyme and Receptor

The ability of purified preparations to generate cGMP from GTP is monitored over increasing substrate concentrations (Table III). The enzyme demonstrated typical linear Michaelis–Menten kinetics as a function of MnGTP. The concentration of substrate yielding half-maximal enzyme velocity is 130 μM of MnGTP, and the V_{max} is 19 μmol cGMP formed/min/mg protein. Similarly, these preparations were tested for their ability to bind labeled ANP over increasing ligand concentrations (Table III). Purified ANP receptor demonstrated classic dose-dependent binding, and Scatchard analysis of the data yielded a single straight line, indicating a

TABLE III
KINETICS OF ANP RECEPTOR AND PARTICULATE GUANYLYL CYCLASE PURIFIED FROM RAT LUNG[a]

Activity	Parameter	Value
ANP binding	K_d[b]	150
	B_{max}[c]	1200
Guanylyl cyclase	K_m[d]	130
	V_{max}[e]	19
	Hill coefficient	1

[a] After Ref. 11.
[b] K_d, pmol ^{125}I-labeled ANP.
[c] B_{max}, pmol ^{125}I-labeled ANP bound/mg protein.
[d] K_m, μmol MnGTP.
[e] V_{max}, μmol cGMP produced/min/mg protein.

single class of binding sites with a K_d of 150 pM and a B_{max} of 1200 pmol/mg protein.

Discussion

The data presented above are the first demonstration that the 130K ANP receptor and particulate guanylyl cyclase resided in the same transmembrane glycoprotein in rat lung.[11] This observation is consistent with previous results demonstrating that the 130K ANP-R1 receptor but not the 66K ANP-R2 receptor is functionally coupled to particulate guanylyl cyclase.[12] Subsequently, ANP receptors have been purified from a variety of sources.[13,21–23] The 66K receptor purified from smooth muscle cells possesses no measurable particulate guanylyl cyclase activity.[13] This is consistent with the observation that it does not bind to the GTP affinity column. Similarly, both the 66K and 130K receptors have been purified from adrenal cells.[22,23] Only the 130K but not the 66K receptor is associated with guanylyl cyclase activity in these studies.[22,23] In addition, proteolytic maps of these different receptors demonstrated no shared peptides, suggesting that these receptors are different gene products.[23] These data support the suggestion that particulate guanylyl cyclase and ANP receptor-binding activities are contained in one protein.[11] Indeed, the cDNA encoding this protein has recently been isolated from rat brain and expressed in mammalian cells.[24] This cDNA encodes a protein which possesses both particulate guanylyl cyclase and ANP receptor-binding activities, confirming the results obtained with protein purification.[24]

Thus, it is readily apparent that there are multiple populations of both ANP receptors and particulate guanylyl cyclase. As described, ANP receptor binding subunits exist as 130K and 66K proteins. Particulate guanylyl cyclase exists as a 130K monomer possessing ANP receptor-binding activity, as described. Two additional forms of particulate guanylyl cyclase have been described. One form is not associated with ANP-binding activity[25] whereas other forms of particulate guanylyl cyclase are tightly coupled to cytoskeletal components in retinal rod outer segments and intesti-

[21] A. K. Paul, R. B. Marala, R. K. Kaiswal, and R. K. Sharma, *Science* **235,** 1224 (1987).

[22] R. Takayanagi, L. M. Snajdar, T. Imada, M. Tamura, K. N. Pandey, K. S. Misono, and T. Inagami, *Biochem. Biophys. Res. Commun.* **144,** 244 (1987).

[23] R. Takayanagi, T. Inagami, R. M. Snajdar, T. Imada, M. Tamura, and K. S. Misono, *J. Biol. Chem.* **262,** 12104 (1987).

[24] M. Chinkers, D. L. Garbers, M. S. Chang, D. G. Lowe, H. Chin, D. V. Goeddel, and S. Schulz, *Nature (London)* **338,** 78 (1989).

[25] S. A. Waldman, D. C. Leitman, L. Y. Chang, and F. Murad, *Biochim. Biophys. Acta* (in press).

nal brush border membranes.[26–28] It is the latter form in intestinal membranes that may be coupled to receptors for the heat-stable enterotoxin produced by *Escherichia coli* which produces travelers' diarrhea.[28] The relationship of these other forms of guanylyl cyclase to the ANP receptor–particulate guanylyl cyclase protein is not known. The role for these various forms of receptors and particulate guanylyl cyclase are currently under investigation.

[26] D. Fleishman and M. Denisevich, *Biochemistry* **18,** 5060 (1979).
[27] D. Fleishman, M. Denisevich, D. Raveed, and R. Pannbacker, *Biochim. Biophys. Acta* **630,** 176 (1980).
[28] S. A. Waldman, T. Kuno, Y. Kamisaki, L. Y. Chang, J. Gariepy, G. Schoolnik, and F. Murad, *Infect. Immun.* **51,** 320 (1986).

[38] Copurification of Atrial Natriuretic Factor Receptors and Guanylyl Cyclase from Adrenal Cortex

By TADASHI INAGAMI, RYOICHI TAKAYANAGI, and RUDOLF M. SNAJDAR

Introduction

Atrial natriuretic factor (ANF) is a peptide of 28 amino acids which is primarily synthesized and stored in atrial myocytes.[1] The ANF peptide has been shown to be a potent circulating natriuretic, diuretic, and vasorelaxant hormone. Very early in the characterization and purification, it was discovered that many of the actions of ANF were paralleled by increases in cGMP levels,[2] and it was suggested that ANF action was in some way coupled to stimulation of guanylyl cyclase. Atrial natriuretic factor has been demonstrated to antagonize the release of such hormones as aldosterone, renin, vasopressin, and norepinephrine. In addition, ANF has been shown not only to inhibit renal tubular reabsorption, centrally mediated water drinking, and salt appetite, but also to stimulate testicular and ovarian steroidogenesis. Such diverse actions of ANF are mediated by ANF receptors that have been identified in numerous sites, including the vascular tissue, kidneys, adrenals, brain, testis, ovary, and adipose tissue.

While receptor-binding studies using radioiodinated ANF revealed a

[1] T. Inagami, *J. Biol. Chem.* **264,** 3043–3046 (1989).
[2] P. Hamet, J. Tremblay, S. C. Pang, R. Garcia, G. Thibault, J. Gutkowska, M. Cantin, and J. Genest, *Biochem. Biophys. Res. Commun.* **123,** 515 (1984).

high-affinity binding site with a single affinity constant of approximately 10^{-10} *M*, photoaffinity or cross-linking experiments followed by sodium dodecyl sulfate-polyacrylamide gel electrophoresis (SDS–PAGE) revealed the presence of at least two types of ANF receptors. In most of the tissues which were examined, specifically labeled protein bands were observed at 60–70 kDa and 120–140 kDa and were shown to possess high affinity for the 28 amino acid ANF. However, in cultured rat adrenocortical tumor cells, only a 180-kDa protein band was identified.[3] In general, the ratio of low to high molecular weight ANF receptor appears to vary from tissue to tissue.[4] Furthermore, ANF receptors have been shown to undergo down-regulation which is inversely proportional to ANF levels. For example, in the brain, aorta, or adrenal cortex of spontaneously hypertensive rats (SHR), the total receptor number was markedly reduced, presumably due to down-regulation because of the markedly increased plasma ANF levels. However, plasma membrane guanylyl cyclase activity was increased.[5] To explain such a discrepancy, the presence of more than one type of receptor has to be postulated. It is possible that several different types of receptors behave differently in the down-regulation and play different roles. The first direct evidence that guanylyl cyclase activity was associated with the 120-kDa ANF receptor was shown by Kuno *et al.*[6] Questions arose as to whether the 60-kDa ANF receptor was a proteolytic product of the guanylyl cyclase-linked ANF receptor or whether the cyclase was a dimer of the smaller receptor. In order to clarify these questions studies were undertaken to purify the two types of ANF receptors and to compare their structural and functional properties.

Preliminary studies have shown that ANF receptor could be readily solubilized and binding activity maintained. In contrast, the associated guanylyl cyclase activity appeared to be more labile. To retain as much receptor binding and guanylyl cyclase activity as possible, we used a detergent–protease inhibitor mixture for solubilization. Furthermore, since the guanylyl cyclase activity was precipitously destroyed below pH 5.0, conditions were optimized, as described below, to minimize loss of activity while maintaining yield.

Atrial natriuretic factor receptor was solubilized with a Triton X-100–protease inhibitor mixture and was applied to ANF-agarose. After

[3] A. K. Paul, R. B. Marala, R. K. Jaiswal, and R. K. Sharma, *Science* **235,** 1224 (1987).

[4] D. C. Leitman, J. W. Andersen, R. M. Catalano, S. A. Waldman, J. J. Tuan, and F. Murad, *J. Biol. Chem.* **263,** 3720 (1988).

[5] R. Takayanagi, T. Imada, R. T. Grammer, K. S. Misono, M. Naruse, and T. Inagami, *J. Hypertens.* **4,** 5303 (1986).

[6] T. Kuno, J. W. Andersen, Y. Kamisaki, L. Y. Chang, S. Sahei, D. C. Leitman, M. Nakane, and F. Murad, *J. Biol. Chem.* **261,** 5817 (1986).

elution two ANF receptor subtypes were effectively separated by passing the mixture successively through guanine triphosphate (GTP)-agarose and wheat germ agglutinin (WGA)-Sepharose chromatography columns as previously described.[7]

Assays

ANF Binding Assay. ANF-binding activity is determined at 25° by incubating 5–50 μl of solubilized membranes for 60 min with 200 pM ^{125}I-labeled ANF(99–126) in 0.5 ml of a reaction mixture consisting of 50 mM Tris-HCl, pH 7.5, 0.15 M NaCl, 5 mM $MgCl_2$, 1 mg/ml bovine serum albumin, 0.5 mg/ml bacitracin, and 0.1% (v/v) Triton X-100. The receptor–^{125}I-labeled ANF complex is separated from free ^{125}I-labeled ANF by adding 0.25 ml of 0.25% bovine γ-globulin and 3 ml of a precipitation solution consisting of 10% polyethylene glycol 8000, 20 mM sodium phosphate buffer, pH 7.5, and 0.15 M NaCl, followed by filtration under vacuum through 0.3% (w/v) polyethyleneimine-treated Whatman GF/B filters. The filters are washed 3 times with 3 ml of the same precipitation solution. The binding reaches equilibrium in 30 min and is stable for 150 min. Data are analyzed by nonlinear curve-fitting by use of the LIGAND computer program.[8] The concentrations of ANF peptides used are determined by amino acid analysis.

Guanylyl Cyclase Assay. Guanylyl cyclase activity is determined by incubating samples (membrane or solubilized preparations) at 37° for 20 min in 0.1 ml of a reaction mixture consisting of 50 mM Tris-HCl, pH 7.6, 10 mM theophylline, 2 mM 1-methyl-3-isobutylxanthine, 7.5 mM creatine phosphate, 3 units of creatine phosphokinase, 2 mM $MnCl_2$, 1 mM GTP, and 1 mg/ml bovine serum albumin. The reaction is initiated by the addition of samples and terminated by the addition of 0.9 ml of 0.1 M sodium acetate, pH 4.0, and heating for 3 min at 100°. Generated cGMP is determined by radioimmunoassay.[9]

The protein concentration is determined by the method of Lowry *et al.* with Peterson's modifications.[10] After affinity chromatography on ANF-agarose, the protein content is measured fluorometrically with fluorescamine and nitrocellulose membrane filters as described by Nakamura and Pisano.[11] Bovine serum albumin is used as standard in each assay.

[7] R. Takayanagi, T. Inagami, R. M. Snajdar, T. Imada, M.Tamura, and K. S. Misono, *J. Biol. Chem.* **262,** 12104 (1987).

[8] P. J. Monson and D. Rodbard, *Anal. Biochem.* **107,** 220 (1980).

[9] D. L. Garbers and F. Murad, *Adv. Cyclic Nucleotide Res.* **10,** 57 (1979).

[10] G. L. Peterson, *Anal. Biochem.* **83,** 346 (1977).

[11] H. Nakamura and J. J. Pisano, *Arch. Biochem. Biophys.* **172,** 102 (1976).

Affinity Columns

Preparation of ANF Affinity Column. Ten milligrams of ANF(99–126) is dissolved in 0.1 *M* HEPES, pH 7.5, containing 1 *M* NaCl and tracer amounts of ^{125}I-labeled ANF(99–126) and mixed gently at 4° overnight in a polypropylene column with 8 ml of Affi-Gel 10 (Bio-Rad, Richmond, CA), afterward, the coupled agarose is treated with ethanolamine, and the gel is washed successively with 2 liters of water, 1 liter of 1 *M* NaCl, 1 liter of 0.1 *M* acetic acid containing 1 *M* NaCl, 1 liter of 2 *M* urea, 2 liters of water, and finally with 2 liters of 50 m*M* Tris-HCl, pH 7.4, 0.15 *M* NaCl. Under these conditions 86% of ANF(99–126) is incorporated into the gel, which contains 0.35 μmol of ANF in 1 ml of packed wet gel. GTP-agarose and WGA-Sepharose 6MB were purchased from Sigma (St. Louis, MO), and Pharmacia P-L Biochemicals (Piscataway, NJ), respectively.

Purification of ANF Receptor

Preparation of Adrenocortical Membranes. All the following procedures are performed at 4°. Fresh bovine adrenal glands are obtained from a local slaughterhouse. The adrenal cortex is sliced (1.0 mm thick) with a Stadie-Riggs microtome, and the outermost slices are used for membrane preparation. Plasma membrane-rich fractions are prepared by stepwise sucrose density gradient according to Glossman *et al.*[12] The membranes are suspended in 50 m*M* Tris-HCl buffer, pH 7.5, 0.15 *M* NaCl, 0.4 m*M* phenylmethylsulfonyl fluoride (PMSF), 2 μg/ml leupeptin, and 1 m*M* EDTA, then frozen in ethanol/dry ice and stored at −70°.

Solubilization of Membranes. Frozen membranes (2.0–2.3 g of protein) are thawed at 4°, and the protein concentration is adjusted to 14 mg/ml in 50 m*M* Tris-HCl buffer, pH 7.5, containing 0.15 *M* NaCl, 0.4 m*M* PMSF, 1 m*M* EDTA, and 1 m*M* EGTA. After the addition of an equal volume of solubilizing buffer, which consists of 50 m*M* Tris-HCl, pH 7.5, 0.15 *M* NaCl, 2% Triton X-100, 1 m*M* EDTA, and 1 m*M* EGTA, 20 μg/ml each leupeptin, aprotinin, and phosphoramidon, 2 μg/ml pepstatin A, 1.0 mg/ml bacitracin, 0.4 m*M* PMSF, and 0.1 m*M* diisopropyl fluorophosphate, the suspension is stirred for 60 min. A clear supernatant is obtained by centrifugation after 90 min at 100,000 *g*.

Chromatography on ANF-Agarose. The Triton X-100-solubilized preparations are diluted by the addition of 3 volumes of 50 m*M* Tris-HCl, pH 7.5, containing 13.4% (v/v) glycerol, 0.0167% (w/v) phosphatidylcholine, 0.15 *M* NaCl, 10 μg/ml each leupeptin, aprotinin, and phosphoramidon,

[12] H. Glossman, A. J. Baukal, and K. J. Catt, *J. Biol. Chem.* **249,** 825 (1974).

0.5 mg/ml bacitracin, 0.4 mM PMSF, 2 μg/ml pepstatin A, 1 mM EGTA, and 14.7 mM $MgCl_2$. The diluted samples are loaded simultaneously at a flow rate of 0.8 ml/min onto two ANF-agarose columns (10 $\times$ 60 mm) equilibrated with 50 mM Tris-HCl, pH 7.5, 0.15 M NaCl, 0.1% Triton X-100, 10% (v/v) glycerol, 0.0125% (w/v) phosphatidylcholine, 0.4 mM PMSF, and 10 mM $MgCl_2$. Each column is washed with 100 ml of 50 mM HEPES buffer, pH 7.5, containing 0.15 M NaCl, 0.1% Triton X-100, 10% glycerol, 0.0125% phosphatidylcholine, 1 mM EGTA, 0.4 mM PMSF, and 11 mM $MgCl_2$ (buffer A), plus 10 μg/ml each leupeptin, aprotinin, and phosphoramidon, 0.5 mg/ml bacitracin, and 2 μg/ml pepstatin A. The columns are then washed successively with 100 ml of buffer A containing 1 M NaCl and the above-mentioned protease inhibitors, 100 ml of buffer A without NaCl, and 100 ml of buffer A containing 0.2% CHAPS or Zwittergent 3-14 instead of 0.1% Triton X-100. The washed gel in each column is then mixed with 5 ml of an elution buffer, consisting of 50 mM sodium acetate buffer, pH 5.5, 1 M NaCl, 0.1% Triton X-100, 20% glycerol, 0.025% phosphatidylcholine, 0.1 mM PMSF, and 0.5 mM EDTA, for 20 min by gentle tumbling of the column. The eluate is collected in a tube containing 0.8 ml of 1 M HEPES buffer, pH 7.5. The elution procedure is repeated 14 times, collecting 5 ml eluate plus 0.8 ml of HEPES buffer in each fraction. Fractions possessing ANF-binding and guanylyl cyclase activities are pooled and immediately dialyzed against 30 volumes of 20 mM HEPES buffer, pH 7.5, containing 0.1% Triton X-100, 0.025% phosphatidylcholine, 20% glycerol, and 0.1 mM PMSF for 12 hr with changes of the dialysis buffer every 3 hr.

Chromatography on GTP-Agarose. Dialyzed samples are adjusted to 3 mM $MnCl_2$ and 2 mM NaN_3, and each 7.5-ml aliquot is added to a Bio-Rad Poly-Prep minicolumn, which contains 3 ml of GTP-agarose equilibrated with 20 mM HEPES buffer, pH 7.5, 0.1% Triton X-100, 0.025% phosphatidylcholine, 20% glycerol, 0.1 mM PMSF, 3 mM $MnCl_2$, and 2 mM NaN_3 (buffer B), and mixed for 12 hr by gentle tumbling of the column. After the pass-through fraction is collected, the column is washed with 60 ml of buffer B. Three milliliters of buffer B containing 5 mM GTP is added, the column is gently tumbled for 20 min, and 3 ml of eluate is collected. An additional 1 ml of buffer B containing 5 mM GTP is added to the column, and 1 ml eluate is collected. This elution procedure is repeated 10 times.

Chromatography on WGA-Sepharose. The pass-through fractions from GTP-agarose as well as the GTP-eluted fractions are separately pooled, adjusted to 0.5 M NaCl and 0.5 mM EDTA, and applied at a flow rate of 0.15 ml/min to 10 $\times$ 110 mm columns of WGA-Sepharose equilibrated with 20 mM HEPES buffer, pH 7.5, containing 0.1% Triton

X-100, 20% glycerol, 0.025% phosphatidylcholine, 0.1 m*M* PMSF, 0.5 m*M* EDTA, and 0.5 *M* NaCl (buffer C). The columns are washed with 100 ml of buffer C and then with 200 ml of 50 m*M* HEPES buffer, pH 7.5, containing 0.02% Triton X-100, 0.0125% phosphatidylcholine, 20% glycerol, 0.1 m*M* PMSF, 0.5 m*M* EDTA, and 0.15 *M* NaCl (buffer D). The columns are allowed to stand with 0.25 ml of 0.4 *M* *N*-acetylglucosamine in buffer D for 30 min and eluted. This elution procedure is repeated 10 times. The active fractions are concentrated about 5-fold with a YM30 Centricon ultrafiltration membrane (Amicon, Danvers, MA) and stored at −70°.

Comments

Solubilization. The Triton X-100-solubilized plasma membranes contained 31–38% of the membrane proteins, 85–90% of the ANF-binding activity, and 240–310% of the guanylyl cyclase activity, when used at a weight ratio of 1.4–1.5 of Triton X-100 to protein. Above this ratio, only greater amounts of protein were extracted without further enrichment of the receptor or the guanylyl cyclase. A similar increase in guanylyl cyclase was reported,[13] apparently due to the exposure to detergents. Addition of phosphatidylcholine and glycerol[6] to the solubilized membranes afforded excellent stability for the ANF-binding and guanylyl cyclase activities during subsequent purification steps as well as storage.

The particulate guanylyl cyclase can be solubilized by Lubrol PX or CHAPS from rat lung or adrenal tumor cells.[3,6] However, these detergents destroy the ANF-induced activation of the cyclase. Triton X-100 maintained some (2.5- to 3.5-fold) response of the cyclase to ANF.[7]

Purification. Figure 1 shows the elution profile of the sequential purification of ANF receptors on the ANF (Fig. 1A), GTP (Fig. 1B), and WGA (Fig. 1C,D) chromatography columns. The ANF column retained 90% of ANF receptor activity and 23% of the particulate guanylyl cyclase activity of the Triton X-100-solubilized plasma membrane. Affinity columns bearing truncated ANF analogs such as ANF(99–121) ANF(103–123) could bind only about 25% of the receptor activity and practically no guanylyl cyclase activity. The carboxyl-terminal sequence, Phe-Arg-Tyr, in ANF seems to be essential for the binding of the particulate guanylyl cyclase. Use of 10 m*M* Mg^{2+} bestowed the best binding capability to the ANF gel,

[13] J. Tremblay, R. Gerzer, S. C. Pang, M. Cantin, J. Genest, and P. Hamet, *FEBS Lett.* **194,** 210 (1985).

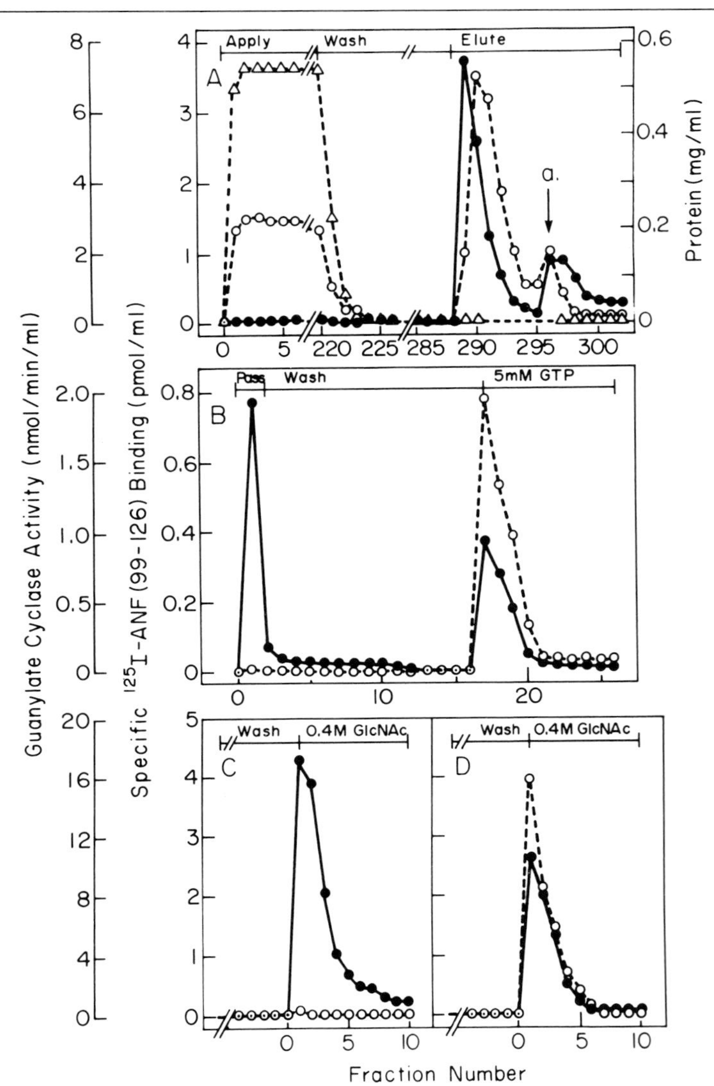

FIG. 1. Elution profiles of the sequential chromatographic steps during purification of ANF receptors with and without particulate guanylyl cyclase activity from bovine adrenal cortex. (△) Protein concentration; (●) specific ^{125}I-ANF (99–126) binding; (○) guanylyl cyclase activity. (A) Chromatographic profile of the ANF column. The Triton X-100-solubilized extract of bovine adrenocortical membranes was applied in a flow-through fashion to an ANF column. The column was extensively washed as described in text and then eluted with 50 m*M* sodium acetate buffer, pH 5.5, containing 1 *M* NaCl, 0.5 m*M* EDTA, 0.025% phosphatidylcholine, and 20% glycerol. Fractions of 3- and 6-ml volume were collected during the application and washing periods, respectively, and 5.5-ml fractions, with addition of 0.8 ml neutralization buffer, were collected during the elution step as described in text.

in agreement with the earlier observation that divalent cations markedly increased the binding affinity of ANF peptides to its receptor.[14]

The elution of the cyclase from the ANF gel was most delicate and demanded foremost attention. Recovery of the ANF-binding capacity and guanylyl cyclase activity was 42 and 39%, respectively. Essential for the elution of the cyclase was the use of phospholipid, without which only a very small amount of protein was released from the column, even in the presence of 1 *M* NaCl and at a pH below 4.5. The phospholipid was also essential for preserving the cyclase activity at an acidic pH. Since the guanylyl cyclase activity was sensitive to exposure to an acidic pH of 5.0 or below (Fig. 2), conditions were devised to elute it at pH 5.5. A mixture containing 50 m*M* sodium acetate (pH 5.5), 1 *M* NaCl, 0.5 m*M* EDTA, 0.025% phosphatidylcholine, and 20% glycerol was effective in eluting the ANF receptor from the affinity gel. At this pH the cyclase activity remained stable for 100 min or longer at 4°.

The ANF-agarose chromatographic step attained 1600- and 6800-fold purification of the cyclase and ANF–receptor activity, respectively. The slight dissociation of the cyclase activity peak and ANF receptor activity peak (as shown in Fig. 1A) indicated the heterogeneity of these activities. This observation was substantiated by the affinity chromatography on GTP-agarose which effectively separated the two ANF-receptor subtypes, guanylyl cyclase-free (first peak in Fig. 1B) and guanylyl cyclase-containing (second peak in Fig. 1B) receptors. About one-half of the receptor distributed to the first and the other half distributed to the second peak.

The sample eluted from the ANF gel had to be dialyzed to remove EDTA and NaCl before application to the GTP-agarose affinity column. With 5 m*M* GTP in the elution solution, about 75% of the applied guanylyl

[14] A. DeLean, J. Gutkowska, N. McNicol, P. Schiller, M. Cantin, and J. Genest, *Life Sci.* **35,** 2311 (1984).

The pH of the elution buffer was changed from 5.5 to 4.5 at the fraction indicated by arrow a. (B) Chromatography on GTP-agarose. Approximately 60 ml of the pooled peak fractions from the ANF-agarose columns (fraction numbers 289–293) was dialyzed and adjusted to 3 m*M* $MnCl_2$ and 2 m*M* NaN_3, and each 7.5-ml aliquot was loaded onto 3 ml of GTP-agarose. After a pass-through fraction (~7.5 ml) was collected, the column was washed with buffer B and eluted with 5 m*M* GTP in buffer B (see text). Four-milliliter fractions were collected during the washing and elution periods. (C and D) WGA-Sepharose chromatography. The passed-through fractions (C) from GTP-agarose columns and the GTP-eluted fractions (D) (fractions 17–19) were separately pooled, adjusted to 0.5 *M* NaCl and 0.5 m*M* EDTA, and loaded onto the WGA column. The column was washed with buffers C and D, as described in text, and eluted with 0.4 *M* *N*-acetylglucosamine, while 2.5-ml fractions were collected. (Reproduced from Ref. 7 with the permission of the *Journal of Biological Chemistry*.)

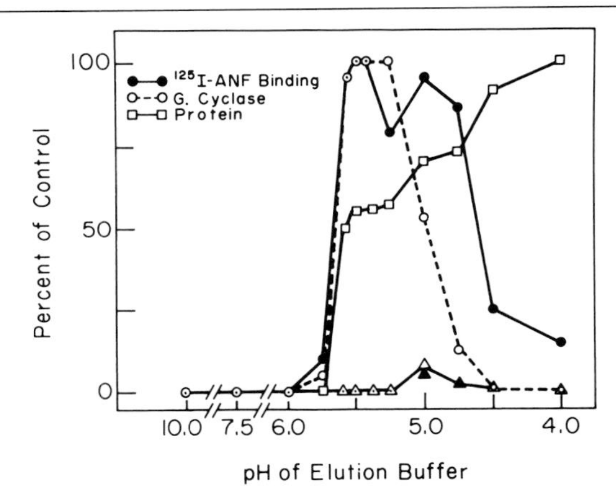

FIG. 2. Effects of pH, salt concentration, and phosphatidylcholine on the elution of the ANF receptor and guanylyl cyclase from the ANF(99–126) column. Bovine adrenocortical membranes were solubilized with Triton X-100 and diluted, as described in text, and 30 ml was applied to 0.8 ml of ANF(99–126)-agarose gel. The column was washed and eluted at various pH values with sodium acetate buffer from pH 4.0 to 6.0, HEPES buffer at pH 7.5, and ethanolamine buffer at pH 10.0, in combination with 1 *M* NaCl and 0.025% phosphatidylcholine. The filled triangle and the open triangle symbols represent the ANF-binding activity and the guanylyl cyclase activity, respectively, when the column was eluted without NaCl or phosphatidylcholine. The values are expressed as percentages of the maximally eluted activities [2.34 pmol for ANF-binding activity, with 200 p*M* ^{125}I-labeled ANF(99–126); 5.34 nmol cGMP formed/min for guanylyl cyclase activity (maximally eluted at pH 5.5)] or amount of protein (1.6 μg protein eluted at pH 4.0). (Reproduced from Ref. 7 with the permission of the *Journal of Biological Chemistry*.)

cyclase activity and ANF-binding activity were recovered from the GTP-agarose column. By this step the guanylyl cyclase was purified by an additional factor of 3. While the subsequent purification on the WGA gel did not increase the specific activity of the guanylyl cyclase, it did result in cleaner product as examined by SDS–PAGE. Recovery of the enzyme activity at this step was approximately 60%.

Starting with 576 g of the outermost layer of bovine adrenal cortex, 3.7 μg of guanylyl cyclase-linked ANF receptor and 8.4 μg of guanylyl cyclase-free ANF receptor were obtained. These preparations were homogeneous as judged by SDS–PAGE under reducing conditions.

Properties

Guanylyl Cyclase-Linked ANF Receptor. The cyclase and receptor activities were stable for 4 months at $-70°$. The molecular weight determined by SDS–PAGE was 135,000 and was not altered by the presence or absence of dithiothreitol. Observations supporting the bifunctionality of this protein containing both guanylyl cyclase and ANF receptor-binding activities are several. The purity of this protein is 92%, as calculated from its molecular weight, the B_{max} of a Scatchard plot of ANF binding, and the assumption of stoichiometric binding of ANF. The specific activity of guanylyl cyclase (23.1 μmol cGMP formed/min/mg protein) is the highest among the guanylyl cyclases purified to date. The guanylyl cyclase from rat lung had an activity of 19 μmol/min/mg protein,[6] that from sea urchin spermatozoa had 15.2 μmol/min/mg protein,[15] and the rat adrenal cortical tumor guanylyl cyclase showed an activity of 1.8 μmol/min/mg protein.[3] The binding and the guanylyl cyclase activities copurified at a constant ratio on the GTP-agarose and WGA-Sepharose affinity chromatography columns.

As to the ANF receptor activity, the receptor does not bind effectively to truncated ANF species such as ANF(103–123), that is, atriopeptin I with 21 amino acid peptide, or the 17 amino acid peptide which is without carboxy- and amino-terminal tails [i.e., ANF(105–121)]. Although the cyclase activity was slightly increased by ANF binding at an early stage of Triton X-100 solubilization, the purified protein lost the response to ANF activation. Whether this is due to partial denaturation or to loss of a second protein needed for the activation is not clear.

Guanylyl Cyclase-Free Receptor. As shown in Fig. 1B (first peak) and Fig. 1C, the guanylyl cyclase-free receptor is devoid of guanylyl cyclase activity and has a high affinity for various truncated forms of ANF.[16] The amino acid sequence of this protein was recently determined[17] and was found to have a tentative transmembrane domain very close to the carboxy terminus, suggesting little intracellular function for this protein. An intriguing hypothesis has been proposed for the role of this protein as that of a clearance receptor, which may bind excess circulating ANF and thereby buffer its actions.[18] Peptide mapping of the two receptors indicated that these receptors are not structurally related and are products of two different genes.[7]

[15] E. W. Radany, R. Gerzer, and D. L. Garbers, *J. Biol. Chem.* **258,** 8346 (1983).

[16] R. M. Scarborough, D. B. Schenk, G. A. McEnroe, A. Arfstein, L. L. Kang, K. Schwartz, and J. A. Lewicki, *J. Biol. Chem.* **261,** 12960 (1986).

[17] F. H. Fuller, J. G. Porter, A. E. Avfsten, J. Miller, J. W. Schilling, R. M. Scarborough, J. A. Lewicki, and D. B. Schenk, *J. Biol. Chem.* **263,** 9395 (1988).

[18] T. Maack, M. Suzuki, F. A. Almedia, D. Nussenzveig, R. M. Scarborough, G. A. McEnroe, and J. A. Lewicki, *Science* **238,** 675 (1987).

[39] Molecular Cloning of Membrane Forms of Guanylyl Cyclase

By SUJAY SINGH and DAVID L. GARBERS

Introduction

The formation of 3′,5′-cyclic GMP (cGMP) from GTP is catalyzed by guanylyl cyclase, an enzyme found in both the soluble and particulate fractions of various cell types.[1] At least two different cellular compartments for the particulate enzyme exist: plasma membrane and cytoskeleton. The soluble and particulate enzymes have different primary structures, and the regulation of each appears to occur by different mechanisms.[2,3] The soluble enzyme has been reported to exist as a heterodimer with subunit molecular weights of 82,000 and 70,000; it appears to contain heme as a prosthetic group and is activated by nitroprusside, nitric oxide, and reactive free radicals.[4–7] The membrane form is a glycoprotein and is composed of a single polypeptide chain of molecular weights ranging from 140,000 to 180,000. A plasma membrane form was first purified from sea urchin spermatozoa[8,9]; later, mammalian plasma membrane forms of guanylyl cyclase were purified from rat lung,[10] rat adrenocarcinoma,[11] and other tissues.[12] Both the sea urchin and mammalian enzymes appear to be similar in size and kinetic properties.[2,3] Sea urchin sperm guanylyl cyclase is regulated by peptides obtained from sea urchin egg-conditioned media in a species-specific manner. Cross-linking studies with a radiolabeled analog of resact, an egg peptide, demonstrated specific interaction with an

[1] J. G. Hardman, J. A. Beavo, J. P. Gray, T. D. Chrisman, W. D. Patterson, and E. W. Sutherland, *Ann. N.Y. Acad. Sci.* **185,** 27 (1971).

[2] D. L. Garbers, J. G. Hardman, and F. G. Rudolph, *Biochemistry* **13,** 4166 (1974).

[3] H. Kimura and F. Murad, *J. Biol. Chem.* **249,** 6910 (1974).

[4] S. C. Tsai, V. C. Manganiello, and M. Vaughn, *J. Biol. Chem.* **253,** 8542 (1978).

[5] P. H. Craven and F. R. DeRubertis, *J. Biol. Chem.* **253,** 8433 (1978).

[6] J. A. Lewicki, H. J. Brandwein, S. A. Waldman, and F. Murad, *J. Cyclic Nucleotide Res.* **6,** 283 (1980).

[7] R. Gerzer, E. Bohme, F. Hoffman, and G. Schultz, *FEBS Lett.* **132,** 71 (1981).

[8] D. L. Garbers, *J. Biol. Chem.* **251,** 4071 (1976).

[9] E. W. Radany, R. Gerzer, and D. L. Garbers, *J. Biol. Chem.* **258,** 8346 (1983).

[10] T. Kuno, J. W. Anderson, Y. Kamisaki, S. A. Waldman, L. Y. Chang, S. Sahaki, D. C. Leitman, M. Nakane, and F. Murad, *J. Biol. Chem.* **261,** 5817 (1986).

[11] A. K. Paul, R. B. Marsala, R. K. Jaiswal, and R. K. Sharma, *Science* **235,** 1224 (1987).

[12] R. Takayanagi, T. Inagami, R. M. Snajadar, T. Imada, M. Tamura, and K. S. Misono, *J. Biol. Chem.* **262,** 12104 (1987).

Arbacia punctulata spermatozoan protein having an apparent molecular weight of 160,000; the protein was subsequently identified as guanylyl cyclase.[13] We have recently cloned the cDNA for *A. punctulata* guanylyl cyclase[14] but were unable to demonstrate resact binding or guanylyl cyclase activity of the expressed product. The reasons for the failure are not yet clear. In mammalian cells two natriuretic peptides, atrial natriuretic peptide (ANP) and brain natriuretic peptide (BNP), have been shown to bind to specific sites on various mammalian tissues and to cause rapid increases in intracellular cGMP.[15] In addition, preparations of partially purified guanylyl cyclase have continued to retain ANP-binding activity.[10–12] The recent cloning of cDNAs for mammalian guanylyl cyclase has demonstrated that ANP and BNP binding sites and cyclase activity reside in the same molecule. These findings also demonstrate diversity within the plasma membrane guanylyl cyclases. Two different forms of the membrane form of the enzyme have been cloned from both human and rat.[16,17] In this chapter, we describe the cloning of guanylyl cyclase cDNA from sea urchin spermatozoa, human placenta, and rat brain.

Molecular Cloning of *Arbacia punctulata* cDNA

Step 1: Purification of Particulate Guanylyl Cyclase. Guanylyl cyclase appears to exist entirely as a membrane-bound glycoprotein in sea urchin spermatozoa. Purification of guanylyl cyclase is based on the observation that, in *A. punctulata*, it is the only polypeptide with an apparent molecular weight of 160,000 on sodium dodecyl sulfate-polyacrylamide gel electrophoresis (SDS–PAGE). Spermatozoa are collected from sea urchins by intracoelomic injection with 0.5 *M* KCl. Twenty grams (wet weight) of intact spermatozoa is sonicated in homogenization buffer (9 volumes) which contains 25 m*M* 2-(*N*-morpholino)ethanesulfonic acid (MES) (pH 6.2), 10 m*M* benzamidine hydrochloride, 10 m*M* NaF, 5 m*M* dithiothreitol, and 0.5% (v/v) Lubrol PX.

The homogenate is centrifuged at 30,000 *g* in a JA-20 rotor (Beckman) for 10 min at 2°, and the supernatant fluid is prepared for electrophoresis by the addition of 5× Laemmli sample buffer with a final concentration

[13] H. Shimomura, L. J. Dangott, and D. L. Garbers, *J. Biol. Chem.* **261,** 15778 (1986).

[14] S. Singh, D. G. Lowe, D. S. Thorpe, H. Roderiguez, W.-J. Kuang, L. J. Dangott, M. Chinkers, D. V. Goeddel, and D. L. Garbers, *Nature (London)* **334,** 708 (1988).

[15] T. Inagami, *J. Biol. Chem.* **264,** 3043 (1989).

[16] M. Chinkers, D. L. Garbers, M.-S. Chang, D. G. Lowe, H. Chin, D. V. Goeddel, and S. Schulz, *Nature (London)* **338,** 78 (1989).

[17] D. G. Lowe, M.-S. Chang, R. Hellmiss, S. Singh, E. Chen, D. L. Garbers, and D. V. Goeddel, *EMBO J.* **8,** 1377 (1989).

of 100 mM Tris (pH 6.8), 1% (w/v) SDS, and 40 mg/ml bromphenol blue and heating at 57° for 10 min. An aliquot of this material is electrophoresed through a 10% SDS–polyacrylamide gel continuously eluted by gel buffer containing 0.05% SDS, 192 mM glycine, and 25 mM Tris (pH 8.8).[18] Aliquots of collected fractions are analyzed on minislab 12.5% SDS–polyacrylamide gels to assess both the presence of the 160,000 molecular weight protein and its purity. Protein is stained by a modification of Wray's technique.[19] Fractions containing the M_r 160,000 protein are pooled and concentrated by the use of an Amicon (Danvers, MA) Diaflo stir cell with a YM30 membrane. Further purification is done (to remove excess glycine) by a passage through a Sephadex G-50 column equilibrated with 0.01% SDS and 0.1 M ammonium bicarbonate. The total yield from 5 g of spermatozoa is approximately 2 nmol of protein.

Step 2: Partial Amino Acid Sequencing. About 1 nmol of M_r 160,000 protein is digested with 200 pmol of *Staphylococcus aureus* V8 (SV8) protease at room temperature for 22 hr. The resulting peptides are purified by reversed-phase high-performance liquid chromatography (HPLC) with an HP 1090 (Hewlett Packard) fitted with an octyldodecylsilane, 0.46 × 25 cm column (Vydac) equilibrated with 5 mM ammonium formate (pH 6.5) and developed with a linear gradient of 1%/min of acetonitrile in 5 mM ammonium formate (pH 6.5). Peptides are also obtained from digestion of guanylyl cyclase with SV8 according to the method of Cleveland *et al.*[20] Peptides are eluted from gel slices and prepared for automated microsequencing. Two sequences (ATLXYNPTVINLDRGR, peptide 1, and LKTSINMAXVYIF, peptide 2) obtained from the microsequencing are used to synthesize oligonucleotide probes (see below). Seven HPLC-purified peptides are sequenced by automated Edman degradation on an Applied Biosystems 470A gas-phase sequenator with an on-line 120A phenylthiohydantoin (PTH) analyzer. The SV8-generated peptides purified by HPLC are first dried in a Speed-Vac and then dissolved in 40% 2-propanol, 0.1% trifluoroacetic acid (TFA) before subjecting 50% of the sample to sequencing. This is done to minimize the effects of ammonium formate on Edman chemistry and subsequent analysis by HPLC. About 75 pmol of the intact protein is used to obtain the amino-terminal sequences. Amino-terminal sequencing of the intact protein yields a sequence ATLHYNPTVINLDRGR, which corresponds to peptide 1. This sequence is helpful in deciding the cleavage site of the signal peptide.

[18] P. P. van Jaarsveld, B. J. van Der Walt, and C. H. LeRoux, *Anal. Biochem.* **75,** 363 (1976).

[19] W. Wray, T. Boulikus, V. P. Wray, and P. Hancock, *Anal. Biochem.* **118,** 197 (1981).

[20] D. W. Cleveland, S. G. Fisher, M. W. Kirschner, and U. K. Laemmli, *J. Biol. Chem.* **252,** 1102 (1977).

Step 3: Preparation of Antibodies. A synthetic peptide corresponding to the amino-terminal sequence YNPTVINLDRGR was synthesized by Peninsula Laboratories (Palo Alto, CA) and conjugated to bovine serum albumin with the bifunctional cross-linking reagent disuccinimidyl suberate. Antiserum was raised in rabbits at East Acres (Southbridge, MA). Antibodies against the intact M_r 160,000 protein are obtained by immunization of rabbits with electroeluted protein from polyacrylamide gel slices. All initial immunizations are done with Freund's complete adjuvant, and booster injections use conjugated peptides or protein in Freund's incomplete adjuvant.

Step 4: Western Blot Analyses. The purity of the isolated protein is tested by SDS–PAGE and Western blot analysis. It has been demonstrated that the egg peptide resact causes a species-specific dephosphorylation of the enzyme in intact spermatozoa, which results in a reduction in the apparent molecular weight on SDS–PAGE.[21] Control or resact-treated plasma membrane fractions are prepared from *A. punctulata* spermatozoa according to the method of Bentley *et al.*[22] and analyzed on immunoblots. When the antiserum is used to probe blots of untreated or resact-treated sea urchin spermatozoa membrane extracts, the antibodies react with both forms of the enzyme. But with resact-treated membranes, the antiserum reacts only with the dephosphorylated low molecular weight form. These data, then, demonstrate that the purified protein for which the partial amino acid sequence was obtained is guanylyl cyclase.

Step 5: Oligonucleotide Probes. Two 144-fold degenerate antisense oligodeoxyribonucleic acid probes based on the amino acid sequences and codon usage frequencies[23] are synthesized. One probe (probe A) [5′-(A/G)CGGCC(A/G)CG(A/G)TC(C/G)AGGTTGATCAC(T/G/A)GT(A/G/T)GGGTTGTA-3′] is derived from the amino-terminal sequence YNPTVINLDRGR, and a second probe (probe B) [5′-GAAGATGTA(C/G)ACTTT(A/G)GCCAT(A/G)TTGAT(A/G)GA(A/G/T)GTCTT-3′] is derived from the amino acid sequence LKTSINMAXVYIF (where X stands for an ambiguous amino acid).

Step 6: Isolation of Poly(A)$^+$ RNA. Poly(A)$^+$ RNA from *A. punctulata* testes is prepared by the guanidinium isothiocyanate and lithium chloride procedure.[24] Eight grams of testes is homogenized in 40 ml of 5 *M* guanid-

[21] N. Suzuki, H. Shimomura, E. W. Radany, C. S. Ramarao, G. E. Ward, J. K. Bentley, and D. L. Garbers, *J. Biol. Chem.* **259,** 14874 (1984).

[22] J. K. Bentley, D. J. Tubb, and D. L. Garbers, *J. Biol. Chem.* **261,** 14859 (1986).

[23] R. Grantham, C. Gautier, M. Gouy, M. Jacobzone, and R. Mercier, *Nucleic Acids Res.* **9,** 43 (1981).

[24] G. Cathala, J.-F. Savourt, B. Mandez, B. L. West, M. Karin, J. A. Martial, and J. D. Baxter, *DNA* **2,** 329 (1983).

ium isothiocyanate in an Omnimix. Cell debris is removed by centrifugation at 3000 rpm for 5 min. Five volumes of 4 *M* lithium chloride solution is added to the supernatant, and, after incubation overnight at 4°, RNA is isolated by centrifugation at 13,000 rpm for 90 min. The pellet containing RNA is washed with 3 *M* lithium chloride and solubilized in 10 m*M* Tris, 1 m*M* EDTA. Poly(A)$^+$ RNA is purified by two passages over an oligo(dT)-Sepharose column (Collaborative Research, Inc., Bedford, MA).

Step 7: cDNA Cloning and Sequencing. A random-primed cDNA library is constructed in bacteriophage λ-ZAP (Stratagene, San Diego, CA). The cDNA is synthesized as described by Gubler and Hoffman,[25] using an Amersham (Arlington Heights, IL) cDNA synthesis kit and random hexanucleotides as primers for first-strand synthesis. Approximately 200,000 independent clones are amplified in *E. coli* strain, BB-4. Four hundred thousand clones from the amplified library are hybridized to ^{32}P-end-labeled oligonucleotide probes [1×10^6 counts per minute (cpm)/ml] in 20% formamide, 5× SSC (20× SSC is 3 *M* NaCl, 0.3 *M* sodium citrate, pH 7.4), 0.1% sodium pyrophosphate, 0.1% SDS, 100 μg/ml denatured salmon sperm DNA, and 5× Denhardt's solution for 12 hr at 42°. After a second round of screening for plaque purification, cloned cDNAs are rescued as Bluescript plasmids, utilizing the automatic excision process of the λ-ZAP cloning system.[26] To identify regions in the cloned cDNAs responsible for hybridizations to probe A and B, *Sau*3AI fragments are shotgun cloned into M13mp18 (*Bam*HI digested), screened by filter hybridization, and sequenced to determine regions of identity with either probe. The nucleotide sequence is determined by the Sanger dideoxy method.[27] In some cases dITP is substituted for dGTP where G–C compression is encountered.

Nucleotide analysis shows that three clones cover 3,444 base pairs (bp) and, based on the assignment of the initiation and termination codons, contain an open reading frame of 986 codons and 5′- and 3′-untranslated regions. The deduced amino acid sequence predicts guanylyl cyclase to be an intrinsic membrane protein of 986 amino acids with a 21 amino acid signal sequence. A single transmembrane domain of 28 amino acids separates the protein into a 478-residue amino-terminal domain and a 459-residue carboxy-terminal domain. The mature protein is showed to be composed of 965 amino acids with a calculated molecular weight

[25] L. Gubler and B. J. Hoffman, *Gene* **25,** 263 (1983).

[26] J. M. Short, J. M. Fernandez, J. A. Sorge, W. D. Huse, *Nucleic Acids Res.* **16,** 7583 (1988).

[27] F. Sanger, S. Nicklen, and A. Coulson, *Proc. Natl. Acad. Sci. U.S.A.* **74,** 5463 (1977).

of 106,150. The size difference between the purified enzyme (150,000–160,000) and the predicted amino acid sequence is possibly due to N-linked and/or O-linked glycosylation or to the phosphorylation state of the native protein. The complete nucleotide sequence has been published.[14]

Amino acid sequence comparison with published sequences shows distinct sequence similarities between sea urchin guanylyl cyclase and the highly conserved region of protein kinases. It also shows sequence similarities between the extracellular domain of the ANP-C receptor[28] and the 70-kDa soluble guanylyl cyclase from bovine and rat lung.[29,30] We were unable to functionally express this protein in transfection of COS-7 cells, even though the protein was synthesized as demonstrated by immunoprecipitation. However, an antipeptide antibody raised against the amino-terminal sequence could specifically precipitate guanylyl cyclase activity from intact sea urchin spermatozoa. These data confirm that we have isolated the membrane form of guanylyl cyclase from *A. punctulata* spermatozoa. By use of this cDNA probe, guanylyl cyclase from another sea urchin, *Strongylocentrotus purpuratus*, has been cloned.[31]

Cloning of Mammalian Plasma Membrane Guanylyl Cyclase

Recently cloned cDNAs encoding two different forms of mammalian membrane guanylyl cyclases, arbitrarily called GC-A and GC-B, from human and rat, have been identified.

Isolation of GC-A cDNA clones

Step 1: Choice of Probes. Based on the assumption that the catalytic domain of guanylyl cyclase would be highly conserved across species, coupled with the observation that antibody against sea urchin guanylyl cyclase immunoprecipitates rat membrane guanylyl cyclase,[32] sea urchin cDNA probes[14] are used to screen human cDNA libraries. As described earlier, sea urchin sperm guanylyl cyclase is a protein with a single trans-

[28] F. Fuller, J. G. Porte, A. E. Arfsten, J. Miller, J. W. Schilling, R. M. Scarborough, J. A. Lewicki, and B. Schenk, *J. Biol. Chem.* **263,** 9395 (1988).

[29] D. Koesling, J. Hertz, H. Gausepohl, F. Nirromand, K.-D. Hirsch, A. Muelsch, E. Boehme, G. Schultz, and R. Frank, *FEBS Lett.* **239,** 29 (1988).

[30] M. Nakane, S. Saheki, T. Kuno, K. Ishii, and F. Murad, *Biochem. Biophys. Res. Commun.* **157,** 1139 (1988).

[31] D. S. Thorpe and D. L. Garbers, *J. Biol. Chem.* **264,** 6545 (1989).

[32] D. L. Garbers, *J. Biol. Chem.* **253,** 1898 (1976).

membrane domain. The cyclase catalytic domain should reside in the cytoplasmic portion of the protein or the 3′-end of the cDNA. Approximately 10^6 recombinant clones from an oligo(dT)-primed human kidney cDNA library in λgt10 are screened under conditions of low stringency hybridization in a solution containing 20% formamide, 5× SSC, 0.1% sodium pyrophosphate, 50 m*M* sodium phosphate (pH 6.5), 5× Denhardt's solution, 50 μg/ml denatured salmon sperm DNA, 10% dextran sulfate, and 10^6 cpm/ml of the 3′-end of the *A. punctulata* guanylyl cyclase cDNA probe. The filters are hybridized for 18 hr at 42° and washed at 52° in 5× SSC, 0.1% SDS. The filters are subjected to autoradiography for 5 hr at −80°. After a second round of screening to obtain pure plaques, the *Eco*RI insert fragments are subcloned into M13mp vectors for dideoxynucleotide sequencing.[27]

Three clones were found to be homologous to the sea urchin probe. Since none of the clones is long enough to represent a full-length cDNA, one of the clones is used to screen a human placental cDNA library, where the longest clone is found to be 3 kb. This cDNA clone also does not contain the entire coding sequence, in that it appeared to lack sequences at the 5′-end. Full-length clones are difficult to obtain, probably due to a (G–C)-rich region in the 5′-end. The inability to obtain a full-length clone could have been due to the failure of avian myeloblastosis virus (AMV) reverse transcriptase to read through the secondary structure of the mRNA. Further 5′-ends of the sequence are obtained by specifically primed and homopolymer tailed cDNA in a modified polymerase chain reaction procedure (PCR). Cloned PCR fragments are screened by oligonucleotide hybridization and sequenced. In this way the entire coding region of human guanylyl cyclase is obtained.

While the screening of human cDNA libraries was in progress, an oligo(dT)-primed rat brain cDNA library, obtained from Dr. Hemin Chin, NIH,[16] was screened with a partial length human cDNA probe. It is worth mentioning here that this cDNA library was created using murine Moloney leukemia virus (MMLV) reverse transcriptase (BRL, Gaithersburg, MD), and was size selected for cDNA inserts of more than 2.5 kb. Several clones are obtained after high-stringency hybridization with the partial length human cDNA clones: insert sizes ranging from 2 to 4 kb are obtained. The longest clone is sequenced, and a complete coding region for this *A. punctulata* guanylyl cyclase and human placenta related protein is obtained. The cDNA clones from rat and human are named GC-A.

Both human and rat GC-A has a 32-residue signal sequence followed by a 441-residue extracellular domain homologous to the 60-kDa ANP-C receptor, a 21-residue transmembrane domain, and a 568-residue cytoplasmic domain with homology to the protein kinase family and to the 70-kDa soluble guanylyl cyclase.

Step 2: Expression of cDNA Clones. To confirm that the cDNA clones from human placenta and rat brain indeed contain guanylyl cyclase activity, COS-7 cells are transfected with mammalian expression vectors containing either human or rat cDNA. The guanylyl cyclase activity in the membrane fractions, ANP binding, and ANP-stimulated cGMP elevation are measured. The detailed experimental protocols have been published.[16,17] The ligand-binding properties of the receptor encoded by human and rat GC-A expression vectors are determined by equilibrium binding of ^{125}I-labeled ANP in the presence of ANP or its truncated analog atriopeptin I (AP I). Competition for ^{125}I-labeled ANP binding with unlabeled ANP gives an EC_{50} in the range of 3 n*M*, while AP I has an EC_{50} in the range of 300 n*M*. Guanylyl cyclase activities in the membrane fractions are 10- to 20-fold higher than the control transfected cells. When intact COS-7 cells transfected with guanylyl cyclase cDNA from either rat or human are incubated with ANP, cellular cGMP concentrations are markedly elevated. In addition, ^{125}I-labeled ANP could be specifically cross-linked in the presence of disuccinimidyl suberate to an expressed protein of apparent molecular weight equal to 130,000. These data established the identity of the cloned cDNAs as a guanylyl cyclase containing cyclase activity as well as ANP-binding activity.

Isolation of GC-B cDNAs

The discovery of GC–ANP receptor complexes raises the question of whether there are other members of the family. A rat brain cDNA library used to isolate GC-A is screened with the 3′ 2.7-kb *Eco*RI restriction fragment of GC-A as probe. Low-stringency hybridization and washing conditions are used to detect GC-A homologous clones. Hybond-N nylon filters are hybridized at 42° in a solution containing 20% formamide, 5× SSC, 10% dextran sulfate, 50 m*M* sodium phosphate (pH 6.5), 5× Denhardt's solution, 0.1% sodium pyrophosphate, 100 μg/ml denatured salmon sperm DNA, and 5×10^5 cpm/ml probe for 18 hr. The filters are washed twice at room temperature and twice at 55°. Following autoradiography, the filters are stripped according to the manufacturer's protocol and hybridized again to the 5′ 1.2-kb *Eco*RI fragment of the GC-A probe. DNA is prepared from the clones which hybridize only to 3′ probe. Several clones greater than 3.4 kb in size without any internal *Eco*RI site are used for subsequent analysis (GC-A has one internal *Eco*RI site).

The amino acid sequence derived from the nucleotide sequences predicts a protein of 1047 amino acids containing a 22-residue signal peptide and 21-amino acid single transmembrane domain. Cleavage of the signal peptide would result in a mature protein of M_r 115,085. Overall, GC-A and GC-B are 62% identical, but the intracellular region containing the kinase and cyclase domains appear to be more highly conserved than the extracellular domain (78 versus 43% identity).[33] COS-7 cells transfected with GC-B are able to be stimulated by both ANP and BNP to produce cGMP. However, cGMP concentrations in cells transfected with GC-A are one-half maximally elevated at 3, 25, and 65 n*M* ANP, BNP, and AP-1, respectively, while 25 μM, 6 μM, and greater than 100 m*M* ANP, BNP, and AP-1, respectively, are required to one-half maximally stimulate GC-B. A homologous GC-B clone from human cDNA library has also been isolated,[33] which has 98% identity with the rat GC-B and exhibits similar enzyme kinetics.

Comments

Sea urchin spermatozoa provided an excellent source of the membrane form of guanylyl cyclase since they are one of the richest sources of the enzyme. Most or all of the guanylyl cyclase activity resides in the membrane fraction. Purification of the sea urchin sperm guanylyl cyclase based on its response to treatment with the egg peptide, resact, was instrumental in the cloning of the *A. punctulata* guanylyl cyclase cDNA. Nucleotide sequence homology between the sea urchin *A. punctulata* cytoplasmic domain and the human guanylyl cyclase was conserved significantly to allow the cloning of the human guanylyl cyclase using sea urchin cDNA as a probe. Subsequently, rat brain guanylyl cyclase was isolated by use of the human cDNA probe. In cloning of the membrane form of GC from mammalian cDNA libraries, we have noticed that use of MMLV reverse transcriptase is more efficient in generating full-length cDNA clones than the AMV reverse transcriptase. Since the cloning of the rat and human GC-A and GC-B clones, we have isolated full-length cDNA clones from a mouse Leydig cell cDNA library created by using MMLV reverse transcriptase,[34] which seems to be the enzyme of choice for generating full-length cDNAs coding for the membrane form of guanylyl cyclase. Alternate methods which keep mRNA in a denatured state during the first strand synthesis also may be used.

We have described the molecular cloning of different forms of mem-

[33] M.-S. Chang, D. G. Lowe, M. Lewis, R. Hellmiss, E. Chen, and D. V. Goeddel, *Nature* (*London*) **341,** 68 (1989).

[34] K. N. Pandey and S. Singh, *J. Biol. Chem.* **265,** 12342 (1990).

brane guanylyl cyclases from sea urchin, human, and rat. These studies have confirmed the observations that membrane guanylyl cyclase is a receptor for egg or natriuretic peptides. ANP and BNP can stimulate cGMP production by specifically binding to the plasma membrane-associated guanylyl cyclase.

[40] Radiation-Inactivation Analysis of Multidomain Proteins: The Case of Particulate Guanylyl Cyclase

By MICHEL POTIER, CÉLINE HUOT, CAROLINE KOCH, PAVEL HAMET, and JOHANNE TREMBLAY

Introduction

Particulate guanylyl cyclase (EC 4.6.1.2, guanylate cyclase), one of two enzyme types synthesizing cyclic GMP (cGMP) (for review, see Refs. 1 and 2), is directly activated by a few peptides, including speract, resact,[3] atrial natriuretic peptide,[4,5] and brain natriuretic peptide (BNP).[6] It is sensitive to *Escherichia coli* toxin in the intestine.[7] This enzyme has been mainly studied in relation to atrial natriuretic peptide (ANP) action. It was initially demonstrated that ANP, either injected into rats[8] or incubated with different tissues or cultured cells,[5,8,9] produces huge increases in cGMP levels. This nucleotide represents a second messenger mediating some of the biological functions of ANP, such as diuresis and vasodilata-

[1] J. Tremblay, R. Gerzer, and P. Hamet, *in* "Advances in Second Messenger and Phosphoprotein Research" (P. Greengard and G. A. Robison, eds.), Vol. 22, p. 319. Raven, New York, 1988.
[2] F. Murad, *J. Clin. Invest.* **78,** 1 (1986).
[3] J. K. Bentley, D. J. Tubb, and D. L. Garbers, *J. Biol. Chem.* **261,** 14859 (1986).
[4] J. Tremblay, R. Gerzer, P. Vinay, S. C. Pang, R. Beliveau, and P. Hamet, *FEBS Lett.* **181,** 17 (1985).
[5] S. A. Waldman, R. M. Rapoport, and F. Murad, *J. Biol. Chem.* **259,** 14332 (1984).
[6] D.-L. Song, K. P. Kohse, and F. Murad, *FEBS Lett.* **232,** 125 (1988).
[7] M. Field, L. H. Graf, W. J. Laird, and P. L. Smith, *Proc. Natl. Acad. Sci. U.S.A.* **75,** 2800 (1978).
[8] P. Hamet, J. Tremblay, S. C. Pang, R. Garcia, G. Thibault, J. Gutkowska, M. Cantin, and J. Genest, *Biochem. Biophys. Res. Commun.* **123,** 515 (1984).
[9] Y. Hirata, M. Tomita, H. Yoshimi, and M. Ikeda, *Biochem. Biophys. Res. Commun.* **125,** 562 (1984).

tion.[10] It was subsequently shown that ANP functions correlate with the distribution of particulate guanylyl cyclase rather than with that of the soluble form of the enzyme,[4] and a direct effect of ANP on cGMP levels was observed in washed membranes of several tissues.[5,11,12] Thus, ANP elevates cGMP levels by exclusively stimulating particulate guanylyl cyclase, which is the predominant cyclase form in ANP target tissues.

Heterogeneity of the ANP receptor system has been demonstrated *in vivo*[13] as well as *in vitro*.[14] One of the ANP receptor types is a 130-kDa protein composed of two 65-kDa disulfide-linked subunits.[15–17] This "C receptor" has been so named because of its potential clearance role.[18] Copurification of the other ANP receptor type with particulate guanylyl cyclase activity has been accomplished by several groups.[19–24] Their results have revealed that one ANP receptor type is coupled to cGMP formation and could be responsible for the biological effects of the peptide. Recently, the isolation, sequencing, and expression of a complete cDNA clone coding for the membrane guanylyl cyclase of rat brain clearly showed that an ANP receptor domain is present on guanylyl cyclase.[25] The ANP

[10] P. Hamet, J. Tremblay, S. C. Pang, R. Skuherska, E. L. Schiffrin, R. Garcia, M. Cantin, J. Genest, R. Palmour, F. R. Ervin, S. Martin, and R. Goldwater, *J. Hypertens.* **4** (Suppl. 2), S49 (1986).

[11] R. J. Winquist, E. P. Faison, S. A. Waldman, K. Schwartz, F. Murad, and R. M. Rapoport, *Proc. Natl. Acad. Sci. U.S.A.* **81,** 7661 (1984).

[12] J. Tremblay, R. Gerzer, S. C. Pang, M. Cantin, J. Genest, and P. Hamet, *FEBS Lett.* **194,** 210 (1986).

[13] R. C. Willenbrock, J. Tremblay, R. Garcia, and P. Hamet, *J. Clin. Invest.* **83,** 482 (1989).

[14] K. N. Pandey, T. Inagami, and K. S. Misono, *Biochem. Biophys. Res. Commun.* **147,** 1146 (1987).

[15] D. B. Schenk, M. N. Phelps, J. G. Porter, F. Fuller, B. Cordell, and J. A. Lewicki, *Proc. Natl. Acad. Sci. U.S.A.* **84,** 1521 (1987).

[16] M. Shimonaka, T. Saheki, H. Hagiwara, M. Ishido, A. Nogi, T. Fujita, K.-I. Wakita, Y. Inada, J. Kondo, and S. Hirose, *J. Biol. Chem.* **262,** 5510 (1987).

[17] F. Fuller, J. G. Porter, B. E. Arfsten, J. Miller, J. W. Schilling, R. M. Scarborough, J. A. Lewicki, and D. B. Schenk, *J. Biol. Chem.* **269,** 9395 (1988).

[18] T. Maack, M. Suzuki, F. A. Almeida, D. Nussenzveig, R. M. Scarborough, G. A. McEnroe, and J. A. Lewicki, *Science* **238,** 675 (1987).

[19] T. Kuno, J. W. Andresen, Y. Kamisaki, S. A. Waldman, L. Y. Chang, S. Saheki, D. C. Leitman, M. Nakane, and F. Murad, *J. Biol. Chem.* **261,** 5817 (1986).

[20] T. Kuno, Y. Kamisaki, S. A. Waldman, J. Gariepy, G. Schoolnik, and F. Murad, *J. Biol. Chem.* **261,** 1470 (1986).

[21] A. K. Paul, R. B. Marala, R. K. Jaiswal, and R. K. Sharma, *Science* **235,** 1224 (1987).

[22] R. Takayanagi, R. M. Snajdar, T. Imada, M. Tamura, K. N. Pandey, K. S. Misono, and T. Inagami, *Biochem. Biophys. Res. Commun.* **144,** 244 (1987).

[23] R. Takayanagi, T. Inagami, R. M. Snajdar, T. Imada, M. Tamura, and K. S. Misono, *J. Biol. Chem.* **262,** 12104 (1987).

[24] S. Meloche, H. Ong, M. Cantin, and A. De Lean, *J. Biol. Chem.* **261,** 1525 (1986).

[25] M. Chinkers, D. L. Garbers, M. S. Chang, D. G. Lowe, H. Chin, D. V. Goeddel, and S. Schulz, *Nature (London)* **338,** 78 (1989).

receptor–guanylyl cyclase molecule is a transmembrane protein that contains an extracellular ANP-binding domain and an intracellular guanylyl cyclase catalytic domain.[25] Another type of guanylyl cyclase receptor, cloned recently, appears to be more specific for BNP than ANP. However, because of the high BNP concentration needed to stimulate guanylyl cyclase activity, it is possible that other natural endogenous agonists of this receptor exist.

The mechanism of particulate guanylyl cyclase activation is not yet fully understood. The degree of stimulation of the enzyme by ANP decreases dramatically in the course of purification. Although purified guanylyl cyclase possesses catalytic activity and ANP-binding capacity, its catalytic activity becomes refractory to ANP stimulation. These results suggest that coupling components are lost during purification of the protein or that integrity of the membrane environment is required for guanylyl cyclase sensitivity to ANP. In this chapter, we apply the radiation-inactivation method to characterize the domains responsible for guanylyl cyclase and ligand-binding functions. Our results demonstrate that particulate guanylyl cyclase behaves as a multidomain protein.

Basic Principle of Radiation-Inactivation Method

Radiation-inactivation analysis has been used extensively to determine the molecular size of proteins (for review, see Refs. 26–29). The unique feature of this method is that it does not require purified or even soluble protein preparations. It is particularly useful with membrane-bound enzymes, receptors, or transport systems which can be studied in their natural membrane environment.

Target Theory

The target theory is employed to analyze the effects of radiation on proteins.[30] This theory is based on the assumption that a protein molecule hit by ionizing radiation completely loses its activity whereas untouched molecules retain full biological activity.[31,32] The Poisson distribution [Eq.

[26] E. S. Kempner and W. Schlegel, *Anal. Biochem.* **92,** 2 (1979).
[27] C. Y. Jung, *in* "Receptor Biochemistry and Methodology" (J. C. Venter and L. C. Harrison, eds.), Vol. 3, p. 193. Alan R. Liss, New York, 1984.
[28] G. Beauregard, A. Maret, R. Salvayre, and M. Potier, *Methods Biochem. Anal.* **32,** 313 (1987).
[29] J. T. Harmon, T. B. Nielsen, and E. S. Kempner, this series, Vol. 117, p. 65.
[30] D. E. Lea, *in* "Actions of Radiations on Living Cells" (D. E. Lea, ed.), 2nd Ed., p. 69. Cambridge Univ. Press, Cambridge, 1955.
[31] C. J. Steer, E. S. Kempner, and G. Ashwell, *J. Biol. Chem.* **256,** 5851 (1981).
[32] D. J. Fluke, *Radiat. Res.* **51,** 56 (1972).

(1)] describes the probability that a protein is not hit and retains biological activity after irradiation, where μ is directly related to the molecular mass of the protein and D is the dose expressed in megarads (Mrad). In terms of residual biological activity after dose D, relationship (1) is as shown in Eq. (2), where A_o is the activity measured in a nonirradiated sample and A_D the activity measured after the sample has received dose D.

$$P(O) = \exp(-\mu D) \tag{1}$$

$$A_D/A_O = \exp(-\mu D) \tag{2}$$

μ contains two parameters: molecular mass, M, and $Q(t)$, which is the amount of energy deposited per mole of protein inactivated at temperature t [see Eq. (3)]. The dependence of $Q(t)$ on temperature is given by the empirical relationship [Eq. (4)], which is valid between $-200°$ and $+200°$.[33] Equation (4) was established after studying various proteins at temperatures between $-200°$ and $+200°$.[34] The probability of a hit on a protein cannot be dependent on temperature. Thus, physicochemical processes following energy absorption are responsible for the temperature dependence of the radiation inactivation of a protein. At 30°, the amount of energy transferred to the protein per inactivation event is around 65 eV, which is enough to break 15 to 20 covalent bonds.[35] Excited electrons are also generated, and both excited electrons and charges travel along the polypeptide chain of the protein.[36,37] The energy absorbed by the protein will be dissipated by breakage of a covalent bond or by a conformational change in the chain, resulting in an irreversible loss of biological function of the molecule.

$$\mu = M/Q(t) \tag{3}$$

$$\log Q(t) = 13.89 - 0.0028t \tag{4}$$

Experimental Conditions

High-energy ionizing radiation sources delivering particles with energies higher than 1 MeV are used in radiation-inactivation experiments to penetrate whole protein samples up to a few millimeters thick. Linear accelerators or van de Graaff generators are capable of high dose rates of the order of 1 to 2 Mrad/min, but these sources are not widely accessible. In our laboratory, we routinely use a Gammacell 220 irradiator (Atomic

[33] G. Beauregard and M. Potier, *Anal. Biochem.* **150,** 117 (1985).
[34] E. S. Kempner and H. T. Haigler, *J. Biol. Chem.* **257,** 13297 (1982).
[35] G. R. Kepner and R. I. Macey, *Biochim. Biophys. Acta* **163,** 188 (1968).
[36] F. Patten and W. Gordy, *Proc. Natl. Acad. Sci. U.S.A.* **46,** 1137 (1960).
[37] A. Muller, *in* "Biological Effects of Ionizing Radiation at the Molecular Level," p. 61. International Atomic Energy Agency, Vienna, 1962.

Energy of Canada Ltd., Ottawa), which contains a cylindrical ^{60}Co source delivering γ rays at energies of 1.17 and 1.33 MeV. With γ irradiators, dose rates of 0.1 to 3 Mrad/hr are obtained.[38] Dosimetry is precise inside the cylindrical irradiation chamber, and the equipment is easy to operate, being generally more accessible than electron accelerators.[39]

Experimental conditions must be chosen to ensure that the fundamental assumptions of the target theory are respected, that is, only direct effects of a radiation hit on the protein should be observed, and secondary inactivation events caused by the reaction of protein with free radicals generated from water or oxygen must be avoided.[26] Thus, irradiation is carried out with lyophilized preparations under a nitrogen atmosphere or with frozen solutions to impair the mobility of free radicals.

Dosimetry is established with ferrous sulfate solutions (Fricke dosimeter),[40] thermoluminescent dosimeters,[41] Perspex dosimeters,[42] or with standard enzymes of known D_{37}.[43–45] In some laboratories, internal standards are added to protein preparations so that any variations in the radiation beam can be corrected.[46] This is particularly useful with electron accelerators, but this method could be laborious and standard proteins should be well characterized. Gamma irradiators deliver constant and uniform radiation fields so the equipment needs only to be calibrated periodically.

Molecular Size Determination

The fundamental equation of radiation inactivation of a protein is shown in Eq. (5). $Q(t)$, which represents the amount of energy absorbed per mole of protein inactivated, is determined empirically by using a series of proteins of known molecular mass. Kepner and Macey[35] proposed the empirical relationship shown in Eq. (6), which is employed by most laboratories when irradiation is undertaken at around 30°. Since radiation-

[38] G. Beauregard and M. Potier, *in* "Receptor Biochemistry and Methodology" (J. C. Venter and C. Y. Jung, eds.), Vol. 10, p. 61. Alan R. Liss, Inc., New York, 1987.

[39] G. Beauregard, S. Giroux, and M. Potier, *Anal. Biochem.* **132,** 362 (1983).

[40] H. Fricke and S. Morse, *Am. J. Roentgenol.* **18,** 426 (1927).

[41] H. Fricke and E. J. Hart, *in* "Radiation Dosimetry" (F. H. Attix and W. C. Roesch, eds.), 2nd Ed., Vol. 2, Instrumentation, p. 167. Academic Press, New York, 1966.

[42] R. J. Berry and C. H. Marshall, *Phys. Med. Biol.* **14,** 585 (1969).

[43] G. Beauregard, M. Potier, and B. D. Roufogalis, *Biochem. Biophys. Res. Commun.* **96,** 1290 (1980).

[44] G. Beauregard and M. Potier, *Anal. Biochem.* **122,** 379 (1982).

[45] M. M. Lo, E. A. Barnard, and J. O. Dolly, *Biochemistry* **21,** 2210 (1982).

[46] S. Schulz, S. Singh, R. A. Bellet, G. Singh, D. J. Tubb, H. Chin, and D. L. Garbers, *Cell (Cambridge, Mass.)* **58,** 1155 (1989).

inactivation experiments are often done at different temperatures and $Q(t)$ can be expressed as a function of temperature by Eq. (4), the general Eq. (7) for the molecular size determination[33] has been used between $-200°$ and $+200°$.

$$M = Q(t)/D_{37,t} \tag{5}$$
$$M = 6.4 \times 10^5/D_{37} \tag{6}$$
$$\log M = 5.89 - \log D_{37,t} - 0.0028t \tag{7}$$

The "molecular mass" of a protein includes bound sugars and lipids. Therefore, this terminology is not satisfactory for values obtained by radiation inactivation since with glycoproteins and membrane proteins only the core polypeptide chain is seen by the ionizing radiation. The energy absorbed by sugars or lipids is not transferred to the protein core. The term radiation-inactivation size (RIS) was therefore proposed when the loss of biological activity is monitored as a function of dose.[28] This value is defined as the mass of 1 mol of protein inactivated by a hit. The term target size (TS) is reserved for size determined by monitoring the loss of a native protein band after separation from radiolysis fragments by sodium dodecyl sulfate (SDS)–polyacrylamide gel electrophoresis. This value is defined as the mass of 1 mol of protein whose polypeptide chain structure (subunit) has been broken by a hit. Equation (7) is thus re-written as Eq. (8).

$$\log(\text{RIS or TS}) = 5.89 - \log D_{37,t} - 0.0028t \tag{8}$$

Radiation Inactivation of Oligomeric Proteins

A great advantage of the radiation-inactivation method is that, with oligomeric proteins, it is possible to establish structure–function relationships in terms of protomer assembly necessary for a given biological function. Beauregard *et al.*[28] described what information can be obtained from radiation-inactivation experiments of oligomers.

When oligomeric proteins are irradiated, the RIS and TS could correspond to either the monomer or oligomer. Three categories of oligomers exist. In the first category, the RIS and TS are identical and yield the monomer value. Among all the oligomeric proteins studied by irradiation, this class is by far the most common.[26] This means that the protomer is active by itself and does not need interaction with other protomers inside the oligomer to express activity. Thus, the protomer itself is the functional unit of the protein, that is, the minimal assembly of structure required for activity.[31] In the second category of oligomers, the RIS corresponds to oligomer size and the TS to the monomer. One monomer is fragmented by a hit, but, as a consequence, all other associated monomers lose their biological activity. It is then concluded that the integrity of each protomer in the oligomer is necessary for the expression of biological activity. In this case, the whole oligomer constitutes the functional unit. The third and last category comprises all oligomers where the RIS and TS both correspond to the whole oligomer. In this situation, the energy absorbed by a hit on one monomer is transferred to all other monomers in the oligomer, causing their fragmentation. The RIS then corresponds to all fragmented subunits which, as a consequence of breaks in the chain, lose biological activity. However, recent studies have revealed that energy transfer between the subunits of an oligomer is probably reserved to a small number of proteins.[47] The determination of TS values of several well-characterized standard oligomeric proteins has allowed tests of energy transfer. The calibration curve obtained by plotting experimentally determined $1/D_{37}$ values as a function of known molecular mass gives a better fit when the monomer mass is used for the plot rather than the whole oligomer (Fig. 1).[47] Therefore, no simultaneous fragmentation of several monomers from a single hit was noted.

Radiation Inactivation of Particulate Guanylyl Cyclase

Radiation inactivation of particulate guanylyl cyclase was done in membranes from bovine adrenal cortex. After several washes in 50 m*M* triethanolamine buffer, the membranes are rapidly frozen in liquid nitrogen and kept in dry ice before and during irradiation. The frozen samples are irradiated with incremental doses of γ rays in a Gammacell model 220 irradiator (Atomic Energy of Canada, Ltd.).[38] They are rapidly thawed for the guanylyl cyclase assay, which is performed at 37° as described previously.[12] The cGMP produced is measured by radioimmunoassay (RIA).[48]

The radiation-inactivation curves of particulate guanylyl cyclase activity are shown in Fig. 2. The three curves represent different conditions of guanylyl cyclase treatment: (1) untreated membranes; (2) membranes preincubated with a fully activating dose of ANP; and (3) Triton X-100-solubilized membranes. For untreated membranes, the relationship between log residual guanylyl cyclase activity (%) versus radiation dose has a downward concave shape (Fig. 2). When the membranes were preincubated with 5×10^{-7} *M* ANP for 10 min at 37° before rapid freezing

[47] M. le Maire, L. Thauvette, B. de Foresta, A. Viel, G. Beauregard, and M. Potier, *Biochem. J.* **267,** 431 (1990).

[48] R. A. Richman, G. S. Kopf, P. Hamet, R. A. Johnson, *J. Cyclic Nucleotide Res.* **6,** 461 (1980).

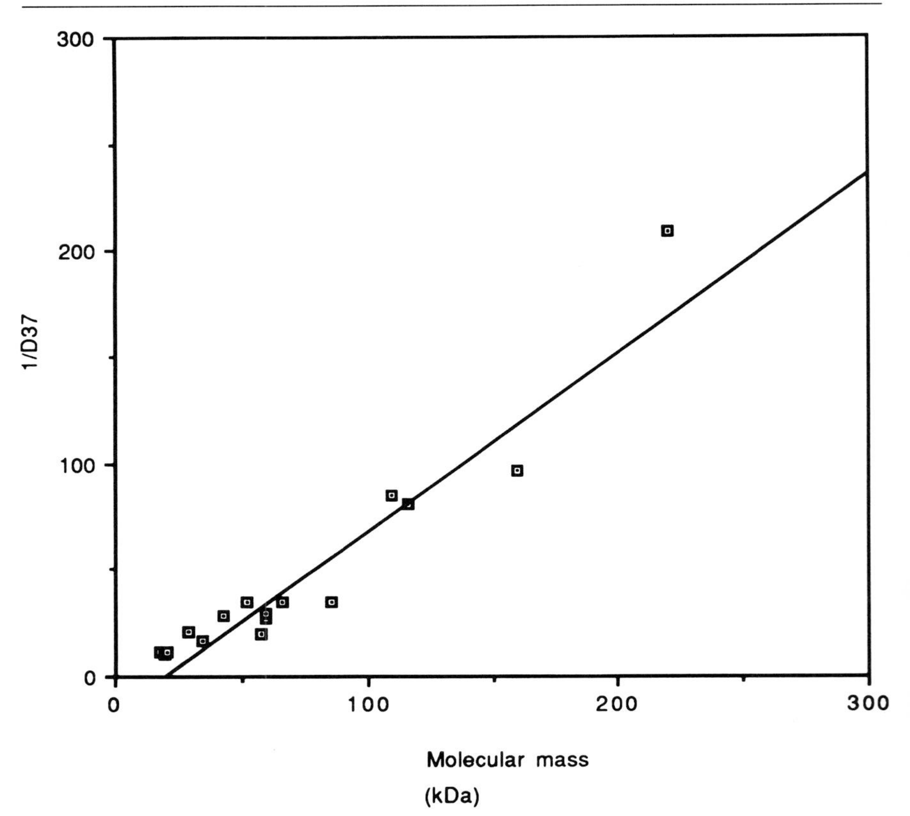

FIG. 1. Calibration curve for target size values of standard purified proteins. The decrease of band intensity corresponding to the native protein was followed as a function of dose to determine D_{37} values. The graph shows a good relationship of monomer molecular mass with experimental D_{37} values between 17.5 and 220 kDa. (Data are from le Maire *et al.*[47])

and irradiation, a linear monoexponential loss of guanylyl cyclase activity occurred. The same phenomenon was observed after membrane solubilization with Triton X-100.

In the models described by Verkman *et al.*,[49] a concave downward log(activity) versus radiation dose curve is expected when the inactive state of the enzyme exists for the oligomeric species that predominates over the monomeric fully active state. The straight line represents the monoexponential inactivation of a monomer. The transition to a linear log(activity) versus dose relation after addition of the agonist (ANP) sug-

[49] A. S. Verkman, K. L. Skorecki, and D. A. Ausiello, *Am. J. Physiol.* **250,** C103 (1986).

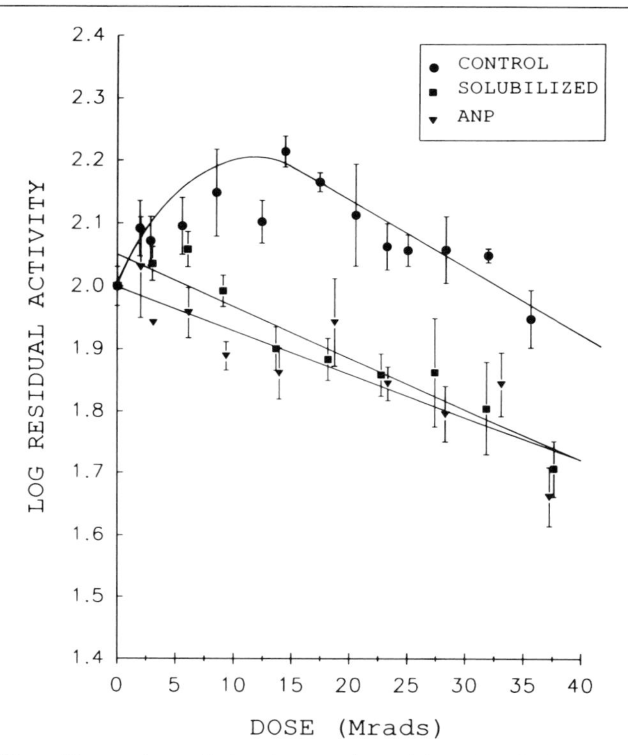

FIG. 2. Effect of increasing radiation dose on the activity of particulate guanylyl cyclase from bovine adrenal cortex membranes. Washed membranes were exposed to increasing radiation doses in the frozen state. Samples were rapidly thawed, and residual guanylyl cyclase activity was measured as previously described.[4] Data were plotted as a semilogarithmic function of residual activity (%) versus radiation dose. Membrane treatment before irradiation was as follows: for control, 10 min at 37° with 0.1 *N* acetic acid before rapid freezing; for ANP treatment, incubation for 10 min at 37° with 5×10^{-7} *M* ANP before freezing; and for solubilized protein, incubation for 1 hr on ice with 3% Triton X-100 followed by centrifugation at 105,000 *g*. [Data are from J. Tremblay, C. Huot, C. Koch, and M. Potier, submitted (1990).]

gests that ANP acts by promoting dissociation from a relatively inactive oligomer to an active monomeric species. The apparent enzymatic activation observed at low irradiation doses may also indicate that, after irradiation, full reequilibration was not achieved, and the number of active monomers produced by oligomer dissociation exceeds the number of destroyed monomers. Alternatively, under conditions of full equilibration, the initial cyclase activation at low doses could be explained by destruction of inhibi-

TABLE I
RADIATION-INACTIVATION SIZE OF GUANYLYL CYCLASE AND ANP RECEPTOR DOMAINS[a]

	Guanylyl cyclase activity	ANP-binding activity
Basal	32 ± 8	15 ± 4
ANP treated	21 ± 2	21 ± 3
Triton X-100 treated	25 ± 5	23 ± 2

[a] Values are expressed in kilodaltons and represent means ± S.E.M. Data are from J. Tremblay, C. Huot, C. Koch, and M. Potier, submitted (1990).

tory components of a size larger than the catalytic site of the enzyme.[50] Thus, the concave downward log(residual activity) versus dose curve shown in Fig. 2 is compatible with the possibility that guanylyl cyclase exists in native membranes of adrenal cortex as an inactive oligomer. The transition to a linear relation caused by ANP may be due to the dissociation of inactive oligomeric guanylyl cyclase form into active monomers. A similar phenomenon is observed when membrane proteins are dispersed by Triton X-100. The two treatments stimulate guanylyl cyclase activity.

Our results suggest that enzyme activation occurs via a dissociative mechanism. Activation by ANP or Triton X-100 produces a shift in the equilibrium between oligomers and monomers in favor of the active monomeric state. The curves obtained can also provide information about the size of the active form of the enzyme. This value can then be compared with the documented molecular mass of the peptidic chain of guanylyl cyclase, which is of the order of 120 kDa.[25]

As shown in Fig. 2, the three curves become parallel at high radiation doses. The slope of these curves is equal to $-1/D_{37}$. This relationship allows the calculation of the RIS of guanylyl cyclase function according to the previously described relationship [Eq. (8)]. Table I summarizes the RIS values calculated under the three conditions. An RIS of 30 kDa was obtained for guanylyl cyclase function. This is significantly lower than the structural mass of 116 kDa for the peptide backbone derived from the deduced amino acid sequence.[25] Thus, the RIS value of the guanylyl cyclase function is dramatically below the structural size of the protein, suggesting that the catalytic site of guanylyl cyclase behaves as an independent domain in the protein. Since bovine adrenal cortex membranes pos-

[50] M. Potier and S. Giroux, *Biochem. J.* **226,** 797 (1985).

sess almost exclusively the ANP receptor type linked to guanylyl cyclase,[51] the RIS was also calculated for the ^{125}I-ANP-binding function of particulate guanylyl cyclase under basal conditions and following Triton X-100 solubilization. As for the cyclase catalytic function, the ANP receptor function appears to be an independent domain for which an RIS of 20 kDa was determined (Table I). This is an important corollary since if one domain behaves independently, it is expected that other domains on the same protein will also behave independently.

Behavior of Independent Domains in Proteins

Loops in the polypeptide chain link secondary structure elements and larger structural domains in the case of multidomain proteins. Therefore, it appears possible that structural domains could be left intact and biologically active if the absorbed energy is dissipated by a break in a loop instead of inside a domain. In this context, a given functional domain may be relatively independent from the rest of the structure of a protein and may need to be hit directly to lose its activity. There are a few examples of such proteins in the literature, including enzymes and receptors. An unexpected finding in the analysis of both immunoglobulin E (IgE) and its receptor is that, in both cases, a smaller size was obtained by radiation inactivation than from the covalent structure of the molecule.[52] The enzymes β-hexosaminidase (EC 3.2.1.52 β-*N*-acetylhexosaminidase) and β-glucuronidase (EC 3.2.1.31, β-*D*-glucuronide glucuronosohydrolase) present in the receptor preparation were used as internal standards and gave the expected molecular size. The authors concluded that the functions for ligand binding in IgE and its receptors reside in a conformationally stable domain and that the radiation-inactivation method is capable of revealing that aspect.

The RIS for membrane-bound Na^+,K^+-ATPase activity from the pig kidney was 264 kDa, whereas for high-affinity binding of ATP, vanadate, and ouabain (with Mg^{2+}, Na^+, and ATP) it was 145 kDa and for ouabain binding (with Mg^{2+}) it was 97 kDa.[53] These results are consistent with a multidomain protein, each domain exerting a function independently of the others and inactivated independently by a direct hit.

NADH : nitrate reductase (EC 1.6.6.1) is perhaps one of the best documented multidomain proteins.[54,55] The enzyme has three distinct functions,

[51] S. Meloche, H. Ong, and A. De Lean, *J. Biol. Chem.* **262,** 10252 (1987).

[52] C. Fewtrell, E. Kempner, G. Poy, and H. Metzger, *Biochemistry* **20,** 6589 (1981).

[53] P. Ottolenghi and C. Ellory, *J. Biol. Chem.* **258,** 14895 (1983).

[54] L. P. Solomonson and M. J. McCreery, *J. Biol. Chem.* **261,** 806 (1986).

[55] L. P. Solomonson, M. J. McCreery, C. J. Kay, and M. J. Barber, *J. Biol. Chem.* **262,** 8934 (1987).

all on a single polypeptide chain of M_r around 100,000. Determination of the RIS for each function gave the following results: NADH : cytochrome *c* reductase has a RIS of 58 kDa; NADH : ferricyanide reductase, 47 kDa; and reduced methyl viologen : nitrate reductase, 28 kDa. NADH : ferricyanide reductase was found to be associated with a 28-kDa proteolytic fragment, which behaves as an independent domain in the protein and retains its activity following its release from the protein by proteolytic digestion. Relatively large but defined radiolysis fragments have also been obtained using SDS–polyacrylamide gel electrophoresis and immunoblotting. Discrete bands of 90, 75, and 40 kDa elicited with increasing doses of radiation probably correspond to intact domains in the proteins.

Random deposition of energy along the polypeptide chain is likely to generate random radiolysis fragments if fragmentation of the polypeptide chain occurs close to the site of energy deposition. Analysis of irradiated proteins by SDS–polyacrylamide gel electrophoresis shows the progressive formation of defined radiolysis fragments as a function of dose. These results suggest that "fragile sites," which exist along the polypeptide chain of proteins, are preferentially broken after energy deposition.[47] The absorbed energy seems to travel from its deposition site along the chain until it reaches a structure which is likely to be broken.

By amino-terminal sequencing and determination of the apparent M_r of radiolysis fragments of the catalytic subunit of aspartate transcarbamylase (EC 2.1.3.2 aspartate carbamoyltransferase) by SDS–polyacrylamide gel electrophoresis, the "fragile sites" were approximately located along the chain.[47] All five major radiolysis fragments corresponding to apparent M_r values of 30,000, 28,000, 27,000, 25,000 and 21,000 on the SDS gel contained the amino-terminal sequence of the native protein. Therefore, in this protein, all fragile sites seem to be localized near the carboxy-terminal end. Since the three-dimensional structure of aspartate transcarbamylase is known at a 2.6 Å resolution,[56] we have been able to localize the breaks in loops of the polypeptide chain. Such a finding was expected since tension and mobility are greater in loops and β turns. If some fragile loops are also localized between protein domains, it is thus possible that, at a low radiation dose, intact domains are released by radiolysis.

In summary, particulate guanylyl cyclase is a membrane protein with at least two known functions: ANP-binding and cGMP-synthesizing activities. Analysis of the radiation effects on these two functions suggest that they reside in two independent domains. These observations are compatible with the reported amino acid sequence of a complementary DNA clone

[56] K. H. Kim, Z. Pan, R. B. Honzatko, H. Ke, and W. N. Lipscomb, *J. Mol. Biol.* **196,** 853 (1987).

of particulate guanylyl cyclase.[25] The most extensive similarity between the intracellular region of particulate guanylyl cyclase and other proteins is within the carboxy terminus of one of the subunits of soluble guanylyl cyclase. If one assumes a mean *M* of 115 Da per amino acid, the size of the 253 amino acid sequence would be a value close to the RIS calculated in our studies for the catalytic function. An additional functional domain on particulate guanylyl cyclase has high homology to protein kinases.[57] Interestingly, this domain appears to function as a regulatory element of the catalytic domain.[58] The authors suggested that the catalytic domain is normally repressed by the kinaselike domain and that this repression may be removed by ANP. Analysis of the curve shape of the radiation inactivation of guanylyl cyclase function under the different conditions shown in Fig. 2 also suggests that ANP binding or solubilization with a detergent dissociates an inhibitory element from the catalytic domain. The RIS of this inhibitory element is presently being determined in our laboratory.

We conclude that the radiation-inactivation method may be useful to determine the size and potential interactions of functional domains in proteins such as particulate guanylyl cyclase.

Acknowledgments

The authors wish to express thanks to Louise Thauvette for technical assistance, Louise Chevrefils for secretarial skills, and Ovid Da Silva for editing the manuscript. Research was supported by Medical Research Council of Canada Grants No. MA-9299 and No. MA-8769 and the Heart and Stroke Foundation of Canada. J. T. and C. H. received a scholarship and fellowship, respectively, from Fonds de la Recherche en Santé du Québec.

[57] S. Singh, D. G. Lowe, D. S. Thorpe, H. Rodriguez, W. J. Kuang, L. J. Dangott, M. Chinkers, D. V. Goeddel, and D. L. Garbers, *Nature (London)* **334,** 708 (1988).

[58] M. Chinkers and D. L. Garbers, *Science* **245,** 1392 (1989).

[41] Identification of Atrial Natriuretic Peptide Receptors in Cultured Cells

By DALE C. LEITMAN, SCOTT A. WALDMAN, and FERID MURAD

Introduction

Atrial natriuretic peptide (ANP) is a 28 amino acid peptide that was initially isolated from the atria of the heart.[1] Recently, it has become clear that ANP is a member of a distinct class of peptides that are distinguished by the presence of a 17 amino acid disulfide ring structure and by their capacity to elicit a spectrum of physiological effects that include hypotension, natriuresis, and diuresis.[2] Presently, the natriuretic peptide family includes ANP, the 26 amino acid brain natriuretic peptide (BNP),[3] and a recently discovered novel atrial peptide.[4] Animal and human studies indicate that ANP is an important regulator of blood pressure and vascular volume during physiological and pathological states.[5] The development of drugs that mimic ANP to treat renal and cardiovascular disorders is a major goal that requires an understanding of the mechanism of action of ANP.

The physiological effects of ANP are mediated by ANP receptors. Scatchard analysis of radioligand binding studies initially indicated that there was only a single class of ANP receptors. However, two functionally and physically distinct classes of ANP receptors were originally identified by cyclic GMP (cGMP) response and cross-linking studies.[6–8] The presence of two ANP receptor subtypes was definitively demonstrated by the purification[9–11] and molecular cloning of the cDNA for both receptor sub-

[1] T. G. Flynn, M. L. de Bold, and A. J. de Bold, *Biochem. Biophys. Res. Commun.* **117,** 859 (1983).

[2] A. de Bold, *Science* **230,** 767 (1985).

[3] T. Sudoh, K. Kangawa, N. Minamino, and H. Matsuo, *Nature* (*London*) **332,** 78 (1988).

[4] T. G. Flynn, A. Brar, L. Tremblay, I. Sarda, C. Lyons, and D. B. Jennings, *Biochem. Biophys. Res. Commun.* **161,** 830 (1989).

[5] S. A. Atlas, *Recent Prog. Horm. Res.* **42,** 207 (1986).

[6] D. C. Leitman and F. Murad, *Biochim. Biophys. Acta* **885,** 74 (1986).

[7] D. C. Leitman, J. W. Andresen, T. Kuno, Y. Kamasaki, J.-K. Chang, and F. Murad, *J. Biol. Chem.* **261,** 11650 (1986).

[8] D. C. Leitman, C. R. Molina, S. A. Waldman, and F. Murad, *UCLA Symp. Mol. Cell. Biol.* **81,** 39 (1988).

[9] T. Kuno, J. W. Andresen, Y. Kamisaki, S. A. Waldman, L. Y. Chang, S. Saheki, D. C. Leitman, M. Nakane, and F. Murad, *J. Biol. Chem.* **261,** 5817 (1986).

types.[12,13] The ANP-R1 receptor is a 130-kDa glycoprotein that contains two functional domains joined by a single transmembrane region. The extracellular domain functions as an ANP receptor, whereas the intracellular domain contains guanylyl cyclase enzymatic activity that converts GTP to cGMP. The precise mechanism whereby the binding of ANP activates the guanylyl cyclase domain is unknown. However, recent studies suggest that accessory proteins may be regulated by ANP, which then may participate in the activation of guanylyl cyclase.[14] Binding of ANP to the ANP-R1 receptor results in the formation of cGMP, which is the second messenger that mediates most of physiological effects of ANP, such as smooth muscle relaxation, natriuresis, and diuresis.[15]

The second ANP receptor, designated ANP-R2, may exist in the membrane in two forms, a 130-kDa dimer composed of two identical 66-kDa subunits joined by disulfide bonds and a 66-kDa monomer.[10,16] Despite the purification[10] and cloning[12] of the cDNA for this receptor subtype, the physiological role of this receptor has remained a mystery. This receptor subtype is devoid of intrinsic guanylyl cyclase activity[10] and apparently is not functionally coupled to the reported inhibition of adenylyl cyclase activity.[17] It has been shown that the ANP-R2 receptor is associated with the activation of phospholipase C and the increased formation of the inositol phosphates and diacylglycerol.[18] However, no physiological effects of ANP have been shown to be mediated by inositol phosphates or diacylglycerol. It has been proposed by Maack and coworkers that the ANP-R2 receptor functions as clearance receptor that binds, internalizes, and degrades ANP.[19] The physiological basis for a

---

[10] D.Shenk, M. N. Phelps, J. G. Porter, R. M. Scarborough, G. A. McEnroe, and J. A. Lewicki, *Proc. Natl. Acad. Sci. U.S.A.* **84,** 1521 (1987).

[11] R. Takayanagi, L. M. Snajdar. T. Imada, M. Tamura, K. N. Pandey, K. S. Misono, and T. Inagami, *Biochem. Biophys. Res. Commun.* **144,** 244 (1987).

[12] F. Fuller, J. G. Porter, A. E. Arfsen, J. Miller, J. W. Schilling, R. M. Scarborough, J. A. Lewicki, and D. B. Shenk, *J. Biol. Chem.* **19,** 9395 (1988).

[13] M. Chinkers, D. L. Garbes, M.-S. Chang, D. G. Lowe, H. Chin, D. V. Goeddel, and S. Schulz, *Nature (London)* **338,** 78 (1989).

[14] C.-H. Chang, K. P. Kohse, B. Chang, M. Hirata, and F. Murad, *FASEB J.* **3,** A1005 (1989).

[15] D. C. Leitman and F. Murad, *Endocrinol. Metab. Clin. North Am.* **16,** 79 (1987).

[16] D. C. Leitman, J. W. Andresen, R. M. Catalano, S. A. Waldman, J. J. Tuan, and F. Murad, *J. Biol. Chem.* **263,** 3720 (1988).

[17] M. B. Anand-Srivastava, D. J. Franks, M. Cantin, and J. Genest, *Biochem. Biophys. Res. Commun.* **121,** 855 (1984).

[18] M. Hirata, C.-H. Chang, and F. Murad, *Biochim. Biophys. Acta* **1010,** 346 (1989).

[19] T. Maack, M. Suzuki, F. A. Almeida, D. Nussenzveig, R. M. Scarborough, G. A. McEnroe, and J. A. Lewicki, *Science* **238,** 675 (1987).

receptor that functions exclusively as a clearance receptor for ANP is unclear.

ANP receptors are present in a variety of different cell types, including cells that are not involved in volume and blood pressure regulation, suggesting that ANP exerts other actions in addition to its well-characterized renal and cardiovascular effects. The properties and the regulation of the ANP receptors can be characterized by radioligand binding studies. Furthermore, the two ANP receptor subtypes can be distinguished by using selective atrial peptide analogs in cross-linking, competition binding, and cGMP response studies. In this chapter, we describe our methods for characterizing ANP receptors and the cGMP response to ANP in cultured cells.[6,7,16,20,21]

Iodination of ANP

The first step in the development of a receptor binding assay is to prepare a radioactive ligand. Most ANP binding studies have used iodinated forms of the 26 amino acid peptide ANP(101–126) or the 28 amino acid peptide ANP(99–126). Both of these peptides contain a single tyrosine residue that is located at the carboxyl terminus and which can be iodinated. In our binding studies, we use the 26 amino acid atrial peptide that is iodinated by the Iodogen method.[22] Prior to iodination, microfuge tubes are coated with Iodogen. One milligram of Iodogen is dissolved in 25 ml dichloromethane, and then 50 μl is added to the bottom of a microfuge tube. The Iodogen is then completely evaporated by a gentle stream of nitrogen gas. Once the tube is free of liquid it is used for the iodination of ANP.

For the iodination reaction, 10 μl of 50 mM sodium phosphate, pH 7.4, is added to a microfuge tube coated with Iodogen. Five microliters of ANP (0.2 nmol) dissolved in 10 mM acetic acid is then added. After the addition of 1 mCi $Na^{125}I$, the reaction is incubated for 10 min at room temperature. The unbound iodide is separated from the labeled peptide with a Sep-Pak C_{18} column. During the iodination reaction a Sep-Pak C_{18} column is prepared by washing the column with 5 ml of 0.1% trifluoroacetic acid in 100% acetonitrile, followed by 10 ml of 0.1% trifluoroacetic acid in water. After the iodination reaction, the sample is removed from the microfuge

[20] D. C. Leitman, V. L. Agnost, J. J. Tuan, J. W. Andresen, and F. Murad, *Biochem. J.* **244,** 69 (1987).

[21] D. C. Leitman, V. L. Agnost, R. M. Catalano, H. Schroder, S. A. Waldman, B. M. Bennett, J. J. Tuan, and F. Murad, *Endocrinology* **122,** 1478 (1988).

[22] P. R. P. Salacinski, C. McLean, J. E. C. Sykes, V. V. Clement-Jones, and P. J. Lowry, *Anal. Biochem.* **117,** 136 (1981).

tube and put at the bottom of a plastic syringe containing the Sep-Pak C_{18} column. The microfuge tube is washed with 500 μl 0.1% trifluoroacetic acid in water, which is then added to the syringe. The free iodide is then slowly eluted. The column is subsequently washed with 20 ml of 0.1% trifluoroacetic acid in water to remove free iodide. The ^{125}I-labeled ANP is eluted into a plastic centrifuge tube with 3 ml 0.1% trifluoroacetic acid in 100% acetonitrile. The sample is concentrated to approximately 1 ml under a continuous stream of nitrogen gas and then stored at 4°. The specific radioactivity of ANP ranges from 700 to 1400 Ci/mmol. Recently, iodinated ANP that is purified by high-performance liquid chromatography (HPLC) has become commercially available. However, we have obtained similar results when comparing ANP that we labeled to purified ANP purchased from companies.

ANP Receptor Assay

We have used the following binding assay to characterize the ANP receptors in a variety of cultured cell types, including aortic endothelial and smooth muscle cells, fibroblasts, and Leydig cells.[6,7,16,20] We perform binding studies at 37° in order to compare the properties of the ANP receptors under the same conditions we used to determine the biological response (i.e., cGMP determinations) of ANP. For binding assays, 25,000 cells are plated into 24-well tissue culture dishes.The cells are then grown until confluency. The cells are washed twice with 1 ml serum-free Dulbecco's modified Eagle's medium (DMEM), pH 7.3, containing 3.7 g/liter sodium bicarbonate. The binding assay is initiated by the addition of 200 μl of DMEM, pH 7.4, containing 25 m*M* HEPES, 2 mg/ml bovine serum albumin, and ^{125}I-labeled ANP. The cells are incubated in a 5% CO_2 incubator maintained at 37° for 15–30 min. The unbound ANP is removed by rapidly washing the cells 4 times with ice-cold Hanks' balanced salt solution (HBSS) containing 2 mg/ml bovine serum albumin. After the cells are washed, they are solubilized by addition of 500 μl of 1 *N* sodium hydroxide. The cell lysate is removed after at least 1 hr of incubation at room temperature. An additional 500 μl of 1 *N* sodium hydroxide is used to wash the wells. The two samples are pooled, and the radioactivity is determined with a gamma counter. To determine nonspecific binding, parallel cultures are exposed to the same concentration of ^{125}I-labeled ANP in the presence of 100 n*M* unlabeled ANP(101–126). Nonspecific binding of ANP to cultured cells usually does not exceed 10% of total binding. Specific binding is calculated by subtracting nonspecific binding from total binding.

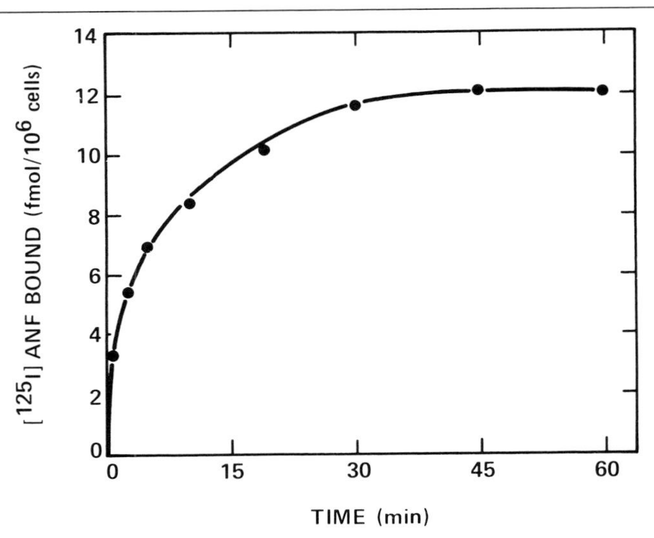

FIG. 1. Time course of specific binding of ^{125}I-labeled ANP to confluent bovine aortic endothelial cells. Endothelial cells were incubated with 150 pM ^{125}I-labeled ANP for various times at 37° in the absence and presence of 100 nM unlabeled ANP(8–33). Nonspecific binding was less than 2% of total binding.

ANP Binding Characteristics

At 37° the binding of ANP to cells is rapid, and equilibrium is reached in about 15 min (Fig. 1). In most cell types, we find that ANP binding is stable for at least 30 min after equilibrium is reached. However, in rat lung fibroblasts maximal ANP binding is maintained until 40 min, after which binding drops sharply.[20] By 60 min specific ANP binding is only 50% of the binding observed at 30 min, probably due to degradation of ANP. These results indicate that ANP binding studies should be restricted to 15–30 min at 37°. The binding of ANP to intact cultured cells displays a typical saturation isotherm. Figure 2 shows the effect of adding increasing concentrations of ANP on total, specific, and nonspecific binding in bovine aortic endothelial cells.[6] Specific ANP binding becomes saturated at approximately 0.5 nM. In contrast, nonspecific binding increases in a linear fashion. Scatchard analysis of the specific ANP binding data results in a straight line, suggesting that these cells contain a single class of binding sites. We later discovered that there are actually two ANP receptors present in cells, but the affinity of the two sites are nearly identical, accounting for the linear Scatchard plot.[7] The binding of ANP to cultured

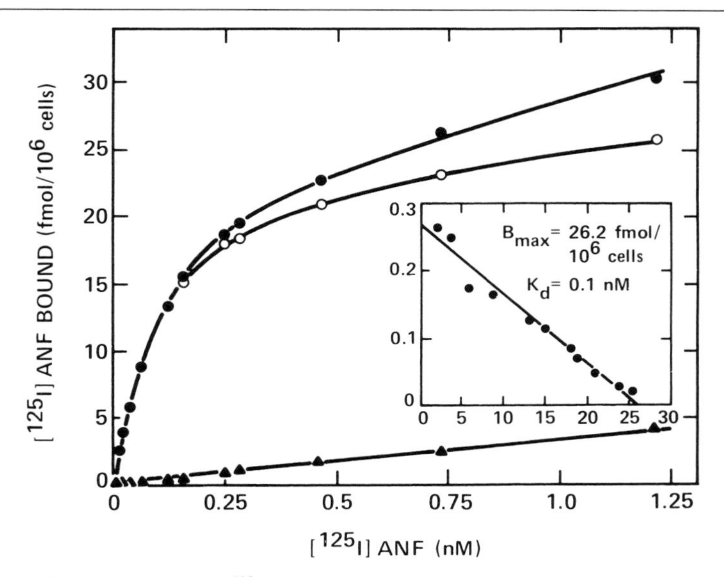

FIG. 2. Saturation curve of ^{125}I-labeled ANP binding to bovine aortic endothelial cells. Confluent endothelial cells were incubated at 37° for 30 min with increasing concentrations of ^{125}I-labeled ANP(101–126), and total (●), specific (○), and nonspecific binding (▲) was measured. (Inset) Scatchard analysis showing bound/free as a function of ^{125}I-labeled ANP specific binding.

cells is of high affinity, with a K_d value in the range of 0.1–2 n*M*.[16] When performing saturation curves, we typically use a range of 0.005–3 n*M* ^{125}I-labeled ANP. Whereas the K_d values are very similar in different cell types, the number of ANP receptors varies substantially (Table I). The richest source of ANP receptors are aortic smooth muscle cells, with approximately 300,000 receptors/cell.[16]

Another feature of ANP receptors is that the binding of ANP can be inhibited by a variety of ANP analogs. Analogs lacking the amino-terminal amino acids are nearly as effective at competing for ANP binding sites as the native ANP. Atrial peptides with a deletion of the carboxy-terminal amino acids phenylalanine-arginine-tyrosine[6] and ANP analogs with amino acid substitutions[23] effectively compete for ANP receptors, despite their having little or no intrinsic biological activity in some cell types.

[23] R. M. Scarborough, D. Shenk, G. A. McEnroe, A. E. Arfsen, L.-L. Kang, K. Schwartz, and J. A. Lewicki, *J. Biol. Chem.* **261,** 12960 (1986).

TABLE I
ANP RECEPTOR BINDING AND EFFECTS OF ANP ON cGMP AND PARTICULATE GUANYLYL CYCLASE ACTIVITY IN CULTURED CELLS[a]

	Stimulation (-fold)			
Cell type	cGMP	Guanylyl cyclase	Receptors/cell	K_d (nM)
BAC	13	1.5	50,000	0.12
HLF	35	3.1	80,000	0.32
MDCK	58	3.2	N.D.	
BASM	60	2.5	310,000	0.82
RME	120	5.0	N.D.	
RL	260	7.0	3000	0.11
MDBK	300	7.8	N.D.	
BAE	475	8.0	14,000	0.09

[a] The values for ANP stimulation of cGMP accumulation and particulate guanylyl cyclase were obtained from the concentration–response curve using an ANP concentration of 100 nM and 1 μM, respectively. The ANP receptor number and affinity were obtained from the saturation curves. The cell types shown are bovine aortic endothelial (BAE), bovine adrenal cortical (BAC), bovine aortic smooth muscle (BASM), human lung fibroblasts (HLF), rat Leydig (RL), rat mammary epithelial (RME), Madin–Darby canine kidney (MDCK), and Madin–Darby bovine kidney (MDBK) cells. N.D., None detectable (i.e., <3000 sites/cell).

ANP Cross-Linking

Cross-linking studies are useful in characterizing the molecular properties of ANP receptors. Specifically, cross-linking of ANP to intact cells has been instrumental for the elucidation of the molecular size and subunit composition of ANP receptors and the identification of multiple ANP receptor subtypes. ANP has been successfully cross-linked to receptors by either chemical affinity or photoaffinity cross-linking techniques. Both methods have yielded similar results. We have used chemical cross-linking techniques since no special equipment is necessary and the same ligand can be used for binding and cross-linking studies. The cross-linking reagent most commonly used to cross-link ANP to its receptor is disuccinimidyl suberate (DSS). Once ANP is covalently bound to its receptors by DSS, the molecular size of ANP-binding proteins can be determined by comparing the mobility of the ANP–receptor complex with known protein standards on sodium dodecyl sulfate-polyacrylamide gel electrophoresis (SDS-PAGE).

Cultured cells are plated at a density of 50,000 cells in 6-well tissue culture plates (35 mm). When the cells reach confluence they are washed

3 times with HBSS containing 10 m*M* HEPES, pH 7.3. The cells are incubated for 1 hr at room temperature with 1 ml HBSS containing 1–2 n*M* ^{125}I-labeled ANP in the absence and presence of unlabeled ANP. Just before the end of the 1-hr incubation time point, DSS is dissolved in dimethyl sulfoxide to a final concentration of 50 m*M*. The 50 m*M* DSS stock solution is diluted into HBSS (4 μl/ml) to a final concentration of 0.2 m*M*. One milliliter of HBSS containing DSS is added to the cells, which are then incubated for an additional 30 min at room temperature. The medium is aspirated, and the cells are washed quickly with 2 ml of 50 m*M* Tris-HCl, pH 7.3. Four hundred microliters of SDS sample buffer (62.5 m*M* Tris-HCl, pH 7.6, 10% glycerol, and 2.3% SDS) is added to the cells, and the plates are placed on top of a thin layer of boiling water for 3 min. The solubilized cells are removed and split into two equal portions. 2-Mercaptoethanol is added to one portion to a final concentration of 5%. The samples are electrophoresed on a 7.5% SDS-polyacrylamide gel. The gel is dried on Whatman 3 MM paper and exposed to Kodak Omat X-ray film at $-70°$.

Figure 3 shows the autoradiograms of ANP that were cross-linked to intact cultured cells. ANP cross-links to 66- and 130-kDa proteins under both nonreducing and reducing conditions.[16] However, the proportion of these two bands varies markedly in the presence of reducing agents such as 2-mercaptoethanol. Under nonreducing conditions approximately 40% of the ANP binding sites have a molecular mass of 130 kDa, whereas the remaining binding sites have a molecular mass of 66 kDa. In contrast, when the proteins are reduced prior to electrophoresis, over 95% of the ANP binding sites consist of the 66-kDa protein. These results suggest that ANP receptors exist in three molecular forms in intact cells: (1) a nonreducible 130-kDa protein that consists of a single polypeptide, (2) a 130-kDa dimer protein that is made up of two identical 66-kDa subunits bridged by disulfide bonds, and (3) a 66-kDa protein that exists as a monomer. We previously designated the 130-kDa nonreducible protein as ANP-R1 and the 130-kDa dimer and the 66-kDa monomer ANP-binding protein as ANP-R2.[7] These two ANP receptors have been purified,[9–11] and the cDNA for each receptor subtype has been isolated.[12,13]

In addition to the different subunit structure and amino acid composition, the two ANP receptors have different pharmacological and functional properties. The ANP-R1 receptor has intrinsic guanylyl cyclase activity and has been shown to mediate many of the physiological responses of ANP. The ANP-R2 receptor is devoid of guanylyl cyclase activity[10] and is not associated with any known physiological effects of ANP. The ANP-R1 receptor has much more rigid requirements for binding of ANP analogs compared to the ANP-R2 receptor. ANP analogs that are truncated at the

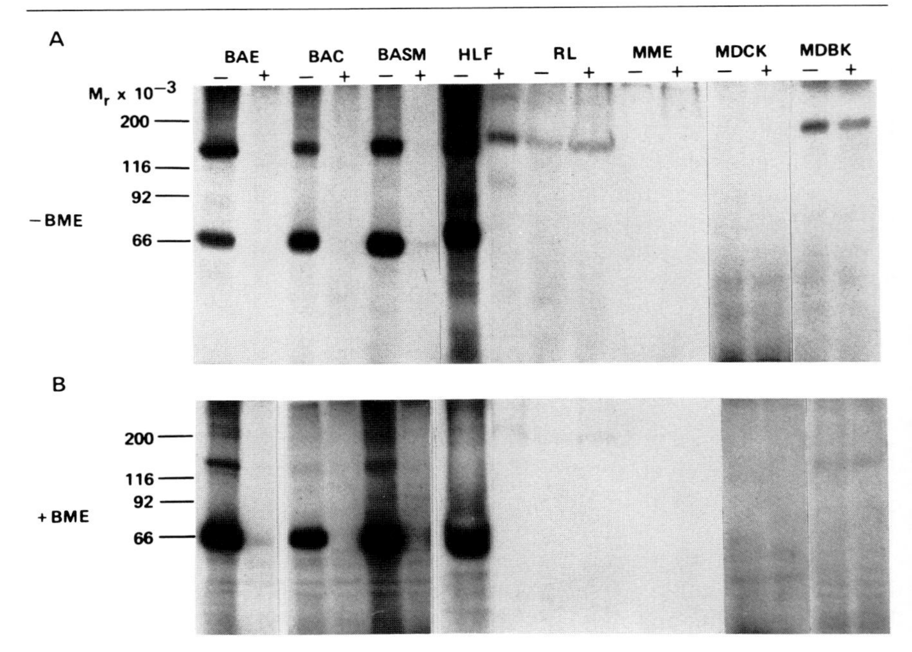

FIG. 3. Autoradiograms of SDS-polyacrylamide gels of cultured cells after binding with ^{125}I-labeled ANP and cross-linking with DSS. Confluent cultures were incubated with 2 n*M* ^{125}I-labeled ANP for 1 hr at room temperature in the absence (−) or presence (+) of 1 μ*M* ANP. The cells were then cross-linked by adding DSS to a final concentration of 0.1 m*M* for 30 min. The cells were solubilized with SDS sample buffer and subjected to SDS-PAGE under nonreducing (A) or reducing (B) conditions. The mobilities of molecular weight standards are given at left. The cell types shown are bovine aortic endothelial (BAE), bovine adrenal cortical (BAC), bovine aortic smooth muscle (BASM), human lung fibroblasts (HLF), rat Leydig (RL), rat mammary epithelial (RME), Madin–Darby canine kidney (MDCK), and Madin–Darby bovine kidney (MDBK) cells.

carboxy terminus and linear peptides lacking the disulfide bridge bind much more effectively to the ANP-R2 receptor compared with the ANP-R1 receptor. Several ANP-R2 selective peptides, such as atriopeptin I, are commercially available. These peptides completely inhibit ANP binding to many cells in competition binding studies, but they do not antagonize the ANP-induced stimulation of cGMP.[6,23,24] Unfortunately, selective agonists or antagonists for the guanylyl cyclase–ANP-R1 receptor are presently not available.

[24] R. M. Scarborough, G. A. McEnroe, A. E. Arfsen, L.-L. Kang, K. Schwartz, and J. A. Lewicki, *J. Biol. Chem.* **263,** 16818 (1988).

Cyclic GMP Determination

We have shown previously that there are two functionally distinct ANP receptor subtypes in cultured cells.[6] Only the nonreducible ANP-R1 receptor mediates the ANP activation of guanylyl cyclase and increased cGMP levels.[7] Table I illustrates that certain cell types which have either no detectable or a small number of ANP receptors can exhibit a much greater biological response to ANP compared with cells with the greatest number of receptors. These results demonstrate that the number of ANP receptors does not reflect the biological response to ANP. These results suggest that the magnitude of the cGMP response may depend only on the number of ANP-R1 receptors or possibly the ratio of the ANP-R1 receptors to ANP-R2 receptors, rather than the total amount of ANP receptors.

We believe that, in addition to performing competition binding studies, it is necessary to determine the stimulation of cGMP levels to evaluate the biological potency of various atrial peptide analogs. Measuring the cGMP response to ANP is also vital when comparing the number of ANP receptors after physiological perturbation and hormone or drug treatment. In different studies, the stimulation of cGMP has decreased, increased, or remain unaltered after the number of ANP receptors declined by ANP or angiotensin II treatment.[25-27] These studies suggest that the two receptor subtypes can be independently regulated.

Culture cells are plated in 6-well dishes at a density of 50,000 cells/35 mm dish. At confluence the cells are washed 2 times with 2 ml serum-free DMEM. The cells are then preincubated for 15 min in 990 μl DMEM, pH 7.3, containing 10 mM HEPES and 0.5 mM 3-isobutyl-1-methylxanthine. After preincubation, 10 μl of ANP is added to the medium, and the cells are incubated for 3 or 5 min. To minimize ANP binding to the tubes, we use polystyrene tubes and dilute ANP in water that contains a carrier such as 0.01% bacitracin or 1 mg/ml bovine serum albumin. The medium is aspirated, and 1 ml of 6% trichloroacetic acid is added to cells. The cell culture plates are stored at $-70°$. The plates are thawed at room temperature, and the cellular lysate is placed in 12×75 mm disposable glass test tubes. After centrifugation at 4000 rpm for 20 min at 4°, 850 μl of the supernatant fraction is placed in 13×100 mm disposable glass test tubes. Fifty microliters of 1 N HCl is added to each sample, and the

[25] Y. Hirata, S. Hirose, S. Takagi, H. Matsubara, and T. Omae, *Eur. J. Pharmacol.* **135,** 439 (1987).

[26] P. Roubert, M.-O. Lonchampt, P. Chabrier, P. Plas, J. Goulin, and P. Braquet, *Biochem. Biophys. Res. Commun.* **26,** 61 (1987).

[27] P. Chabrier, P. Roubert, M.-O. Lonchampt, P. Plas, and P. Braquet, *J. Biol. Chem.* **26,** 13199 (1988).

samples are extracted 5 times with 2 ml of water-saturated ether. The residual ether is evaporated by incubating the samples in a 55° water bath for 30 min. The samples are neutralized by the addition of 50 μl of 1 *N* sodium hydroxide and 50 μl of 1 *M* sodium acetate, pH 4.0. To acetylate the samples,[28] 500 μl of the sample is removed to 12 × 75 mm test tubes. Twenty microliters triethylamine is added, and the tube is vortexed. Then 10 μl acetic anhydride is immediately added, and the tube is again vortexed. The cGMP content in the sample is measured by radioimmunoassay (see [30], this volume).

ANP Stimulation of Cyclic GMP

In the presence of the phosphodiesterase inhibitor, 3-isobutyl-1-methylxanthine, ANP produces a maximal increase in intracellular cGMP content in 3–10 min at 37°.[6,16] The magnitude of the increase in cGMP levels and the concentration–response curves vary markedly in different cell types. Table I shows that ANP increases cGMP from 15- to 500-fold in various cell types. The most sensitive cell type that we have studied is bovine aortic endothelial cells. An approximate 5-fold increase in cGMP levels can be detected at 1 p*M*. The exquisite sensitivity of the BAE cells to ANP has permitted us to develop a bioassay for ANP in human plasma. The concentration–response curves for vascular smooth muscle, fibroblasts, and kidney cells are shifted to the right by 10- to 100-fold compared to BAE cells.[16] The EC_{50} for cGMP formation ranges from 100 p*M* in BAE cells to greater than 1 n*M* in some cell types, such as fibroblasts and vascular smooth muscle cells.

[28] J. F. Harper and G. Brooker, *J. Cyclic Nucleotide Res.* **1,** 207 (1975).

[42] Evaluating Atrial Natriuretic Peptide-Induced cGMP Production by Particulate Guanylyl Cyclase Stimulation *in Vitro* and *in Vivo*

By PAVEL HAMET and JOHANNE TREMBLAY

Introduction

The discovery of atrial natriuretic peptide (ANP) by Adolfo de Bold in 1982,[1] which was soon followed by observations suggesting that cyclic GMP (cGMP) is the second messenger and biological marker of this peptide,[2,3] and that particulate guanylyl cyclase (EC 4.6.1.2, guanylate cyclase) is the target enzyme for the biological actions of ANP[4,5] renewed interest in cGMP among researchers from different fields of biology. Several reviews have dealt with the role of cGMP in cell function.[6,7] This chapter specifically considers the methods used to measure cGMP in intracellular and extracellular media as well as some aspects of particulate guanylyl cyclase modulation in relation to the expression of actions by ANP.

cGMP Determination

Intracellular cGMP

cGMP is usually measured by radioimmunoassay using antibodies and labeled antigen available from several commercial sources or investigators. General guidelines for the measurement of cyclic nucleotide levels

[1] A. de Bold, *Proc. Soc. Exp. Biol. Med.* **170,** 133 (1982).

[2] P. Hamet, J. Tremblay, S. C. Pang, R. Garcia, G. Thibault, J. Gutkowska, M. Cantin, and J. Genest, *Biochem. Biophys. Res. Commun.* **123,** 515 (1984).

[3] P. Hamet, J. Tremblay, S. C. Pang, R. Skuherska, E. L. Schiffrin, R. Garcia, M. Cantin, J. Genest, R. Palmour, F. R. Ervin, S. Martin, and R. Goldwater, *J. Hypertens.* **4** (Suppl. 2), S49 (1986).

[4] J. Tremblay, R. Gerzer, P. Vinay, S. C. Pang, R. Beliveau, and P. Hamet, *FEBS Lett.* **181,** 17 (1985).

[5] S. A. Waldman, R. M. Rapoport, and F. Murad, *J. Biol. Chem.* **259,** 14332 (1984).

[6] F. Murad, *J. Clin. Invest.* **78,** 1 (1986).

[7] J. Tremblay, R. Gerzer, and P. Hamet, *in* "Advances in Second Messenger and Phosphoprotein Research" (P. Greengard and G. A. Robison, eds.), p. 319. Raven, New York, 1988.

have been well described in the past,[8,9] but since many new investigators who are not fully up-to-date on cyclic nucleotide research are now using cGMP levels to assess the biological activities of ANP, it is probably imperative to recall several aspects of cGMP assays. One of them is the importance of rapidly terminating cGMP metabolic reactions to evaluate levels of the nucleotide accurately. It is usually quite acceptable to add up to 2 *N* perchloric acid to isolated cells to arrest the activities of the synthesizing enzyme, guanylyl cyclase, and the catabolizing enzyme, cGMP phosphodiesterase. However, when the assay is performed in whole organs or tissue, it is necessary to freeze the tissue rapidly in liquid nitrogen and to use tools such as the Wollenberger clamp[10] to preserve the levels of cGMP reached. Tissues have to be quickly homogenized thereafter in the presence of perchloric acid and, to benefit from the heat and acid stability of the cyclic nucleotide, the sample can be boiled to destroy cyclic nucleotide phosphodiesterase activity, which can significantly and rapidly lower nucleotide levels. These basic rules are well recognized in the field of cyclic nucleotides.

The use of alkyl-xanthines, such as 3-isobutyl-1-methylxanthine (IBMX), throughout the incubation period to inhibit phosphodiesterase activity is also often recommended. Exact requirements of IBMX concentrations have to be assessed while working with ANP, taking into consideration the specific activity of cGMP phosphodiesterase, the tissue partition of alkylxanthine,[11,12] and the potential metabolic effects of this phosphodiesterase inhibitor on the tissue under study. Thus, in some animal species and in certain tissues, the addition of IBMX will not significantly modify cGMP responsiveness to ANP; an example is the glomerulus from dog kidney, where increases of cGMP in response to ANP are obtained in the absence or presence of IBMX[4] (Fig. 1). However, in the same organ isolated from rats, IBMX is an absolute necessity since without phosphodiesterase inhibition the cGMP increase is almost undetectable, while in the presence of 500 μM IBMX a 15-fold elevation of cGMP can easily be obtained in less than 2 min of incubation (Fig. 1). In the latter tissue usually a 15-min preincubation of glomeruli with IBMX is needed to preserve the nucleotide levels achieved[13] (Fig. 1).

[8] A. L. Steiner, this series, Vol. 38, p. 96.

[9] R. A. Richman, G. S. Kopf, P. Hamet, and R. A. Johnson, *J. Cyclic Nucleotide Res.* **6,** 461 (1980).

[10] S. E. Mayer, J. T. Stull, and W. B. Wastila, this series, Vol. 38, p. 3, (1974).

[11] J. A. Beavo, *Adv. Second Messenger Phosphoprotein Res.* **22,** 1 (1988).

[12] J. N. Wells and J. R. Miller, this series, Vol. 159, p. 489, (1988).

[13] B. J. Ballermann, R. L. Hoover, M. J. Karnovsky, and B. M. Brenner, *J. Clin. Invest.* **76,** 2049 (1985).

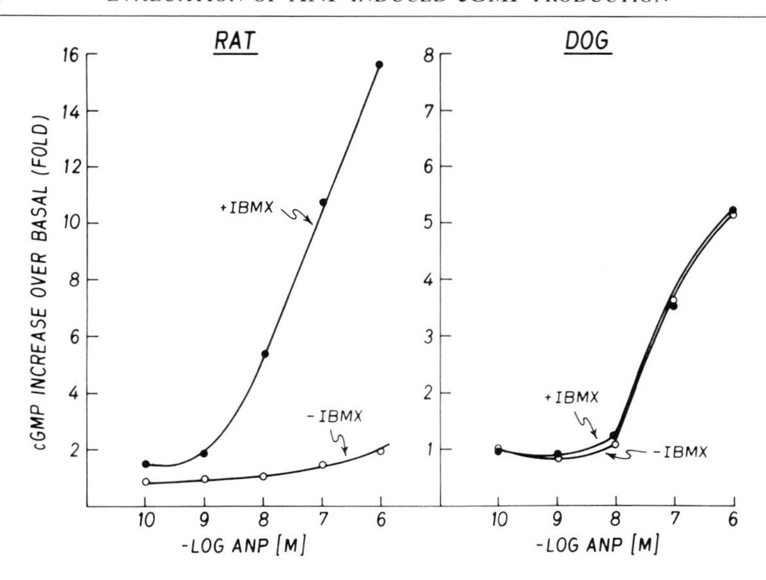

FIG. 1. Effect of IBMX on ANP-induced cGMP production in isolated rat and dog glomeruli. Isolated glomeruli were preincubated for 15 min at 37° without or with 500 μM IBMX. Increasing concentrations of ANP were then added, and the incubation was arrested after 90 sec by the addition of 1 N perchloric acid. cGMP levels were measured by radioimmunoassay following extraction as described in Ref. 4.

Extracellular cGMP

Egression. It is now fully acknowledged that cGMP increases in the extracellular milieu, that is, in tissue culture media and body fluids, are a faithful marker of the biological activity of ANP, particularly when acute elevations of ANP levels occur.[3,14] The teleological implications of cGMP egression from the cell are presently unknown, but it has been demonstrated that egression is an active process.[15] An understanding of the characteristics of cGMP egression is important to fully appreciate the usefulness of extracellular cGMP as a marker of ANP activity. While intracellular cGMP levels, under control of the synthesizing enzyme, particulate guanylyl cyclase, are subjected to rapid destruction by cyclic nucleotide phosphodiesterases, an entirely different metabolic fate is reserved for cGMP released into the extracellular milieu. Rapid secretion and hydrolysis of the intracellularly synthesized nucleotide result in a

[14] R. Gerzer, H. Witzgall, J. Tremblay, J. Gutkowska, and P. Hamet, *J. Clin. Endocrinol. Metab.* **61,** 1217 (1985).

[15] P. Hamet, S. C. Pang, and J. Tremblay, *J. Biol. Chem.* **264,** 12364 (1989).

short-lived intracellular peak, usually followed by a new steady state.[15] Quite distinctly in extracellular environments, such as cell culture media, cGMP continues to rise for minutes and hours after stimulation. The extracellular activity of membrane-associated cGMP phosphodiesterase is usually weak enough to allow a progressive accumulation of cGMP in the extracellular medium. For this reason, cGMP accumulation in the extracellular milieu provides a useful method for assessing ANP activity. Extracellular fluids can therefore be collected and assayed for cGMP levels without prior purification or extraction. However, care must be taken to avoid cell damage, which may lead to phosphodiesterase leakage and nucleotide destruction.

The process of cGMP egression is temperature dependent. Thus, decrements of cGMP egression are observed at the low temperature at which cGMP accumulates intracellularly, leading to a modification of the extracellular/intracellular ratio. IBMX, used for its ability to inhibit phosphodiesterase activity and to preserve intracellular cyclic nucleotide levels, indeed enhances accumulation in the intracellular compartment, resulting in an apparently increased overall production due to a blockade of nucleotide catabolism without modification of the extracellular/intracellular ratio.

The biochemical nature of cGMP transport is currently unknown but appears to share, at least in part, some of the properties of cAMP transport. This transport can be suppressed by amino acid transport inhibitors, such as probenecid. In cultured vascular smooth muscle cells (VSMC), cAMP levels rise in response to the adenylyl cyclase agonist, forskolin, while the addition of ANP leads to a progressive accumulation of cGMP in the extracellular environment (Fig. 2). When both agonists are added simultaneously, the augmented levels of cGMP by ANP do not interfere with cAMP egression in the extracellular medium (Fig. 2, right), since the intracellular cGMP levels attained in maximal situations are still two orders of magnitude lower than those of cAMP. However, the elevated intracellular cAMP reaches values which interfere with cGMP, leading to its decreased egression into the extracellular milieu (Fig. 2, left) with a parallel accumulation of intracellular cGMP.[15] These results suggest that cAMP and cGMP share the same transport system, which may have significant physiological consequences in situations where both cyclic nucleotides are modified. It must be kept in mind that, although cAMP levels in the intracellular milieu are higher than those of cGMP by several orders of magnitude, the circulating levels of these cyclic nucleotides are similar.[15]

cGMP egression of up to 7 orders of magnitude also occurs against an extracellular/intracellular gradient. This has been demonstrated after prelabeling the endogenous GTP pool with [^{14}C]guanine, as described

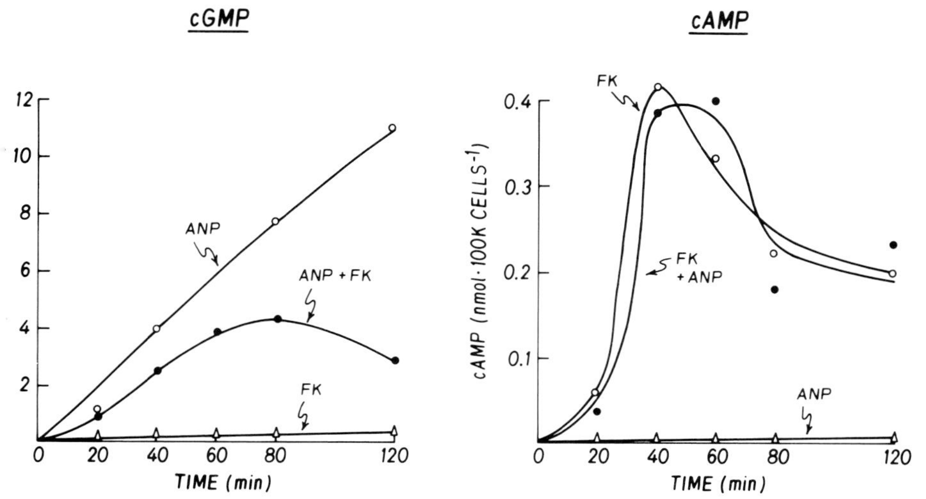

FIG. 2. Time course of extracellular cGMP and cAMP accumulation in rat mesenteric smooth muscle cells after the addition of 0.1 μM ANP, 0.1 mM forskolin (FK), or the two agents together. Conditions of cell culture are described in Ref. 15.

elsewhere.[15,16] After achieving a steady-state level of labeled [^{14}C]GTP within 60 to 90 min, the addition of ANP leads to [^{14}C]cGMP egression from cells, independently of extracellular cGMP content. In fact, even millimolar concentrations of cGMP in the extracellular environment do not impede egression of the cyclic nucleotide.[15]

Thus, cGMP egression is an active process which can be modulated by intracellular events, such as increased cAMP levels, or by extracellular transport modifiers, such as probenecid. This suggests that cGMP egression is also a regulated process, which may participate in modulating the signal of the biological activity of ANP.

cGMP in Body Fluids. cGMP concentrations in body fluids, plasma, and urine can be used as a marker of both pharmacological or physiological increases of ANP. Specific methodologic requirements for plasma measurements include cGMP purification or extraction from plasma to eliminate interference by plasma proteins and electrolytes in most radioimmunoassays. Determination of urinary cGMP is quite satisfactory after significant dilution since micromolar cGMP levels in urine exceed the sensitivity of the radioimmunoassay while nanomolar plasma levels preclude diluting it significantly. The appearance of cGMP in plasma fulfills

[16] R. J. Haslam, M. M. L. Davidson, T. Davies, J. A. Lynham, and M. D. McClenaghan, *Adv. Cyclic Nucleotide Res.* **9,** 533 (1978).

Sutherland's requirements for a second messenger candidate.[17] One of these requirements is that the increase of cyclic nucleotides (second messenger) should be observable with doses and at times preceding the expression of biological activity of the agonist (first messenger). Such is indeed the case with cGMP increases in plasma, which appear with bolus ANP doses lower than those required for the appearance of the diuretic and natriuretic actions of ANP.[2,18]

Reflection of Biological Activity of ANP. Initial studies have demonstrated that injections of atrial extracts, boluses, or continuous infusions of purified and synthetic ANP increase urinary and plasma cGMP.[2] Although plasma cGMP levels usually correlate significantly with ANP increases, the best reflection of the effectiveness of ANP on natriuresis is obtained by measuring urinary cGMP excretion. This is observed particularly with high doses of administered ANP, which lead to prolonged increments of plasma and urinary cGMP as well as sustained natriuresis and diuresis.[19] In these situations, cGMP levels remain elevated well beyond any increase of plasma ANP. This phenomenon may be related to the relative irreversibility of stimulation of the target enzyme, as will be discussed in the following sections. Good correlations between cGMP and ANP increases have been recorded in humans,[14] dogs,[20] rats,[2] and monkeys.[3]

Chronic ANP administration produces effects that are different from those of acute injection. Although the profile of modulation of cGMP levels is yet unknown, it has been demonstrated that, at the end of prolonged ANP infusion, plasma cGMP is actually significantly decreased, despite the persistence of the biological activities of ANP, such as lowering of blood pressure.[21] Chronic ANP administration may lead to new steady state cGMP levels where particulate guanylyl cyclase and phosphodiesterase remain activated. Such a modification of phosphodiesterase activity following augmented cAMP levels has been demonstrated in other situations.[22]

[17] E. W. Sutherland, *in* "Cyclic AMP" (G. A. Robison, R. W. Butcher, E. W. Sutherland, eds.), p. 36. Academic Press, New York, 1971.

[18] P. Larochelle, J. Cusson, P. Hamet, P. DuSouich, E. L. Schiffrin, J. Genest, and M. Cantin, *in* "Biologically Active Atrial Peptides" (B. M. Brenner and J. H. Laragh, eds.), p. 451. Raven, New York, 1987.

[19] J. R. Cusson, P. Hamet, J. Gutkowska, O. Kuchel, J. Genest, M. Cantin, and P. Larochelle, *in* "Biologically Active Atrial Peptides" (B. M. Brenner and J. H. Laragh, eds.), Vol. 1, p. 602. Raven, New York, 1987.

[20] E. H. Blaine, A. A. Seymour, E. A. Marsh, and M. A. Napier, *Fed. Proc., Fed. Am. Soc. Exp. Biol.* **45,** 2122 (1986).

[21] P. Hamet, E. Testaert, R. Palmour, P. Larochelle, M. Cantin, S. Martin, F. Ervin, and J. Tremblay, *Am. J. Hypertens.* **2,** 690 (1989).

[22] J. Tremblay, B. Lachance, and P. Hamet, *J. Cyclic Nucleotide Protein Phosphorylation Res.* **10,** 397 (1985).

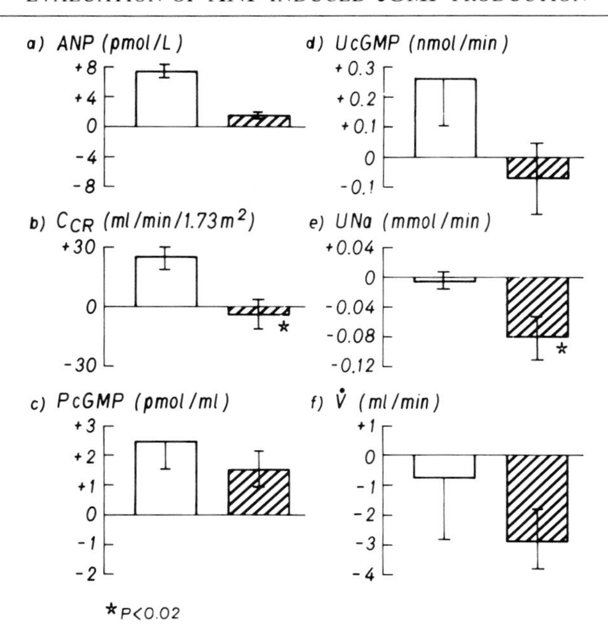

FIG. 3. Effect of exercise on (a) plasma ANP levels, (b) creatinine clearance, (c) plasma cGMP levels, (d) urinary cGMP levels, (e) urinary sodium excretion, and (f) diuresis in two groups of human volunteers. Plasma ANP levels were increased by exercise in one group (open bars) but not in other subjects (hatched bars).

In addition to this cGMP increment after exogenous ANP administration, more recent studies have shown that cGMP also increases with manipulation of endogenous ANP levels. Thus, cGMP is augmented after the application of ANP secretory stimuli, including volume expansion,[23–26] head-out water immersion, and exercise.[27,28] Figure 3 illustrates the effect of exercise on ANP and cGMP levels in human volunteers. Urinary cGMP excretion can be used here as a marker of ANP secretion since its modifi-

[23] J. V. Anderson, N. D. Christofides, and S. R. Bloom, *J. Endocrinol.* **109,** 9 (1986).

[24] A. S. Hollister, I. Tanaka, T. Imada, J. Onrot, I. Biaggioni, D. Robertson, and T. Inagami, *Hypertension* **8** (Suppl. II), II106 (1986).

[25] M. F. J. Walsh, S. N. Barakat, M. B. Zemel, S. M. Gualdoni, and J. R. Sowers, *in* "Biologically Active Atrial Peptides" (B. M. Brenner and J. H. Laragh, eds.), Vol. 1, p. 574. Raven, New York, 1987.

[26] P. Larose, S. Meloche, P. du Souich, A. De Lean, and H. Ong, *Biochem. Biophys. Res. Commun.* **130,** 553 (1985).

[27] V. K. Somers, J. V. Anderson, J. Conway, P. Sleight, and S. R. Bloom, *Horm. Metab. Res.* **18,** 871 (1986).

[28] A. M. Richards, G. Tonolo, J. G. F. Cleland, G. D. McIntyre, B. J. Leckie, H. J. Dargie, S. G. Ball, and J. I. S. Robertson, *Clin. Sci.* **72,** 159 (1987).

cation is observed only in subjects with heightened ANP levels and not in those showing no ANP response to exercise. It is further related to changes in renal function.[29] cGMP is also significantly increased while assuming a recumbent position, again in parallel to an increment of plasma ANP levels.[30–32] Dietary manipulations are also known to modify ANP secretion and, consequently, cGMP levels, particularly when the diets result in alterations of body volume and enhanced sodium intake.[33] It is therefore methodologically an absolute requirement to assess carefully the conditions such as volume status, exercise, posture, and diet when measuring cGMP or ANP levels in clinical studies. In both humans and experimental animals, major modifications of cGMP levels are observed in such pathological situations as renal failure[34] and heart failure,[35] and an abnormal responsiveness of cGMP to ANP has been described in hypertension,[19] which are considerations beyond the scope of this chapter (for review, see Ref. 36).

Circulating cGMP Levels. The cellular origin of plasma cGMP is unknown, but experimental evidence suggests that most of it derives from endothelial cells.[37] This possibility is supported by experiments in which human subjects alternatively received sodium nitroprusside (SNP) or ANP infusions in doses leading to a similar degree of vasodilation.[38] Both agonists exercised their biological activity via the cGMP system: ANP activates particulate guanylyl cyclase whereas SNP stimulates the soluble form of the enzyme. Both forms of the enzyme are present in VSMC, whereas endothelial cells possess only the particulate form of guanylyl

[29] D. Blanchard, M. Verdy, J. Gutkowska, J. Tremblay, and P. Hamet, *Clin. Invest. Med.* **11,** C34 (1988).

[30] M. L. Pearce, E. V. Newman, and M. R. Birmingham, *J. Clin. Invest.* **33,** 1089 (1954).

[31] F. H. Epstein, A. V. N. Goodyer, F. D. Lawrison, and A. S. Relman, *J. Clin. Invest.* **30,** 63 (1951).

[32] J. R. Cusson, P. Larochelle, J. Gutkowska, J. Tremblay, and P. Hamet, *Clin. Invest. Med.* **11:C63** (1988).

[33] N. E. Bruun, M. D. Nielsen, P. Skott, J. Giese, A. Leth, H. J. Schutten, and S. Rasmussen, *J. Hypertens.* **7,** 287 (1989).

[34] K. Hasegawa, Y. Matsushita, T. Inoue, H. Morii, M. Ishibashi, and T. Yamaji, *J. Clin. Endocrinol. Metab.* **63,** 819 (1986).

[35] M. Cantin, G. Thibault, J. Ding, J. Gutkowska, R. Garcia, G. Jasmin, P. Hamet, and J. Genest, *Am. J. Pathol.* **130,** 552 (1988).

[36] P. Hamet and J. Tremblay, *in* "Atrial Natriuretic Peptides" (W. K. Samson and R. Quirion, eds.), p. 95. CRC Press, Boca Raton, Florida, 1989.

[37] J. Tremblay, R. Willenbrock, J. R. Cusson, P. Larochelle, P. W. Schiller, F. H. H. Leenen, R. Palmour, F. Ervin, E. Testaert, and P. Hamet, *in* "Biological and Molecular Aspects of Atrial Factors" (P. Needleman, ed.), Vol. 81, p. 97. Alan R. Liss, New York, 1988.

[38] L. F. Roy, R. I. Ogilvie, P. Larochelle, P. Hamet, and F. H. H. Leenen, *Circulation* **79,** 383 (1989).

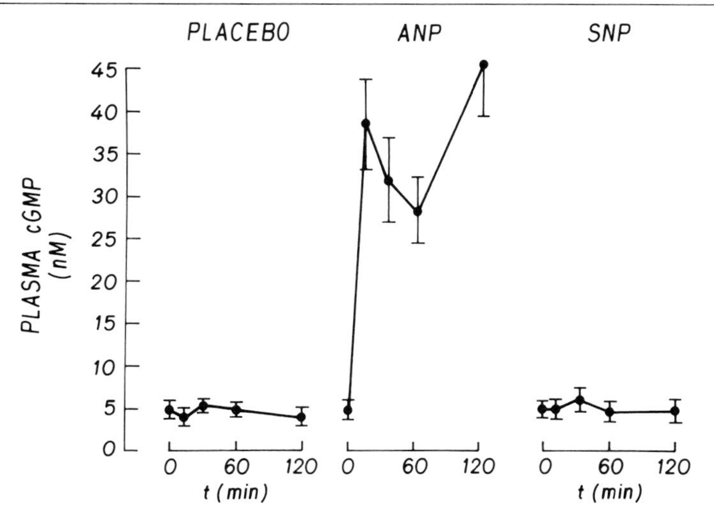

FIG. 4. Effect of ANP or SNP injection on plasma cGMP levels in human volunteers. (Data are from Roy *et al.*[38])

cyclase. Figure 4 illustrates the effects of the two agonists in human volunteers. Only ANP injections lead to an increment of cGMP levels in plasma. cGMP remains unchanged by SNP infusion. These observations support the notion that circulating cGMP reflects the effect of ANP selectively at the level of the endothelium. This methodological consideration has to be borne in mind when evaluating the vasodilatory activity of ANP *in vivo*.

The proportion of circulating cGMP levels which are due to the effect of ANP is difficult to assess, but current evidence suggests that a significant percentage of plasma cGMP is indeed dependent on ANP activity. Thus, Hardman *et al.*[39] demonstrated a long time ago that hypophysectomized animals have low plasma cGMP levels. It is now recognized that this significant fall in cGMP is preceded by a decrease of ANP secretion and plasma ANP levels, perhaps related to the modulatory effect of gluco- and mineralocorticoids on ANP gene expression in the atria and peptide secretion (J. Gutkowska, 1989, unpublished results).

Urinary cGMP Levels. Another methodological consideration concerns the assessment of urinary cGMP excretion in response to ANP. The classic studies of Broadus *et al.*[40] and Steiner[8] have demonstrated that

[39] J. G. Hardman, J. W. Davis, and E. W. Sutherland, *J. Biol. Chem.* **244,** 6354 (1969).
[40] A. E. Broadus, N. I. Kaminsky, J. G. Hardman, E. W. Sutherland, and G. W. Liddle, *J. Clin. Invest.* **49,** 2222 (1970).

TABLE I
DISTRIBUTION OF PARTICULATE AND SOLUBLE GUANYLYL CYCLASE AND EFFECTS OF ANP AND SNP ON cGMP IN VARIOUS TISSUES

Tissue	Guanylyl cyclase (%)		cGMP levels	
	Particulate	Soluble	ANP (0.1 μM)	SNP (0.1 mM)
Glomeruli	83	17	30	3
Collecting duct	30	70	3	10
Vascular smooth muscle	50	50	10	10
Platelets	5	95	N.D.	100

cyclic nucleotides are filtered by the kidney, and, in the case of cAMP, a fraction is added by tubular secretion under the influence of parathyroid hormone (PTH). Thus, PTH activity can be assessed by measuring nephrogenous fractions of urinary cAMP. In contrast, the effect of antidiuretic hormone (ADH), although clearly due to adenylyl cyclase stimulation at the level of the distal tubule, cannot be evaluated in urine since ADH does not increase urinary cAMP secretion. A somewhat analogous situation occurs with cGMP in response to ANP. By the use of Met^{110}-O-ANP, an oxidized analog of human ANP, dissociation of the natriuretic and diuretic activities of ANP is demonstrable.[41] Injection of Met^{110}-O-ANP in rats leads to significant diuresis which is not accompanied by natriuresis and urinary cGMP increases. In addition to the suggested heterogeneity of the cGMP effector system of ANP in the kidney, this result also reveals a compartmentalization of ANP receptor subtypes. The cGMP increment in some parts of the nephron, such as glomeruli, is reflected by increases of urinary cGMP, while the effects of ANP on other parts, such as the distal tubule, are only poorly assessed by measurements of urinary cGMP.

Particulate Guanylyl Cyclase as a Target Enzyme of ANP

The availability of ANP has brought about rapid developments in the study of the hormonal regulation of particulate guanylyl cyclase.

Tissue Distribution

It was noted quite early that ANP does not increase cGMP in all tissues, and cGMP increments are seen only in tissues rich in particulate guanylyl cyclase. Table I illustrates the concomitance between increments of cGMP

[41] R. C. Willenbrock, J. Tremblay, R. Garcia, and P. Hamet, *J. Clin. Invest.* **83,** 482 (1989).

by 0.1 μM ANP and the presence of particulate guanylyl cyclase.[42,43] Extreme examples of this distribution are glomeruli rich in particulate guanylyl cyclase. In this tissue, the agonist of soluble cyclase, SNP, only slightly elevates cGMP levels. The other extreme appears to be platelets, in which SNP increases cGMP, resulting in inhibition of aggregation, while ANP has no significant effect on platelet aggregation despite the presence of its receptor type, devoid of particulate guanylyl cyclase activity.[3] Thus, the biological activity of ANP is related to the presence of its receptor(s) linked with particulate guanylyl cyclase.

ANP Receptors

To date, two classes of ANP receptors have been identified, purified and cloned, one being a 120-kDa homodimer linked by a disulfide bridge (C-receptor), the other (A- and B-receptors) representing glycoproteins in which the extracellular domain consists of an ANP receptor linked via a transmembrane domain to the particulate guanylyl cyclase catalytic subunit (for review, see Ref. 44). It is of significant methodological relevance that the relative density of these ANP receptor classes varies from species to species and from tissue to tissue. At one extreme are the zona glomerulosa of the adrenals[45] and the kidney papillae,[46] where guanylyl cyclase-related receptors are predominant, and at the other extreme are the VSMC in which the C-receptor type predominates (Fig. 5). The only function of the C-receptor suggested at present is a potential role in ANP clearance from the circulation.[47] Many tissues possess the two classes of receptors in about an equal proportion. This heterogeneity in the ratio of ANP receptor classes has to be kept in mind. In tissues in which the C-receptor is absent, lower concentrations of ANP appear to be needed to increase cGMP levels.[46,48] Furthermore, blockade of the C-receptor class with the C-ANP analog C-ANP(102–121) produces a nonparallel shift to the left of the dose–response curve of the cGMP increase by ANP in cultured VSMC (Fig. 6). Similarly, a shift of several orders of magnitude to the left is observed directly on guanylyl cyclase stimulation by ANP in

[42] P. A. Craven and F. R. DeRubertis, *Biochemistry* **15,** 5131 (1976).

[43] T. P. Dousa, S. V. Shah, and H. E. Abboud, *Adv. Cyclic Nucleotide Res.* **12,** 285 (1980).

[44] S. Schulz, M. Chinkers, and D. L. Garbers, *FASEB J.* **3,** 2026 (1989).

[45] S. Meloche, H. Ong, M. Cantin, and A. De Lean, *Mol. Pharmacol.* **30,** 537 (1986).

[46] E. Nuglozeh, G. Gauquelin, R. Garcia, J. Tremblay, and E. L. Schiffrin, *Am. J. Physiol.* **259,** F130 (1990).

[47] T. Maack, M. Suzuki, F. A. Almeida, D. Nussenzveig, R. M. Scarborough, G. A. McEnroe, and J. A. Lewicki, *Science* **238,** 675 (1987).

[48] A. Kurtz, R. Della Bruna, J. Pfeilschifter, R. Taugner, and C. Bauer, *Proc. Natl. Acad. Sci. U.S.A.* **83,** 4769 (1986).

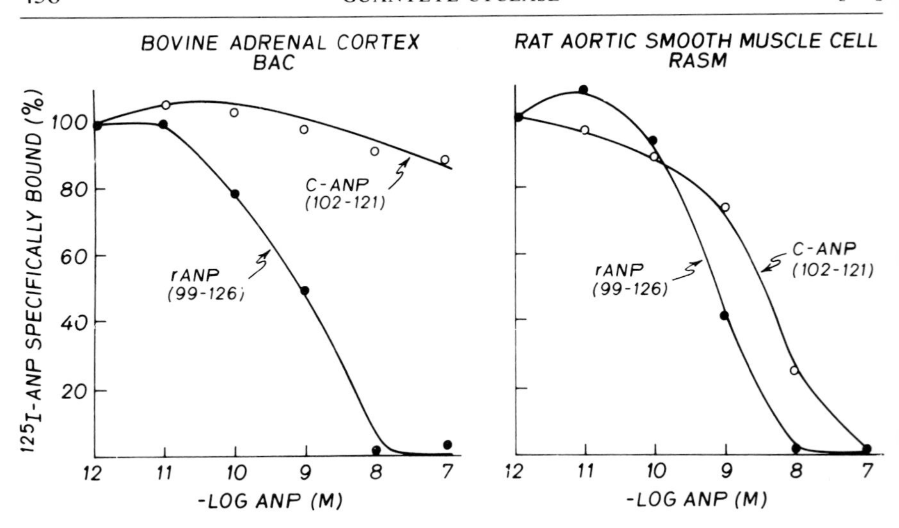

FIG. 5. Evaluation of the ratio of different classes of ANP receptors in bovine adrenal cortex and rat aortic smooth muscle cell membranes by competitive binding studies using rat ANP(99–126) or an analog specific for the C-receptor subtype C-ANP(102–121).

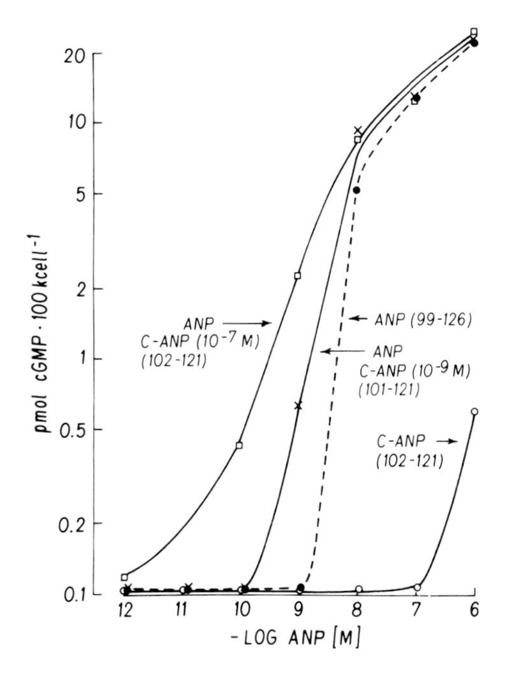

FIG. 6. Effect of two concentrations of C-ANP(102–121) (10^{-9} and 10^{-7} *M*) on the dose–response curve of cGMP increase in rat aortic smooth muscle cells by ANP.

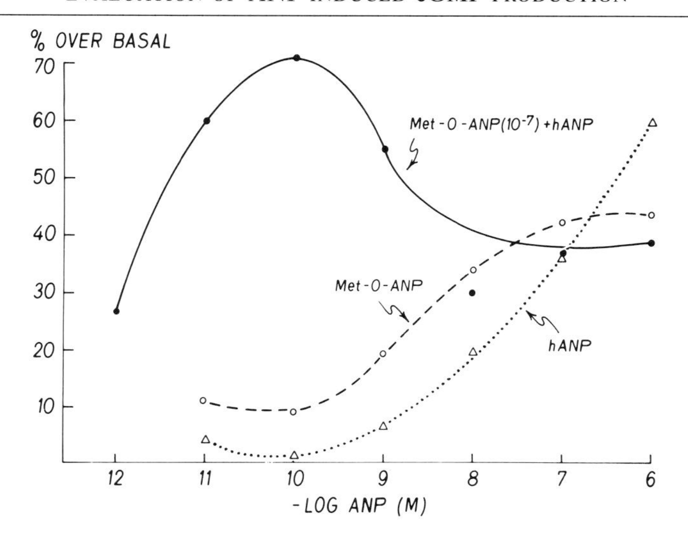

FIG. 7. Potentiation and inhibition of ANP-stimulated guanylyl cyclase activity by simultaneous addition of Met[110]-O-ANP (10^{-7} *M*).

placental membranes (J. Tremblay, 1989, unpublished results). These data suggest an inhibitory action of C-receptors on guanylyl cyclase stimulability by ANP.

Both potentiation and inhibition of the guanylyl cyclase stimulability by ANP have also been demonstrated *in vitro*. Figure 7 is an example of such an experiment. It can be seen that ANP or Met[110]-O-ANP significantly stimulates particulate guanylyl cyclase, with a threshold concentration of the order of 1 n*M*. However, prior addition of Met[110]-O-ANP (10^{-7} *M*) to increasing concentrations of ANP results in a dose–response curve of guanylyl cyclase activity with two components. The stimulatory component is displaced by several orders of magnitude to the left, followed by an inhibitory component, compatible with the partial agonistic activity of the analog on guanylyl cyclase. In its stimulatory component, the effective doses of ANP *in vitro* are similar to those in *in vivo* situations. It is therefore possible to speculate that, *in vivo*, endogenous levels of ANP facilitate particulate guanylyl cyclase stimulation as a result of preoccupancy of the C-receptor type. Alternatively, the explanation of this potentiation could reside in an interaction between the two ANP receptor types, clustering of receptors, etc. The stimulation followed by inhibition of activity by increasing ligand concentration may be attributed to the increasing size of receptor clusters following the "bifurcation theory" of

Schiffmann.[49] In this theory, the formation of microclusters is responsible for the increased activity, but when the clusters get too large the changes in membrane geometry result in decreased activity. Direct proof of such a situation with ANP receptors remains to be demonstrated.

Irreversibility of Guanylyl Cyclase Stimulation by ANP

Another distinctive feature of the biological activity of ANP is its long-lasting effect on natriuresis, diuresis, and cGMP excretion following a bolus injection in different species.[3,37,50] Furthermore, in Green Vervet monkeys, ANP produces a fall in blood pressure that persists for 48 hr after cessation of chronic infusion.[21] This phenomenon may also be related to the functional irreversibility of particulate guanylyl cyclase documented *in vitro*. Exposure of cultured VSMC to ANP for 2 min leads to prolonged cGMP accumulation in the extracellular medium, in spite of several washes of the cells with fresh medium without ANP.[50] This is quite distinct from the stimulation of adenylyl cyclase[15] and somewhat similar to some of the actions of agents such as insulin.[51] Irreversible activation of particulate guanylyl cyclase by ANP is also demonstrable in washed membranes.[50] Preexposure of the membranes to ANP results in an irreversibly activated guanylyl cyclase despite thorough washes of the membrane fractions and undetectability of added labeled ANP molecules in the supernatant. These properties are seen mainly at higher doses of ANP (100 n*M*) and could be due to either a turn-on of the enzyme or irreversible ANP binding at the receptor site. Nevertheless, the properties appear to have both methodological and biological consequences.

It is noteworthy that the stimulation of guanylyl cyclase activity by ANP disappears after isolation of the enzyme. The protein purified by many groups[52–55] demonstrates both high-affinity binding of ^{125}I-labeled ANP and significant guanylyl cyclase activity. In all cases, however, the enzyme becomes refractory to ANP stimulation. This repeatedly observed behavior suggests that integrity of the membrane environment is necessary for the full response of guanylyl cyclase to ANP. In many circumstances,

[49] Y. Schiffmann, *Prog. Biophys. Mol. Biol.* **36,** 87 (1980).

[50] J. Tremblay and P. Hamet, *Can. J. Physiol. Pharmacol.* **67,** 1124 (1989).

[51] R. M. Denton, *Adv. Cyclic Nucleotides Protein Phosphorylation Res.* **20,** 293 (1986)

[52] T. Kuno, J. W. Andresen, Y. Kamisaki, S. A. Waldman, L. Y. Chang, S. Saheki, D. C. Leitman, M. Nakane, and F. Murad, *J. Biol. Chem.* **261,** 5817 (1986).

[53] R. Takayanagi, R. M. Snajdar, T. Imada, M. Tamura, K. N. Pandey, K. S. Misono, and T. Inagami, *Biochem. Biophys. Res. Commun.* **144,** 244 (1987).

[54] A. K. Paul, R. B. Marala, R. K. Jaiswal, and R. K. Sharma, *Science* **235,** 1224 (1987).

[55] S. Meloche, N. McNicoll, B. Liu, H. Ong, and A. De Lean, *Biochemistry* **27,** 8151 (1988).

the degree of cGMP increment reached in intact cells is greatly reduced in washed membranes and completely lost when the enzyme is isolated. This may be a consequence of removal of an accessory protein or loss of positive receptor interaction. Alternatively, the purified enzyme may be in a fully activated state due to elimination of an inhibitory component.

[43] Phosphorylation and Dephosphorylation of Sea Urchin Sperm Cell Guanylyl Cyclase

By J. KELLEY BENTLEY

Introduction

The sea urchin spermatozoan guanylyl cyclase [GTP pyrophosphate-lyase (cyclizing); EC 4.6.1.2, guanylate cyclase] is a plasma membrane glycoprotein regulated by species-specific egg peptides. Both the resact (Cys-Val-Thr-Gly-Ala-Pro-Gly-Cys-Val-Gly-Gly-Gly-Arg-Leu-NH_2) and speract (Gly-Phe-Asp-Leu-Asn-Gly-Gly-Gly-Val-Gly) egg peptides increase spermatozoan respiration and motility rates as well as cyclic AMP (cAMP) and cGMP concentrations in homologous spermatozoa.[1-6] Resact is also a spermatozoan chemoattractant.[7] At high ($\geq$100 n*M*) concentrations, egg peptides cause a dephosphorylation of serine residues on guanylyl cyclase; the loss of phosphate decreases the specific activity of the enzyme.[3,7-15] The dephosphorylation coincides with a decrease in the apparent molecular weight (M_r) of guanylyl cyclase from M_r 160,000 to

[1] J. R. Hansbrough and D. L. Garbers, *J. Biol. Chem.* **256,** 1447 (1981).
[2] J. R. Hansbrough and D. L. Garbers, *J. Biol. Chem.* **256,** 2235 (1981).
[3] N. Suzuki, H. Shimomura, E. W. Radany, C. S. Ramarao, G. E. Ward, J. K. Bentley, and D. L. Garbers, *J. Biol. Chem.* **259,** 14874 (1984).
[4] H. Shimomura and D. L. Garbers, *Biochemistry* **25,** 3405 (1986).
[5] J. K. Bentley, D. J. Tubb, and D. L. Garbers, *J. Biol. Chem.* **261,** 14859 (1986).
[6] J. K. Bentley, A. S. Khatra, and D. L. Garbers, *J. Biol. Chem.* **262,** 15708 (1987).
[7] G. E. Ward, C. J. Brokaw, D. L. Garbers, and V. D. Vacquier, *J. Cell Biol.* **101,** 2324 (1985).
[8] J. K. Bentley and D. L. Garbers, *Biol. Reprod.* **35,** 1249 (1986).
[9] G. E. Ward and V. D. Vacquier, *Proc. Natl. Acad. Sci. U.S.A.* **80,** 5578 (1983).
[10] G. E. Ward, G. W. Moy, and V. D. Vacquier, *J. Cell Biol.* **103,** 95 (1986).
[11] G. E. Ward, D. L. Garbers, and V. D. Vacquier, *Science* **227,** 768 (1985).
[12] C. S. Ramarao and D. L. Garbers, *J. Biol. Chem.* **260,** 8390 (1985).
[13] V. D. Vacquier and G. W. Moy, *Biochem. Biophys. Res. Commun.* **137,** 1148 (1987).
[14] J. K. Bentley, H. Shimomura, and D. L. Garbers, *Cell (Cambridge, Mass.)* **45,** 281 (1986).
[15] C. S. Ramarao and D. L. Garbers, *J. Biol. Chem.* **263,** 1524 (1988).

150,000 and with a lowered rate of egg peptide-stimulated cGMP formation.[5] After guanylyl cyclase dephosphorylation, sperm cells no longer swim to a resact source.[7]

The cDNA-derived sequence of the enzyme is now known, and the sequence shares some homology with the tyrosine kinase receptor families.[16,17] When cross-linked to sperm cells with disuccinimidyl suberate, resact is covalently bound to guanylyl cyclase (M_r 160,000–150,000) and to a different protein of M_r 80,000–75,000, so guanylyl cyclase is closely apposed to if not an actual resact receptor.[18] However, such an experiment cross-links speract only to a M_r 77,000 protein,[19,20] and speract stimulation of guanylyl cyclase in solubilized preparations is lost on dilution or fractionation of the detergent extract.[21] Therefore, it is possible that guanylyl cyclase is regulated by another egg peptide receptor. Here, methods for the phosphorylation, activation, dephosphorylation, and desensitization of guanylyl cyclase in whole or broken cell preparations are described.

Materials

Common reagents of the highest available quality are sufficient. However, sodium dodecyl sulfate (SDS) obtained from Sigma (St. Louis, MO) as the 95% grade (66% lauryl) is needed in order to observe the M_r shift upon gel electrophoresis.[3] When highly purified SDS is used, guanylyl cyclase migrates as a single electrophoretic band regardless of its phosphorylation state.

The gametes of different sea urchin species have unique peptide receptors and biochemical characteristics. Only *Arbacia punctulata* spermatozoa incorporate ortho[^{32}P]phosphate from the surrounding medium into guanylyl cyclase; resact (but not speract) activates spermatozoa of this species.[3,9] The sperm cells of *Strongylocentrotus purpuratus* respond to speract (and not resact) with guanylyl cyclase activation and desensitization, but speract causes no guanylyl cyclase M_r shift in this species.[12] *Lytechinus pictus* spermatozoa activate guanylyl cyclase in response to speract (not resact) and desensitize with a guanylyl cyclase M_r shift from M_r 150,000 to 140,000.[5,8]

[16] S. Singh, D. L. Lowe, D. S. Thorpe, H. Rodriguez, W.-J. Kuang, L. J. Dangott, M. Chinkers, D. V. Goeddel, and D. L. Garbers, *Nature (London)* **334,** 708 (1988).

[17] S. Singh and D. L. Garbers, this volume, [39].

[18] H. Shimomura, L. J. Dangott, and D. L. Garbers, *J. Biol. Chem.* **261,** 15778 (1986).

[19] L. J. Dangott and D. L. Garbers, *J. Biol. Chem.* **259,** 13712 (1984).

[20] J. K. Bentley and D. L. Garbers, *Biol. Reprod.* **34,** 413 (1986).

[21] J. K. Bentley, A. S. Khatra, and D. L. Garbers, *Biol. Reprod.* **39,** 639 (1988).

^{32}P Incorporation into *Arbacia punctulata* Sperm Cell Guanylyl Cyclase

Sperm cells collected "dry" are radiolabeled most effectively in a pH 6.5 buffer.[6] Sperm cells are suspended at 100 mg wet weight/ml in 0.5 *M* NaCl, 50 m*M* $MgCl_2$, 10 m*M* benzamidine, 20 m*M* 2-(*N*-morpholino) ethanesulfonic acid (MES), pH 6.5 (abbreviated as NMBM buffer), with 20 mCi/ml carrier-free ortho[^{32}P]phosphate and incubated for 1 h at 21°. Afterward, spermatozoa are incubated at 4° for 10 min and washed twice in 4° NMBM buffer. Washing spermatozoa is accomplished by centrifugation for 5 min at 1000 *g* and resuspension in fresh buffer. Sperm suspensions remain responsive to resact for at least 8 hr when stored at 4°.

The incorporation of ^{32}P into cellular guanylyl cyclase is assessed by reducing SDS–polyacrylamide gel electrophoresis, since several other membrane proteins also incorporate ^{32}P from [γ-^{32}P]ATP.[6] Guanylyl cyclase dephosphorylation must be blocked in order to prepare samples for electrophoresis. Although trichloroacetic acid precipitates guanylyl cyclase and halts its dephosphorylation,[3] the recovery of protein from the acid precipitate varies. Guanylyl cyclase dephosphorylation and proteolytic degradation may be halted without acid precipitation. Subsequent membrane solubilization for electrophoresis gives uniform protein recovery. A solution of 1 *M* tris(hydroxymethyl)aminomethane (Tris) base, 0.3 *M* ethylenediaminetetraacetic acid (EDTA), 0.1 *M* benzamidine hydrochloride, and 0.1 *M* NaF (abbreviated TEBF buffer) is first prepared. The pH of the solution is adjusted to 6.5 with acetic acid. In order to halt guanylyl cyclase dephosphorylation, at least 9 volumes of the TEBF buffer at 2°–4° is added to a cell or membrane vesicle suspension. If concentration is needed, diluted samples are washed by centrifugation and resuspension to a desired volume in TEBF buffer at 2°–4°. An equal volume of a concentrated sample buffer consisting of 20% glycerol, 10% 2-mercaptoethanol, 5% SDS, 200 m*M* Tris, 60 m*M* EDTA, 20 m*M* benzamidine, and 20 m*M* NaF at pH 6.5 is added to solubilize the sample for electrophoresis. The solubilization of whole spermatozoa requires shearing DNA by homogenization before electrophoresis. Samples in the buffer are incubated at 95°–100° for 5 min before electrophoresis in a Laemmli gel system.[8,14,22] Autoradiography is performed after silver staining for protein.[3,23]

Sea urchin sperm guanylyl cyclase is entirely membrane bound.[10,20] Therefore, membranes with the proteins that regulate the enzyme provide an ideal starting point for *in vitro* studies. Membranes prepared from spermatozoa of all three species listed above respond to their homologous egg peptide with guanylyl cyclase activation and desensitization.[5–7]

[22] U. K. Laemmli, *Nature (London)* **227,** 600 (1970).

[23] J. H. Morrissey, *Anal. Biochem.* **117,** 307 (1981).

Membrane Vesicle Preparation

In vitro studies of phosphorylated guanylyl cyclase require its protection from spermatozoan phosphatase and protease activities. At all times in the preparation, sperm cells or particles are kept at 4° (i.e., kept on ice). Whole sperm are suspended in at least 5 volumes of the following solution: 0.4 *M* NaCl, 0.1 *M* NaF, 20 m*M* disodium EDTA, 10 m*M* Na_2MoO_4, 10 m*M* benzamidine hydrochloride, 10 m*M* KCl, and 20 m*M* 4-(2-hydroxyethyl)-1-piperazinethanesulfonic acid (HEPES) buffered as the Na^+ salt to pH 6.5. Cells must be broken under isosmotic conditions with protease and phosphatase inhibitors to obtain a phosphorylated guanylyl cyclase and functional resact or speract receptor.[8,14,20] After incubating 1 hr at 4°, cells are dispersed with 10 passes in a glass–Teflon homogenizer and then subjected to nitrogen cavitation for 20 min at 200 psi. The homogenate is centrifuged for 30 min at 30,000 *g* to obtain a membrane-enriched supernatant fluid. The pellet is dispersed again and centrifuged to increase the yield of membranes. The 30,000 *g* supernatant fluid is centrifuged for 1 hr at 200,000 *g*. Pelleted particles from the 200,000 *g* centrifugation are dispersed in the same solution with an additional 20 m*M* MES but at pH 6.0. The particles are layered over a 33% (w/v) sucrose cushion made in the pH 6.0 buffer and are centrifuged for 2 hr at 100,000 *g* in a swinging-bucket rotor. The membrane-enriched material at the 0–33% sucrose interface is removed, diluted 5-fold in the pH 6.0 buffer, and centrifuged again at 200,000 *g* for 1 hr to obtain a membrane vesicle pellet.[5]

Membranes may be stored at −70° after resuspension in the pH 6.0 buffer with 33% sucrose. However, repeated freeze–thaw cycles destroy guanylyl cyclase activity. Thawed membranes kept at 4° should be used within a few hours before enzyme activity and egg peptide-binding capacity decay.

^{32}P Incorporation into Guanylyl Cyclase in Membrane Vesicles

Spermatozoan membranes are thawed at 4° and washed. Membranes are washed by dilution with 2 volumes of 4° NMBM buffer and centrifugation for 1 hr at 200,000 *g*. Membranes are resuspended in the same buffer at 4°. Guanylyl cyclase radiolabeling is initiated by diluting membrane vesicles into a solution with 25 m*M* NaCl, 2.5 m*M* $MgCl_2$, 1 m*M* dithiothreitol (DTT), 1 m*M* benzamidine hydrochloride, 0.4 m*M* 1-methyl-3-isobutylxanthine (MIX), 0.2 m*M* ethylene bis(oxyethylenenitrilo)tetraacetic acid (EGTA), 1 m*M* Na_2HPO_4, 10 m*M* MES, pH 6.5, and 100 μM [γ-^{32}P]ATP at a specific activity of 2000–4000 counts per minute (cpm)/pmol.[6] The phosphorylation proceeds for 10 min at 15°. After washing membranes in NMBM buffer, samples are kept at 4° until the dephosphorylation reaction is initiated. In order to assess ^{32}P incorporation into guanylyl cyclase, TEBF buffer is added to the sample. The sample is then processed for polyacrylamide gel electrophoresis.

Dephosphorylation of Guanylyl Cyclase

In order to dephosphorylate guanylyl cyclase by interaction with an egg peptide, cells or membranes are diluted 10-fold into 21° artificial seawater (450 m*M* NaCl, 50 m*M* $MgCl_2$, 10 m*M* $CaCl_2$, 10 m*M* KCl, 50 m*M* sodium HEPES, pH 8.0) with at least 100 n*M* egg peptide concentrations.[8,14] The egg peptide-stimulated dephosphorylation of membrane guanylyl cyclase is essentially complete after 1–5 min at 21°.[5] Incubating membranes at a pH no less than 7.5 for 1 hr without egg peptide also dephosphorylates guanylyl cyclase.[5] Dephosphorylation is halted with a 10-fold dilution into TEBF buffer. Samples are processed for electrophoresis as described above. The progress of the interaction should also be assessed by measuring guanylyl cyclase activity.

Guanylyl Cyclase Activity with Different Phosphorylation States

Several characteristics of guanylyl cyclase are worth considering before initiating studies. The basal specific activity of guanylyl cyclase is lower in the absence of detergent, but membrane-bound guanylyl cyclase retains positive cooperativity upon speract activation.[5] Radioimmunoassay methods are therefore used to monitor cGMP formation at lower (<100 μM) GTP concentrations.[24,25] Mn^{2+} also supports a greater specific activity of the enzyme, but, owing to an accelerated enzyme dephosphorylation in the presence of Mn^{2+}, speract stimulation of guanylyl cyclase is lost when Mn^{2+} is substituted for Mg^{2+}.[21]

When an homologous egg peptide activates and desensitizes spermatozoan guanylyl cyclase, the stimulated rate of cGMP production declines as the enzyme is dephosphorylated.[5-7] In order to measure directly the receptor-mediated activation and desensitization of guanylyl cyclase, 10–50 μg membranes in NMBM are diluted into 100 μl of 50 m*M* NaCl, 5 m*M* $MgCl_2$, 1 m*M* benzamidine hydrochloride, 2 m*M* GTP ([α-^{32}P]GTP at a specific activity of 20,000 cpm/nmol), 1 m*M* dithiothreitol (DTT), 0.4 m*M* MIX, 1 m*M* cGMP, 0.2 m*M* EGTA, and 40 m*M* Tris, pH 7.6, at 21° with at least 100 n*M* egg peptide concentration.[5] Added cGMP should be omitted when radioimmunoassays are used to monitor cGMP formation. Rates of guanylyl cyclase activation by an egg peptide are maximal at no

[24] D. L. Garbers and F. Murad, *Adv. Cyclic Nucleotide Res.* **10,** 57 (1979).
[25] S. E. Domino, D. J. Tubb, and D. L. Garbers, this volume, [30].

later than 1 min at pH 7.6. Guanylyl cyclase activation by speract is sustained without dephosphorylation when 40 mM MES, pH 6.5, is substituted for Tris and 100 μM ATP is included in the assay.[6,21] Reactions are run at 15° and stopped at timed intervals over 10 min with the addition of 250 μl of 0.2 M zinc acetate.[24] The α-^{32}P-labeled guanine nucleotide tri- and diphosphates are precipitated with Zn^{2+} by adding 250 μl of 0.2 M Na_2CO_3. [α-^{32}P]cGMP is purified from the supernatant fluid by chromatography over neutral alumina and Dowex.[24]

In order to monitor only guanylyl cyclase inactivation, dephosphorylation reactions in buffers at pH above about 7.5 and at 21° are halted by a further 10-fold dilution into a 4° solution of 0.5% (w/v) Lubrol PX, 25 mM KF, 100 μM Na_2VO_4 and 25 mM MES at pH 6.0 (abbreviated as LMFV buffer).[12,14] The samples are incubated for 1 hr at 4° for solubilization. The guanylyl cyclase activity is assayed by further dilution of about 1 μg protein into 100 μl of 0.4 mM MIX, 10 mM $MnCl_2$, 2 mM GTP [α-^{32}P]GTP specific activity of 2500 cpm/nmol), and 1 mM cGMP in LMFV buffer. Reactions are carried out at 30° for 15 min and terminated by 250 μl of 0.2 M zinc acetate as described above.[24]

Acknowledgments

Thanks are due to David L. Garbers, for his counsel. This work was supported by the Howard Hughes Medical Institute, and National Institutes of Health Grants HD10254 and HD05797.

[44] Calcium-Regulated Guanylyl Cyclases from *Paramecium* and *Tetrahymena*

By JOACHIM E. SCHULTZ and SUSANNE KLUMPP

Introduction

In metazoans, distinct guanylyl cyclase (EC 4.6.1.2, guanylate cyclase) activities are found in cytosolic as well as in particulate fractions. In contrast, in the unicellular protozoans *Paramecium* and *Tetrahymena* all activity is membrane bound. In addition, there are other major differences indicative of the uniqueness of protozoan guanylyl cyclases[1]: (1) no effect of nitroso compounds or fatty acids (mammalian soluble guanylyl cyclases

[1] J. E. Schultz and S. Klumpp, *Adv. Cyclic Nucleotide Protein Phosphorylation Res.* **17,** 275 (1984).

are activated); (2) no effect of atrial natriuretic peptides or porphyrins (mammalian particulate guanylyl cyclases are activated); (3) Mg^{2+} as metal cofactor is as effective as Mn^{2+} (mammalian guanylyl cyclases are about 10-fold less active with Mg^{2+}); (4) micromolar concentrations of Ca^{2+} are absolutely required for full activity (mammalian guanylyl cyclases are unaffected); and (5) the Ca^{2+} sensitivity of the protozoan guanylyl cyclases is mediated by calmodulin. These particular properties of guanylyl cyclases from *Paramecium* and *Tetrahymena* make the study of the protozoan enzymes interesting from a functional and from an evolutionary point of view. Guanylyl cyclases in other protozoans have not been investigated so far.

Materials and Reagents

Skim milk powder is purchased from Nestle (Frankfurt, FRG), proteose peptone and yeast extract from Difco (Detroit, MI), yeast RNA from Sigma (St. Louis, MO), Dowex 50W-X4 (100–200 mesh) from Serva (Heidelberg, FRG), neutral aluminum oxide "90 active" from Merck (Darmstadt, FRG), and Chelex-100 from Biorad (Munich, FRG). [α-^{32}P]GTP (400–600 Ci/mmol) and [8-^{3}H]cGMP (16 Ci/mmol) are from Amersham (Amersham, UK).

Guanylyl Cyclase Assay

Guanylyl cyclase activity is determined at 37° in a total volume of 50 μl containing 30 m*M* Tris-HCl, pH 7.5, 3 m*M* $MgSO_4$, 1.5 m*M* phosphoenolpyruvate, 1 U pyruvate kinase, 1 m*M* cGMP (including 5 nCi [8-^{3}H]cGMP to monitor phosphodiesterase activity during the assay and yield upon subsequent purification), and up to 30 μg protein. The reaction is started by addition of 0.5 μCi [α-^{32}P]GTP (0.4 m*M* final). After 10 min, the incubation is terminated with 200 μl of a solution containing 1% sodium dodecyl sulfate (SDS), 1 m*M* GTP, and 5 m*M* cGMP, pH 7.2. Water (750 μl) is added, and the samples are applied to Dowex 50 columns (5 ml; pretreated with 4 ml of 2 *N* HCl and 10 ml water). The columns are washed with 3 ml water, and cGMP is eluted with 2 ml water onto alumina columns (2 ml; activated with 4 ml of 0.1 *M* Tris-HCl, pH 7.5). The alumina columns are washed with 2 ml of Tris buffer, and cGMP is eluted into scintillation vials with 4 ml of this buffer. Phosphodiesterase activity is usually negligible when purified membranes are used as the enzyme source. The yield of [8-^{3}H]cGMP after separation from [α-^{32}P]GTP by the two-column method is around 60%. With 0.5 μCi [α-^{32}P]GTP in the assay, a background of 50 counts per minute (cpm) is obtained in control incubations (see also this volume, [30]).

TABLE I
COMPONENTS OF MILK POWDER MEDIUM FOR AXENIC CULTIVATION OF *Paramecium tetraurelia*

Component	Concentration (mg/liter)	Component	Concentration (mg/liter)
Skim milk powder	8500	Folic acid	2.5
Ribonucleic acid	1000	Thiamin-HCl	15
$Mg_2SO_4 \cdot 7H_2O$	500	Biotin	0.00625
Phosphatidylcholine	250	Lipoic acid	0.05
Stigmasterol	5	Palmitic acid	7.5
Calcium pantothenate	5	Stearic acid	5
Nicotinamide	5	Oleic acid	1
Riboflavin	5	Linoleic acid	0.5
Pyridoxamine-HCl	5.8	Linolenic acid	0.15

Mass Cultivation

Axenic mass production of *Paramecium* has only been accomplished with *Paramecium tetraurelia* 51s, and several mutants derived therefrom, and with strain 299s.[2] The medium constituents are listed in Table I. The vitamins are suspended in water, and the lipids are dissolved in ethanol as 200- and 400-fold concentrated stocks, respectively, and stored at $-20°$. For preparation of the medium, the major solid constituents (skim milk powder, ribonucleic acid, $MgSO_4$, and phosphatidylcholine) are mixed with demineralized water. After addition of the appropriate amounts of vitamin and lipid stock solutions, the pH of the highly turbid, yellowish suspension is adjusted to pH 7.2 by 6 *N* NaOH, and the medium is autoclaved. Failure to bring the pH above 7.1 results in massive coagulation of milk protein upon sterilization.[2]

Cells are cultured at 25° either in Erlenmeyer flasks (preferably with two baffles) filled to one-third with medium or in airlift bioreactors of up to 250 liters. For Erlenmeyer cultures, optimal cell densities (up to 3×10^4/ml) and generation times (12–14 hr) are obtained at 70–80 rpm on a rotary shaker (diameter of gyration 2.5 cm). In bioreactors, the aeration rate has to be adjusted according to volume and size of the vessel. An oxygen tension of 65%, measured with a calibrated oxygen electrode, yields optimal growth. Because of the size of *Paramecium* ($100 \times 30\ \mu m$), the organism is very sensitive to shearing forces, and aeration conditions must be carefully optimized. Cells usually reach the late logarithmic

[2] U. Schönefeld, A. W. Alfermann, and J. E. Schultz, *J. Protozool.* **33,** 222 (1986).

growth stage within 4–5 days. The turbid medium clears with progressing culture age (exoproteases), and, after an initial drop, the pH reaches 7.3–7.5. In addition to the cell count, the pH can reliably be taken as measure for culture progression.[2]

For mass production of *Tetrahymena* the medium contains 1% protease peptone, 0.1% yeast extract, 0.2% glucose, and 5 μg/ml Fe^{2+}. With a 10% inoculum cells have a doubling time of 3–6 hr and grow to densities of 10^6/ml within 2–3 days. The technical parameters for culture are identical to those detailed above for *Paramecium*.

Paramecium and *Tetrahymena* alike are harvested by centrifugation at room temperature (400 *g*, 5 min). Large cultures can be concentrated with a cream separator operated at a flow rate of 500 ml/min and a rotor speed corresponding to 300 *g*.

Separation of Cilia and Cell Bodies[3]

Solutions

Dryl's solution: 2 m*M* sodium phosphate, 2 m*M* sodium citrate, 1.5 m*M* $CaCl_2$, pH 7

Solution A: 150 m*M* sucrose, 4 m*M* KCl, 1 m*M* $CaCl_2$, 1 m*M* MOPS-Na, pH 7.2

Solution B: 150 m*M* sucrose, 4 m*M* KCl, 625 m*M* glycerol, 15 m*M* MOPS-Na, pH 8.3

MOPS buffer: 10 m*M* MOPS-Na, pH 7.2

Procedure. The volumes given refer to a 30-liter culture of *Paramecium tetraurelia* wild-type 51s. Packed cells (100 ml) are washed at 20° with 2.4 liters of Dryl's solution (400 *g*, 5 min). All subsequent steps are carried out at 4°. The concentrated cells are suspended as a thick slurry in 240 ml solution A and poured into 1.5 liters solution B. After 5 min, 15 ml of a 3 *M* $CaCl_2$ solution is added, and the suspension is vigorously shaken by hand. Deciliation is usually complete within 5 min, as checked by phase-contrast microscopy. The bulk of the cell bodies is removed as a pellet by centrifugation at 100 *g* (15 min). The supernatant is centrifuged again (370 *g*, 10 min). The small pellet is discarded, and cilia are collected from the supernatant at 5900 *g* (30 min) and suspended in 80 ml solution A. The whitish suspension is centrifuged again at 270 *g* (10 min), and the pellet is discarded. From the supernatant, the cilia are concentrated at 7700 *g* (30 min). The purified cilia are suspended in 15 ml MOPS buffer and stored at −80°. From 30 liters of a late-log culture 150 mg of ciliary protein is

[3] J. Thiele, S. Klumpp, J. E. Schultz, and C. F. Bardele, *Eur. J. Cell Biol.* **28,** 3 (1982).

obtained. The fraction of deciliated cell bodies constitutes about 3000 mg protein. The procedure described above is also applicable for *Tetrahymena*.

Subcellular Distribution of Guanylyl Cyclase

Ninety percent of guanylyl cyclase activity is associated with cell bodies (specific activity 0.1–0.2 mU/mg). The cilia, comprising 5% of the cell protein, contain 10% of the guanylyl cyclase with an average specific activity of 0.3 mU/mg. Cilia are disintegrated by vigorous shaking on a vortex (3 min), and ciliary membranes are isolated free from axonemes by a discontinuous sucrose density gradient[4] (25 mg cilia in 2 ml MOPS buffer on top of 2 ml 20%, 5 ml 45%, and 3 ml 55% sucrose in MOPS buffer; 135,000 *g*, 160 min). Pure ciliary membranes are recovered at the 20–45% sucrose interface; axonemes are in the pellet. Essentially all ciliary guanylyl cyclase is membrane bound.[3]

In *Tetrahymena*, the distribution of guanylyl cyclase between cilia and cell bodies has been reported to vary depending on the growth stage at which the cultures are harvested.[5] The total and specific activity in the cell body appears to increase with progressing culture age.[5] All guanylyl cyclase activity in cilia is membrane bound.[6]

Solubilization

Guanylyl cyclase from cilia of *Paramecium* can be solubilized (see the scheme in Fig. 1). Cilia [10 mg protein/ml in 20 m*M* Tris-HCl, pH 7.5, 1% Brij 35] are vigorously vortexed 3 times for 1 min each time at 0°, and centrifuged at 48,000 *g* (30 min). The supernatant, which contains 30% of the protein but very little guanylyl cyclase activity, is discarded. The pellet is resuspended to two-thirds of the original volume in 20 m*M* Tris-HCl, pH 7.5, 1% thioglycerol, 20% (v/v) glycerol. At this point, 20% Lubrol PX (freshly dissolved in MOPS buffer) is added to give a final detergent concentration of 2%, and the suspension is sonicated in 1-ml aliquots for 20 sec using a microtip and an output setting of 5 (Branson Sonifier). The supernatant of a 100,000 *g* (60 min) spin contains 40% of the total guanylyl cyclase activity. This procedure is repeated twice with the pellet, and altogether 60% of guanylyl cyclase activity can be solubilized (Fig. 1).

[4] A. Adoutte, R. Ramanathan, R. M. Lewis, R. R. Dute, K.-L. Ling, C. Kung, and D. L. Nelson, *J. Cell Biol.* **84,** 717 (1980).

[5] Y. Muto, S. Kudo, S. Nagao, and Y. Nozawa, *Exp. Cell Res.* **159,** 267 (1985).

[6] J. E. Schultz, U. Schönefeld, and S. Klumpp, *Eur. J. Biochem.* **137,** 89 (1983).

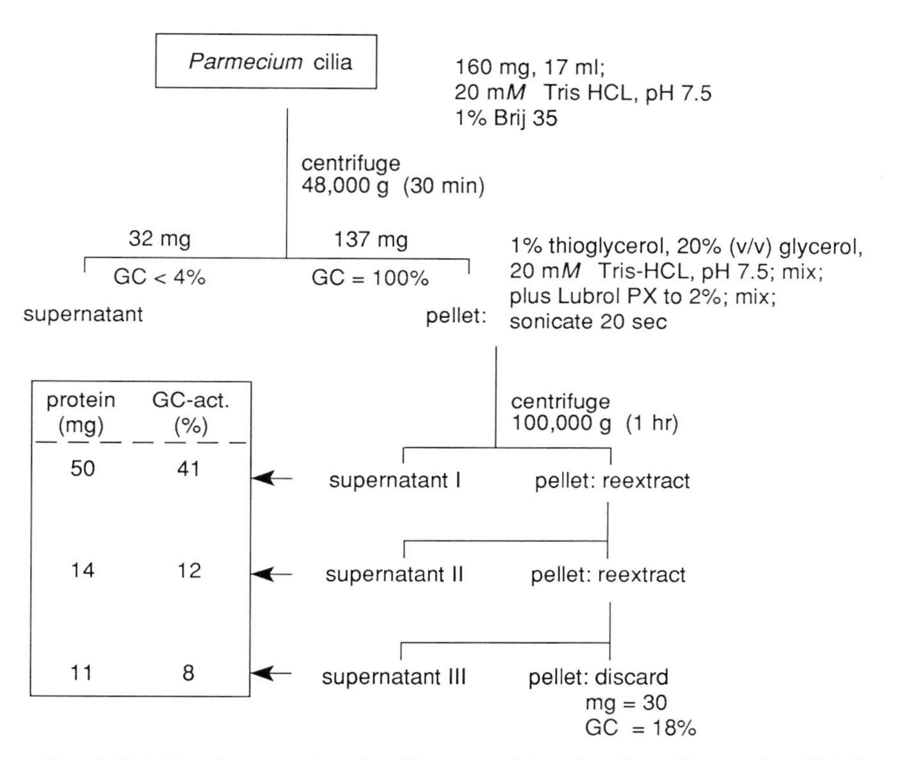

FIG. 1. Solubilization procedure for ciliary guanylyl cyclase from *Paramecium*. Details are given in the text.

The procedure is not successful with *Tetrahymena* guanylyl cyclase because 1% Brij and 0.5% Lubrol are potent inhibitors. Guanylyl cyclase from *Tetrahymena* plasma membranes is solubilized at a yield of 22% using 0.5% digitonin in the presence of 0.2 m*M* $CaCl_2$ and 20% glycerol.[7]

Characterization

Paramecium guanylyl cyclase has a broad pH optimum of 7–9, it uses MnGTP equally well as MgGTP as substrate (K_m for GTP 71.5 and 36 μM, respectively[8]). The temperature optimum is at 37°, the activation energy 55 kJ/mol.[8] Guanylyl cyclase is inhibited by SH-group blocking agents,

[7] S. Nagao and Y. Nozawa, *Arch. Biochem. Biophys.* **252,** 179 (1987).

[8] S. Klumpp and J. E. Schultz, *Eur. J. Biochem.* **124,** 317 (1982).

for example, N-ethylmaleimide (1 mM gives 50% inhibition) and p-chloromercuribenzoate (10 μM gives 90% inhibition).[8] It is permanently activated by several polypeptide antibiotics (e.g., alamethicin, suzukacillin) or by melittin (6-fold at 30 μM).[9]

Guanylyl cyclase from *Tetrahymena* also uses MgGTP and MnGTP equally well as substrate. Maximal activities are obtained with 3 mM Mg^{2+} or 0.5 mM Mn^{2+} as metal cofactors.[10] The K_m values for GTP are 50 and 20 μM, respectively.[10]

Physiological Regulation

Guanylyl cyclase activity in ciliary membranes from *Paramecium* assayed without addition of Ca^{2+} is 0.3 mU/mg. The enzyme is inhibited half-maximally by 20 μM EGTA or 400 μM EDTA.[8] Removal of Ca^{2+} from the membranes by washing once with 60 μM EGTA, and from assay reagents by treatment with Chelex-100, reduces activity by 80%.[8] Addition of Ca^{2+} dose-dependently restores activity (Fig. 2). V_{max} is increased, but K_m is not altered.[8] Half-maximal activation of guanylyl cyclase is observed at 8 μM Ca^{2+}. Ca^{2+} concentrations above 100 μM are inhibitory. Ba^{2+} neither activates guanylyl cyclase nor inhibits the activation by Ca^{2+} (Fig. 2). Paramecium guanylyl cyclase is specifically inhibited by addition of 15 μl antiserum raised against purified calmodulins from *Paramecium, Tetrahymena,* bovine brain, *Dictyostelium,* and soybean (70% reduction of activity).[1] An inhibition of 80% is observed with affinity-purified antibodies against *Paramecium* calmodulin. While addition of calmodulins from various metazoan sources is without any effect on basal guanylyl cyclase activity, calmodulins from *Tetrahymena* and *Paramecium* reproducibly cause a 10–30% activation.[1] The effects of protozoan calmodulins and calmodulin antibodies described above are also found with guanylyl cyclase from cell bodies.

La^{3+} completely inhibits guanylyl cyclase at 10 μM.[8,11] An inactive guanylyl cyclase is prepared by incubation of ciliary membranes with 13 μM La^{3+} for 15 min at 0°. La^{3+} is then removed by centrifugation and membranes are washed once with 60 μM EGTA. Guanylyl cyclase activity is reduced by 85%.[11] Reactivation is accomplished by concomitant addition of Ca^{2+} and calmodulins (from protozoan and metazoan sources), or by

[9] S. Klumpp, G. Jung, and J. E. Schultz, *Biochim. Biophys. Acta* **800,** 145 (1984).

[10] K. Nakazawa, H. Shimonaka, S. Nagao, S. Kudo, and Y. Nozawa, *J. Biochem.* (*Tokyo*) **86,** 321 (1979).

[11] S. Klumpp, G. Kleefeld, and J. E. Schultz, *J. Biol. Chem.* **258,** 12455 (1983).

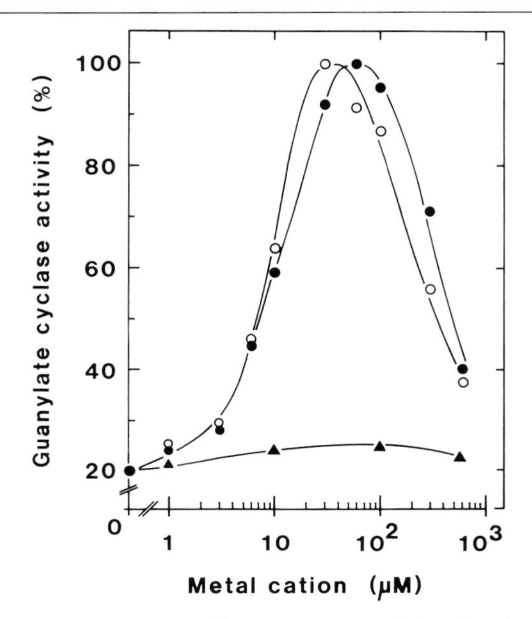

FIG. 2. Dose–response curve for Ca^{2+}-stimulated guanylyl cyclase in ciliary membranes from *Paramecium*. Membranes were washed with 60 μM EGTA followed by two washings with MOPS buffer to remove the chelator. All reagents were passed over Chelex 100 to remove Ca^{2+}. The remaining Ca^{2+} concentration was below 1 μM. Activity was determined at indicated concentrations of Ba^{2+} (▲), Ca^{2+} (●), or Ca^{2+} in the presence of 1 mM Ba^{2+} (○). An activity level of 100% corresponds to 1.6 mU/mg.

equimolar EGTA– and EDTA–strontium complexes. Guanylyl cyclase reactivated with chelator–strontium complexes can still be inhibited by addition of antisera against calmodulins. The molecular events of this reactivation are not yet fully understood.

Basically similar yet quantitatively different results are obtained with guanylyl cyclase from *Tetrahymena* cell bodies and ciliary membranes.[1,12] Basal guanylyl cyclase activity, usually between 0.1 and 0.3 mU/mg, is stimulated up to 20-fold by addition of calmodulin from *Tetrahymena* or *Paramecium*.[5,6,12] Maximal activation is observed with 6 μM of the protozoan calmodulins.[12] Calmodulins from several metazoan sources, for example, bovine brain, sea anemone, soybean, scallops, are inactive.[12] As already found with the enzyme from *Paramecium*, basal guanylyl cyclase

[12] S. Kakiuchi, K. Sobue, R. Yamazaki, S. Nagao, S. Umeki, Y. Nozawa, M. Yazawa, and K. Yagi, *J. Biol. Chem.* **256,** 19 (1981).

activity from *Tetrahymena* is inhibited by 70% when incubated with 15 μl antiserum or with affinity-purified antibodies against *Paramecium* calmodulin.[1] The relationship between the regulatory function of that calmodulin, which obviously is present as an endogenous subunit of guanylyl cyclase, and the stimulation by the exogenous protozoan calmodulins is not understood.

NOTE ADDED IN PROOF: The procedure for solubilization of the guanylyl cyclase from *Paramecium* can be equally well applied to the adenylyl cyclase activity present in the ciliary membrane.

Author Index

Numbers in parentheses are footnote reference numbers and indicate that an author's work is referred to although the name is not cited in the text.

A

B

D

E

F

G

H

I

J

K

L

M

N

O

P

Q

R

S

Y

Z

Subject Index

A

B

C

G

H

I

K

L

M

N

S